Two-variable (bivariate) procedures

Measurement level of second variable

Nominal (c categories)	Ordinal	Interval and ratio
. . . .	. . . .	. . . .
Chi square, V, C, T, τ_b, λ_b Chap. 15	. . . .	. . . .
Analysis of variance with ranks Chap. 16	Rank-order correlation, Kendall's tau, gamma, r_s, d_{yx} Chap. 18	. . . .
Analysis of variance, E^2, intraclass correlation Chap. 16	. . . .	Correlation and regression Chaps. 17, 18

Page
68   Good discussion of mean and median

Social Statistics

McGraw-Hill Series in Sociology

Consulting Editor:

Otto N. Larsen   University of Washington

Social Statistics

Second Edition

Hubert M. Blalock, Jr.

Professor of Sociology
University of Washington

McGraw-Hill Book Company

New York   St. Louis   San Francisco   Düsseldorf   Johannesburg
Kuala Lumpur   London   Mexico   Montreal   New Delhi   Panama
Rio de Janeiro   Singapore   Sydney   Toronto

Library of Congress Catalog Card Number
76-37090
07-005751-6
 234567890 DODO 798765432

This book was set in Modern 8A by
The Maple Press Company, and printed
and bound by R. R. Donnelley & Sons Company.
The designer was J + M Condon, Inc.;
the drawings were done by John Cordes,
J. & R. Technical Services, Inc. The
editors were David Edwards and
Claudia A. Hepburn. Peter D. Guilmette
supervised production.

To Ann, Susie, and Katie

Contents

Preface

This text is written primarily for those students of sociology, both advanced undergraduates and graduate students, who actually intend to engage in social research.

In the twelve years that have elapsed since the publication of the first edition, the level of training and the sophistication in applied statistics has undergone considerable improvement not only in sociology but also in political science, anthropology, geography, and social work. Nevertheless, the overwhelming majority of students and practitioners in these fields still lack the necessary mathematical backgrounds to take full advantage of the rapidly accumulating technical literature in mathematical statistics and econometrics. With these basic facts in mind, this text has been written so as to avoid mathematical derivations insofar as possible, and only a quick review of certain algebraic principles listed in Appendix 1 should therefore be sufficient preparation for the average student. Although it is not necessary in a first course in statistics to stress mathematical derivations, the author is convinced that certain basic and fundamental ideas underlying the principles of statistical inference must be thoroughly understood if one is to obtain more than a mere "cookbook" knowledge of statistics. Therefore, there is a relatively heavy emphasis on the underlying logic of statistical inference, including a chapter on probability, with relatively less attention being given to some of the more or less routine topics ordinarily discussed in elementary texts.

One of the most difficult problems encountered in the teaching of applied statistics is that of motivating students, both in enabling them to overcome their fears of mathematics and in learning to apply statistics to their own field of interest. It is for the latter reason that the author has not attempted to cover a wide range of applications but has selected examples of primary interest to sociologists. To some extent, examples have also been chosen from disciplines which border on sociology: fields such as social psychology, social work, and political behavior. In most instances each new topic has been illustrated by a single example, under the assumption that most students will lose track of the basic line of thought if too many examples are used to illustrate the same point. Additional examples are therefore given in the form of exercises at the end of each chapter. In general, the author has tried to strike a reasonable compromise between the desirability of stating basic principles as clearly and concisely as possible and the necessity of repeating some of the more difficult ideas each time a new topic is discussed. Insofar as possible, new ideas have been introduced gradually, and, equally important, an effort has been made to relate each new topic to those which have preceded it. In so doing, the major goal has been to give an appreciation of the basic similarities underlying many of the most commonly used tests and measures.

Almost all the suggestions I have received from those wishing to help improve the first edition have implied additions to the book, rather than subtractions, and they have also implied that many of the topics originally treated should be discussed more technically. My own position is that sociologists and political scientists, in particular, need greater exposure to the more technical literature on experimental designs and on the use of simultaneous-equation procedures in connection with nonexperimental research. Yet, it became clear that if these materials were added to the original text, it would lose its appeal as an introductory text appropriate for advanced undergraduates majoring in the social sciences. It was therefore decided to treat experimental designs, factor analysis, and simultaneous-equation procedures, as well as other more advanced topics, in a separate text to be written with two of my colleagues, Lewis F. Carter and Krishnan Namboodiri.

Included in the text are a number of sections, paragraphs, and exercises which are either conceptually difficult or which presuppose that the student is reasonably familiar with topics ordinarily covered in courses on research methods. These portions of the text have been marked with an asterisk (*) and may be skimmed on first reading or omitted entirely. Instructors using the text for a one-semester course may wish to indicate that students should omit these materials.

There has been a slight upgrading in terms of technical coverage, but the basic structure of the text has not changed. There are a few modifications in the section on descriptive statistics. The author has added discussions of assumptions and basic concepts, hoping to clarify the relationship between the statistical models and the real world with which the social scientist must deal. In addition to these changes, this edition contains discussions of several procedures, tests, and measures that have been increasingly used during the decade of the 1960s.

Chapter 9 on probability has been expanded to include discussions of permutations, tree diagrams, Bayes' theorem, calculations involving conditional probabilities, and the notion of expected values. And to Chapter 10, which includes a discussion on the binomial distribution, the author has added brief discussions of the multinomial distribution, the hypergeometric distribution, and the Poisson distribution. These additional topics should make the transition to texts specifically oriented to nonparametric statistics a much simpler task.

The author has also extended the coverage of nonparametric techniques to include the Friedman test for two-way analysis of variance with ranks; gamma and d_{yx} as measures of ordinal association; a test for interaction involving the difference of differences of proportions; and standardization in the case of nominal-scale procedures. There is also an expanded discussion of the properties of the several ordinal measures and partialling techniques with ordinal scales.

Similarly, the discussion of parametric approaches has been expanded to include discussion of the assumptions for the general linear model and coverage of the dummy variable approach as an alternative perspective on analysis of covariance. The basic essentials of the theory underlying the use of linear combinations have also been introduced and applied to the discussions of the standard error of the mean, difference of means, difference of differences of proportions, and the use of orthogonal comparisons in the case of multiple samples.

In an effort to help the reader to see the overall picture, the author has included a summary table of tests and measures, which appears on the front, inside cover, and extended summaries at the end of Chapters 2, 14, and 20.

Numerous persons have contributed to this revision, but the author would especially like to thank Richard G. Ames, Erica Borden, and Louis Goodman for providing extensive comments on the entire manuscript.

For assistance in the preparation of the first edition, I would especially like to thank those students and colleagues at the University of Michigan who read and suggested improvements in various drafts of the book. To Richard T. LaPiere, Sanford Dornbusch, Robert Ellis, Santo Camilleri, and Theodore Anderson my appreciation for reading and criticizing the original manuscript. For proofing, typing, and checking computations, I would like to thank Ann Blalock, Diane Etzel, Ann Laux, and Doris Slesinger. My thanks also to Daniel O. Price, who deserves the major credit for stimulating my interest in statistics.

I am indebted to Professor Sir Ronald A. Fisher, Cambridge, to Dr.

Frank Yates, Rothamsted, and to Messrs. Oliver and Boyd Ltd., Edinburgh, for permission to reprint Tables III, IV, and V from their book *Statistical Tables for Biological, Agricultural and Medical Research*. I am equally grateful to those other publishers and authors, acknowledged in the appropriate places, who kindly gave their permission for use of various tables and computing forms.

Hubert M. Blalock, Jr.

Social
Statistics

Introduction   Part 1

Introduction: Purposes and Limitations of Statistics

The field of statistics has widespread applications as indicated by the fact that statistics courses are offered in such dissimilar subjects as dentistry and sociology, business administration and zoology, and public health and education. In spite of this fact, there are many misconceptions concerning the nature of this rapidly developing discipline. The layman's conception of statistics is apt to be very different from that of the professional statistician. A statistician is sometimes thought of as a person who manipulates numbers in order to prove his point. On the other hand, some students of sociology or other social sciences have tended to worship the statistician as someone who, with the aid of his magical computing machine, can make almost any study "scientific." Possibly because of the awe many persons have for anything with a hint of mathematics, students often find it difficult to approach a course in statistics with other than mixed feelings. Although they may be terrified at the prospect of working with numbers, they may also come to expect too much of a discipline that appears so formidable. Before jumping into the subject too quickly and thereby losing perspective, let us first ask ourselves just what statistics is and what it can and cannot do.

It is perhaps easiest to begin by stating what statistics is not. Statistics first of all is not a method by which one can prove almost anything he wants to prove. In fact, we shall find statisticians very carefully laying down rules of the game to ensure that interpretations do not go beyond the limits of the data. There is nothing inherent in statistical methods to prevent the careless or intellectually dishonest individual from drawing his own conclusions in spite of the data, however, and one of the most important functions of an introductory course in statistics is to place the student on guard against possible misuses of this tool.

Statistics is not simply a collection of facts. If it were, there would hardly be much point in studying the subject. Nor is statistics a substitute for abstract theoretical thinking or for careful examination of exceptional cases. In some of the older textbooks on research methods one used to find lengthy discussions on the relative merits of the case-study method versus the statistical method. It is now clearly recognized that statistical methods are in no sense opposed to the qualitative analysis of case studies but that the two approaches are complementary. It is not even true that statistics is applicable only when there are a large number of cases or that it cannot be used in exploratory studies. Finally, statistics is not a substitute for measurement or the careful construction of an interview schedule or other instrument of data collection. This last point will receive further attention toward the end of this chapter and in Chap. 2.

Having indicated what statistics is not, can we say definitely what it is? Unfortunately, persons who call themselves statisticians seem to disagree somewhat as to exactly what is covered under the general heading of statistics. Taking a pragmatic approach to the problem, we shall say that statistics has two very broad functions. The first of these functions is description, the summarizing of information in such a manner as to make it more usable. The second function is induction, which involves either making generalizations about some population, on the basis of a sample drawn from this population, or formulating general laws on the basis of repeated observations. Each of these functions will be discussed in turn.

1.1 Functions of Statistics

Descriptive statistics   Quite frequently in social research a person will find himself in the position of having so much data that he cannot adequately absorb all his information. He may have collected 200 questionnaires and be in the embarrassing position of having to ask, "What do I do with it all?" With so much information it would be exceedingly difficult for any but the most photographic minds to grasp intuitively what is in the data. The information must somehow or other be boiled down to a point at which the researcher can see what is in it; it must be summarized. By computing measures such as percentages, means, standard deviations, and correlation coefficients it may be possible to reduce the data to manageable proportions. In summarizing data by substituting a very few measures for many numbers, certain information is inevitably lost, and, more serious, it is very possible to obtain results that are misleading unless cautiously interpreted. Therefore, the limitations of each summarizing measure must be clearly indicated.

Descriptive statistics is especially useful in instances where the investigator finds it necessary to handle interrelationships among more than two variables. For example, suppose one needed to use eight or ten variables to help explain delinquency rates, and furthermore suppose these explanatory or *independent* variables were themselves highly intercorrelated. If one wished to isolate the effects of one or two of these variables, controlling for the remainder, how would he proceed? And what kinds of assumptions would be needed? Questions of this degree of complexity arise in a branch of statistics referred to as *multivariate analysis*. Relatively simple problems of multivariate analysis will be discussed in Chaps. 15, 16, 19, and 20, but more complex questions must be reserved for a second volume.

Inductive statistics An equally important function of statistics, and certainly the one which will occupy most of our attention in this text, is that of *induction* or inferring properties of a population on the basis of known sample results. *Statistical inference*, as the process is called, involves rather complex reasoning, but when properly used and understood becomes a very important tool in the development of a scientific discipline. Inductive statistics is based directly on probability theory, a branch of mathematics. We thus have a purely deductive discipline providing a rational basis for inductive reasoning. To the writer's knowledge there is no other rational basis for induction. This general point will be discussed in more detail in Chap. 8.

There are several practical reasons why it is often necessary to attempt to generalize on the basis of limited information. The most obvious is the time-cost factor. It would be wholly impractical, if not prohibitively costly, to ask every voter how he intended to vote in order to predict a national election. Nor can the ordinary researcher afford to tap every resident of a large city in order to study prejudice, social mobility, or any other phenomenon. He first decides upon the exact nature of the group about which he wishes to generalize (the population). For example, he may select all citizens of voting age or all white males over eighteen residing within the city limits of Detroit. He then usually draws a sample consisting of a relatively small proportion of these people, being primarily interested, however, not in this particular sample but in the larger population from which the sample was drawn. For example, he may find that within his particular sample of 200 white males there is a negative relationship between education and prejudice. Recognizing that had another set of 200 individuals been sampled the results might have been different, he would nevertheless like to make certain inferences as to the nature of the relationship had the entire population of adult white males in Detroit been studied.

Another reason for generalizing on the basis of limited information is that it may be impossible to make use of the entire population simply because the population is infinite or not easily defined. In replicating an experiment in the natural or social sciences the goal always seems to be some kind of generalization which it is hoped will apply under similar circumstances. Or a social scientist may have collected data on all of the cases available to him. For example, he may have used all 50 states as his units of analysis in studying internal migration. Nevertheless, he may want to generalize about migration under *similar* conditions. In each of these instances the situation calls for inductive statistics.

At this point you may ask a question of the following sort: "If statistics is so important, why is it that sciences such as physics and chemistry have been able to get along so well without the extensive use of statistical techniques? Is there anything different about these sciences?" Quite obviously there is. Some of the natural sciences have developed for centuries without the use of inductive statistics. But this seems to be primarily a matter of good fortune or, to give these scientists credit for their own efforts, a relatively satisfactory control over disturbing elements in the environment. As will become apparent in later chapters, to the degree that carefully controlled laboratory conditions prevail, there is less practical need for statistical techniques. In this sense statistics is a poor man's substitute for contrived laboratory experiments in which all important relevant variables have been controlled. It should be emphasized, however, that many of the same statistical principles apply to laboratory experiments in physics, to somewhat less precise agricultural experiments, and to social surveys. For example, if an experiment in physics has been replicated 37 times with similar results, it is nevertheless conceivable that subsequent trials will yield different outcomes. The scientist must therefore generalize on the basis of a limited number of experiments, and the inferences he makes are essentially statistical in nature. Also, the problem of measurement error can be conceived in statistical terms. No matter how precise the measuring instrument, the scientist never obtains exactly the same results with each replication. He may attribute these differences either to measurement error or to disturbing effects of uncontrolled variables. Statistics becomes especially necessary whenever there is so much variation from one replication to the next that differences cannot be ignored or attributed to measurement error. Basically, then, statistical inference underlies all scientific generalizations, although the need for statistical training and the use of sophisticated statistical techniques vary considerably from one field to another.

1.2 The Place of Statistics in the Research Process

The importance of statistics in the research process is sometimes exaggerated by the emphasis given to it in graduate curricula. Statistics proper does not include measurement problems such as the construction of indices or the scoring of items on a questionnaire. Rather, statistics involves a manipulation of numbers under the assumption that certain requirements have been met in the measurement procedure. Actually, statistical considerations enter only into the analysis stage of the research process, after all data have been collected, and near the beginning when initial plans for analysis are made and when a sample is to be drawn.

While the above statement to the effect that statistics enters only into the analysis and sampling stages of the research process may be technically correct, it is very likely to be misleading unless qualified. It certainly does *not* mean that a social scientist can plan and carry out his entire research without any knowledge of statistics and then dump the whole project into the lap of a statistician saying, "Now I've done my job. You analyze it." If he did this, the results would probably be disappointing if not completely useless. Obviously, problems that will be encountered in analysis have to be anticipated in every stage of the research process, and in this sense statistical considerations may be involved throughout. A highly sophisticated statistical analysis can rarely if ever compensate for a poorly conceived project or a poorly constructed data-collection instrument, however. This last point deserves special note. It means that statistics may be an aid to, but never a substitute for, good sound thinking. From the standpoint of the social scientist it is merely a tool.

Having said this, I should add that statistics is a much more flexible and useful tool for exploratory analyses than might be imagined. Most social research must be based on highly tentative theoretical ideas that usually do not imply precise guidelines in terms of peculiar relationships to be anticipated, the variables that should be controlled in the analysis, or even the priorities and sequences of analysis steps that should be followed. Students are commonly surprised at the complexity of data analysis, as soon as one introduces as many as a half-dozen variables into the picture. It is in these instances, especially, where a knowledge of the statistical theory of experimental designs or of simultaneous-equation estimation techniques become an invaluable tool for enabling the social scientist to disentangle highly complex interrelationships. Verbal or intuitive approaches are far too inadequate. The topics of experimental designs and multivariate analysis can only be introduced in a general text such as this, but it is important to realize that there are numerous more advanced subjects that have proven to be invaluable even in exploratory research, the aim of which is to assess the relative importance of numerous factors, to narrow down the range of alternatives in a systematic way, and to generate more precise hypotheses for later research.

1.3 A Word of Advice

Some students experience a degree of fear ranging from mild apprehension to an extreme mental block whenever they see a number or mathematical equation. If you happen to belong to this category, you should especially try to set aside any notion you may have to the effect that "statistics is something I know I'll never be able to understand." The level of mathematics required in this text is such that several years of high school algebra plus a quick review of a few elementary algebraic operations given in Appendix 1 should be ample preparation. It must be remembered, however, that mathematics and statistics texts cannot be expected to read like novels. Material is usually presented in a highly condensed form. Therefore careful rereading and an active rather than passive orientation to the material presented will be required. For this reason, there is no substitute for daily preparation and the working of practice problems found at the end of each chapter.

References

1. Downie, N. M., and R. W. Heath: *Basic Statistical Methods*, 2d ed., Harper and Row, Publishers, Incorporated, New York, 1965, chaps. 1 and 2.
2. Hagood, M. J., and D. O. Price: *Statistics for Sociologists*, Henry Holt and Company, Inc., New York, 1952, chaps. 1 and 2.
3. Hammond, K. R., and J. E. Householder: *Introduction to the Statistical Method*, Alfred A. Knopf, Inc., New York, 1962, chap. 1.
4. Hays, W. L.: *Statistics*, Holt, Rinehart and Winston, Inc., New York, 1963, pp. 1–12.
5. Tippett, L. H. C.: *Statistics*, 2d ed., Oxford University Press, New York, 1956.
6. Walker, H. M.: *Mathematics Essential for Elementary Statistics*, Henry Holt and Company, Inc., New York, 1951.
7. Wallis, W. A., and H. V. Roberts: *Statistics: A New Approach*, The Free Press of Glencoe, Ill., Chicago, 1956, chaps. 1–3.

Theory, Measurement, and Mathematics

2

The purpose of this chapter is to outline the relationships among theoretical propositions, empirical hypotheses, measurement, and mathematical models. Many of the problems dealt with in this chapter are not usually discussed in connection with courses in statistics, partly because of the regrettable tendency to compartmentalize courses under the headings of *theory, research methods,* and *statistics.* This often means that the nature of the interrelations among these areas becomes obscured. In order to place statistics in its proper perspective it will be helpful to give at least some attention to the relationships between theoretical propositions and research hypotheses and also to those between research hypotheses and mathematical models.

One commonly hears statements to the effect that it is the purpose of research to test hypotheses developed theoretically and that statistical methods enable us to make such tests. It must be realized, however, that the processes involved in getting from theory to actual research hypotheses and from these hypotheses to probability statements of the kind used in statistical inference are by no means direct. In both cases certain decisions have to be made, decisions which may lead to considerable controversy. Let us first examine the nature of those decisions required in developing testable hypotheses from theoretical propositions.

*2.1 Theory and Hypotheses: Operational Definitions

The moment we begin to design a research project the aim of which is to test a proposition of the sort that may appear in a theoretical work, it becomes very evident that a number of things must be done before the

* An asterisk preceding a section, paragraph, or exercise indicates that the material involved either is conceptually difficult or deals with subjects which are likely to be unfamiliar to students with limited backgrounds in research methodology. These materials can safely be omitted or skimmed by the beginning student since they are not really essential for an understanding of subsequent materials. An asterisk preceding a section heading indicates that the entire section may be omitted if desired.

test can be made. As a concrete example let us take the proposition "The higher a white person's social status, the lower his prejudice toward blacks." Suppose *social status* has been defined as one's position relative to others in the status hierarchy and *prejudice* as an underlying tendency to discriminate against a minority or as a negative attitude based on prejudgment. Although you may prefer to substitute other definitions of these two concepts, you will undoubtedly discover that no matter what definitions you choose, it will not be possible to use them directly to enable you to decide exactly what Jones's status or prejudice level may be.

The reason for this is that most ordinary definitions are theoretical rather than operational definitions. In a theoretical definition a concept is defined in terms of other concepts which supposedly are already understood. In the ideal model of the completely deductive system, certain concepts would be taken as undefined (primitive), and all other concepts would be defined in terms of these. In Euclidean geometry, for example, the concepts *point* and *line* may be taken as undefined. The notions of *angle*, *triangle*, or *rectangle* can then be defined in terms of these primitive concepts. Although the choice of undefined concepts is to a certain extent arbitrary, the fact that there must always be some primitive concepts is a reflection of the necessity of defining theoretical concepts in terms of each other.

Operational definitions, on the other hand, are definitions that actually spell out the procedures used in measurement ([8], pp. 58 to 65). An operational definition of *length* would indicate exactly how the length of a body is to be measured. An example of an operational definition of prejudice would include a test such as the Bogardus social-distance scale or, perhaps, a 24-item list of antiblack stereotypes, together with detailed instructions for collecting the data, scoring the items, and so forth. Since all measurement involves classification as a minimal requirement, an operational definition can be considered to be a detailed set of instructions enabling one to classify individuals unambiguously. The notion of reliability is thus built into this conception of the operational definition. The definition should be sufficiently precise that all persons using the procedure will achieve the same results. The theoretical definitions of prejudice and social status given above would not directly permit this to be done.

We are thus arguing that two different kinds of definitions are used in any science. Several alternative ways of viewing the relationship between theory and research lead to essentially the same conclusion. Northrop refers to what we have called theoretical definitions as "concepts by postulation" and to operational definitions as "concepts by intuition" [9]. We have used a terminology that seems to imply that there are two distinct ways of defining the *same* concept, whereas Northrop chooses to refer to two different kinds of concepts. Others prefer to think in terms of indices rather than operational definitions. The concept *index* usually implies that the procedure used gives only an imperfect indicator of some underlying variable which is not directly measurable. According to this perspective, then, there is both an underlying variable and an indicator of this variable. Regardless of the perspective one prefers, it is necessary to understand the nature of the linkage between the two kinds of definitions, concepts, or variables. We may ask whether or not there is any purely *logical* method of associating the two kinds of definitions. An alternative way of phrasing the question would be to ask if there is any logical way of determining whether a given operational definition (or index) *really* measures the theoretically defined concept or variable. The answer to both questions seems to be in the negative.

Northrop essentially argues that there is no method of associating the two kinds of concepts or definitions except by convention or common agreement. Persons simply agree that a given operational definition should be used as a measure of a certain concept if the operations seem reasonable on the basis of the theoretical definition. Presumably, if several operational definitions are possible, those operations will be selected which seem most appropriate and at the same time most reliable. Appropriateness must inevitably be judged on the basis of one's understanding of the theoretical definition. The term *validity* is sometimes used to refer to the appropriateness of the index or operational definition ([11], p. 165). Ideally, as Bridgman points out, operations and theoretical definitions should be associated on a one-to-one basis ([2], pp. 23 ff.). In other words, if we change the operation, we should use a new concept. Such an ideal is perhaps an unrealistic one for the social sciences in their present stage of development, however. Its application would undoubtedly lead either to a rigidity capable of stifling further methodological advances or to a proliferation of theoretical concepts [1].

What can be done then? We can admit the possibility of having a number of different operations or indices associated with each theoretical concept. But then we may run into a common difficulty: these procedures may yield different results. One procedure for measuring prejudice may lead to results which indicate that our hypothesis has been confirmed. In another case a different procedure may lead to the opposite conclusion. In a sense this is the way progress is made, provided it does not lead to endless bickering about which of the procedures is really measuring prejudice (the essence of which is presumed understood). In order to avoid confusion it is important to realize that *the actual test is made in terms of the concepts as operationally defined. Propositions involving concepts defined theoretically are therefore not directly testable.* Thus, if there are two distinct operational definitions of prejudice, there will be two distinct hypotheses being tested.

It has been admitted that it may be desirable to have more than one operation associated with any given theoretical concept, and it has been pointed out that such operations may lead to different results. We are now in a position to give a working, pragmatic criterion for an empirically satisfactory theoretical definition of a concept. Let us imagine that we have a concept defined theoretically and several operational definitions which might possibly be associated with this theoretical definition. On the basis of this latter definition, most scientists in the field will probably agree that certain of the operations should be eliminated as not tapping what is implied in the theoretical definition. For example, they may decide that items referring to delinquent tendencies or musical tastes should not be used to measure prejudice. But there may be several operations which have more or less equal status in the opinion of these judges. In other words, on the basis of the theoretical definition, experts may not be able to agree that one operational procedure should be selected in preference to the others. We may then say that *to the extent that these several procedures yield different results* (under similar circumstances), *the theoretical definition is unsatisfactory* in the sense that it is probably in need of revision or clarification. For example, the concept *prejudice* may have been defined in such a manner that it is too vague. Perhaps it will be found necessary to distinguish between several kinds or dimensions of prejudice and to associate different operations with each. In a manner such as this—whether it is explicitly recognized or not—the research process may be used to help clarify theoretical concepts.

There thus seem to be two distinct languages linked by a kind of dictionary, arrived at by consensus, which enables one to associate concepts in the one language with those in the other. Scientists think in the theoretical language and make tests in the operational language. It is not necessary to associate operations with all concepts in the theoretical language. It is important to realize, however, that concepts which have not been operationally defined should not ordinarily be permitted to appear in statements purporting to be testable hypotheses. If this occurs, the questions raised by the hypotheses will usually be operationally meaningless and may lead to endless debate.

2.2 Level of Measurement: Nominal, Ordinal, and Interval Scales

We have just seen that the process of going from theoretically defined concepts to operationally defined concepts is by no means direct. Certain decisions have to be made in associating one type of concept with the other. Likewise, the process of selecting the appropriate mathematical or statistical model to be used with a given research technique or operational procedure also involves a number of important decisions. It might be thought that once a phenomenon had been measured, the choice of a mathematical system would be a routine one. This all depends on what one means by the word *measure*. If we use the term to refer only to those types of measurement ordinarily used in a science such as physics (e.g., the measurement of length, time, or mass), there is little or no problem in the choice of a mathematical system. But if we broaden the concept of measurement to include certain categorization procedures ordinarily used in the social sciences, as will be done in this text, the whole problem becomes more complex. We can then distinguish among several levels of measurement, and we shall find different statistical models appropriate to each.[1]

Nominal scales The basic and most simple operation in any science is that of classification. In classifying we attempt to sort elements with respect to a certain characteristic, making decisions about which elements are most similar and which most different. Our aim is to sort them into categories that are as homogeneous as possible as compared with differences between categories. If the classification is a useful one, the categories will also be found to be homogeneous with respect to *other* variables [10]. For example, we sort persons according to religion (Methodists, Presbyterians, Catholics, etc.) and then see if religion is related to prejudice or political conservatism. We might find that Presbyterians tend to be more conservative than Catholics, the Presbyterians having uniformly high scores as compared with the Catholics. Had individuals been sorted as to hair color, a perfectly proper criterion for classification, probably no significant differences would have been found among hair-

---

[1] For further discussions of these different levels of measurement see [5], [7], [12], and [13].

color classes with respect to other variables studied. In other words, differences *among* hair-color classes would have been slight as compared with differences *within* each category.

Classification is fundamental to any science. All other levels of measurement, no matter how precise, basically involve classification as a minimal operation. We therefore can consider classification to be the lowest level of measurement as the term is used in its broadest sense. We arbitrarily give names to the categories as convenient tags, with no assumptions about relationships between categories. For example, we place Presbyterians and Catholics in distinct categories, but we do not imply that one is greater than or better than the other. As long as the categories are exhaustive (include all cases) and nonoverlapping or mutually exclusive (no case in more than one category), we have the minimal conditions necessary for the application of statistical procedures. The term *nominal scale* has been used to refer to this simplest level of measurement. Formally, nominal scales possess the properties of symmetry and transitivity. By symmetry we mean that a relation holding between A and B also holds between B and A. By transitivity we mean that if $A = B$ and $B = C$, then $A = C$. Put together, this simply means that if A is in the same class as B, B is in the same class as A, and that if A and B are in the same class and B and C in the same class, then A and C must be in the same class.

It should be pointed out that numbers may be arbitrarily used as tags for different categories, but this fact in no way justifies the use of the usual arithmetic operations on these numbers. The functions of numbers in this case is exactly the same as that of names, i.e., the labeling of categories. It would obviously make absolutely no sense to add social security numbers or room numbers in a hotel. Although we would never be tempted to carry out such a ridiculous operation, there are other instances in social science research where the absurdity is by no means as obvious. Thus although numerical values may arbitrarily be assigned to various categories, the use of certain of the more common mathematical operations (addition, subtraction, multiplication, and division) requires also that certain *methodological* operations be carried out in the classification procedure. We shall have occasion shortly to see what the nature of these operations must be.

Ordinal scales   It is frequently possible to order categories with respect to the degree to which they possess a certain characteristic, and yet we may not be able to say exactly how much they possess. We thus imagine a single continuum along which individuals may be ordered. Perhaps we can rank individuals so precisely that no two are located at the same point on the continuum. Usually, however, there will be a number of ties. In such instances, we are unable to distinguish among certain of the individuals, and we have lumped them together into a single category. We are able to say, however, that these individuals all have higher scores than certain other individuals. Thus we may classify families according to socioeconomic status: *upper, upper-middle, lower-middle,* and *lower.* We might even have only two categories, *upper* and *lower.*

The kind of measurement we are discussing is obviously at a somewhat higher level than that used in obtaining a nominal scale since we are able not only to group individuals into separate categories, but to order the categories as well. We refer to this level of measurement as an *ordinal scale.* In addition to having the symmetrical properties of the nominal scale, an ordinal scale is asymmetrical in the sense that certain special relationships may hold between A and B which do *not* hold for B and A. For example, the relationship *greater than* ($>$) is asymmetric in that if $A > B$, it cannot be true that $B > A$. Transitivity still holds: if $A > B$ and $B > C$, then $A > C$. It is these properties, of course, which enable us to place $A, B, C, \ldots$ along a single continuum.

It is important to recognize that an ordinal level of measurement does not supply any information about the *magnitude* of the differences between elements. We know only that A is greater than B but cannot say how much greater. Nor can we say that the difference between A and B is less than that between C and D.[2] We therefore cannot add or subtract distances except in a very restricted sense. For example, if we had the following relationships

we can say that the distance

$$\overline{AD} = \overline{AB} + \overline{BC} + \overline{CD}$$

but we cannot attempt to compare the distances $\overline{AB}$ and $\overline{CD}$. In other words, when we translate order relations into mathematical operations, we cannot, in general, use the usual operations of addition, subtraction, multiplication, and division. We can, however, use the operations *greater than* and *less than* if these prove useful.

---

[2] The term *ordered-metric* has been used to refer to scales in which it is possible to rank the size of *differences* between elements. See [7].

Interval and ratio scales In the narrow sense of the word, the term *measurement* may be used to refer to instances in which we are able not only to rank objects with respect to the degree to which they possess a certain characteristic but also to indicate the exact distances between them. If this is possible, we can obtain what is referred to as an *interval scale.* It should be readily apparent that an interval-scale level of measurement requires the establishment of some sort of physical unit of measurement which can be agreed upon as a common standard and which is replicable, i.e., can be applied over and over again with the same results. Length is measured in terms of feet or meters, time in seconds, temperature in degrees Fahrenheit or centigrade, weight in pounds or grams, and income in dollars. On the other hand, there are no such units of intelligence, authoritarianism, or prestige which can be agreed upon by all social scientists and which can be assumed to be constant from one situation to the next. Given a unit of measurement, it is possible to say that the difference between two scores is twenty units or that one difference is twice as large as a second. This means that it is possible to add or subtract scores in an analogous manner to the way we can add weights on a balance or subtract 6 inches from a board by sawing it in two ([3], pages 296 to 298). Similarly, we can add the incomes of husband and wife, whereas it makes no sense to add their IQ scores.

If it is also possible to locate an absolute or nonarbitrary zero point on the scale, we have a somewhat higher level of measurement referred to as a *ratio scale.* In this case we are able to compare scores by taking their ratios. For example, we can say that one score is twice as high as another. Had the zero point been arbitrary, such as in the case of the centigrade or Fahrenheit scales, this would not have been legitimate. Thus we do not say that 70°F is twice as hot as 35°F, although we can say that the *difference* between these temperatures is the same as that between 105°F and 70°F. In practically all instances known to the writer this distinction between interval and ratio scales is purely academic, however, as it is extremely difficult to find a legitimate interval scale which is not also a ratio scale. This is due to the fact that if the size of the unit is established, it is practically always possible to conceive of zero units, even though we may never be able to find a body with no length or mass or to obtain a temperature of absolute zero. Thus in practically all instances where a unit is available it will be legitimate to use all of the ordinary operations of arithmetic, including square roots, powers, and logarithms.

*Certain important questions arise concerning the legitimacy of using interval scales in the case of a number of sociological and social-psychological variables. Unfortunately, it will be impossible to discuss these questions in detail in a general text such as this, although some can be mentioned briefly. It is sometimes argued that a variable such as income, if measured in dollars, is not really an interval scale since a difference of $1,000 will have different psychological meanings depending on whether the difference is between incomes of $2,000 and $3,000 or between incomes of $30,000 and $31,000. This argument, it seems, confuses the issue. What is actually being said here is that income as measured in dollars and "psychological income" (assuming this could be measured in terms of some unit) are not related in a straight-line or linear fashion. This is a question of fact which is irrelevant to the question of whether or not there is a legitimate unit of measurement.

*Many ratio scales are obtained through the operation of enumeration of behavioral acts, persons, occupations, or groups of various kinds. For example, crime rates are obtained by counting the number of recorded criminal acts and standardizing for the population base. Most of our census data for cities, counties, states, or regions are obtained by counting various kinds of people and dividing by the population base: per cent urban, percentage of the labor force that is unemployed, the average size of families, per cent nonwhite, and so forth. Complexity of the division of labor may be measured in terms of the number of distinct occupations, or one may obtain an index of organizational complexity by counting the number of branch offices. Arguments sometimes arise as to whether such measures are legitimate ratio scales (see Coleman [4] for an excellent statement of the problem). If one takes a strictly operational point of view that the measure used constitutes the definition of the variable of interest, there can be little question that a legitimate ratio scale has been attained, since there are distinct units being counted and these units are being taken as equivalent (and therefore interchangeable). Thus if we add 1,000 blacks to a city's population and take away 1,000 whites, we make the fundamental assumption that, in terms of the measure used, it makes no difference *which* blacks or whites are involved. Furthermore, the zero point is well defined. A statement that the percentage of nonwhites in a city is zero is unambiguous.

*Whenever disputes arise over the adequacy of enumeration measures, and whether or not they legitimize the assumption of ratio-level measurement, I strongly suspect that the basic issue is of an entirely different sort, namely that of the linkage between the measure being used and the theoretical construct it is intended to measure. For example, unemployment rates may be used as an indicator of the malfunctioning of the economy, a minority percentage as an indicator of a threat posed by the minority, or per cent urban as an indicator of the influence of urban values. In all such instances, statistics per se will never be able to resolve the controversies, and we shall have to sidestep the basic issues by assuming that we are always dealing with the variable we are intending to measure.

*Another question can be raised as to whether or not it is possible to attain an interval scale in the area of attitude measurement. Several attempts have been made to approximate this goal. In the Thurstone method of equal-appearing intervals, judges are asked to sort items into a number of piles which are equally far apart along the attitude continuum ([11], pages 359 to 365). It is reasoned essentially that if there is a high degree of consensus among the judges, an interval scale can legitimately be used. This process, it is argued, is similar in essence to that used in other disciplines to obtain interval scales. The argument seems to be a legitimate one provided there actually is a high degree of consensus among judges and provided judges are given a large number of piles into which they can sort the items. For example, if they were forced to place items into one of three or four piles, we would expect a high degree of consensus simply due to the crudity of the measuring instrument. There would be such a wide range of variability within each pile that one could hardly claim that the items in the various piles were equal distances apart. Even assuming perfect agreement and maximal freedom in sorting the items into piles, Thurstone's method still presents difficulties in conceptualizing the unit concerned. It becomes necessary to postulate that it is the existence of such a unit which makes agreement among judges possible. It seems safe to say that at this point in the development of attitude measurement most techniques yield very poor approximations to interval scales. Many probably should not even be considered to give legitimate ordinal scales. The implications of this for statistical analysis will become apparent as we proceed.

2.3 Measurement and Statistics

We have seen that there are several distinct levels of measurement, each with its own properties. It should be noted that these various levels of measurement themselves form a cumulative scale. An ordinal scale possesses all the properties of a nominal scale plus ordinality. An interval scale has all the qualities of both nominal and ordinal scales plus a unit of measurement, and a ratio scale represents the highest level since it has not only a unit of measurement but an absolute zero as well. The cumulative nature of these scales means that it is always legitimate to drop back one or more levels of measurement in analyzing our data. If we have an interval scale, we also have an ordinal scale and can make use of this fact in our statistical analysis. Sometimes this will be necessary when statistical techniques are either unavailable or otherwise unsatisfactory in handling the variable as an interval scale. We lose information in doing so, however. Thus if we know that Jones has an income of $11,000 and Smith an income of $6,000 and if we make use only of the fact that Jones has the higher of the two incomes, we in effect throw away the information that the difference in incomes is $5,000. Therefore in general we shall find it to our advantage to make use of the highest level of measurement that can legitimately be assumed.

What can be said of the reverse process of going up the scale of measurement from, say, an ordinal to an interval scale? We are often tempted to do so since we would then be able to make use of more powerful statistical techniques. It is even possible that we may do so without being at all aware of exactly what has occurred. It is important to realize that there is nothing in the statistical or mathematical procedures we ultimately use that will enable us to check upon the legitimacy of our research methods. *The use of a particular mathematical model presupposes that a certain level of measurement has been attained.* The responsibility rests squarely on the shoulders of the researcher to determine whether or not his operational procedures permit the use of certain mathematical operations. He must first make a decision as to the appropriate level of measurement, and this in turn will determine the appropriate mathematical system. In other words, a particular mathematical model can be associated with a certain level of measurement according to the considerations discussed in the previous section. The common arithmetic operations, for example, can ordinarily be used only with interval and ratio scales.

*Again, we are faced with a problem of translating from the one language to the other. The operational language involves certain physical operations such as the use of a unit of measurement. The mathematical language involves a completely abstract set of symbols and mathematical operations and is useful not only because it is precise and highly developed, but also because its abstract nature permits its application to a variety of empirical problems. Mathematics makes use of deductive reasoning in which one goes from a set of definitions, assumptions, and rules of operation to a set of conclusions by means of purely logical reasoning. Mathematics per se tells us nothing new about reality since all the conclusions are built into the original definitions, assumptions, and rules and are not determined empirically. These mathematical conclusions then have to be translated back into the operational and theoretical languages if they are to be useful to the scientist [5].

We are arguing, then, that it is not legitimate to use a mathematical system involving the operations of addition or subtraction when this is not warranted by the method of measurement. Although the meaning of this fact will not become fully apparent until we actually begin using the various scales of measurement, we are in effect saying that we cannot legitimately travel up the measurement hierarchy unless the measurement process itself has been improved. No mathematical manipulations can make this possible. How do we decide what level of measurement is legitimate, then? Unfortunately the problem is not quite so simple as might be assumed. A few brief examples should be sufficient to warn you of the complexity of the problem.

*To illustrate one such problem, it is necessary to distinguish ordinal and interval scales from the *partially ordered scale* resulting from the combination of two or more ordinal (or interval) scales into a single index. It is frequently the case in sociology and other social sciences that what first appears to be a simple ordinal (interval) scale is in reality a combination of several ordinal (interval) scales, the result being that one cannot obtain an unambiguous ranking of individuals unless he is willing to make some further decisions. Let us take the example of socio-economic status. Usually, we determine a person's status by examining a number of distinct criteria such as income, occupation, education, family background, or area of residence. If A has a higher rank than B on each of these criteria, then clearly A can be rated higher as to general status. But what if A has the higher income but B the more prominent family name? Which has the higher general status? We have several alternatives. The first is to drop the notion of general status and to think in terms of separate dimensions of status, each of which may admit of an ordinal level of measurement. We then end up with not one but a number of ordinal scales, and it becomes an empirical question as to how highly intercorrelated the different dimensions may be. Of course if there is a perfect relationship among all dimensions, the question becomes an academic one since A will be higher than B on all dimensions if he is higher on any one dimension. In practice, of course, this never happens.

*Our second alternative is to try to "force" an ordinal scale on the data by making some decisions about the relative weightings of each dimension and the equivalences involved. For example, if we can assume that a year's extra education is equivalent to $1,338.49 in extra income, we can then translate educational units into income units and arrive at a unidimensional scale. Obviously, the problem of translating family background or area of residence into income is even more complicated. The method of measurement we are discussing here involves a type of index construction. Suffice it to say that such index construction usually involves arbitrary decisions about relative weighting. If the weighting system can be justified, an ordinal scale can be used; if it cannot, then one is in real doubt about whether or not individuals can legitimately be rank-ordered.

*One commonly used method of obtaining an ordinal scale is to make use of one or more judges to rank-order individuals according to some criterion such as power or prestige. Let us assume for simplicity that there is only one judge and that he has been forced to rank-order individuals according to their "social standing" in the community. Assuming the person cooperates, the method used assures us of obtaining an ordinal scale *regardless* of how individuals actually compare in the eyes of the judge. It is possible that had some other method been used, an ordinal scale would not have been obtained. Had a paired-comparisons technique been used in which judgments were made between every pair combination, the judge might have rated Smith higher than Brown, Brown over Jones, but Jones over Smith, thus violating the transitivity property of ordinal scales. The researcher is now faced with a choice. He can reach the conclusion that there is a partially ordered scale of some kind. Or he may consider the judge to be inconsistent or in error. As Coombs points out, this problem of what to call measurement error is a basic dilemma faced by the social scientist ([7], pages 485 to 488). Generally speaking, he may assume a high level of measurement and conceive of deviations of the above sort as measurement errors, or he may drop down to a lower level of measurement.

*The same dilemma may be illustrated in the case of *Guttman scaling*. In the perfect Guttman scale, items have a cumulative property which justifies the assumption of an ordinal scale [14]. Items may be arranged from least to most extreme so that the exact response pattern of an individual can be reproduced from his total score. For example, if there are five arithmetic problems ranging from least to most difficult, a person who solves the most difficult problem correctly will also be able to solve the others. If he gets three correct, he will solve the three easiest ones and miss the last two. In a perfect social-distance scale, prejudice items may be arranged according to the degree of intimacy of contact with the minority. A white person willing to marry a black will obviously be willing to live on the same street with one; if he will accept him as a neighbor, he will be willing to ride on a bus next to one. Thus we can say in the case of the perfect Guttman scale that a person endorsing four items will have endorsed exactly the same items as a person with a score of three, *plus one more*. If the scale were only partially ordered, it would be possible to say that A is in some respects more prejudiced than B but in other respects less so since the two individuals have endorsed different sets of items.

*Seldom if ever do we attain a perfect Guttman scale in practice,

however. There are always certain persons whose response patterns
deviate from the ideal. The question then arises as to whether or not
to call these persons in error. Are they actually inconsistent because
they may accept a black as a neighbor but refuse to ride on a bus next
to him? Perhaps. But then again, perhaps not. Unless the researcher
is willing to *assume* that he has a legitimate ordinal scale, he cannot claim
the individual to be in error. If the number of "errors" is at all large,
we begin to suspect our scale. On the other hand, we are usually willing
to tolerate a relatively small number of errors. It is this principle which
is behind the decision to accept the Guttman scale as an ordinal scale if
the number of errors, as measured by the coefficient of reproducibility, is
very small. One should be well aware of the fact that his decision is
somewhat arbitrary, however, and that ultimately he will be faced with
the problem of what to call error.

*These examples should be sufficient to indicate that it is often not a
simple matter to decide what type of scale can legitimately be used.
Ideally, one should make use of a data-gathering technique that permits
the lowest levels of measurement, if these are all the data will yield, rather
than using techniques which force a scale on the data. Thus, the method
of paired comparisons will produce an ordinal scale only if the judge is in
fact actually able to rank the individuals. On the other hand, if he is
required to place them in a definite rank order, he will have to do so
whether he believes this can be achieved legitimately or not. Having
used this latter method of data collection and being unable to demonstrate
empirically that individuals can be ordered without doing an injustice to
data, one then has to *assume* a single continuum.

In order to emphasize that any given statistical technique presupposes
a specific level of measurement, we shall develop the habit of always
indicating the level of measurement required for each procedure. In
choosing among alternative procedures one of the important questions
that should always be asked is, "Is it legitimate to assume the level of
measurement required for a given technique?" If it is not, then perhaps
an alternative procedure should be found. If level of measurement were
the only consideration, the problem of choice among alternative procedures would be relatively simple. However, we often find that procedures
that do not require strong measurement assumptions, and which are
therefore preferable on these grounds alone, are also less satisfactory
with respect to other desired characteristics. Therefore one is often faced
with difficult decisions involving the necessity of evaluating the relative
seriousness of violated assumptions of various kinds. In such instances
it may be desirable to analyze one's data by several different methods to
see if the conclusions differ in important ways.

At this point, our discussion of these different levels of measurement and problems of choice among alternative tests and measures may not be too meaningful. One of the dangers of "cookbook" statistics is the tendency to oversimplify the criteria and problems involved in making basic decisions in data analysis. It is impossible to overemphasize the important point that, in using any statistical technique, one must be aware of the underlying assumptions that the procedure requires. In the context of the present discussion, *one* of the first questions that must always be asked concerns the level of measurement that can legitimately be assumed.

2.4 Organization of the Book

The organization of the remaining chapters is determined by a number of considerations, the first being the objective of presenting the simplest ideas first and gradually building to greater complexity. Therefore, since each section generally presupposes the previous materials, it will be to your advantage to follow this organization, skipping only the starred paragraphs and sections as appropriate. An exception is Chap. 14, which may be skipped altogether or considered in conjunction with the "nonparametric" tests and procedures of Chaps. 16 and 18. Chapter 21 on sampling may be read in conjunction with Chap. 9 on probability, though the sampling chapter contains several sections that can only be understood after reading Chaps. 11, 13, and 16. The bulk of Chap. 17 may be understood without first taking up Chap. 16 on analysis of variance. Otherwise, however, it is recommended that topics be studied in the order in which they have been presented.

Statistical tools are not easily grouped under only one or two headings, and therefore the labels for the major sections of the book are only partly appropriate and indicate only the primary focus of attention. In Part 2 there is an exclusive focus on descriptive statistics, whereas in Parts 3 and 4 the primary though not exclusive focus is on induction, tests of hypotheses, and estimation of population parameters on the basis of sample data. In Parts 2 and 3 we will be concerned almost completely with procedures that involve a single variable at a time, whereas in Part 4 we move to the more challenging problem of handling two or more variables at once.

Cutting across these distinctions between description and induction, and between *univariate* and *bivariate* or *multivariate* statistics, is a third organizational principle, namely, one involving the levels of measurements for each of the variables. Many of the chapter titles indicate this level of measurement, but perhaps the best way to obtain a summary perspective of the contents is to refer to the table of tests and measures that appears on the inside cover. Procedures for use with single variables are indicated in the first column of the table. We see that in Chap. 3 we shall be concerned with the very simple measures (percentages, proportions, and ratios) used in connection with both dichotomies and general nominal scales with more than two categories. Tests of hypotheses involving single nominal scales will be considered in Chaps. 10, 11, and 12. Measures (the median, quartile deviation) appropriate for use with a single ordinal scale are taken up very briefly in Chaps. 5 and 6, and a very simple test (the binomial) that can be applied with ordinal data will be considered in Chap. 10. Interval and ratio scales will occupy relatively more of our attention, being considered in Chaps. 4 to 7 that deal with univariate descriptive procedures and again in Chaps. 11 and 12 in Part 3 on inductive statistics.

Beginning with Chap. 13 we turn our attention to relationships among two or more variables, which of course implies that we must be concerned with the level of measurement of the second (and additional) variable as well as that of the first. Columns 2 to 5 of the table allow for varying combinations with respect to the level of measurement of the two variables. For example, the top cell of column 2 refers to situations in which two dichotomies are being interrelated (e.g., sex versus political preference). In the second row of column 2 we allow for the possibility that the first nominal scale has more than two categories (e.g., Protestant, Catholic, and Jew). In the third row one variable is a dichotomy (e.g., sex) whereas the second is an ordinal scale, and so forth. There is only one cell that is not filled in, namely, that for which one variable is measured at the ordinal level and the second at the interval or ratio level. Although we may indeed handle such situations, we lack really satisfactory tools that do not require a loss of information by reducing the level of measurement in either one of the two variables. There is of course no need to fill in the cells above the diagonal cells in the table, since these are already covered by the cells below the diagonal.

At this point it is premature to introduce a discussion of each of the possibilities listed in the table. The main point to note is that the level of measurements involved is one of the important considerations in one's choice among procedures. As long as one is confined to only two variables, this choice is usually relatively (though not completely) straightforward. It is much more difficult in the case of multivariate analysis, where one is often working with five or even as many as fifteen or twenty variables at once, and where it is extremely unlikely that all will be measured at the same level, and where it is often undesirable to use too many different kinds of tests and measures. Some of these problems of multivariate analysis are discussed in Chaps. 15, 16, 19, and 20. At several points, particularly at the ends of Chaps. 14 and 20, you will find summary statements concerning some of the basic considerations involved in selecting among these alternative procedures.

It will be seen that not all the possible combinations are handled with the same degree of thoroughness in this text. This is the case not only because of space limitations and the necessity of dwelling upon fundamental ideas, but also because the theory of statistics is much more highly developed in some respects than others. In particular, much more work has been done in the area of so-called "parametric statistics" involving interval and ratio scales than with ordinal procedures. Thus our tools for handling interval and ratio scales are much better developed, particularly in the case of multivariate analysis. The distinction between interval and ratio scales is also not exploited in statistical theory, at least at the level with which we shall be concerned. The basic reason is that the statistical models with which we generally work are based upon a general linear equation that is additive, rather than involving ratios of variables. Therefore, for all practical purposes, it is not necessary to keep this distinction in mind as you proceed. Periodic referrals to the table on the inside cover, however, will probably be necessary.

Glossary
You should develop the habit of explaining in your own words the meanings of important concepts. New concepts introduced in this chapter are:

Interval scale
Nominal scale
*Operational definition
Ordinal scale
Ratio scale

References
1. Blalock, H. M.: "The Measurement Problem: A Gap between the Languages of Theory and Research," in H. M. Blalock and Ann B. Blalock (eds.), *Methodology in Social Research*, McGraw-Hill Book Company, New York, 1968, chap. 1.
2. Bridgman, P. W.: *The Logic of Modern Physics*, The Macmillan Company, New York, 1938, pp. 1–39.
3. Cohen, M. R., and E. Nagel: *An Introduction to Logic and Scientific Method*, Harcourt, Brace and Company, Inc., New York, 1937, chaps. 12 and 15.
4. Coleman, James S.: *Introduction to Mathematical Sociology*, The Free Press, New York, 1964, chap. 2.
5. Coombs, C. H., H. Raiffa, and R. M. Thrall: "Some Views on Mathematical Models and Measurement Theory," *Psychological Review*, vol. 61, pp. 132–144, March, 1954.
6. Coombs, C. H.: *A Theory of Data*, John Wiley & Sons, Inc., New York, 1964.

7. Coombs, C. H.: "Theory and Methods of Social Measurement," in L. Festinger and D. Katz (eds.), *Research Methods in the Behavioral Sciences*, The Dryden Press, Inc., New York, 1953, pp. 471–535.

8. Lundberg, G. A.: *Foundations of Sociology*, The Macmillan Company, New York, 1939, chaps. 1 and 2.

9. Northrop, F. S. C.: *The Logic of the Sciences and the Humanities*, The Macmillan Company, New York, 1947, chaps. 5–7.

10. Radcliffe-Brown, A. R.: *A Natural Science of Society*, The Free Press of Glencoe, Inc., New York, 1957, pp. 28–42.

11. Selltiz, C., M. Jahoda, M. Deutsch, and S. W. Cook: *Research Methods in Social Relations*, Henry Holt and Company, Inc., New York, 1959, chaps. 5 and 10.

12. Senders, V. L.: *Measurement and Statistics*, Oxford University Press, New York, 1958, chap. 2.

13. Stevens, S. S.: "Mathematics, Measurement, and Psychophysics," in S. S. Stevens (ed.), *Handbook of Experimental Psychology*, John Wiley & Sons, Inc., New York, 1951, pp. 1–49.

14. Stouffer, S. A., et al.: *Measurement and Prediction*, Princeton University Press, Princeton, N.J., 1950, chaps. 1 and 3.

15. Weiss, R. S.: *Statistics in Social Research*, John Wiley & Sons, Inc., New York, 1968, chap. 2.

Univariate Descriptive Statistics

Part 2

Nominal Scales: Proportions, Percentages, and Ratios

3

It is a much simpler matter to summarize data involving nominal scales than is the case when interval scales are used. The basic arithmetic operation is that of counting the number of cases within each category and then noting their relative sizes. A given group may consist of 36 males and 24 females, or 25 Protestants, 20 Catholics, and 15 Jews. In order to make comparisons with other groups, however, it is necessary to take the number of cases in the two groups into consideration. The measures to be discussed in the present chapter permit comparisons among several groups by essentially standardizing or controlling for size. At least two of these measures, proportions and percentages, are undoubtedly already familiar.

3.1 Proportions

In order to make use of proportions we must assume that the method of classification has been such that categories are mutually exclusive and exhaustive. In other words, any given individual has been placed in one and only one category. For sake of simplicity, let us take a nominal scale consisting of four categories with N_1, N_2, N_3, and N_4 cases, respectively. Let the total number of cases be N. The proportion of cases in any given category is defined as the number in the category divided by the total number of cases. Therefore the proportion of individuals in the first category is given by the quantity N_1/N, and the proportions in the remaining categories are N_2/N, N_3/N, and N_4/N, respectively. Obviously, the value of a proportion cannot be greater than unity. Since

$$N_1 + N_2 + N_3 + N_4 = N$$

we have

$$\frac{N_1}{N} + \frac{N_2}{N} + \frac{N_3}{N} + \frac{N_4}{N} = \frac{N}{N} = 1$$

Thus if we add the proportions of cases in all (mutually exclusive) categories, the result is unity. This is an important property of proportions and can easily be generalized to any number of categories.

Let us illustrate the use of proportions with the data given in Table 3.1.

Table 3.1   Numbers of delinquents and nondelinquents in two hypothetical communities

Subjects	Community 1	Community 2
Delinquents		
First offenders	58	68
Repeaters	43	137
Nondelinquents	481	1081
Total	582	1286

It is rather difficult to tell which community has the larger relative number of delinquents because of their different sizes. Expressing the data in terms of proportions enables us to make a direct comparison, however. The proportion of first offenders in community 1 is $58/582$ or .100; the comparable figure for the second community is 68/1,286 or .053. Other proportions can be computed in a similar fashion and the results summarized in tabular form (Table 3.2). From this table we see that the relative numbers of delinquents are quite similar for the two communities but that the second community contains a substantially lower proportion of first offenders and a higher proportion of repeaters.

Table 3.2   Proportions of delinquents and nondelinquents in two hypothetical communities

Subjects	Community 1	Community 2
Delinquents		
First offenders	.100	.053
Repeaters	.074	.107
Nondelinquents	.826	.841
Total	1.000	1.001

The sum of the proportions in community 2 is not exactly unity because of rounding errors. Sometimes it is desirable to present the data in such a manner that the sums are exactly equal to 1.000. This may require adjustment of some of the category proportions, and by convention we usually modify the figures for those categories having the largest proportions of cases.[1] The argument in favor of this procedure is that a change in the last decimal place in a larger proportion is relatively less significant than the same change in a smaller figure. Thus the proportion of non-delinquents in community 2 could be changed to .840 so that the resulting sum is unity.

Table 3.2 involves proportions of the total number of cases in each community. Suppose, however, that interest was mainly centered on the delinquents, and we wanted to know the proportion of repeaters *among the delinquents*. The total numbers of delinquents in the two communities are 101 and 205, respectively. Therefore, among the delinquents the proportions of repeaters are $43/101$ or .426 and $137/205$ or .668. At first glance these figures may give a slightly different impression from the first set of proportions. We would especially want to be on guard against concluding that the second sample is "more delinquent" than the first. This last set of proportions, of course, tells us absolutely nothing about the relative numbers of nondelinquents in the two samples. Obviously there is no substitute for a careful reading of the tables. It is a good idea to develop the habit of always determining the categories that are being included in the total number of cases used in the denominator of the proportion. You should always ask, *"Of what* is this a proportion?" The answer should be clear from the context.

3.2 Percentages

Percentages can be obtained from proportions by simply multiplying by 100. The word *per cent* means *per hundred*. Therefore in using percentages we are standardizing for size by calculating the number of individuals who would be in a given category if the total number of cases were 100 and if the proportion in each category remained unchanged. Since proportions must add to unity, it is obvious that percentages will sum to 100 unless the categories are not mutually exclusive or exhaustive.

Percentages are much more frequently used in reporting results than are proportions. The figures in Table 3.2 could just as well have been expressed in terms of percentages. Rather than make use of these same data, let us take another table which can be used to illustrate several new points. Suppose there are three family service agencies having a distribution of cases as indicated in Table 3.3.

As is customary, percentages have been given to the nearest decimal and adjustments made in the last digits so that totals are exactly 100.0. Here there are a sufficient number of cases for each agency to justify the

[1] Exactly the same procedure can be used in the case of percentages.

computation of percentages. It would have been misleading to give percentages had the number of cases been much smaller, however. Suppose agency C had handled only 25 cases in all. If there had been four unwed mothers and seven engaged couples, the percentages in these categories would have been 16 per cent and 28 per cent, respectively. Since many persons are accustomed to looking only at percentages and not at the actual number of cases involved, the impression that there were many more engaged couples than unwed mothers could easily be

Table 3.3   Numbers and percentage distributions of cases handled by three hypothetical family service agencies

Type of case	Agency A		Agency B		Agency C		Total	
	No.	%	No.	%	No.	%	No.	%
Married couples	63	47.3	88	45.5	41	36.6	192	43.8
Divorced persons	19	14.3	37	19.2	26	23.2	82	18.7
Engaged couples	27	20.3	20	10.4	15	13.4	62	14.2
Unwed mothers	13	9.8	32	16.6	21	18.8	66	15.1
Other	11	8.3	16	8.3	9	8.0	36	8.2
Total	133	100.0	193	100.0	112	100.0	438	100.0

conveyed. As will be seen when we come to inductive statistics, a difference between four and seven cases could very possibly be due to the operation of chance factors. The use of percentages and proportions usually implies considerably more stability of numbers. For this reason two rules of thumb are important: (1) *always report the number of cases along with percentages or proportions*, and (2) *never compute a percentage unless the number of cases on which the percentage is based is in the neighborhood of 50 or more*. If the number of cases is very small, it would be preferable to give the actual number in each category rather than percentages. For example, we would simply state that agency C handled four unwed mothers and seven engaged couples.

Now look at the total column which gives the breakdown of percentages for all three agencies. These figures were obtained by adding the number of cases of each type and also the total number of cases handled by the three agencies. An N of 438 was then used as a base in computing total percentages. Suppose, however, that the number of cases had not been given to us in the body of the table but had been presented as in Table 3.4. There might be some temptation to obtain the total percentages by taking a straightforward arithmetic mean of the three percentages in each row. Such a procedure would not take into consideration the fact that the agencies handled different numbers of cases; it would be justified only if the numbers of cases were in fact identical. The correct procedure would be to weight each percentage by the proper number of cases. One way to do this would be to work backwards to obtain the actual number of cases in each cell. This could be accomplished by multiplying the total number of cases handled by the agency by the *proportion* in a given category. For example, $(133)(.473) = 63$.

Notice that the percentages as given in Tables 3.3 and 3.4 are designed to answer certain questions but not others. They enable us to examine each agency separately and see the breakdown of cases handled. They also permit comparison of agencies with respect to type of cases handled. For example, agencies B and C handled relatively more unwed mothers

Table 3.4 Percentage distributions of cases handled by three hypothetical family service agencies, with percentages computed down the table

Type of case	Agency A ($N = 133$), %	Agency B ($N = 193$), %	Agency C ($N = 112$), %
Married couples	47.3	45.5	36.6
Divorced persons	14.3	19.2	23.2
Engaged couples	20.3	10.4	13.4
Unwed mothers	9.8	16.6	18.8
Other	8.3	8.3	8.0
Total	100.0	100.0	100.0

and divorced persons than did agency A. Suppose, however, that we had been primarily interested in cases of a given type and in the relative numbers handled by each agency. For example, we might wish to know the percentage of all married couples who went to agency B. It would be more helpful in this instance to compute percentages across the table. We could take the total number of married couples and determine what percentages of this category were handled by agencies A, B, and C, respectively. Percentages would then add to 100 across the table, but not down, and the results would be summarized as in Table 3.5.

Percentages may thus be computed in either direction, and careful attention should be given to each table to determine exactly how each percentage or proportion has been obtained. In instances where one's theory dictates which variable is taken to be causally dependent, and which is to be considered causally prior or independent, a simple rule of thumb usually suffices. If we follow the practice of placing the independent variable at the top of the table and the dependent variable on the left-hand side, then we should compute percentages to add to 100 *down*, making our comparisons *across*. Thus in our example comparing delinquency rates in two communities, we would ordinarily expect that certain community characteristics may affect delinquency, rather than vice versa. When we compute percentages to add to 100 down, we are in effect standardizing or controlling for size of community, since we recognize that factors affecting the relative sizes of these communities, or our samples from within each community, are *not* causally dependent on

Table 3.5 Percentage distributions of cases handled by three hypothetical family service agencies, with percentages computed across the table

Type of case	Agency A ($N = 133$), %	Agency B ($N = 193$), %	Agency C ($N = 112$), %	Total ($N = 438$), %
Married couples ($N = 192$)	32.8	45.8	21.4	100.0
Divorced persons ($N = 82$)	23.2	45.1	31.7	100.0
Engaged couples ($N = 62$)	43.5	32.3	24.2	100.0
Unwed mothers ($N = 66$)	19.7	48.5	31.8	100.0
Other ($N = 36$)	—*	—*	—*	—*

* Percentages not computed where base is less than 50.

their delinquency rates. In computing percentages down we are therefore controlling for these factors that affect the sizes of our two samples. This point should become more understandable once we have considered the notion of the slope of a straight line, where one of the variables is taken as dependent on the other (see Chap. 17). It turns out that percentages computed in the suggested direction can be considered special cases of these slopes.

3.3 Ratios

The ratio of one number A to another number B is defined as A divided by B. Here the key term is the word *to*. Whatever quantity precedes this word is placed in the numerator, while the quantity following it becomes the denominator. Suppose that there are 365 Republicans, 420 Democrats, and 130 Independents registered as voters in a local election. Then the ratio of Republicans *to* Democrats is $^{365}/_{420}$; the ratio of Republicans and Democrats *to* Independents is $(365 + 420)/130$. Notice that unlike a proportion, a ratio can take on a value greater than unity. We also see that the expression preceding or following the word *to* can consist of several separate quantities (e.g., Republicans and

Democrats). Usually a ratio is reduced to its simplest form by canceling common factors in the numerator and denominator. Thus the ratio of Democrats to Independents would be written as $42\!\!/\!_{13}$ or equivalently as 42:13. Sometimes it is desirable to express a ratio in terms of a denominator of unity. For example, the ratio of Democrats to Independents might be written as 3.23 to 1.

Proportions obviously represent a special type of ratio in which the denominator is the total number of cases and the numerator a certain fraction of this number. Ordinarily the term *ratio* is used to refer to instances in which the A and B represent separate and distinct categories, however. We might obtain the ratio of delinquents to nondelinquents or of married couples to engaged couples. It is evident that with four or five categories, the number of possible ratios that could be computed is quite large. Therefore, unless interest is centered primarily on one or several pairs of categories, it will usually be more economical and less confusing to the reader to make use of percentages or proportions. Notice that if there are only two categories, it will be possible to compute a proportion directly from a ratio or vice versa. For example, if we know that the ratio of males to females is $3:2$, then out of every five persons there must be an average of three males and two females. Therefore the proportion of males is $3\!\!/\!_5$ or .6.

Ratios can be expressed in terms of any base that happens to be convenient. The base of a ratio is indicated by the magnitude of the denominator. For instance sex ratios are conventionally given in terms of the number of males per 100 females. A sex ratio of 94 would therefore indicate slightly fewer males than females, whereas a sex ratio of 108 would mean a preponderance of males. Bases involving large numbers such as 1,000 or 100,000 are often used in computing *rates*, another type of ratio, whenever the use of proportions or percentages might result in small decimal values. Birth rates, for example, are usually given in terms of the number of live births per 1,000 females of childbearing age. Murder rates may be given in terms of the number of murders per 100,000 population.

Rates of increase are another common type of ratio. In computing such a rate we take the actual increase during the period divided by the size at the *beginning* of the period. Thus if the population of a city increases from 50,000 to 65,000 between 1940 and 1950, the rate of increase for the decade would be

$$\frac{65,000 - 50,000}{50,000} = .30$$

or 30 per cent. In the case of rates of increase, percentages may obviously go well over 100 per cent and will be negative if the city has actually decreased in size.

Glossary
Percentage
Proportion
Rate
Ratio

Exercises
1. Suppose you are given the following table showing the relationship between church attendance and class standing at a certain university:

Church attendance	Class standing				Total
	Freshmen	Sophomores	Juniors	Seniors	
Regular attenders	83	71	82	59	295
Irregular attenders	31	44	61	78	214
Total	114	115	143	137	509

    a. What is the percentage of regular attenders in the total sample? (*Ans.* 57.96%)
    b. What is the ratio of freshmen to seniors?
    c. Among the regular attenders, what is the ratio of underclassmen (freshmen and sophomores) to upperclassmen? (*Ans.* 1.09 to 1)
    d. What is the proportion of irregular attenders who are seniors? The proportion of seniors who are irregular attenders? (*Ans.* .364; .569)
    e. Are there relatively more irregular attenders among freshmen and sophomores than among juniors and seniors? Express results in terms of percentages.
    f. Summarize the data in several sentences.

2. A social psychologist studying the relationship between industrial productivity and type of group leadership obtains the following data indicating the productivity levels of individuals within each of three types of leadership groups:

Productivity	Group-leadership type			Total
	Democratic	Laissez-faire	Authoritarian	
High	37	36	13	86
Medium	26	12	71	109
Low	24	20	29	73
Total	87	68	113	268

a. In which direction would you prefer to compute percentages? Why?

b. Compute these percentages and summarize the data briefly.

c. What is the ratio of high to low producers in each of the three groups? For these particular data, would these three ratios adequately summarize the data? Explain.

3. If the ratio of whites to nonwhites in a particular community is $\frac{8}{5}$, what is the proportion of nonwhites? Suppose the ratio of whites to blacks were $\frac{8}{5}$. Could you obtain the proportion of blacks in the same manner? Why or why not?

4. If a city had a population of 153,468 in 1960 and of 176,118 in 1970, what was the rate of growth (expressed as a percentage) between 1960 and 1970? (*Ans.* 14.76%)

5. If there are 12,160 males and 11,913 females in a certain county, what is the sex ratio (expressed in terms of the number of males per 100 females)?

References

1. Anderson, T. R., and M. Zelditch: *A Basic Course in Statistics,* 2d ed., Holt, Rinehart and Winston, Inc., New York, 1968, pp. 24–31.

2. Freeman, L. C.: *Elementary Applied Statistics,* John Wiley & Sons, Inc., New York, 1965, chap. 4.

3. Hagood, M. J., and D. O. Price: *Statistics for Sociologists,* Henry Holt and Company, Inc., New York, 1952, chap. 7.

4. Weiss, R. S.: *Statistics in Social Research,* John Wiley & Sons, Inc., New York, 1968, chap. 4.

5. Zeisel, Hans: *Say It with Figures,* 5th ed., Harper and Row, Publishers, Incorporated, New York, 1968, chaps. 1 and 2.

Interval Scales: **4**
Frequency
Distributions
and Graphic
Presentation

In the present chapter our concern will be with methods of summarizing data which are very similar to those discussed in the previous chapter. We shall group the interval-scale data into categories, order these categories, and then make use of these groupings to give an overall picture of the distribution of cases. In so doing, we can boil down information about a very large number of cases into a simple form that will enable a reader to picture the way the cases are distributed. Later we shall find that by grouping our data, we can also simplify considerably certain computations. In the following two chapters we shall be concerned with methods of summarizing data in a more compact manner so that they may be described by several numbers representing measures of typicality and degree of homogeneity.

4.1 Frequency Distributions: Grouping the Data

In the previous chapter we encountered few, if any, major decisions in summarizing our data. This was due to the fact that, presumably, the classes were already determined and all that was necessary was to count the number of cases in each class and then standardize for the number of cases in the total sample by computing a proportion, percentage, or ratio. If interval-scale data are to be summarized in a similar manner, however, an initial decision must be made as to the categories that will be used. Since the data will ordinarily be distributed in a continuous fashion, with few or no large gaps between adjacent scores, the classification scheme may be somewhat arbitrary. It will be necessary to decide how many categories to use and where to establish the cutting points. Unfortunately, there are no simple rules for accomplishing this since the decision will depend on the purposes served by the classification. Let us take up a specific example in order to illustrate the nature of the problem. Suppose the numbers given below represent the percentage of eligible voters voting in a school board election for 93 census tracts within a particular city.

39.2%	11.6%	36.3%	26.3%	37.1%	15.3%	27.3%	23.5%	13.3%
28.1	26.3	27.1	35.1	23.0	26.1	31.0	36.3	27.3
22.8	33.4	25.6	21.6	46.8	7.1	16.8	26.9	46.6
44.3	58.1	33.1	13.4	27.8	33.4	22.1	42.7	33.0
36.3	20.7	9.3	26.3	29.9	39.4	5.3	24.3	17.8
18.2	37.1	21.6	17.5	12.3	23.6	37.2	37.1	25.1
27.1	28.8	27.8	33.6	26.5	28.3	26.9	24.8	41.0
33.6	19.3	43.7	28.2	19.9	83.6	47.1	4.8	9.7
39.5	32.3	22.4	15.1	26.3	26.1	29.2	14.3	14.6
21.6	37.9	37.1	24.9	10.0	20.7	11.8	22.9	36.0
46.1	21.5	13.3						

Raw data presented in this manner are almost useless in giving the reader a clear picture of what is taking place. This will be especially true if the number of cases is very large. Suppose we wanted to compare this community and another with respect to voting turnout. A quick glance at the data indicates that most tracts had between 20 and 40 per cent turnout, but that there was one with an extremely high record. But it is indeed difficult to get a clear mental picture of the total distribution.

Number and size of intervals In order to visualize this total distribution we shall find it helpful to classify similar scores within the same category. We are immediately confronted with a problem, however. How many intervals should be used in grouping these data? How big should they be? Obviously, there is no point in using intervals with peculiar widths or limits. Thus we would probably select intervals of width 5, 10, or 20 rather than those of width 4.16. Also our end points, or class limits as they are usually called, would ordinarily involve round numbers such as 5.0 or 10.0. If we are in doubt about the intervals ultimately to be used, it is a good idea to tally the scores using a large number of rather narrow intervals. The reason is obvious; having used narrow intervals we can always later combine cases into wider intervals. But if we have started with a small number of more crude categories, we cannot subdivide unless we go through the counting operation all over again. Therefore, we might decide to classify the data into intervals of width 5 per cent as in Table 4.1.

Examining the frequencies in each category we see that the picture presented is a rather jagged or uneven one. Probably we can account for the variations between adjacent categories in terms of chance fluctuations. Had there been more cases, we would have expected a smoother distribution. The reasoning behind this intuitive judgment will become more apparent in later chapters. Suffice it to say that empirically it always

Table 4.1 Frequency distribution, with data grouped into intervals of 5 per cent

Interval	Frequency, f	Interval	Frequency, f
0.0– 4.9	1	45.0–49.9	4
5.0– 9.9	4	50.0–54.9	0
10.0–14.9	9	55.0–59.9	1
15.0–19.9	8	60.0–64.9	0
20.0–24.9	16	65.0–69.9	0
25.0–29.9	23	70.0–74.9	0
30.0–34.9	8	75.0–79.9	0
35.0–39.9	14	80.0–84.9	1
40.0–44.9	4		93

seems to happen this way. Given our particular N of 93 tracts, however, the best we can do to obtain a smoother-looking distribution is to make use of a smaller number of wider intervals. Using intervals of width 10 we obtain Table 4.2.

Table 4.2 Frequency distribution, with data grouped into intervals of 10 per cent

Interval	Frequency, f
0.0– 9.9	5
10.0–19.9	17
20.0–29.9	39
30.0–39.9	22
40.0–49.9	8
50.0–59.9	1
60.0–69.9	0
70.0–79.9	0
80.0–89.9	1
	93

If we had used still wider intervals, say of width 20, the picture would have been as shown in Table 4.3. Here, we are beginning to obscure most of our original information. We know only that about two-thirds

Table 4.3 Frequency distribution, with data grouped into intervals of 20 per cent

Interval	Frequency, f
0.0–19.9	22
20.0–39.9	61
40.0–59.9	9
60.0–79.9	0
80.0–99.9	1
	93

of the cases lie between 20.0 and 39.9, but by looking at the data in this last form we cannot say much about where the bulk of the cases are within this very large interval. Clearly, we have to compromise between using so many intervals that the picture remains too detailed or irregular and using so few categories that too much information is lost. Incidentally, we see that in summarizing interval-scale data *some important information is practically always lost.* On the other hand, to report all the information is to present so much detail as to confuse rather than enlighten.

Although mathematical formulas have been given that may serve as guides to the number of intervals to use, such formulas often give the impression of exactness where the best decision is usually that based on common sense and a knowledge of the use to which the frequency table will be put. Regardless of the number of cases or smoothness of the frequency distribution, it is wise to follow the rule of thumb that an interval should be no wider than the amount of difference between values that can safely be ignored. A $5 difference between house prices is insignificant, but not so for the prices of shirts. Therefore, an interval should contain cases whose actual values can for practical purposes be considered alike.

Another problem presents itself with the above data. What about the single tract with 83.6 per cent turnout? Using intervals of width 10 leaves several empty classes with this single tract off to itself, so to speak. Of course this is essentially what must be done if the data are to be summarized accurately. This tract *is* unique. On the other hand it may be desirable under some circumstances to shorten the table. If percentages could go well above 100 and if there were several extremes strung out over 10 or more intervals, we should be posed with a more difficult decision. Several alternatives present themselves. First, we may use intervals of *different widths*, permitting the extreme intervals to be much wider than the others. Thus we might use a single interval 50.0 to 89.9 which would include the largest two scores. In so doing, we of course lose information since we now have much less precise an indication of the scores of the two extreme cases.

Secondly, we might use an *open interval* to take care of the extreme cases. The last category might then read, "50 per cent or more." Here, we retain even less information about the data although in this particular example we happen to know that percentages cannot run over 100. Had the data involved income and had the last interval been "$50,000 and over," the reader would have had absolutely no way of guessing from the table alone what the highest incomes might have been. It should be noted, however, that under some circumstances it may not be at all important what these highest incomes happen to be. In this case the simplifications introduced by using open intervals may outweigh the disadvantages. With distributions having a small number of very extreme scores, there may be no satisfactory alternative. If one wishes to indicate the incomes of the most wealthy citizens without distorting his table, he may find it easier to do so in his textual presentation. As we shall see in the following chapters, open intervals should not be used if the primary purpose of grouping data is to simplify *computations* rather than to *display* the data in a meaningful way.

True limits You may have noticed that in indicating the intervals, the class limits have been stated in such a way that there is no overlap between classes. In fact, there is a slight gap between them. Limits are usually stated in this manner in order to avoid ambiguity for the reader. Had they been stated as 10 to 20, 20 to 30, etc., the question would have arisen as to what we do with a score that is exactly 20.0. Actually there will always be ambiguities no matter how the intervals are stated, as can be seen if we ask what would be done with a case that is between 19.9 and 20.0. We of course notice that there are no such cases, but a little thought will convince us that this is due to the fact that the data have been *rounded* to the nearest tenth of a per cent. We therefore have to ask the following question: "Which cases actually belong in a particular interval, given the fact that data have been rounded?" We see immediately that the *true* class limits are not the same as the *stated* limits. If we had followed conventional rounding rules, a tract with a percentage turnout slightly over 19.95 would have been rounded up to 20.0 and placed in the interval 20.0 to 29.9. Had the percentage been ever so slightly below 19.95, we would have rounded the score down to 19.9 and placed the tract in the next lower category. Therefore the true limits actually used run as follows:

$$-0.05-\ 9.95$$
$$9.95-19.95$$
$$19.95-29.95$$
$$\text{etc.}$$

We see that by using the true limits each interval is exactly of width 10.0 (rather than 9.9) and that the upper limit of one interval exactly coincides with the lower limit of the following interval.[1] Had a score

[1] Where the lowest interval involves zero and where values cannot be negative (as with percentages), we still consider all intervals to be of equal width, imagining that the lower limit of the first interval is actually -0.05 and that scores have been rounded up to 0.00.

been exactly 9.95000 we would have followed the conventional procedure and rounded up, since the digit immediately preceding the last five is odd.[2] We can therefore assign every case unambiguously to the proper interval. Notice that if rounding has been to the *nearest* figure, as is usually the case, the true limits will always involve splitting the difference between the stated limits of adjacent intervals. For example, if we split the difference between 19.9 and 20.0, we get 19.95. The convention is to report figures in such a manner as to indicate the degree of accuracy in measurement, e.g., 10.45 indicates accuracy to two decimal places, 10.450 to three, and 10.4 to one. Measurement accuracy should therefore be reported in the stated intervals so that the reader can ascertain the true limits if he wishes to make use of these limits in his computations. For example, if limits are stated to be 10.00 to 19.99 we know that measurement is accurate to two decimal places, that rounding has been to the nearest $\frac{1}{100}$ of 1 per cent, and that therefore the true limits run from 9.995 to 19.995. Had limits been given as 10 to 19, true limits would of course have been 9.5 to 19.5.

In a few instances, such as age *last* birthday, data may not have been rounded in the conventional way. If we ask ourselves which interval any given case really belongs in, however, the answer should always be apparent. Since a person who will be aged 20 tomorrow lists his age as 19 today, it is clear that an interval stated as 15 to 19 has as its true limits the values 15 and 20. Although we may appear to be splitting hairs in distinguishing true limits from stated limits, we shall see in the following chapters that true limits must be used in computations even though they are not usually explicitly given when data are presented in the form of a frequency distribution.

Discrete and continuous data   The data we have been using are *continuous* in the sense that any value might theoretically have been obtained for a percentage provided that measurement accuracy were sufficiently precise and that tracts were very large. Thus, the value 17.4531 per cent is as possible as 17.0000 per cent. Certain other kinds of data are *discrete* in that not all values are possible. A woman may have exactly 0, 1, 2, or even 17 children, but she cannot have 2.31 children. Income and city size are theoretically discrete variables in that it is impossible to earn $3,219.5618 or for a city to have a population of 43,635.7 people. Because

---

[2] Notice that in the case of the intervals we have been using, there would be a very slight bias since cases exactly between intervals would always be placed in the higher category. For most practical purposes such a bias can be ignored.

of the limitations of any measuring instrument and the consequent necessity of rounding at some point, empirical data always come in discrete form, but in many cases we are at least able to imagine a continuous distribution that could be obtained by the perfect measuring instrument. As we shall see in the chapter on the normal curve, mathematicians often have to develop theoretical distributions that assume a continuous variable.

In some cases, such as income or city size, it is not too difficult to imagine that data are continuous even though actually there are very small units (pennies, persons) that cannot be subdivided. But what about the number of children in a family? Here we would seem to be doing grave injustice to the data to assume continuity. In presenting data in a frequency distribution we would obviously not be so foolish as to use intervals running from 0.5 to 2.4 or 2.5 to 4.4 children. We would simply list intervals as 0 to 2, 3 to 4, etc., and there would be no ambiguity concerning the gaps between intervals. For some computations, however, it will be necessary, for pragmatic reasons, to treat the data as continuous and to spread discrete scores over small intervals. As peculiar as it may seem, we may wish to consider mothers with one child as ranging in number of children from 0.5 to 1.5. For most purposes we shall get essentially the same results as would be obtained by keeping the data in discrete form. A compromise with reality in this and other instances may be necessary in order to fit the models set up by the mathematician. Provided we realize what we are doing, little or no confusion will result.

4.2 Cumulative Frequency Distributions

For some purposes it is desirable to present data in a somewhat different form. Instead of giving the number of cases in each interval, we may indicate the number of scores that are less than (or greater than) a given value. In the case of the data we have been using, there are obviously no tracts having less than zero turnout; there are five with less than 9.95 per cent; 22 with less than 19.95 per cent; and all 93 have less than 89.95 per cent turnout. We can thus represent the data in a cumulative fashion as indicated in Table 4.4. Notice that we can cumulate downward as well as upward by asking how many cases are *above* a certain value. Cumulative frequencies are usually indicated by using a capital F rather than the lowercase letter. If we wish, we can convert actual frequencies into percentages. We shall have occasion to make use of cumulative distributions in Chap. 5 when computing medians and again in Chap. 14.

Table 4.4   Cumulative frequency distributions

Cumulating up			Cumulating down		
Number of cases below	Cumulative frequency, F	Per cent	Number of cases above	Cumulative frequency, F	Per cent
0.0	0	0.0	0.0	93	100.0
9.95	5	5.4	9.95	88	94.6
19.95	22	23.7	19.95	71	76.3
29.95	61	65.6	29.95	32	34.4
39.95	83	89.2	39.95	10	10.8
49.95	91	97.8	49.95	2	2.2
59.95	92	98.9	59.95	1	1.1
69.95	92	98.9	69.95	1	1.1
79.95	92	98.9	79.95	1	1.1
89.95	93	100.0	89.95	0	0.0

4.3   Graphic Presentation: Histograms, Frequency Polygons, and Ogives

There are always some persons who hesitate to read tables and who obtain a better understanding of materials if these are presented in graphic or visual form. One of the simplest and most useful ways of presenting data so that differences among frequencies readily stand out is to make use of figures which have areas or heights proportional to the frequencies in each category. For example, a bar can be used to represent each category, the height of the bar indicating its relative size. If the scale is nominal, the actual ordering of bars will be irrelevant. For ordinal and interval scales, the bars can be arranged in their proper order, giving a good visual indication of the frequency distribution. The resulting figure is referred to as a *histogram*. Either the absolute frequencies or the proportion of cases in each interval may be indicated along the ordinate as in Fig. 4.1.

It should be noted that if *heights* of bars are taken to be proportional to frequencies in each class interval, the visual picture can be very misleading unless all intervals are closed and of equal width. Suppose, for example, that one of the middle intervals had been of width 20 instead of 10. We would therefore find a larger number of cases in the interval, and the result would be as in Fig. 4.2. Obviously, if we wish to obtain a histogram which represents the data more faithfully, we should make the bar only half as high since we have doubled the width and have on the average included twice as many cases in the wider interval as would be in either of the two intervals of regular size. This would give us a histogram (see Fig. 4.3) much more similar to the one originally obtained. A little thought will convince you that if we were to think in terms of areas rather than heights, we could more readily handle data involving unequal intervals. In other words, we let the *areas* of the rectangles be proportional to the number of cases. In the important special case where all

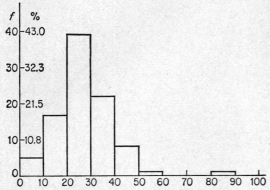

Figure 4.1 Histogram with equal intervals.

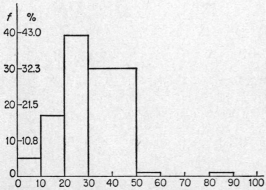

Figure 4.2 Histogram with unequal intervals and heights proportional to frequencies.

intervals are of equal width, the heights will of course also be proportional to the frequencies. If the width of each rectangle is taken to be one unit and if heights are represented as proportions, the total area under the histogram will be unity. Thus

$$1(\tfrac{5}{93}) + 1(\tfrac{17}{93}) + 1(\tfrac{39}{93}) + \cdots + 1(\tfrac{1}{93}) = 1$$

When we take up the normal curve in Chap. 7, we shall find it necessary to deal with areas rather than heights, and it will be convenient to take the total area under the histogram to be unity.

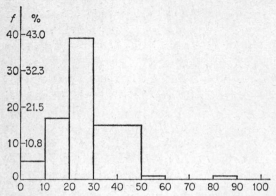

Figure 4.3 Histogram with unequal intervals and areas proportional to frequencies.

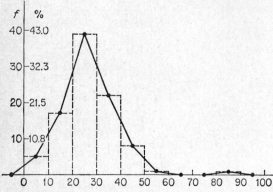

Figure 4.4 Frequency polygon.

Another very similar way of presenting a frequency distribution graphically is by means of the *frequency polygon*. To obtain a frequency polygon we simply connect the midpoints of the tops of each rectangle with straight lines and then blot out the rectangles as in Fig. 4.4. Note that the end points of the frequency polygon have been placed on the base line (horizontal axis) at the midpoints of the intervals on either side of the two extreme intervals. Ordinarily we would not make use of both types of figures, but by superimposing the frequency polygon on the histogram we see that the area under the two figures must be identical. This is true because for every small triangle which lies within the frequency polygon but outside the histogram, there is an identical triangle beneath the histogram but above the frequency polygon. Thus we can also take the area under a frequency polygon to be unity. Notice, however, that we have merely connected a number of points with straight lines. The points themselves may represent the number of cases in each interval, but we should be careful not to infer that there are a given number of cases at any other single point along the continuum. For example, we should *not* infer that there are approximately 28 cases with scores of exactly 20.

Frequency polygons can also be used to represent cumulative frequency distributions. The resulting figure is referred to as an *ogive*. Along the ordinate or *Y* axis we can indicate frequencies or percentages. We place scores of the interval-scale variable along the *X* axis (abscissa) as before, with the understanding that the frequencies represented indicate the number of cases that are *less than* the value of the *X* axis. For example, in Fig. 4.5 we see that approximately 75 per cent of the scores are less

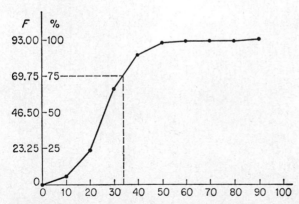

Figure 4.5　Ogive representing cumulative frequency distribution.

than 34. Ogives can therefore be used as a graphic method of determining the number of cases below or above a certain value. Obviously, the form of the ogive must be such that it is either always increasing or always decreasing depending on whether one is cumulating up or down. The curve will be horizontal in those intervals that are empty. If the frequency distribution is of the type given by our data, with the largest number of cases in those intervals near the center of the distribution, the ogive will be S-shaped, with the steepest slope in the vicinity of those intervals containing the largest number of cases.

Glossary
Continuous and discrete data
Cumulative distribution
Frequency distribution
Frequency polygon
Histogram
Ogive
True limits

Exercises
1. Suppose the following represent the annual incomes of a sample of community residents:

$ 5,760	$ 6,850	$ 6,340	$ 6,890	$ 5,860
7,340	7,360	7,350	14,740	7,350
8,210	5,140	5,610	6,560	10,310
6,410	6,330	11,190	5,740	6,550
7,570	10,810	7,250	10,110	7,210
12,300	8,340	6,460	13,300	8,490
6,320	5,970	22,310	7,440	5,110
4,790	7,140	5,670	6,370	26,400
7,560	6,000	6,100	8,170	6,760
6,800	4,610	8,130	6,160	7,170
16,460	7,570	4,710	5,800	9,170
8,210	4,940	7,320	6,180	5,350
5,690	5,780	12,830	7,240	11,340

a. Construct a frequency distribution and cumulative distribution.
b. What are the true limits?
c. Draw a histogram, frequency polygon, and ogive.

2. In a survey of visiting patterns among close friends and relatives, 81 respondents are asked to indicate the number of such friends and relatives with whom they ordinarily visit at least once every month. The results are as follows (figures indicating actual number of persons usually visited):

3	5	2	3	3	4	1	8	4
2	4	2	5	3	3	3	0	3
5	6	4	3	2	2	6	3	5
4	14	3	5	6	3	4	2	4
9	4	1	4	2	4	3	5	0
4	3	5	7	3	5	6	2	2
5	4	2	3	6	1	3	16	5
3	11	4	5	19	4	5	2	2
4	3	14	5	2	1	4	3	4

    a. Construct a frequency distribution and cumulative distribution.
    b. Justify as well as you can your choice of intervals.
    c. Draw a histogram, frequency polygon, and ogive.

3. Indicate the true limits in each of the following:

a. 1,000–1,900	*c.* 1.000–1.999 *(Ans.* 0.9995–1.9995)
2,000–2,900	2.000–2.999
b. 1,000–1,999	*d.* .010–.019
2,000–2,999	.020–.029

What have you assumed about the method of rounding in each of the above?

References

    1. Anderson, T. R., and M. Zelditch: *A Basic Course in Statistics,* 2d ed., Holt, Rinehart and Winston, Inc., New York, 1968, chap. 4.

    2. Downie, N. M., and R. W. Heath: *Basic Statistical Methods,* 2d ed., Harper and Row, Publishers, Incorporated, New York, 1965, chap. 3.

    3. Hagood, M. J., and D. O. Price: *Statistics for Sociologists,* Henry Holt and Company, Inc., New York, 1952, chaps. 4 and 5.

    4. McCollough, C., and L. Van Atta: *Introduction to Descriptive Statistics and Correlation,* McGraw-Hill Book Company, New York, 1965, chap. 1.

    5. Mueller, J. H., K. Schuessler, and H. L. Costner: *Statistical Reasoning in Sociology,* 2d ed., Houghton Mifflin Company, Boston, 1970, chap. 4.

    6. Weiss, R. S.: *Statistics in Social Research,* John Wiley & Sons, Inc., New York, 1968, chap. 5.

Interval Scales: Measures of Central Tendency

We have seen that nominal scales can easily be summarized in terms of percentages, proportions, or ratios and that these summarizing measures are essentially interchangeable. In other words, a single kind of measure is sufficient to describe the data. In the case of interval scales, we have seen that the data may be described by means of a frequency distribution. We can also make use of several distinct types of measures, the most important of which are measures of typicality or *central tendency* and measures of heterogeneity or *dispersion*. In each case we shall find that there are a number of distinct measures from which to choose, each measure having somewhat different properties, advantages, and disadvantages. Summarizing interval scales is therefore somewhat less straightforward than was the case with nominal scales. The present chapter is concerned with measures of typicality; in the following chapter we shall discuss measures of dispersion. Taken together, these two types of measures will ordinarily be adequate for describing interval-scale data.

The layman's conception of the term *average* is likely to be rather vague or ambiguous. In fact, he may not realize that there are several distinct measures of typicality and that under some circumstances these measures may yield very different results. The fact that it is possible to obtain such different measures of central tendency means that it is necessary to understand the advantages and disadvantages of each measure. It is also important to learn the circumstances under which each is appropriate. Why does the Census Bureau report median incomes rather than mean incomes? Would it make sense to tell the layman that the "average" family has 2.3 children and lives in a 4.8 room house? Under what circumstances will it make little difference which measure is used? These are some of the questions that can be raised about the kind of average we should compute.

5.1 The Arithmetic Mean

There are two important measures of central tendency used in sociological research, the arithmetic mean (hereafter referred to simply as the mean) and the median. The mean is by far the more common of the two and is defined as the sum of the scores divided by the total number of cases involved. The symbol $\bar{X}$ is conventionally used to indicate the mean, although occasionally the letter M may be used. The formula for the arithmetic mean is therefore

$$\bar{X} = \frac{X_1 + X_2 + \cdots + X_N}{N} = \frac{\sum_{i=1}^{N} X_i}{N} \tag{5.1}$$

where X_1 represents the score of the first individual, X_2 that of the second, and X_i that of the general individual.[1] If there are no ambiguities, we may drop the subscripts and write simply

$$\bar{X} = \frac{\Sigma X}{N}$$

where it is understood that all quantities are being summed.

The mean has the algebraic property that the sum of the deviations of each score from the mean will always be zero. Symbolically, this property can be expressed by the equation

$$\sum_{i=1}^{N} (X_i - \bar{X}) = 0$$

This fact should not be at all surprising when we realize how the mean has been defined. The proof is simple. Since we have a sum of numbers each of which is actually a difference, we can break the above expression into the difference of two sums. Thus

$$\sum_{i=1}^{N} (X_i - \bar{X}) = \sum_{i=1}^{N} X_i - \sum_{i=1}^{N} \bar{X}$$

But since $\bar{X}$ is a constant we have

$$\sum_{i=1}^{N} \bar{X} = N\bar{X} = N \frac{\sum_{i=1}^{N} X_i}{N} = \sum_{i=1}^{N} X_i$$

[1] For a discussion of the summation notation see Appendix 1.

and we see immediately that the difference between $\sum\limits_{i=1}^{N} X_i$ and $\sum\limits_{i=1}^{N} \bar{X}$ is zero.

The above property can be used to simplify the computation of the mean. For example, suppose we wish to compute a mean from the numbers 72, 81, 86, 69, and 57. Adding and dividing by five we get an $\bar{X}$ of 73.0. Subtracting this mean from each of the figures and then adding, we verify that the resulting sum is zero.

X	$X - 73$	$X - 70$
72	−1	2
81	8	11
86	13	16
69	−4	−1
57	−16	−13
	0	15

Suppose, however, that we had guessed a mean of 70 and then had subtracted this guessed mean from each of the figures. The resulting sum is not zero, but we notice that each of the new differences is three units larger (in the positive direction) than the original differences. We thus see that we guessed a mean that is three too small. If we now add a correction factor of three to our guessed mean, we obtain the correct mean. In actual practice we would not compare the two sets of differences in this manner. We notice, however, that the sum of the second set of differences is +15. Since there are five scores, this indicates that on the average we were below the true mean by $15\!/\!5$ or 3.0 units. As can easily be verified, had we guessed a value that was too high, the resulting sum of the differences would have been negative, and we would have had to subtract from the guessed mean in order to obtain the correct figure. If $\bar{X}'$ represents the guessed mean, we may write a formula for the mean in terms of the guessed mean and a correction factor

$$\bar{X} = \bar{X}' + \frac{\sum\limits_{i=1}^{N} (X_i - \bar{X}')}{N} \qquad (5.2)$$

or, in words,

True mean = guessed mean + $\dfrac{\text{sum of deviations from guessed mean}}{\text{number of cases}}$

In order to verify the correctness of this formula, we expand the right-hand side, getting

$$\bar{X}' + \frac{\sum\limits_{i=1}^{N} (X_i - \bar{X}')}{N} = \bar{X}' + \frac{\sum\limits_{i=1}^{N} X_i}{N} - \frac{\sum\limits_{i=1}^{N} \bar{X}'}{N}$$

$$= \bar{X}' + \frac{\sum\limits_{i=1}^{N} X_i}{N} - \frac{N\bar{X}'}{N}$$

$$= \frac{\sum\limits_{i=1}^{N} X_i}{N} = \bar{X}$$

Although we seem to have gone to a lot of trouble computing $\bar{X}$ in a roundabout manner, it is sometimes possible by this method to save considerable work when desk calculators are not available. Making use of a guessed mean ordinarily reduces the size of the numbers to be added. The closer the guessed mean to the correct value, the smaller in magnitude will be the resulting differences. This principle will be especially useful when we take up the computation of means from grouped data.

A second property of the mean can be stated as follows: The sum of the *squared* deviations of each score from the mean is less than the sum of the squared deviations about any other number. In other words

$$\sum_{i=1}^{N} (X_i - \bar{X})^2 = \text{minimum}$$

*The proof of this property is very simple. Consider deviations of the X_i about any other number $\bar{X}'$, which we have previously treated as a guessed mean. Adding and subtracting the true mean $\bar{X}$ from each such expression we may write

$$X_i - \bar{X}' = (X_i - \bar{X}) + (\bar{X} - \bar{X}')$$

Squaring both sides we get

$$(X_i - \bar{X}')^2 = (X_i - \bar{X})^2 + 2(X_i - \bar{X})(\bar{X} - \bar{X}') + (\bar{X} - \bar{X}')^2$$

Summing over all N cases we obtain

$$\sum_{i=1}^{N} (X_i - \bar{X}')^2 = \sum_{i=1}^{N} (X_i - \bar{X})^2$$

$$+ 2(\bar{X} - \bar{X}') \sum_{i=1}^{N} (X_i - \bar{X}) + \sum_{i=1}^{N} (\bar{X} - \bar{X}')^2$$

where it has been possible to write the quantity $2(\bar{X} - \bar{X}')$ in front of the summation sign in the second term since it is just a constant. We immediately see that the entire second term must be zero since we have just shown that $\sum_{i=1}^{N} (X_i - \bar{X}) = 0$. Also, the last term consists of N terms that are all equal to $(\bar{X} - \bar{X}')^2$. We therefore have

$$\sum_{i=1}^{N} (X_i - \bar{X}')^2 = \sum_{i=1}^{N} (X_i - \bar{X})^2 + N(\bar{X} - \bar{X}')^2$$

and we thus see that the sum of the squared deviations about $\bar{X}'$ is equal to the sum of the squared deviations about the true mean *plus* a squared term that can never be negative. The greater the difference between $\bar{X}'$ and $\bar{X}$, the greater the second term on the right-hand side.

We shall have frequent occasion to use this least-squares property of the mean, and the quantity $\sum_{i=1}^{N} (X_i - \bar{X})^2$ will appear in much of our later work as a measure of the total variation or heterogeneity.

5.2 The Median

We sometimes want to locate the position of the middle case when data have been ranked from high to low. Or we may divide a group of students into percentiles by locating the individuals who have exactly 10 per cent of the class below them, exactly 32 per cent below them, etc. Measures of this sort are often referred to as *positional measures* since they locate the position of some typical (or atypical) case relative to those of other individuals. The median is perhaps the most important of such positional measures. We define the median as a number which has the property of having the same number of scores with smaller values as there are with larger values. Ordinarily, the median thus divides the scores in half. If the number of cases is odd, the median will simply be the score of the middle case. If N is even there will be no middle case and, in fact, any number between the values of the two middle cases will have the property of dividing the scores into two equal groups. Thus the median is ambiguously defined if N is even. By convention we take as the unique value of the median the arithmetic mean of the two middle cases.

If we had the numbers 72, 81, 86, 69, and 57, the median would be 72 (as compared with a mean of 73). Had there been a sixth score, say 55, the two middle scores would be 69 and 72, and we would take as the median $(69 + 72)/2$ or 70.5. If the two middle cases happen to have the same score, the median will of course be that score itself. Notice that if N is odd, the middle case will be the $(N + 1)/2$nd case. When the number of cases is even, the median will be halfway between the $N/2$nd and the $(N/2 + 1)$st case. Thus if $N = 251$, the median will be the score of the 126th case; if $N = 106$, we take a value halfway between the scores of the fifty-third and fifty-fourth cases. These formulas will ordinarily be helpful when N is fairly large.

We have seen that the mean has the properties

$$\sum_{i=1}^{N} (X_i - \bar{X}) = 0$$

and
$$\sum_{i=1}^{N} (X_i - \bar{X})^2 = \text{minimum}$$

The reason that the first property holds is essentially that when the mean is subtracted from each score, the values of the resulting differences are such that the negative scores exactly balance the positive ones. But suppose we had ignored the signs altogether and considered all differences as positive? It can be shown that had the *median* been subtracted from each score, the sign of the difference ignored, and the results summed, we would have obtained a sum which would be less than the comparable figure for any other measure of central tendency. In symbols

$$\sum_{i=1}^{N} |X_i - \text{Md}| = \text{minimum}$$

where Md stands for the median, and the bars around the expression $(X_i - \text{Md})$ indicate that the positive (or absolute) value of each difference is to be used. Although this property of the median is perhaps of some interest, it does not seem to have any direct applications to sociological research.

*5.3 Computation of Mean and Median from Grouped Data

Long method for computing mean When the number of cases becomes large, the computation of the mean or median may become tedious if hand calculations are used. Most social scientists now have available computer programs that handle these and other computations with great ease. It is generally preferable to use such programs, when convenient, since they will minimize the risk of computational and rounding errors, as well as saving considerable time and money. Nevertheless, you should know how to compute various measures without the aid of such programs since it will often be inconvenient to put the data in such a form that they can be handled by high-speed computers. In these instances it is useful to group the data into categories and to compute the mean or median from the resulting frequency distributions. Sometimes, of course, we find data already given to us in grouped form, and it will be either impossible or impractical to go back to the original data for purposes of computation. Census data are usually given in grouped form, for example. We would know only that there are a certain number of persons aged 0 to 4 or 5 to 9, but the exact age of each individual would be unknown.

As we shall see below, the use of data in grouped form may simplify one's work considerably. On the other hand, we inevitably lose information when grouping data into categories. We may know only that there are 17 persons having incomes between $2,000 and $2,900, but we do not know exactly how they are distributed within this interval. In order to compute either a mean or a median from such grouped data, we have to make some simplifying assumptions about the location of individuals within each category. In the case of the mean we shall treat all cases as though they were concentrated at the midpoints of their respective intervals. In computing the median we imagine cases to be spread at equal distances within each interval. These simplifications will of course lead to certain inaccuracies. We would not expect to get exactly the same results as those obtained from the raw data. On the other hand, if the number of cases is large, the distortions introduced will usually be very minor and well worth the saving of time. Obviously, the narrower the intervals, the less information we have lost, and the greater the accuracy. For example, if we know that there are 17 cases between $2,000 and $2,900 and 26 cases between $3,000 and $3,900, we can obtain more accurate results by imagining the 17 cases to be at the midpoint of the first interval and the 26 persons to be at the midpoint of the second than if we were to place all 43 scores at the midpoint of the larger interval, $2,000 to $3,900. These simplifications are most likely to lead to errors in the case of extreme intervals since the scores in these intervals may be skewed toward the center of the total distribution. Thus if there are 17 cases in the lowest interval, most of these scores may be in the upper half of the interval. However, if the number of individuals in these intervals is relatively small, as is usually the case, the distortion introduced is likely to be insignificant.

In computing the mean from grouped data we therefore treat all cases as though they were located at the midpoints of their respective intervals. If we preferred, we could take them as spread at equal distances throughout the interval, but, as can easily be verified, this would lead to the same results since the mean score within each interval would be exactly at the midpoint of the interval. Since all cases within each interval are treated as having the same value, we can multiply the number of cases in the interval by their common value instead of adding the scores separately. For example, if we have placed 26 cases at the value 3,450, the product 26 × 3,450 will be equal to the sum of 26 separate scores of 3,450. If we do this for all intervals, sum the products, and divide by the total number of cases, we obtain the arithmetic mean. The formula for the mean then becomes

$$\bar{X} = \frac{\sum_{i=1}^{k} f_i m_i}{N} = \frac{\sum_{i=1}^{k} f_i m_i}{\Sigma f_i} \tag{5.3}$$

where f_i = number of cases in ith category, with $\Sigma f_i = N$
    m_i = midpoint of ith category
    k = number of categories
The example worked in Table 5.1 should make the process clear.

In Table 5.1 all intervals are of equal width. This is not essential as long as the correct midpoints are used. It is necessary, however, to make use of closed intervals. Suppose the last interval had been $7,000 and above. What would we use as the midpoint? We have absolutely no basis on which to judge unless we can go back to the original data. Sometimes it is feasible to do this since extreme categories often contain relatively few cases. In such instances it usually makes more sense to use the actual mean of the cases in the extreme category rather than the midpoint of some wider interval. In instances where it is impossible to go back to the original data, it will be necessary to make an enlightened guess as to a reasonable value for the midpoint. It is therefore clearly to our advantage to use closed intervals whenever a mean is to be computed. As we shall see in the next chapter, this is also true in computing the standard deviation, the most commonly used measure of dispersion.

Short method for computing mean The above method will usually involve the multiplication of fairly large numbers (e.g., 2,450 × 17) unless the midpoints turn out to be simple numbers. With a desk calculator or computer such products can easily be computed and accumulated. But if computations must be made by hand, there is a much simpler method of computing the mean from grouped data. This so-called "short method" seems, at first glance, to involve more work than the "longer" one, but once it has been mastered it will prove much the simpler of the two. Basically, the short method involves guessing a mean and thereby making use of smaller numbers in the multiplication process. A correction factor is then added to the guessed mean as before.

Table 5.1 Computation of mean from grouped data, using long method

Stated limits	True limits	Mid-points (m_i)	f_i	$f_i m_i$
$2,000–2,900	$1,950–2,950	$2,450	17	$ 41,650
3,000–3,900	2,950–3,950	3,450	26	89,700
4,000–4,900	3,950–4,950	4,450	38	169,100
5,000–5,900	4,950–5,950	5,450	51	277,950
6,000–6,900	5,950–6,950	6,450	36	232,200
7,000–7,900	6,950–7,950	7,450	21	156,450
Totals			189	$967,050

$$\bar{X} = \frac{\sum_{i=1}^{k} f_i m_i}{N} = \frac{967,050}{189} = \$5,117$$

In order to simplify our computations let us take as our guessed mean a midpoint of one of the intervals. In the above example we can see by inspection that the mean will be somewhat less than $5,450, the midpoint of the fourth interval. The advantage of using a midpoint as our guessed mean is obvious. All of the other scores will then be a certain number of intervals away from the guessed mean since each score is taken to be at one or another of the midpoints. If we now subtract the guessed mean from each score, we will obtain differences of exactly $1,000, $2,000, or $3,000 in either direction. We then multiply these *differences* by the appropriate frequencies to obtain the correction factor to be added to the guessed mean. In other words, there will be 17 cases which have scores that are exactly $3,000 less than the guessed mean; there will be 26 cases with a difference of $2,000, etc. If we make use of a column d_i representing the difference between actual scores and the *guessed* mean, we can modify (5.2) and write the formula for the mean as

$$\bar{X} = \bar{X}' + \frac{\sum\limits_{i=1}^{k} f_i d_i}{N} \tag{5.4}$$

where
$$d_i = X_i - \bar{X}'$$

and we can set up our computations in a table such as Table 5.2. Again, the correction factor involves taking the total deviation from the guessed mean (here $-63{,}000$) and then dividing by the number of cases to obtain the average amount the guessed mean deviates from the true mean.

Table 5.2   Computation of mean from grouped data, using short method

True limits	Mid-points	f_i	d_i	$f_i d_i$
\$1,950–2,950	\$2,450	17	\$ −3,000	\$ −51,000
2,950–3,950	3,450	26	−2,000	−52,000
3,950–4,950	4,450	38	−1,000	−38,000
4,950–5,950	5,450	51	0	0
5,950–6,950	6,450	36	1,000	36,000
6,950–7,950	7,450	21	2,000	42,000
Totals		189		\$ −63,000

$$\bar{X} = \bar{X}' + \frac{\sum\limits_{i=1}^{k} f_i d_i}{N}$$

$$= 5{,}450 + \frac{-63{,}000}{189} = 5{,}450 - 333$$

$$= \$5{,}117$$

In this example the correction factor turned out to be negative, indicating that the guessed mean was too high. It should be noted that had we guessed any other value for the mean, we would have obtained the same result. Selecting the midpoint of the third interval (\$4,450) as the guessed mean yields a correction factor of \$667 which, when *added* to \$4,450, gives the correct result. Incidentally, this serves as a useful check on our work. Notice that had we selected the midpoint of any other interval, we would have made more work for ourselves since the numbers to be added in the $f_i d_i$ column would have been numerically larger. Had we failed to use a midpoint, deviations from the guessed mean would have involved much less simple numbers, and we would not have saved ourselves any work. Once the process becomes clearly understood, it will be possible to omit the midpoints column from the computing table.

You have undoubtedly already noticed that each of the deviations from the guessed mean in the above example is an exact multiple of 1,000, the size of the interval used. This will always be the case provided all intervals are exactly the same width. We can therefore factor out the width of the interval in each of the products $f_i d_i$, and then multiply by the interval width when we have finished summing. In other words, we could have obtained the sum $-63,000$ as follows:

$$-63,000 = 1,000(-51 - 52 - 38 + 0 + 36 + 42)$$

In what amounts to the same thing, we could have expressed the original deviations in terms of the number of *intervals* (or *step deviations*) from the guessed mean. We then determine how many intervals the guessed mean is from the true mean and, finally, we translate the amount of error back into the original units by multiplying this correction factor by the size of the interval. Referring to the deviation in interval widths as d', we can revise our table as in Table 5.3.

Table 5.3 Computation of mean from grouped data, using short method and step deviations

True limits	Mid-points	f_i	d'_i	$f_i d'_i$
$1,950–2,950	$2,450	17	−3	−51
2,950–3,950	3,450	26	−2	−52
3,950–4,950	4,450	38	−1	−38
4,950–5,950	5,450	51	0	0
5,950–6,950	6,450	36	1	36
6,950–7,950	7,450	21	2	42
Totals		189		−63

The modified formula for the mean now becomes

$$\bar{X} = \bar{X}' + \frac{\sum\limits_{i=1}^{k} f_i d'_i}{N} i \qquad (5.5)$$

where i is the interval width. Therefore

$$\bar{X} = 5{,}450 + \frac{-63}{189}\,1{,}000 = 5{,}117$$

If unequal intervals have been used, it will be necessary to modify this second form of the short method. Some persons may find it easier to go back to the earlier method, using d_i instead of d'_i, and writing in the actual differences in the original units. Alternatively, if only one or two intervals differ from the rest in width, we may take as the interval width i the width of the majority of class intervals. The deviations of the midpoints of the remaining intervals from the guessed mean can then be expressed in terms of fractions of whole intervals. For example, had the last interval been \$6,950 to \$8,950 instead of \$6,950 to \$7,950, the midpoint would have been \$7,950 instead of \$7,450. The deviation from the guessed mean would therefore be \$2,500 or 2.5 interval widths. Had the interval run to \$9,950, the d'_i value would have been 3.0 as can easily be verified.

Computation of median In computing the median from grouped data we shall treat all cases within a given interval as though they were distributed at equal distances throughout the interval. We first locate the interval containing the middle case, and we then interpolate to find the exact position of the median. In determining the interval containing the median it is usually desirable to obtain the cumulative frequency distribution. Although not absolutely necessary, it is preferable to develop the habit of writing down the entire cumulative distribution as well as indicating in a separate column the meaning of each of the figures in the cumulative column (F). The cumulative distribution for the above data is given in Table 5.4. As a check on our adding, we note that all 189 cases must be less than \$7,950.

Table 5.4 Computation of median from grouped data

True limits	f	F	No. of cases less than
\$1,950–2,950	17	17	\$2,950
2,950–3,950	26	43	3,950
3,950–4,950	38	81⎫	⎧4,950
4,950–5,950	51	132⎭	⎩5,950
5,950–6,950	36	168	6,950
6,950–7,950	21	189	7,950
Total	189		

Next, we locate the interval containing the middle or N/2nd case. Here $18\frac{9}{2} = 94.5$, so that we are looking for the interval containing the ninety-fourth and ninety-fifth cases. Note that had data been ungrouped, we would have located the $(N + 1)/2$nd or ninety-fifth case. The reason for this apparent inconsistency will be discussed below. Since there are 81 cases less than \$4,950 and 132 less than \$5,950, the median must lie somewhere in the interval \$4,950 to \$5,950. It is a good idea to indicate this interval with brackets since there is sometimes a tendency to look across from the figure 81, obtaining the incorrect interval \$3,950 to \$4,950.

Let us now look more closely at the interval containing the median. There are 51 cases within this interval, and therefore we shall divide the entire interval into 51 subintervals, each of width \$1,000/51 or \$19.61. We locate each of the 51 cases at the midpoint of its proper subinterval. The eighty-first case will be located in the last subinterval of the interval \$3,950 to \$4,950, and the 132nd case will be just slightly less than the upper limit of the interval containing the median. We now simply count subintervals until we have come to the median. Had the data been ungrouped, we should have located the score of the $(N + 1)/2$nd or ninety-fifth case. According to our convention, the ninety-fifth case would be located at the *midpoint* of the fourteenth subinterval or ex-

actly 13.5 subintervals from the lower limit of the interval. Notice that this same value of 13.5 could have been obtained by subtracting 81 from 94.5 or $N/2$. It is because we are dealing with midpoints of small intervals that we count exactly $N/2$ intervals in order to locate the position of the $(N + 1)/2$nd case.

The value of the median can now be obtained by simply multiplying the number of subintervals covered by the size of each subinterval, adding the result to the lower limit of the interval. The whole process can be summarized by the formula

$$\text{Md} = l + \frac{N/2 - F}{f}\, i \tag{5.6}$$

where F = cumulative frequency corresponding to lower limit
    f = number of cases in interval containing median
    l = lower limit of interval containing median
    i = width of interval containing median

The quantity i/f represents the size of each subinterval, and $N/2 - F$ gives the distance (in subintervals) between the lower limit of the interval and the median. In our problem we get

$$\text{Md} = 4{,}950 + \frac{94.5 - 81}{51} \, 1{,}000 = 4{,}950 + 13.5 \, \frac{1{,}000}{51}$$
$$= 4{,}950 + 265 = \$5{,}215$$

There is an alternative but equivalent way of visualizing the process of obtaining the median. Instead of finding the width of each subinterval and then multiplying by the number of these subintervals, we can reason that since there are 51 cases in the entire interval and since we need to go 13.5 of these smaller intervals in order to get to the median, we must travel 13.5/51 of the entire interval. Therefore, if we multiply the width of the interval (1,000) by the fraction of the total distance we must go, we obtain the desired result, a process called *interpolation*. When using the formula, it is, of course, irrelevant which explanation we find most satisfactory. Lest one become too dependent on formulas, it is best to reason out the process each time, using the formula as a check, until it is thoroughly understood. As another check it should be noted that the median could have been computed by *subtracting* a certain quantity from the *upper* limit u. As can easily be shown, the formula then becomes

$$\text{Md} = u - \frac{F - N/2}{f} \, i \qquad\qquad (5.7)$$

where F now represents the cumulative frequency corresponding to the upper limit of the interval. Numerically

$$\text{Md} = 5{,}950 - \frac{132 - 94.5}{51} \, 1{,}000 = \$5{,}215$$

5.4 Comparison of Mean and Median

Having discussed the computational procedures used in obtaining the mean and median from both grouped and ungrouped data, we now need to compare their properties. Several differences between the two measures are immediately apparent. First, the mean uses more information than the median in the sense that all the exact scores are used in computing the mean, whereas the median only uses the relative positions of the scores. Returning to the scores 72, 81, 86, 69, and 57 we see that had the highest score been 126 instead of 86, the median would have been unaffected, but the mean would have been increased substantially. Likewise, had the lowest score been zero, the mean would have been lowered, but the median would again be unchanged. We therefore may state a very important difference between the two measures: *the mean is affected by changes in extreme values whereas the median will be unaffected unless the value of the middle case is also changed.* In our example, as long as 72 remains the third case when data have been reranked, the median will be unchanged.

This important difference between the two measures enables us to decide in most instances which will be the more appropriate. Ordinarily, we desire our measure to make use of all information available. We somehow have more intuitive faith in such a measure. Although at this point it is impossible to bolster such a faith with a sound statistical argument, some justification for the preference for the mean under ordinary circumstances can be given. It turns out that the mean is generally a more stable measure than the median in the sense that it varies less from sample to sample. When we turn our attention to inductive statistics, we shall see that the researcher is ordinarily much more interested in generalizing about a population than he is in his particular sample. He is well aware of the fact that had another sample been taken, the results would not have been quite the same. Had a very large number of samples of the same size been drawn, he would be able to see just how much the sample means differed among themselves. What we are saying, here, is that the sample medians will differ from one sample to the next more than will the means. Since, in actuality, we usually draw only one sample, it is important to know that the measure we use will give reliable results in that there will be minimal variability from one sample to the next. We can therefore state the following rule of thumb: *when in doubt, use the mean in preference to the median.*

Because of the fact that it uses all the data, whereas the median does not depend upon extreme values, the mean may give very misleading results under some circumstances. We must keep in mind that in making use of a measure of central tendency we are attempting to obtain a simple description of what is *typical* of our scores. Suppose, to take an extreme case, that the highest score in the series of five numbers had been 962. The median would remain at 72, but the mean would become 1,241/5 or 248.2. Is this value in any sense typical of the scores? Certainly not. It is nowhere near the score of any of the five cases. It is true, of course, that in such an extreme example no single measure could be used to describe adequately the typical case, but since four out of the five scores are around 72, the use of the median would obviously be less misleading. We may say, then, that *whenever a distribution is highly skewed*, i.e., whenever there are considerably more extreme cases in one direction than the other, *the median will generally be more appropriate than the mean.*

The relationship between skewness and the relative positions of the mean and median is indicated in Fig. 5.1. Since the mean can be very much affected by a few extreme values, the mean will be pulled in the

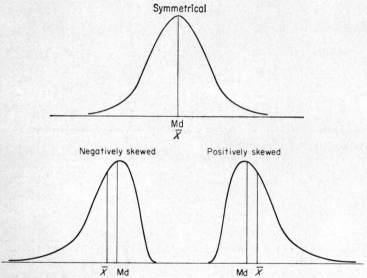

Figure 5.1   Relationship between skewness and the relative positions of the mean and median.

direction of skewness, i.e., toward the tail. If the distribution is perfectly symmetrical, the mean and median will coincide. We know that income distributions are usually skewed toward the higher incomes, with a very few extremely high incomes. It might therefore be misleading to report mean incomes within a corporation or community. For this reason income data are usually reported using the median rather than the mean. If a distribution is highly skewed, of course, this fact should be mentioned in reporting the data. In such instances it might be of some value to report both the mean and the median, although this is seldom done in practice.

The mean has a second property not possessed by the median: it is more easily manipulated algebraically. For example, it is sometimes necessary to obtain a weighted average from several sets of data. Suppose that we have the following mean incomes for three low-income rural communities, A, B, and C:

Community	Size	Mean
A	10,000	$3,518
B	5,000	4,760
C	8,000	4,122

If the sizes of all three communities were exactly the same, we could take the mean of these three scores as the overall mean. But community A is twice as large as community B. In other words, the figure $3,518 represents twice as many cases as does $4,760. Had all 23,000 persons been thrown together and the overall mean computed, the resulting figure would have reflected this fact. To obtain the correct mean we must weight each separate mean by the proper number of cases and then sum, dividing finally by the total number of cases (23,000). We thus get

$$\bar{X} = \frac{\sum_{i=1}^{k} N_i \bar{X}_i}{N} \tag{5.8}$$

where N_i and $\bar{X}_i$ represent the number of cases and the mean for the ith category, and k indicates the number of categories. Therefore

$$\bar{X} = \frac{10,000(3,518) + 5,000(4,760) + 8,000(4,122)}{23,000}$$

$$= \frac{91,956,000}{23,000} = \$3,998.09$$

We can easily justify this weighting procedure by noting that the mean of the ith category was actually obtained by adding scores and dividing by N_i.[2] Therefore the product $N_i \bar{X}_i$ represents the *sum* of all scores in this category. Adding products and dividing by the total N thus gives us the same result as would have been obtained had categories been com-

[2] More generally we may weight the $\bar{X}_i$ by weights w_i, forming as our weighted average the expression $\Sigma w_i \bar{X}_i / \Sigma w_i$. Ordinarily we construct the weights in such a way that they sum to a convenient quantity such as unity (that is, $\Sigma w_i = 1$) or the total sample size N, as in the above example.

pletely ignored. This kind of algebraic manipulation of the mean is sometimes very useful. It should be readily apparent that the overall median for the combined data cannot be obtained in such a manner. If we knew the values of the middle cases of each of the separate categories, we would still not know the value of the middle case for the combined data.

A final important difference between the mean and the median should be noted. The computation of the mean requires an interval scale. Without an interval scale it would be meaningless, of course, to talk about summing scores. It is obviously necessary to assume, for example, that the sum of the numbers 30 and 45 is equivalent to the sum of 20 and 55 since both pairs possess the same mean. The median, on the other hand, can be used for ordinal as well as interval scales. The actual numerical score of the median will be meaningless unless we have an interval scale, but it will certainly be possible to *locate* the middle score. This means that, among other things, we can separate the cases into one of two categories according to whether they are above or below the median. Positional measures can therefore be used with ordinal scales, a fact that is very useful in developing tests which do not require interval scales.

5.5 Other Measures of Central Tendency

There are several additional measures of central tendency, none of which are very commonly used in sociological research. One such measure is the *mode*, which is simply the *most frequent* score or scores. If we take the following series of numbers

(1)   71, 75, 83, 75, 61, 68
(2)   71, 75, 83, 74, 61, 68
(3)   71, 75, 83, 75, 83, 68

we may say that the first has a mode of 75 since there are two individuals with this score and no other score appears twice. There is no mode in the second series of numbers, but there are two modes in the third (75 and 83). The mode is perhaps more useful when there are a larger number of cases and when data have been grouped. We then sometimes speak of a modal category, taking the midpoint of this category as the mode. In the grouped data we have been using, the modal category would be $5,000 to $5,900. In a frequency distribution the mode will be indicated by the highest point on the curve. In a symmetrical distribution with a single mode at the center, the mean, median, and mode will of course all be identical. We also can distinguish between *unimodal* and *bimodal* distributions, the latter taking a form as in Fig. 5.2. In referring to bimodal distributions we usually do not assume that both peaks are of exactly the same height, as would be implied by the definition.

It should be emphasized that since the mode refers to the category with the largest number of cases, we can make use of this concept in describing nominal as well as ordinal and interval scales. Thus in the case of nominal scales the modal category can be considered a type of *central tendency*, as long as it is clearly recognized that no ordering of categories is implied.

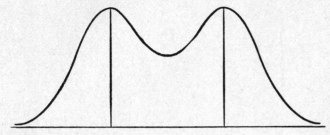

Figure 5.2 A bimodal distribution.

Two other measures of central tendency which are practically never seen in the sociological literature are the *harmonic mean* and the *geometric mean*. They are defined by the formulas

$$\text{Harmonic mean} = \frac{N}{\sum\limits_{i=1}^{N} \frac{1}{X_i}}$$

$$\text{Geometric mean} = \sqrt[N]{(X_1)(X_2) \cdots (X_N)}$$

In the latter formula, the N over the radical indicates that we are taking the Nth root of the product of the N scores.

5.6 Deciles, Quartiles, and Percentiles

In discussing the median, we pointed out that there are certain other positional measures such as percentiles that can be used to locate the position of scores which are larger than a given proportion of cases. These measures, although not necessarily measures of typicality or central tendency, are directly analogous to the median. Thus instead of finding a number which has half of the scores above or below it, we may wish to determine the value of the first quartile, which has the property that one-fourth of the scores are of lesser magnitude. Similarly, the third quartile represents the score having three-quarters of the cases below it in magnitude. If one prefers, he may divide the distribution into 10 deciles by locating scores that have one-tenth, two-tenths, or nine-tenths of the cases with lower values. You are perhaps more familiar with percentiles, which divide the distribution into 100 portions of equal size. Thus a student falling at the ninety-first percentile on an examination knows that 91 per cent of the other students had lower scores than himself.

The computation of deciles, quartiles, and percentiles is directly analogous to the computation of the median. In the case of grouped data, we would first determine the interval within which the desired positional measure falls. Using the data of Table 5.4, we would obtain the first quartile by locating the position of the $N/4$th or the 47.25th case. From our cumulative frequency column we see that the first quartile must lie somewhere within the interval \$3,950 to \$4,950. Since there are 38 cases within this interval, we must go $(47.25 - 43)/38$ of this distance. Thus, the value of the first quartile Q_1 would be

$$Q_1 = 3,950 + \frac{47.25 - 43}{38} \, 1,000 = 3,950 + 112 = \$4,062$$

Other positional measures can be computed in a similar manner. Notice, incidentally, that the median is by definition equivalent to the second quartile, the fifth decile, and the fiftieth percentile. Although deciles, quartiles, and percentiles are seldom used in sociological research, their meanings should at least be familiar.

Glossary
Decile
Mean
Median
Mode
Percentile
Quartile
Skewed distribution

Exercises
1. Give the mean, median, and mode of the following numbers: 26, 37, 43, 21, 58, 26, 33, and 45. (*Ans.* 36.1; 35; 26)

2. Compute a mean and median for the data you grouped in Exercise 1, Chap. 4. Do the same for Exercise 2, Chap. 4.

3. Compute the third quartile, the fourth decile, and the seventy-first percentile for the data of Exercise 1, Chap. 4.

4. The following (hypothetical) data show the distribution of the percentage of farm families in 60 counties. Compute the mean and median. (*Ans.* 32.83; 32.83)

Interval, %	Frequency
10–19	7
20–29	16
30–39	21
40–49	12
50–59	4
	60

5. Using the data for the previous example, indicate how the mean and median would be affected (raised, lowered, or remain the same) if
   a. The last interval were widened to read 50 to 69, with frequencies remaining the same. (*Ans.* raised; same)
   b. Ten per cent were added to each interval (making intervals 20 to 29, 30 to 39, etc.), with frequencies remaining the same.
   c. Intervals remained unchanged, but two cases in the 20 to 29 category were put in the 30 to 39 category (making frequencies 7, 14, 23, 12, and 4).
   d. Intervals remained unchanged, but each frequency were doubled.

6. A group of 10 boys and 7 girls take an algebra quiz. Suppose the mean score for the boys is 84, their median being 74. Both the mean and the median for the girls turn out to be 79. The teacher concludes that on this test the boys did better than the girls. Is her conclusion justified? Why or why not? How might you account for the large difference between the mean and median scores for the boys?

7. Suppose one finds the mean age of the 50 governors to be 51.6, the mean age of 100 senators to be 62.3, and the mean age of 435 representatives to be 44.7. What is the mean age of all of these politicians? Suppose the above figures represented median ages. Could you obtain the overall median in the same manner? Why or why not?

References
   1. Anderson, T. R., and M. Zelditch: *A Basic Course in Statistics*, 2d ed., Holt, Rinehart and Winston, Inc., New York, 1968, chap. 5.
   2. Downie, N. M., and R. W. Heath: *Basic Statistical Methods*, 2d ed., Harper and Row, Publishers, Incorporated, New York, 1965, chap. 4.
   3. Hagood, M. J., and D. O. Price: *Statistics for Sociologists*, Henry Holt and Company, Inc., New York, 1952, chap. 8.
   4. McCollough, C., and L. Van Atta: *Introduction to Descriptive Statistics and Correlation*, McGraw-Hill Book Company, New York, 1965, chap. 2.
   5. Mueller, J. H., K. Schuessler, and H. L. Costner: *Statistical Reasoning in Sociology*, 2d ed., Houghton Mifflin Company, Boston, 1970, chap. 5.
   6. Weinberg, G. H., and J. A. Schumaker: *Statistics: An Intuitive Approach*, Wadsworth Publishing Company, Inc., Belmont, Calif. 1962, chaps. 2 and 6.

Interval Scales: 6
Measures
of Dispersion

In many instances in sociological research the focus of attention is on measures of central tendency. For example, we may wish to compare several religious denominations with respect to average attendance or income level. We may also wish to obtain measures of homogeneity, however. Perhaps we have hypothesized that one of the denominations will be more likely than the others to attract its membership from a single social stratum. But even if we are primarily interested in comparing measures of central tendency, we still may need to know something about the dispersion in each group. We realize intuitively that if each denomination were extremely heterogeneous with respect to income background or church attendance, then a given difference between their means (say $2,000) would not be as important or indicative as would be the case if each group were more homogeneous. When we come to inductive statistics, we shall be in a position to justify this intuition and to see why measures of dispersion are so important. In the present chapter we must concentrate on mechanics. In the following chapter will be given an interpretation for the most important measure of dispersion, the standard deviation.

6.1 The Range

Of the several measures of dispersion to be discussed in this chapter, the range is by far the simplest. The range is defined as the difference between the highest and lowest scores. Thus, for the data given in the previous chapter (72, 81, 86, 69, and 57) the range would be the difference between 86 and 57, or 29. We usually indicate the range either by the actual difference (29) or by giving the two extreme scores, e.g., 57 and 86. If data have been grouped, we take as the range the difference between the *midpoints* of the extreme categories. Thus, if the midpoint of the lowest interval is 2,450 and that of the highest is 7,450, the range will be 5,000.

The extreme simplicity of the range as a measure of dispersion is both an advantage and a disadvantage. The range may prove useful if it is desirable to obtain some very quick calculations which can give a rough indication of dispersion or if computations must be made by persons unacquainted with statistics. If data are to be presented to a relatively unsophisticated audience, the range may be the only measure of dispersion that will be readily interpreted. The level of sophistication of social scientists is rapidly reaching the point, however, where we can assume that more satisfactory measures will be understood. The disadvantage of the range is obvious: it is based on only two cases, the two extreme cases at that. Since extremes are likely to be the rare or unusual cases in most empirical problems, we recognize that it is usually a matter of chance if we happen to get one or two in our sample. Suppose, for example, that there is one millionaire in the community sampled. If we choose 10 persons at random, he will probably not be included. But suppose he is. The range in incomes will then be extremely large and very misleading as a measure of dispersion. If we use the range as our measure, we know nothing about the variability of scores between the two extreme values except that the scores lie somewhere within this range. And, as implied in the above example, the range will vary considerably from one sample to the next. Furthermore, the range will ordinarily be greater for large samples than small ones simply because in large samples we have a better chance of including the most extreme individuals. For these reasons, the range is not ordinarily used in sociology except at the most exploratory levels.

Another exceedingly simple measure, the *variation ratio*, can be used in the case of grouped data and is especially appropriate in the case of nominal scales. Basically, it is a measure of the degree to which the cases are concentrated in the modal category, rather than being distributed more evenly throughout all categories. It is defined as follows:

$$\text{V.R.} = 1 - f_{\text{modal}}/N$$

where f_{modal} refers to the number of cases in the modal category and N to the total number of cases. Obviously, the measure is insensitive to the distribution of cases in the nonmodal categories and is of course dependent on the categorization process. Its advantage lies in its extreme simplicity and intuitive appeal, as well as the fact that in the case of nominal scales one cannot make use of the ordering of categories to construct more refined measures.

6.2 The Quartile Deviation

Another measure that is sometimes used in the fields of psychology and education but that seldom appears in the sociological literature is the quartile deviation or semi-interquartile range. The quartile deviation Q is a type of range, but instead of representing the difference between extreme values, it is arbitrarily defined as half the distance between the first and third quartiles. Symbolically,

$$Q = \frac{Q_3 - Q_1}{2} \tag{6.1}$$

where Q_1 and Q_3 represent the first and third quartiles respectively. Notice that the quartile deviation is one half of the range covered by the middle half of the cases. Since Q_1 and Q_3 will vary less from sample to sample than the most extreme cases, the quartile deviation is a far more stable measure than the range. But it does not take advantage of all the information. We are not measuring the variability among the middle cases nor are we taking into consideration what is happening at the extremes of the distribution. We shall therefore turn our attention to two measures that do have this desirable property.

6.3 The Mean Deviation

If we wish to make use of all scores, common sense would suggest that we take the deviations of each score from some measure of central tendency and then compute some kind of average of these deviations in order to control for the number of cases involved. It would be possible to use the median or mode as our measure of central tendency, but we ordinarily take the mean since this is the most satisfactory single measure under most circumstances. Suppose we were simply to sum the actual deviations from the mean. Unfortunately, as we know, the result would always be zero since the positive and negative differences would cancel each other. This suggests that in order to obtain a measure of dispersion about the mean, we must somehow or other get rid of the negative signs. Two methods immediately occur to us: (1) ignore signs, taking the absolute value of the differences, or (2) square the differences. These two methods lead to the two remaining measures of dispersion to be discussed in this chapter, the mean deviation and the standard deviation.

The mean deviation is defined as the arithmetic mean of the absolute differences of each score from the mean. In symbols

$$\text{Mean deviation} = \frac{\sum_{i=1}^{N} |X_i - \bar{X}|}{N} \qquad (6.2)$$

The mean of the numbers 72, 81, 86, 69, and 57 is 73.0. Subtracting 73.0 from each of these numbers, ignoring the sign, and then adding the results and dividing by 5 we get

$$\frac{\sum_{i=1}^{N} |X_i - \bar{X}|}{N} = \frac{1 + 8 + 13 + 4 + 16}{5} = \frac{42}{5} = 8.4$$

We may therefore say that on the average the scores differ from the mean by 8.4.

Although the mean deviation has a more direct intuitive interpretation than the standard deviation, it has several serious disadvantages. First, absolute values are not easily manipulated algebraically. Second, and more important, the mean deviation is not as easily interpreted theoretically nor does it lead to as simple mathematical results. For purely descriptive purposes, the mean deviation may be adequate although, as we shall see, the standard deviation can be interpreted more readily in terms of the normal curve. When we come to inductive statistics, we shall see that the standard deviation is used almost exclusively because of its theoretical superiority. For this reason, we seldom see references to the mean deviation in the sociological literature.

6.4 The Standard Deviation

Having more or less eliminated several other measures of dispersion, we can turn our attention to the most useful and frequently used measure, the standard deviation. The *standard deviation* is defined as the square root of the arithmetic mean of the squared deviations from the mean. In symbols

$$s = \sqrt{\frac{\sum_{i=1}^{N} (X_i - \bar{X})^2}{N}} \qquad (6.3)$$

where s is used to represent the standard deviation.[1] In words, we take the deviation of each score from the mean, square each difference, sum the results, divide by the number of cases, and then take the square root. If we are to get the correct answer, it is essential that the operations be carried out in exactly this order. In our numerical example the standard deviation could be obtained as follows:

X_i	$(X_i - \bar{X})$	$(X_i - \bar{X})^2$
72	-1	1
81	8	64
86	13	169
69	-4	16
57	-16	256
$\bar{X} = 73.0$	0	506

dominators

$$s = \sqrt{506/5} = \sqrt{101.2} = 10.06$$

The intuitive meaning of a standard deviation of 10.06 will not be apparent until later when we make use of s to give us areas under the normal curve. For the present, we simply accept it as an abstract number. Several properties of the standard deviation are readily apparent, however. We notice that the greater the spread about the mean, the larger the standard deviation. Had all five values been the same, the deviations about the mean would all have been zero, and s would also be zero. Furthermore, we see that extreme deviations from the mean have by far the greatest weight in determining the value of the standard deviation. The values 169 and 256 dominate the other three squared deviations. In squaring the deviations, even though we later take a square root, we are in effect giving even more relative weight to extreme values than was the case in computing the mean. This suggests that we must qualify our initial enthusiasm about the standard deviation as the single "best" measure of dispersion. Certainly if there are several extreme cases, we want our measure to indicate this fact. But if the distribution has a few very extreme, cases, the standard deviation can give misleading results in that it may be unusually large. In such instances we would probably use the median as our measure of central tendency and, perhaps, the quartile deviation as a measure of dispersion. For most data the standard deviation will be appropriate, however.

Extreme Cases

p. 79

[1] Some texts define s with $N - 1$ in the denominator rather than N. The reason for this will not be apparent until Chap. 11.

It is reasonable to ask, "Why bother to take the square root in computing a measure of dispersion?" One easy but unsatisfactory answer would be that this is the way the standard deviation is defined. It would be possible to justify taking the square root by pointing out that since we have squared each deviation, we are essentially compensating for this earlier step. It makes more sense, however, to justify taking a square root in terms of its practicality. Since we shall make considerable use of the normal curve later on, the standard deviation, as defined, turns out to be a very useful measure. For other purposes we shall make use of the square of the standard deviation or *variance* which is defined as

$$\text{Variance} = s^2 = \frac{\sum\limits_{i=1}^{N} (X_i - \bar{X})^2}{N}$$

Mathematical statisticians have found the concept of variance of more theoretical value than the standard deviation. Beginning with Chap. 16 we shall make increasing use of the variance, but for the present we can confine our attention to the standard deviation. The two concepts are so easily interchangeable that we can pass readily from the one to the other. Whether one defines the variance as the square of the standard deviation or the standard deviation as the square root of the variance is unimportant.

Computation of standard deviation from ungrouped data   Although the standard deviation can always be computed from the basic formula given above, it is often simpler to make use of computing formulas that do not require the subtraction of the mean from each separate score. Not only will the mean not ordinarily be a whole number, but rounding errors will usually be made if the above formula is used. In order to see how we can simplify the computations, let us expand the expression under the radical. We get

$$\frac{\sum\limits_{i=1}^{N} (X_i - \bar{X})^2}{N} = \frac{\sum\limits_{i=1}^{N} (X_i^2 - 2X_i\bar{X} + \bar{X}^2)}{N} = \frac{\sum\limits_{i=1}^{N} X_i^2 - 2\bar{X} \sum\limits_{i=1}^{N} X_i + N\bar{X}^2}{N}$$

Notice that since $\bar{X}$ is a constant, we were able to take it in front of the summation sign in the second term of the numerator. In the third term, we have made use of the fact that for any constant k

$$\sum_{i=1}^{N} k = Nk$$

But since $N\bar{X} = \sum_{i=1}^{N} X_i$, the middle term of the numerator reduces to $-2N\bar{X}^2$ and we may write

$$\frac{\sum_{i=1}^{N} (X_i - \bar{X})^2}{N} = \frac{\sum_{i=1}^{N} X_i^2}{N} - 2\bar{X}^2 + \bar{X}^2 = \frac{\sum_{i=1}^{N} X_i^2}{N} - \bar{X}^2$$

Therefore
$$s = \sqrt{\frac{\sum_{i=1}^{N} X_i^2}{N} - \bar{X}^2} \qquad (6.4)$$

Some alternative computing formulas are as follows:

$$s = \sqrt{\frac{\sum_{i=1}^{N} X_i^2}{N} - \left(\frac{\sum_{i=1}^{N} X_i}{N}\right)^2} \qquad (6.5)$$

$$= \sqrt{\frac{\sum_{i=1}^{N} X_i^2 - \frac{\left(\sum_{i=1}^{N} X_i\right)^2}{N}}{N}} \qquad (6.6)^2$$

$$= \frac{1}{N}\sqrt{N \sum_{i=1}^{N} X_i^2 - \left(\sum_{i=1}^{N} X_i\right)^2} \qquad (6.7)$$

Although any of the above forms may be used as computing formulas, Eq. (6.7) will involve the fewest rounding errors and is therefore recommended.

---

[2] The derivation of Eqs. (6.6) and (6.7) from Eq. (6.5) is left as an exercise.

Let us make use of one of these computing formulas, Eq. (6.7), in the above problem where $N = 5$.

X_i	X_i^2
72	5,184
81	6,561
86	7,396
69	4,761
57	3,249
365	27,151

In addition to the total number of cases, the two quantities needed are $\sum_{i=1}^{N} X_i$ and $\sum_{i=1}^{N} X_i^2$. Both sums can be accumulated simultaneously on modern desk calculators. We now compute s from Eq. (6.7).

$$s = \tfrac{1}{5} \sqrt{5(27,151) - (365)^2} = \tfrac{1}{5} \sqrt{135,755 - 133,225} = 10.06$$

We have made use of this very simple problem in order to illustrate that the computing formula gives the same numerical result as the basic formula, Eq. (6.3). Since $\bar{X}$ turned out to be a whole number, the computing formula actually involved more work than did the original formula. Usually, of course, this will not be the case.

*Computation of standard deviation from grouped data   When data have been grouped, we may simplify our work considerably by treating each case as though it were at the midpoint of an interval and by making use of a guessed mean. Of course, we thereby introduce certain inaccuracies, but the saving in time will be substantial if computations must be carried out by hand. Following a common convention, suppose we let

$$x_i = X_i - \bar{X}$$

The small x's therefore represent deviations from the mean, and the basic formula for the standard deviation becomes

$$s = \sqrt{\frac{\sum_{i=1}^{N} x_i^2}{N}}$$

We can now modify the formula to take into consideration the fact that there will be a large number of cases all treated as having the same value, i.e., one of the midpoints. If we multiply the number of cases in each class by the proper midpoint and then sum these products, we can save ourselves the work of adding up all N cases. The formula for the standard deviation then becomes

$$s = \sqrt{\frac{\sum_{i=1}^{k} f_i x_i^2}{N}} \tag{6.8}$$

where f_i is the number of cases in the ith interval and k is the number of intervals.[3]

Now suppose we were to guess a mean and to take deviations from the guessed mean instead of the true mean. We have shown in the previous chapter that the sum of the squared deviations from the mean will be less than the sum of the squared deviations from any other value. In particular, the sum of the squared deviations from the guessed mean will be larger than the figure obtained using the true mean unless, of course, the guessed mean actually equals the true mean. It can also be shown that the closer the guessed mean is to the true mean, the smaller the sum of the squared deviations from the guessed mean. In other words, if we use a guessed mean, we expect to find a sum of squares that is too large. We can, as before, make use of a correction factor which we then subtract from the value obtained using the guessed mean. The formula for the standard deviation then becomes

$$s = \sqrt{\frac{\sum_{i=1}^{k} f_i d_i^2}{N} - \left(\frac{\sum_{i=1}^{k} f_i d_i}{N}\right)^2} \tag{6.9}$$

where the d_i represent the differences between each score and the guessed mean and are directly analogous to the x_i in Eq. (6.8).

Before taking up a numerical example, let us examine the above formula more carefully. The second term under the radical represents the correction factor to be subtracted from the mean of the squared deviations from the guessed mean. Recalling the formula for the mean expressed in terms

[3] Notice that we do *not* square the frequencies f_i in the numerator of the expression under the radical.

of the guessed mean, that is,

$$\bar{X} = \bar{X}' + \frac{\sum\limits_{i=1}^{k} f_i d_i}{N}$$

we see that
$$\frac{\sum\limits_{i=1}^{k} f_i d_i}{N} = \bar{X} - \bar{X}'$$

and therefore
$$\left(\frac{\sum\limits_{i=1}^{k} f_i d_i}{N} \right)^2 = (\bar{X} - \bar{X}')^2$$

Thus the correction factor turns out to be the square of the difference between the true and guessed means. We see immediately that had we guessed the mean exactly, the correction factor would have been zero. Also, the greater the difference between the true and guessed means, the larger the correction factor. A poor guess will always lead to the correct result but will involve larger numerical scores for both terms in the formula.

The formula can be modified still further if we prefer to think in terms of step deviations d_i'. As in Chap. 5, we factor out the width of the interval from each d_i, multiplying the final result by i when we have finished the process. The formula then becomes

$$s = \sqrt{\frac{\sum\limits_{i=1}^{k} f_i d_i{}^2}{N} - \left(\frac{\sum\limits_{i=1}^{k} f_i d_i}{N} \right)^2} = \sqrt{\frac{i^2 \sum\limits_{i=1}^{k} f_i d_i'{}^2}{N} - \left(\frac{i \sum\limits_{i=1}^{k} f_i d_i'}{N} \right)^2}$$

Therefore
$$s = i \sqrt{\frac{\sum\limits_{i=1}^{k} f_i d_i'{}^2}{N} - \left(\frac{\sum\limits_{i=1}^{k} f_i d_i'}{N} \right)^2} \qquad (6.10)$$

$$= i \sqrt{\frac{\sum\limits_{i=1}^{k} f_i d_i'{}^2 - \dfrac{\left(\sum\limits_{i=1}^{k} f_i d_i' \right)^2}{N}}{N}} \qquad (6.11)$$

$$= \frac{i}{N} \sqrt{N \sum_{i=1}^{k} f_i d_i'^2 - \left(\sum_{i=1}^{k} f_i d_i' \right)^2} \qquad (6.12)$$

Notice that in effect we have merely taken the interval width i outside the radical.

When computing the standard deviation from grouped data, we can now extend the procedure used for the mean by adding the column $f_i d_i'^2$. Although we could actually obtain the squared deviations $d_i'^2$ and then multiply by f_i, it will be much simpler to multiply the last two columns used in obtaining the mean (that is, $d_i' \times f_i d_i'$). Having in effect multiplied d_i' by itself, we see that all negative numbers now become positive.[4] Let us now compute the standard deviation for the grouped data used in the previous chapter. For illustrative purposes we shall use Eq. (6.10), although Eq. (6.12) will ordinarily involve fewer rounding errors.

We thus have obtained a mean of $5,117 and a standard deviation of $1,444. These two numbers can now be used to summarize the data or to compare them with data from another sample. As will be seen later,

Table 6.1 Computation of standard deviation, using grouped data

True limits	Mid-points	f_i	d_i'	$f_i d_i'$	$f_i d_i'^2$
$1,950–2,950	$2,450	17	−3	−51	153
2,950–3,950	3,450	26	−2	−52	104
3,950–4,950	4,450	38	−1	−38	38
4,950–5,950	5,450	51	0	0	0
5,950–6,950	6,450	36	1	36	36
6,950–7,950	7,450	21	2	42	84
Totals		189		−63	415

$$s = i \sqrt{\frac{\sum_{i=1}^{k} f_i d_i'^2}{N} - \left(\frac{\sum_{i=1}^{k} f_i d_i'}{N} \right)^2}$$

$$= 1,000 \sqrt{\frac{415}{189} - \left(\frac{-63}{189} \right)^2} = 1,000 \sqrt{2.196 - .111}$$

$$= 1,444$$

they can also be used to test hypotheses or to estimate population measures.

[4] Note specifically that the last column in Table 6.1 is *not* obtained by squaring the column $f_i d_i'$ since this would involve squaring the f_i as well as the d_i'.

6.5   The Coefficient of Variation

It is sometimes desirable to compare several groups with respect to their relative homogeneity in instances where the groups have very different means. It therefore might be somewhat misleading to compare the absolute magnitudes of the standard deviations. One might expect that with a very large mean he would find at least a fairly large standard deviation. He might therefore be primarily interested in the size of the standard deviation *relative to that of the mean*. This suggests that we can obtain a measure of the relative variability by dividing the standard deviation by the mean. The result has been termed the *coefficient of variation*, denoted by *V*. Thus

$$V = \frac{s}{\bar{X}}$$

To illustrate the advantages of the coefficient of variation over the standard deviation, suppose a social psychologist is attempting to show that for all practical purposes two groups are equally homogeneous with respect to age. In the one group the mean age is 26 with a standard deviation of 3. In the other, the mean age is 38 with a standard deviation of 5. The coefficients of variation for the two groups are therefore $\frac{3}{26} = .115$ and $\frac{5}{38} = .132$, a much smaller difference than that between the two standard deviations. In view of the fact that exact age usually becomes less important in determining interests, abilities, and social status as the average age of group members is increased, a comparison of the two coefficients of variation in this instance might very well be much less misleading than if the standard deviations were used.

It is also possible to make use of a relative variance if one so desires. However, these relative measures of dispersion are rather infrequently reported in the sociological literature. More commonly, one finds the means and standard deviations listed in adjacent columns.

6.6   Other Summarizing Measures

We have discussed only two types of summarizing measures, measures of central tendency and measures of dispersion. Other kinds of measures are possible, although they are seldom used in sociological research. Of course we often find the entire frequency distribution given, but this is not a single summarizing measure. It is sometimes desirable to indicate the degree of skewness in a distribution. One measure of skewness takes advantage of the fact that the greater the skewness the larger the difference between the mean and the median. This measure is given by the formula

$$\text{Skewness} = \frac{3(\bar{X} - Md)}{s}$$

If the distribution is skewed to the right (large positive scores), the mean will be greater than the median, and the result will be a positive number. A distribution skewed to the left will produce a negative result.

Very infrequently in sociology we also find references to the general peakedness of a distribution. The term *kurtosis* is used to refer to this type of measure which will be discussed very briefly after we have taken up the normal curve. Statistics texts written primarily for students of economics usually go more deeply into both skewness and kurtosis. Perhaps as we begin to attain greater precision in describing the exact forms of distributions of sociological variables, we may find more use for these other descriptive measures.

Glossary
Coefficient of variation
Mean deviation
Quartile deviation
Range
Standard deviation
Variance

Exercises
1. Compute the mean deviation and standard deviation for the data given in Exercise 1, Chap. 5. (*Ans.* 9.62; 11.59)

2. Compute the standard deviation and the quartile deviation for the data grouped in Exercise 1, Chap. 4. Do the same for Exercise 2, Chap. 4.

3. Compute the standard deviation for the data given in Exercise 4, Chap. 5. Check your computations by selecting a different guessed mean and a different computing formula. (*Ans.* 10.83)

4. Indicate how the standard deviation would be affected by each of the changes indicated in Exercise 5, Chap. 5.

References
1. Anderson, T. R., and M. Zelditch: *A Basic Course in Statistics*, 2d ed., Holt, Rinehart and Winston, Inc., New York, 1968, pp. 76–84.
2. Downie, N. M., and R. W. Heath: *Basic Statistical Methods*, 2d ed., Harper and Row, Publishers, Incorporated, New York, 1965, chap. 5.

3. Hagood, M. J., and D. O. Price: *Statistics for Sociologists*, Henry Holt and Company, Inc., New York, 1952, chap. 9.

4. McCollough, C., and L. Van Atta: *Introduction to Descriptive Statistics and Correlation*, McGraw-Hill Book Company, New York, 1965, chap. 3.

5. Mueller, J. H., K. Schuessler, and H. L. Costner: *Statistical Reasoning in Sociology*, 2d ed., Houghton Mifflin Company, Boston, 1970, chap. 6.

6. Weinberg, G. H., and J. A. Schumaker: *Statistics: An Intuitive Approach*, Wadsworth Publishing Company, Inc., Belmont, Calif., 1962, chap. 3.

7. Weiss, R. S.: *Statistics in Social Research*, John Wiley & Sons, Inc., New York, 1968, chap. 7.

The Normal Distribution

7

The notion of a frequency distribution is already a familiar one. The present chapter is concerned with a very important kind of frequency distribution, the normal curve. This distribution is useful not only because a large number of empirical distributions are found to be approximately normal but also because of its theoretical significance in inductive statistics. At this point you should not be concerned about applications in which the normal curve is used. The purpose of this chapter is to indicate the properties of the normal curve and to enable you to gain facility in using tables based on the normal curve. This distribution is discussed under descriptive rather than inductive statistics for two principal reasons. First, the normal curve can be used to provide an interpretation for the standard deviation. Second, you will find it helpful to become familiar with the normal distribution several chapters before you are exposed to statistical tests that require facility with it. Therefore, the better your understanding of the materials in the present chapter, the less difficulty you will experience later on.

7.1 Finite Versus Infinite Frequency Distributions

Frequency distributions discussed up to this point have involved a finite number of cases. Actually, of course, all empirical distributions necessarily involve a finite although perhaps very large number of cases. Mathematicians frequently find it useful to think in terms of distributions based on an indefinitely large number of cases, however. Rather than dealing with empirical distributions having a jagged appearance, as exemplified by a histogram or frequency polygon, it is possible to conceive of smooth curves which are based on an indefinitely large number of cases and which can be expressed in terms of relatively simple mathematical equations. The normal distribution is one such curve. Before studying this specific distribution, it will be necessary to examine the nature of the process by which such a smooth curve is developed.

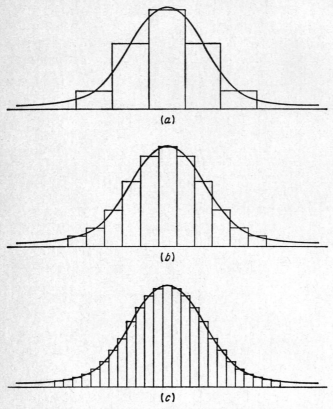

Figure 7.1 Comparison of smooth curves with histograms having different interval widths.

Let us start with a histogram involving five intervals (Fig. 7.1a). For simplicity we shall assume the frequency distribution to be symmetrical. We have already seen that if the number of intervals is increased without changing N, the form of the histogram is likely to become irregular. Suppose, however, that the number of cases is also increased. Then, as in Fig. 7.1b, it will be possible to make use of a larger number of narrower intervals, each of which has a sufficiently large number of cases to maintain regularity. If the number of cases is further increased, still more rectangles can be used while retaining the regular pattern (Fig. 7.1c). Smooth curves have been drawn through the midpoints of the top of each rectangle. It is clear that the rectangles form better and better approximations to the smooth curve as the number of rectangles increases, i.e., as the width of each interval decreases. We now imagine an ever-increasing number of cases and correspondingly narrower intervals until the rectangles approximate the smooth curve so closely that we can no longer see any difference. We refer to the smooth curve approximated by the ever-narrowing rectangles as the *limiting* frequency distribution.[1]

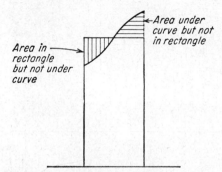

Figure 7.2 Comparison of areas under smooth curve and under rectangle.

Although we cannot possibly imagine an infinite number of cases, we can conceive of such a large number that the rectangles approximate the smooth curve to within any desired degree of accuracy.

It will be remembered that the *area* of each rectangle can be used to represent the proportion of cases within the interval. As indicated in Chap. 4, it is customary to set the total area of all rectangles equal to unity. Thus if the proportion of cases in the first interval is .10, then this same number represents the actual area of the first rectangle. We now notice that the area under the smooth curve within any given interval can be approximated by the area of the corresponding rectangle. This is indicated in Fig. 7.2. As the number of rectangles is increased, the total area of the rectangles becomes a better and better approximation to the area under the smooth curve. This can be seen by noting that the shaded areas become relatively smaller and smaller. In the limit, then,

[1] The notion of a limit is also discussed in Sec. 9.1.

the area under the smooth curve can be obtained by summing the areas of an indefinitely large number of rectangles. Since the area under the rectangles is unity, the area under the smooth curve will also be unity. The process we have just described is exactly the kind of process underlying that branch of mathematics referred to as the calculus.

7.2 General Form of the Normal Curve

The normal curve is a special type of symmetrical smooth curve. Since the normal curve is smooth, perfectly symmetrical, and based on an indefinitely large number of cases, it can only be approximated by frequency distributions involving actual data. It is bell-shaped in form

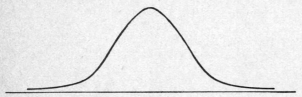

Figure 7.3   General form oɪ the normal curve.

and has a number of remarkable mathematical properties, some of which will be indicated shortly. Since it is symmetrical and unimodal, its mean, median, and mode all coincide. The general form of the normal distribution is indicated in Fig. 7.3.

*The mathematical equation for the normal curve is relatively simple by mathematicians' standards. Although you will never have to use this formula for computations since tables have been constructed for this purpose, it will be instructive to use it to point out and verify some of the properties of this theoretical distribution. The formula is as follows:

$$Y = \frac{1}{s\sqrt{2\pi}} \, e^{-(X-\bar{X})^2/2s^2}$$

where Y is the height of the curve for any given value of X. Since both π and e are constants (approximately equal to 3.14 and 2.72, respectively), the formula involves only two summarizing measures, the mean $\bar{X}$ and the standard deviation s.[2] Therefore, the exact form of the normal curve

---

[2] Other notation for the mean and standard deviation will be introduced when we come to inductive statistics. The formula for the normal curve is usually written in terms of a mean of μ (mu) and a standard deviation of σ (sigma).

will be known if we are given the values of both the mean and standard deviation. In other words, there are many different normal curves, one for every combination of mean and standard deviation.

*Recalling that a quantity with a negative exponent can be written as the reciprocal of that quantity raised to a positive power, we may rewrite the formula as follows:

$$Y = \frac{1}{s\sqrt{2\pi}}\left(\frac{1}{2.72^{(X-\bar{X})^2/2s^2}}\right)$$

where the constant e has been replaced by its approximate numerical value. Let us assume that s has a fixed value and find the value of X for

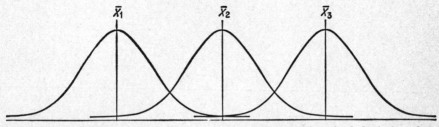

Figure 7.4 Comparison of normal curves with the same standard deviations but different means.

which Y will be a maximum. Clearly, Y will be at its maximum when the denominator within the parentheses is a minimum. But this denominator consists of a positive number greater than unity, raised to a power which cannot be negative, since a squared real number can never be less than zero. The denominator will therefore take on its minimum value when the exponent is zero. This will occur when X takes on the value $\bar{X}$ since we will then have $X - \bar{X} = 0$. This shows that the mode (and therefore the mean and median) is actually $\bar{X}$, a fact that has already been pointed out but not demonstrated. Also, we can see that the equation yields a curve that is symmetrical about $\bar{X}$. Since the quantity $X - \bar{X}$ is squared and therefore cannot be negative, deviations from $\bar{X}$ in either direction will produce identical values of Y.

The specific equation for any particular normal curve can be obtained by using the proper values of $\bar{X}$ and s. Normal curves having the same standard deviations but different means are drawn in Fig. 7.4. On the other hand, curves having different standard deviations will vary in peakedness as indicated in Fig. 7.5. The smaller the standard deviation the more peaked the curve.

It should be pointed out that not all symmetrical bell-shaped curves are normal. Although the curves in Fig. 7.5 differ with respect to peakedness, this is due only to differences in their standard deviations. They

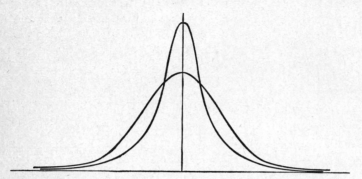

Figure 7.5 Comparison of two normal curves with the same means but different standard deviations.

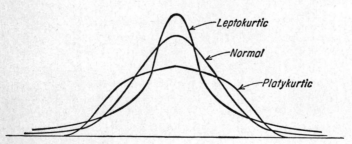

Figure 7.6 Comparison of normal curve with curves having the same standard deviation but which differ with respect to peakedness.

are both normal in form. In general, unimodal symmetrical curves may be either more peaked or more flat than the normal curve even though their standard deviations are all the same. Several such curves are drawn in Fig. 7.6. Curves which are more peaked than the normal curve are referred to as *leptokurtic*, and those which are flatter than normal as *platykurtic*. Unlike the normal curve, the equations of leptokurtic and platykurtic curves will involve summarizing measures in addition to the mean and standard deviation.

7.3 Areas under the Normal Curve

It is frequently necessary to determine the proportion of cases falling within a given interval. Fortunately, the normal curve has an important property that makes this task a relatively simple one. It turns out that regardless of the particular mean or standard deviation a normal curve may have, there will be a *constant area* (or proportion of cases) *between the mean and an ordinate which is a given distance from the mean in terms of standard deviation units.* Figure 7.7 helps to illustrate the meaning of the above statement.

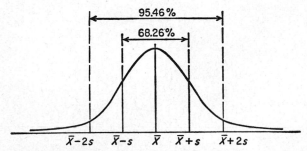

Figure 7.7   Areas under the normal curve.

If we go one standard deviation to the right of the mean, we shall always find .3413 of the area included between the mean and an ordinate drawn at this point. Therefore twice this area, or .6826, will be included between the two ordinates which are located one standard deviation on either side of the mean. In other words, slightly over two-thirds of the cases will always be within one standard deviation of the mean. Similarly, the area between the mean and an ordinate two standard deviations away will always be .4773, and therefore slightly over 95 per cent of the area will be included between the pair of ordinates that are two standard deviations on either side of the mean. Practically all the cases will be within three standard deviations of the mean even though the normal curve theoretically extends infinitely far in either direction. Distances from the mean need not always be exact multiples of the standard deviation, of course. By means of a procedure to be described shortly it is possible to determine the areas between any two ordinates. For example, if we go out 1.96 standard deviations on either side of the mean, we shall include almost exactly 95 per cent of the area; 99 per cent of the area will be included between ordinates that are 2.58 standard deviations from the mean.

This property of the normal curve affords an interpretation for the standard deviation and a method of visualizing the meaning of this measure of dispersion. A number of empirical frequency distributions. are sufficiently similar to the normal distribution that these relationships between areas and the standard deviation will hold up reasonably well. Even in the case of income distributions which are likely to be skewed in the direction of high incomes, we usually find approximately two-thirds of the cases within one standard deviation of the mean. It should be kept in mind that although the normal curve provides an *interpretation* for the standard deviation, this property cannot be used to *define* what is meant by a standard deviation. The definition is in terms of its formula. The above property holds only in the case of normal or approximately normal distributions.

It is possible to take any particular normal curve and to transform numerical values for this curve into such a form that a single table can be used to evaluate the proportion of cases within any desired interval. We can illustrate the process with a numerical example. Suppose we have a normal curve with mean 50 and standard deviation 10. Let us find the proportion of cases within the interval 50 to 65. We first determine how many standard deviations 65 is from the mean 50. In order to do this, we take the difference between these two values, i.e., 15, and divide by the size of the standard deviation. In this case the result is 1.5. Generally, we can make use of the formula

$$Z = \frac{X - \bar{X}}{s}$$
$$= \frac{65 - 50}{10} = 1.5$$

where X is the value of the ordinate and where Z represents the deviation from the mean in standard-deviation units.

*Before discussing how the numerical value of Z can be used to determine the proportion of cases between the mean and the ordinate corresponding to Z, let us give an alternative interpretation of Z. We can think in terms of an actual transformation from the X variable to a new variable Z. Whereas the distribution of the X variable is normal with a mean $\bar{X}$ and standard deviation s, the new variable is normal with a mean of zero and a standard deviation of one.[3] A normal distribution

---

[3] The verification of this fact is left as an exercise. (See Exercise 3.)

with mean zero and standard deviation one is referred to as the *standard form*, and the Z is often called a *standard score*. The transformation of variables is illustrated in Fig. 7.8. From every X we subtract the constant $\bar{X}$. In subtracting this constant value (here, 50) from each X, we have shifted every original score 50 units to the left and have therefore in effect moved the original normal curve to a position directly over the origin. This takes care of the numerator in the expression for Z. We now divide each difference $X - \bar{X}$ by the size of the standard deviation. In so doing we either squeeze the original curve or spread it out depending on whether or not its standard deviation is greater or less than unity. We can thus think of having first shifted the position of the original

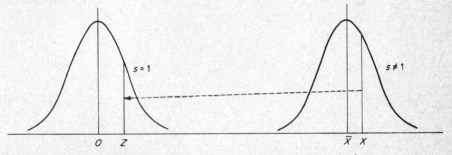

Figure 7.8 Comparison of standard and general forms of the normal curve.

normal curve and then having changed the size of the standard deviation so as to superimpose it on the standard form. In dividing by the standard deviation of 10 we have essentially changed units along the horizontal axis so that a distance of 10 along the X axis corresponds to a distance of 1 along the Z axis.

Regardless of which interpretation is given, a value of $Z = 1.5$ indicates that the ordinate is 1.5 standard deviations from the mean. In the case of the standard form this of course means that the ordinate itself falls at the value 1.5 on the Z scale. Tables showing exact areas have been constructed for the standard form of the normal curve. Table C in Appendix 2 is one such table. The values of Z are given down the left-hand margin and across the top. The first two digits of Z are obtained by reading down and the third by reading across. Figures within the body of the table indicate the proportion of the area between the mean (that is, 0) and the ordinate corresponding to Z. In the above example, we see that .4332 of the area is within these limits. Had Z turned out to be 1.52, the corresponding area would have been .4357.

7.4   Further Illustrations of the Use of the Normal Table

Suppose we wish to find the shaded area in the normal curve indicated by Fig. 7.9. The value of Z in this case is

$$Z = \frac{143 - 168}{12} = \frac{-25}{12} = -2.08$$

The fact that Z is negative simply indicates that the shaded area is to the left of the mean. The sign of Z can be ignored when using the normal table since the curve is perfectly symmetrical. From the table we see that

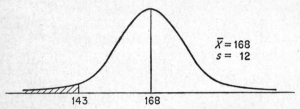

Figure 7.9 Normal curve, with shaded portion representing the area in a single tail.

the area between the mean and a Z of 2.08 is .4812. Since the total area is unity, the area to the left of the mean must be .5 (by symmetry). The shaded area therefore can be obtained by subtracting the area between the mean and the ordinate from the total area to the left of the mean. Thus,

(Proportion of cases $\leq$ 143) = .5000 $-$ .4812 = .0188

Therefore, fewer than 2 per cent of the cases have scores less than or equal to 143.[4] The type of problem illustrated in this example is a very common one because of the fact that tests of hypotheses practically always

---

[4] In a continuous distribution the proportion of cases which are exactly 143.0 will be zero. This can be seen if we imagine two ordinates extremely close together. The proportion of cases between these ordinates will also be very small. Now if we allow these ordinates to move closer and closer together, the proportion of cases will become infinitely small. A mathematical line, it will be remembered, has no width. In practice there may be some cases with scores of 143.0 owing to measurement inaccuracies. Since we are dealing with a theoretical distribution, however, it makes no difference whether or not the ordinate itself is included within the interval. Henceforth, we shall simply refer to the area between (but not including) two ordinates or the area less than a given value.

involve the tails of a frequency distribution. We therefore are often interested in the areas of one or both of these tails. Had we wanted to obtain the total area outside of the region defined by 168 ± 25 (as indicated by the shaded areas in Fig. 7.10), we would simply have doubled the above result since the two shaded areas are exactly equal in size.

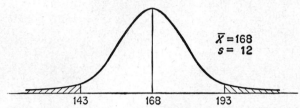

Figure 7.10 Normal curve, with shaded portions representing areas in both tails.

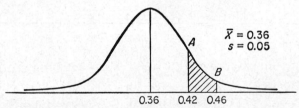

Figure 7.11 Normal curve, with shaded portion representing the area between two ordinates.

To take another example, suppose we need to obtain the shaded area indicated in Fig. 7.11. This area is calculated by first finding the proportion of cases between the mean and ordinate B and subtracting from this number the proportion of cases between the mean and ordinate A. The Z's corresponding to B and A are 2.0 and 1.2, respectively. Thus we have

Proportion between B and mean	.4773
Proportion between A and mean	.3849
Proportion between A and B	.0924

Therefore slightly over 9 per cent of the cases fall between .42 and .46. Notice that had we desired the area between two ordinates on opposite sides of the mean, the result could have been obtained by addition rather than subtraction.

Glossary
Leptokurtic
Limiting-frequency distribution
Normal curve
Platykurtic
Standard score

Exercises

1. You have already computed the mean and standard deviation for the data given in Exercise 1, Chap. 4. What proportion of the 65 cases were within one standard deviation of the mean? Two standard deviations? Three standard deviations? How well do these figures approximate those we would expect if the distribution were exactly normal? Answer these same questions for Exercise 2, Chap. 4. Contrast and account for any differences between the results for the two sets of data.

2. If the mean of a normal distribution is 80 and its standard deviation 12:
   a. What proportion of the cases is between 80 and 93? (*Ans.* .3606)
   b. What proportion of the cases is between 90 and 105? 70 and 105? (*Ans.* .1838)
   c. What proportion of the cases is less than 68?
   d. How many standard deviations on either side of the mean would you need to go in order to obtain two tails each containing 2 per cent of the total area? Ten per cent of the total area? (*Ans.* 2.054)
   e. What score has 4 per cent of the cases above it? (In other words, locate the ninety-sixth percentile.)

*3. Verify that the standard form of the normal curve has a mean of zero and a standard deviation of unity. [*Hint:* Rewrite the formula for the normal curve in terms of Z, using the fact that $Z = (X - \bar{X})/s$.]

4. Raw scores on various aptitude and attitudinal tests are often treated as interval scales by psychologists. These raw scores are then often converted to standard scores with convenient means and standard deviations. Suppose the mean raw score on a college entrance examination is 117 with a standard deviation of 28.5. Assume that these raw scores are distributed normally.
   a. What is the proportion of raw scores above 131? Below 79?
   b. What are the raw scores corresponding to the first, second, and third quartiles?
   c. On college entrance exams, raw scores may be standardized so that the mean of the normal distribution is exactly 500 and the standard deviation 100. Specifically, how would one go about standardizing the above set of scores to obtain a mean of 500 and a standard deviation of 100? (*Hint:* How would you standardize to get a mean of zero and a standard deviation of unity?)

References
1. Downie, N. M., and R. W. Heath: *Basic Statistical Methods*, 2d ed., Harper and Row, Publishers, Incorporated, New York, 1965, chap. 6.

2. Hagood, M. J., and D. O. Price: *Statistics for Sociologists*, Henry Holt and Company, Inc., New York, 1952, chap. 14.

3. Mueller, J. H., K. Schuessler, and H. L. Costner: *Statistical Reasoning in Sociology*, 2d ed., Houghton Mifflin Company, Boston, 1970, chap. 6.

4. Weinberg, G. H., and J. A. Schumaker: *Statistics: An Intuitive Approach*, Wadsworth Publishing Company, Inc., Belmont, Calif., 1962, chap. 8.

5. Weiss, R. S.: *Statistics in Social Research*, John Wiley & Sons, Inc., New York, 1968, pp. 147–156.

Inductive Statistics

Part 3

Introduction to Inductive Statistics

8

The purpose of this brief chapter is to give a general overview of inductive statistics and in particular of the logic underlying the testing of statistical hypotheses. It is very easy to become so overwhelmed with the details of each particular test encountered that one is unable to see the similarities underlying all tests. The learning of statistics can then become a mere "cookbook" exercise in memorizing formulas and procedures. Therefore this chapter is a very important one and should be reread carefully after you have been exposed to two or three specific tests.[1]

8.1 Statistics and Parameters

The purpose of statistical generalizations is to say something about various characteristics of the population studied on the basis of known facts about a sample drawn from that population or universe.[2] We shall refer to the characteristics of the population as *parameters* as contrasted with characteristics of a sample, which are called *statistics*. You are already familiar with a number of parameters and statistics: means, medians, proportions, standard deviations, etc. At this point you should learn to make a careful distinction between those characteristics which refer to the population and those which refer to a sample. Greek letters are usually used to refer to population characteristics; Roman letters indicate sample characteristics.[3] Thus, the population mean will henceforth be designated by μ and the sample mean by $\bar{X}$, the population standard deviation by σ and the sample standard deviation by s.

---

[1] A good point for review would be after Chap. 11.
[2] The terms *population* and *universe* are generally used interchangeably in the statistical literature.
[3] Unfortunately, there are a number of exceptions to this rule.

An important distinction between parameters and statistics can be made. Parameters are *fixed* values referring to the population and are generally *unknown*.[4] For example, at any given time the mean age or grade-point average of all students at Harvard University may be unknown but presumably would be found to be the same value by all observers. Statistics, on the other hand, vary from one sample to the next. If 10 different samples of college students were selected at random, we would not expect them all to have exactly the same mean ages. In fact, we would be very suspicious if they did. In contrast to parameters, the values of statistics for a particular sample are known or can be computed. We do *not* know, however, how representative the sample actually is of the population or how closely the statistic obtained approximates the comparable unknown parameter.

It is the population, rather than any particular sample, in which we are really interested. We have selected a sample as a matter of convenience, and our goal is practically always to make inferences about various population parameters on the basis of known, but intrinsically unimportant, sample statistics. In tests of hypotheses, we make assumptions about the unknown parameters and then ask how likely our sample statistics would be if these assumptions were actually true. In so doing, we attempt to make a rational decision as to whether or not the assumed values of these parameters are reasonable in view of the evidence at hand. Hypothesis testing can thus be viewed as a special type of decision-making process. Since the basic logic underlying the testing of hypotheses is rather complex, it will be helpful at this point to discuss this logic briefly. In later chapters you will see how it is applied to specific tests.

8.2 Steps in Testing an Hypothesis

In the social sciences the term *hypothesis* is used in a number of different senses. Sometimes it is used to refer to a theoretical proposition which has some remote possibility of being tested indirectly. At other times it is used to refer to the kind of statement which can actually be tested statistically. In order to minimize confusion it will therefore be

---

[4] Parameters will always be treated as fixed even though they may actually vary over time. Thus the median age of a population will change from one moment to the next. For this reason, you should conceptualize the notion of repeated sampling in terms of a large number of samples drawn simultaneously rather than in temporal sequence. Actually, in many instances our scientific objective is to infer the nature of the causal processes that generate population values, here assumed to be fixed. However, in learning statistics it seems wise to restrict oneself initially to the simpler notion of generalizing to fixed populations.

necessary to specify how the term will be used in this text. The criteria used in defining what we shall mean by a test of an hypothesis are rather strict and would rule out many of the so-called "tests" made in the current social-science literature. They are, however, consistent with the rather rigid requirements laid down by the statistician. As such, they represent an ideal against which the adequacy of any actual test can be compared.

An hypothesis is a statement about a future event, or an event the outcome of which is unknown at the time of the prediction, set forth in such a way that it can be rejected. In more precise language, let us say that we have tested an hypothesis if the following steps have been taken:

1. All possible outcomes of the experiment or observation were antici-pated *in advance of the test.*[5]
2. Agreement was reached prior to the test on the operations or procedures used in determining which of the outcomes actually occurred.
3. It was decided in advance which of the outcomes, should they occur, would result in the rejection of the hypothesis and which in its nonrejection. As implied above, rejection must have been a possible result.
4. The experiment was performed, or the event observed, the out-comes noted, and a decision made whether or not to reject the hypothesis.

The steps outlined above are very general ones. Statistical inference is primarily concerned with steps 3 and 4 since the statistician must assume that the first two steps have already been accomplished. We shall have occasion to see how the last two steps become more specific in a statistical test. Perhaps the most important general implication in the above list is that all decisions must be made prior to the test. All possible outcomes are divided into two classes: those that will result in rejection and those that will not. If this is not done prior to the test, it becomes possible to retain an hypothesis by simply changing the rules as one goes along. This is directly analogous to the child flipping a coin to decide whether or not to go to the movies. He decides, "Heads I go, tails I don't." If the coin turns up heads he goes. If it comes up tails, he decides on the best two out of three and continues to flip. In this way he always ends up going to the movies unless he loses the coin (an unan-ticipated outcome).

---

[5] The term *experiment* is often used in a very broad sense by the statistician. For example, an experiment might consist of interviewing a housewife and recording a "yes" or "no" to a specific question.

*It was mentioned in Chap. 2 that a test can only be made on a proposition which has been stated in terms of concepts that have been operationally defined. Step 2 indicates that operational definitions must be agreed upon in advance of the test. Unless this has been done, it is always possible to retain an hypothesis regardless of the outcome of the experiment by rejecting the methods used. Suppose one states as his hypothesis that "the higher one's social class position, the less likely that he will be highly ethnocentric." If the results do not seem to confirm this proposition, he may claim that the measure of "social class" or "ethnocentrism" was not really measuring what he intended it to measure and that some other index (which happens to confirm his theory) is a better one. It thus seems desirable to reserve the term *hypothesis* to refer to statements which are on the operational level and are clearly rejectable. If agreement cannot be reached beforehand on the procedures to be used, then agreement on the outcomes can hardly be expected. As indicated in Chap. 2, this point of view does not deny the importance of theory nor does it imply that operational definitions are the only definitions necessary for the development of a science.

The third step is a crucial one since the decision made will usually involve certain risks of error. In some instances the problem is relatively simple. Not all tests of hypotheses involve induction. An hypothesis may be formulated concerning the outcome of a specific event such as a football game. We may predict, for example, that team A will beat team B. As long as there are criteria for determining whether or not the procedures agreed upon have been adequately carried out, there is little chance of error in deciding whether or not to reject this kind of hypothesis. When information is based on a sample of events taken from a larger population, there is a greater risk of error, however. We reject or fail to reject the hypothesis realizing that, since our judgment is based only on a sample, we always have to admit the possibility of error due to the lack of representativeness of the sample. It is probability theory that enables us to evaluate the risks of error and to take these risks into consideration in deciding upon the criteria to be used in rejecting the hypothesis. In the next sections, two kinds of possible errors will be discussed. We can then return to the question of the role played by statistics in tests of inductive hypotheses.

8.3 The Fallacy of Affirming the Consequent

There is often no direct way of checking up on our most important propositions or theories. Instead, we derive from these a number of consequences which should occur if the original proposition or theory were true, and it is the validity of these consequences which can be determined by empirical methods.[6] Thus, evidence for the original theory is indirect. The theory A implies certain consequences B or, written symbolically, $A \Rightarrow B$. It should be emphasized that purely logical or deductive reasoning rather than empirical evidence is used in going from A to B. Therefore, if A is true, B must also be true provided our reasoning in deducing B from A is valid. We then look to see whether or not B has occurred; if B has not occurred (B false), then we know that theory A must also be false.

But what if B turns out to be true? Can we conclude that A *must* be true? We cannot. If we do, we shall be committing the fallacy of affirming the consequent, as logicians refer to it. If B is true, we can say that A *may* be true, but there could be any number of alternative theories that also predict B. We cannot be assured that A is *necessarily* true unless we can also show that there is no valid alternative theory C for which $C \Rightarrow B$. Unfortunately, it is practically never possible to do this, and therefore we have to proceed by *eliminating* theories rather than by definitely establishing them. A good theory is one which successfully resists elimination—provided, of course, that it is stated in such a way that it can be eliminated.[7] In other words, it must lead to hypotheses which themselves can be eliminated. If we fail to reject A when B is true, we run the risk of making an error since A may actually be false. In statistics this type of error, *the error of failing to reject an hypothesis when it is actually false*, is referred to as a *type II* or β *error*.

Perhaps a simple example will make the above argument seem less abstract. Suppose we have a theory A consisting of the following three propositions: (1) All persons will conform to all norms of their society; (2) it is a norm of society X not to steal; and (3) Jones is a member of society X. If all portions of the theory are correct, we may conclude B, that Jones will not steal. Suppose that for some reason we are unable to verify the truth or falsity of A directly, but we are able to ascertain Jones's behavior. Clearly, if Jones does steal, the theory must be at least in part incorrect. Therefore, if B is false, we reject A. But, certainly, if we learn that Jones does not steal, we would not want to conclude that theory A is correct. Perhaps Jones is simply more honest than others.

[6] Strictly speaking, this statement is not quite accurate since a purely deductive theory does not lead *directly* to testable hypotheses. See [2].

[7] The role of the crucial experiment is to enable the scientist to choose among several alternative theories, each of which has previously resisted elimination. For example, theories A and A' may both predict events B_1, B_2, . . . , B_k, all of which occur. But A may predict that B_{k+1} will be true whereas A' predicts it will be false. If B_{k+1} actually is false, then A may be eliminated and A' retained for the time being.

Or he may not even be a member of society X. In this case if we were to accept the theory as true, we would be running a considerable risk of error. We would probably conclude that, although this particular individual may be honest, it would be best to withhold judgment.

The absurdity of the above example should not be allowed to obscure the main point that whenever we are in the position of having a theory which implies certain consequences, and these consequences but not the theory are subject to verification, we are in the logical position of being able to reject the theory, whereas we cannot accept it without running the risk of making an error.

8.4   The Form of Statistical Hypotheses

In the social sciences we do not find propositions of the sort used in the above example for the simple reason that theories about the real world do not imply certainty. Instead of holding that if A is true, B *must* follow, we claim only that if A is true, B will *probably* also be true. We thus have to admit the possibility that B may be false even when A is true. If we follow the rule of rejecting A whenever B is false, we also run the risk of making another kind of error, that of *rejecting a true hypothesis*. We refer to this kind of error as a *type I* or α *error*. Using the above example our propositions would be modified to read, "*Most* members will conform to societal norms," and "Jones will *probably* not steal." If Jones does steal, we reject the revised theory with some risk of error since it may actually be true, Jones being one of the few dishonest members.

There are thus two kinds of error which must be taken into consideration. The first type discussed (type II) stems from the purely logical fallacy of affirming the consequent. When we introduce probability statements into our theory, we admit of an additional type of error (type I). Although we have as yet said nothing about inductive as contrasted with deductive reasoning, it is because of the necessity of generalizing beyond the limits of one's data that we are required to make use of such probability statements.

What specific forms do statistical hypotheses take? What do the A and B look like? Actually the theory A consists of a number of assumptions about the nature of the population and the sampling procedures used, together with the mathematical reasoning necessary to make probability statements concerning the likelihood of particular sample results if the assumptions made are in fact true. By means of these probability statements we decide ahead of time which results are so likely that we would reject the assumptions A should these outcomes B *not* occur. We are reasoning, in effect, that if the assumptions are correct, then most of the time our sample results will fall within a specified range of outcomes. Of course we draw only one sample, but if our particular result happens to fall outside of this range into what is called the *critical region*, we shall reject the assumptions, running the risk of making a type I error. The *B*, then, is represented by a certain range of sample results. If the results are outside this range, then *B* is false and the hypothesis rejected. In deciding on how big a range to include under *B* we must (ideally) take into consideration the risk of errors of types I and II.

To illustrate the process, suppose we wish to compare samples of white-collar and blue-collar workers with respect to the percentage desiring a college education for their children. If we actually wish to show that there is a difference between these two groups, we proceed by trying to eliminate the alternative hypothesis that there is no difference. This seems like an extremely devious way of proceeding, but we must remember that we shall not be in a position to establish directly that there is a difference. To avoid the fallacy of affirming the consequent, we must proceed by the elimination of false hypotheses. In this case there are logically only two possibilities: there either is or is not a difference. If the latter possibility can be eliminated, we can then conclude that some difference in fact exists.

We therefore hypothesize that the percentage desiring a college education is the same for both populations. We might then show mathematically that for 99 per cent of all possible pairs of such samples, the differences between the two sample percentages would be less than 10 per cent *if the assumptions were in fact true*. In other words, *B* consists of sample differences that are less than 10 per cent. If there are actually no differences between the two populations, it is highly likely that the percentages for the two samples will be within 10 per cent of each other. It can therefore be decided that if the difference between sample percentages turns out to be 10 per cent or more, the assumptions *A* will be rejected. This is done with the knowledge that 1 per cent of the time a difference of this magnitude could occur even though *A* were true. In other words, the risk of making a type I error (rejecting a true hypothesis) would be one chance in a hundred.

Let us return to the original list of steps necessary in the testing of hypotheses. It has been pointed out that statistical inference is basically concerned with steps 3 and 4 in the process. The researcher is anticipating all possible sample results and is dividing these into two classes: those for which he can reject and those for which he cannot reject his hypothesis. Actually, all that probability theory does is to provide the criteria to be used in dividing the outcomes into the two classes. Outcomes are put in one or the other of the two classes according to the risks one is willing to take of making types I and II errors. The major advantage of statistical procedures over intuitive methods is in the knowledge they provide about these risks of error.

Explained this way, statistics hardly seems to be worth the trouble. But step 3 is by no means an easy one to accomplish by any other method. Imagine, for example, an experiment consisting of 25 tosses of a coin, the honesty of which is called into question. Suppose we try to decide upon the outcomes which, if they should occur, would result in our challenging the person doing the flipping. Would we reject the hypothesis that the coin is honest if 15 or more heads turned up? More than 18? Only if all tosses resulted in heads? If there were 10 heads in a row regardless of the results on other tosses? Probability theory enables us to evaluate the probabilities of getting any particular set of results if the coin were actually an honest one. We could then select those outcomes which would be very unlikely under this assumption.

It is not expected that a student exposed for the first time to statistical inference will understand on first reading everything that has been said about the logic of testing hypotheses. The process is admittedly an involved one, and one which seems to give students more difficulty than any other part of statistics. You should therefore make a special effort to understand this logic by looking for the basic similarities among all statistical tests. Once the underlying logic is thoroughly understood, the learning of statistics is greatly simplified.

Glossary
Hypothesis
Parameter
Population
Statistic
Types I and II errors

References
1. Ackoff, R. L.: *The Design of Social Research*, University of Chicago Press, Chicago, 1953, chap. 5.

2. Northrop, F. S. C.: *The Logic of the Sciences and the Humanities*, The Macmillan Company, New York, 1947, chaps. 7 and 8.

3. Weiss, R. S.: *Statistics in Social Research*, John Wiley & Sons, Inc., New York, 1968, chap. 13.

Probability 9

We all undoubtedly have some intuitive notion of what is meant by the concept *probability* even though we may not be able to give the term a precise definition. There are a number of words and phrases used almost interchangeably with the concept probability in ordinary language: words such as *likelihood, chance, odds*, etc. These concepts are frequently used in a number of very different senses. A few illustrations are sufficient to indicate some of these diverse usages. We ask, "What is the probability that it will rain today?" referring to a single event (raining today) which may or may not occur in the future. The statement "Jones probably did not murder his mother-in-law" is similar to the above example but refers to an event that has already occurred but about which we lack sufficient information to make a statement of certainty. Or we can refer to what may happen in the long run: "If you gamble, you will probably lose your shirt." Here, the reference is presumably not to losing one's shirt in a single throw of the dice but to what will happen if the experiment is repeated a large number of times. "A male baby born in the United States of native-white parentage will probably live at least 65 years." Such a statement seems to refer to the kind of generalized baby that exists in actuarial tables rather than to a concrete Jimmy Brown.

Obviously, if we are to talk intelligently about probability and especially if the mathematician is to be brought into the picture, the concept must be defined with sufficient precision that we can all use it in the same sense. Unfortunately, however, it is no simple matter to obtain a definition which at the same time satisfies the mathematician and also our intuitive notion of what we ordinarily mean by the term. As we shall see, the mathematician finds it necessary to think in terms of a priori probabilities that cannot actually be obtained empirically and that are not dependent upon any particular sample data. In the next sections the concept probability will be defined in mathematical language and some of its important mathematical properties discussed. At the same time, there will be an attempt to make this definition and these mathematical properties seem reasonable in the light of everyday usage and experience.

9.1   A Priori Probabilities

In statistics we are concerned with generalizing to a population ordinarily made up of a large number of individuals. Such a population may be a finite existing one which is clearly delimitable—such as the population of the United States or native-white males over 65. In such a case we would ordinarily take some sort of a sample from the population, and interest would be primarily in the population itself (or some subpopulation) rather than in those individuals who happen to appear in one particular sample. The population may also be a hypothetical one involving, say, an unlimited number of experiments performed "under similar conditions." The statistician is therefore not interested in the single event or individual except insofar as this event or individual may help him obtain information about the population. Since this is a text in statistics, we shall use the term probability to refer not to single events (raining today, Jones a murderer, etc.) but to a large number of events or to what happens in the long run.[1]

How can we approach probability from the point of view of repeated events? First, it is necessary to think in terms of an idealized experiment that can be carried out a large number of times under similar conditions. Of course in reality conditions change, but it should at least be possible to imagine that they do not. All outcomes must be anticipated in each of these perfect experiments. Thus we must learn to think in terms of an ideal coin being flipped a very large number of times under identical circumstances and with only two outcomes (H or T) possible on each flip. We ignore the fact that a real coin might become worn unevenly in the process of being flipped or that occasionally it might stand on end. We learn to conceive of a perfectly shuffled deck of cards, none of which tend to stick together, even though such a deck could never be found in real life.

Let us call any outcome or set of outcomes of an experiment an *event*. An event can be simple (nondecomposable) or compound (a combination

---

[1] It is possible to approach probabilities from the standpoint of the single event and still make use of the same mathematical properties as discussed in the next section (see [8]). This latter approach presents at least as many conceptual difficulties as the one used in this text, however.

of simple events). Thus, event *A* may be a 6 on a single toss of a die; event *B* (compound) may consist of the outcomes 2, 4, or 6 on a single toss of a die; event *C* (also compound) may be the obtaining of a 7 in two tosses. It is conventional to use the term *success* whenever the event under consideration occurs and *failure* when it does not occur.[2] The experiment can then be performed a very large number of times, and the proportion of times any particular event occurs can be obtained.

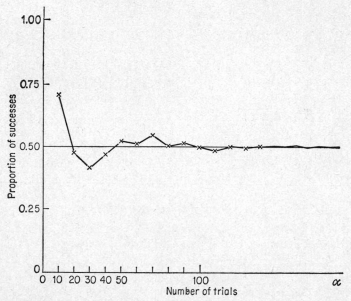

Figure 9.1 Oscillation of the proportion of successes, approaching a limit of .50.

We are not yet quite ready to give a formal definition of probability. First it is necessary to appeal to your knowledge of what happens empirically when an experiment such as the flipping of a coin is repeated a large number of times. Let us suppose that we begin to flip, and with every tenth flip we record the ratio of successes (say, heads) to total number of trials. The results obtained are likely to be similar to those indicated in Fig. 9.1.

On the first 10 flips we would usually not expect to get exactly 5 heads, even with an honest coin. Perhaps there will be 7 heads. The next set

[2] This technical use of the terms *success* and *failure* need not conform to general usage. Thus *success* may indicate the contraction of polio or the election of a demagogue.

of trials may contain a long sequence of tails, so that at the end of 20 trials the proportion of heads may be .45. The next sequence may also result in more tails than heads, the next slightly more heads than tails, and so forth. After we have made 100 trials with an honest coin, we would expect the proportion of successes to be in the neighborhood of .5; after 1,000 trials it should be even closer to this figure. Thus we would expect the ratio of successes to *total* number of trials to settle down in the sense that it ceases to fluctuate very much from one sequence of 10 flips to another. After 10,000 trials, even if we were to get 20 successive tails (an extremely unlikely event), the effect on this ratio would be negligible.[3] Had this happened on the third and fourth sequences the effect would have been pronounced. Therefore, the larger the number of trials, the closer and closer the ratio approaches a given value which mathematicians call a *limit*. If we can imagine the experiment being carried on indefinitely, we can probably also imagine the ratio becoming exactly this limiting value, say .5. Since we are becoming involved with the notion of infinity, and since mathematicians have found this to be a notoriously ambiguous concept, it may be preferable to think in terms of an extremely large number of trials.

*The notion of a limit can be defined somewhat more precisely. We say that the ratio approaches a limit if, having previously decided how close an approximation we want, we can flip the coin a finite number of times until we can be virtually sure that the obtained ratio approximates the limit within the desired degree of accuracy. In other words, we first choose a very small number ϵ representing the degree of approximation desired. Suppose we set $\epsilon = .0001$. If a limit exists, there is a finite large number of flips N such that we can be almost positive that the obtained proportion of successes will be within $\pm.0001$ of the true probability.[4] Furthermore, no matter how small an ϵ we select we can always find a finite number of flips for which this is true. If a limit does not exist, then this will not in general be possible.

It is by no means a *logical* necessity that the ratios obtained in this manner settle down to a limiting value. Indeed, it is at least conceivable

[3] Notice that it is not being claimed that the *absolute numbers* of heads and tails will be nearly equal or that, if there is initially an excess of heads, the tails will eventually catch up. There may continue to be an excess of heads indefinitely, but the *ratio* will approach .5 even when this is the case. Thus, if we had 35 heads and 15 tails on the first trials, the proportion of heads would be .7. An excess of 20 heads in 100 trials (i.e., 60 heads) gives a proportion of .6; the same excess in 200 flips (i.e., 110 heads) gives a value of .55.

[4] The discussion of confidence intervals (Chap. 12) should help to indicate why we can never be absolutely certain that the true probability is within the interval obtained.

that such ratios continue to fluctuate indefinitely. If this were actually the case, we could not speak of a single probability of heads associated with the coin. When such a limit does exist, however, we can define a *probability as a limit of the ratio of successes to total number of trials.* Put more crudely, a probability is the proportion of successes "in the long run."

It will be convenient in later discussions to speak as though we were thinking in terms of the probabilities of single events. Thus we may ask, "What is the probability of getting a 6 on a single toss of a die or of getting a red ace in a single drawing from a deck of cards?" Actually, in using the phrase "a single toss of a die," we are merely attempting to avoid the use of clumsy phrasing. What we really mean is, "What proportion of times in the long run would we expect to get a 6 if a single die were tossed repeatedly?" As a convenience, then, we shall refer to a single toss when we actually mean an indefinitely large number of single tosses of the same die.

Several points need to be made before we proceed with a discussion of the mathematical properties of probabilities. Real-life experiments, when repeated, actually seem to follow the general pattern discussed above and diagramed in Fig. 9.1. That is, a limit is actually approached and can be estimated. This leads us to speak of "the law of averages" and to expect that most coins will come up heads about half the time or that good bridge hands will be mixed with poor ones. A word of caution is necessary regarding this law of averages, however. Some persons have interpreted such a law to mean that if a coin comes up heads 10 times in a row, then the next time it is more likely to come up tails "because of the law of averages." Such an interpretation involves a prediction about a single event (i.e., the result of the eleventh flip). As will be discussed below, we usually assume that what has happened on previous flips is absolutely irrelevant to what follows.[5] A coin is possessed of neither a memory nor a conscience. As a matter of intelligent strategy, if a player were to witness 10 successive heads in 10 trials, he would do well to predict heads on the eleventh under the assumption that the coin is dishonest.

It should be perfectly clear that a priori probabilities, as defined in this section, cannot be obtained *exactly* by empirical methods although they may be *estimated*. This is not only because we have had to imagine idealized experiments but also because no experiment can be repeated indefinitely. With a sufficient number of trials, however, a probability can be estimated to any degree of accuracy. The mathematical rules

[5] This cannot be assumed in the case of human beings, a fact which must be kept in mind whenever repeated measurements are taken on humans or other animals. See Sec. 9.5.

given in the next section and all the mathematical reasoning underlying statistical inference are concerned with a priori probabilities rather than the kinds of probabilities that can actually be obtained by the researcher.[6]

In applying statistical reasoning to any science dealing with the real world, we are thus going to find ourselves in the logical position described in Chap. 8. We have to assume some a priori probability in order to apply mathematical reasoning. We can then say that *if* this is the correct a priori probability, *then* certain empirical results are likely (or unlikely). Thus A is the mathematical theory, B the predicted empirical results, and there is no way of testing the theory directly. If B turns out to be false, we can reject A; but if B is true, some other theory C, involving different a priori probabilities, may also account for the results. If we wish to avoid the fallacy of affirming the consequent, it will be necessary to assume probabilities which we really suspect are false and to proceed by elimination. In the next chapter we shall take up specific examples where this will be done.

9.2   Mathematical Properties of Probabilities

Although you may never again have to calculate probabilities, it is important to realize that underlying every table you will use to make tests of hypotheses, there are a number of fairly simple properties of probabilities. It is not possible in a text such as this to go very deeply into probability theory. The purpose of the discussion that follows is merely to give some insight into the way mathematicians operate with probabilities in laying the foundations for statistical inference. We can begin by identifying three mathematical properties of a priori probabilities.

The first property hardly requires much comment. Since we can obtain no fewer than zero successes and no more than N successes in N trials, it follows that for any event A the probability of A occurring [written $P(A)$] must be greater than or equal to 0 and less than or equal to 1. Thus

$$0 \leq P(A) \leq 1$$

where the symbol $\leq$ should be read as "less than or equal to." If $P(A) = 1$, the event A is certain to occur; if $P(A) = 0$, then A cannot possibly occur.

The addition rule   The second property of probabilities is a more interesting one. Because of its simplicity, we shall first take up a special case of

[6] Strictly speaking, the researcher can only obtain *proportions* since the number of trials or cases will always be finite.

the addition rule which can be stated as follows: *If events A and B are mutually exclusive, the probability of getting either A or B* [written $P(A$ or $B)$] *is equal to the probability of A plus the probability of B,* that is,

$$P(A \text{ or } B) = P(A) + P(B) \quad \text{(if } A \text{ and } B \text{ are mutually exclusive)} \quad (9.1)$$

By mutually exclusive, we mean that A and B cannot possibly occur simultaneously in the same experiment. Thus it is impossible to get both an ace and a king in a single draw from an ordinary deck of cards. Therefore, applying the addition rule to a hypothetical perfect deck we have

$$P(A \text{ or } K) = P(A) + P(K) = \tfrac{1}{13} + \tfrac{1}{13} = \tfrac{2}{13}$$

Of course, we could have obtained this same result by noting that there are eight aces and kings in a deck, and with equal probability of selection the probability of getting one of these cards would be $\tfrac{8}{52}$ or $\tfrac{2}{13}$. Similarly, the probability of getting either a 5 or 6 in a single throw of a die would be $\tfrac{1}{6} + \tfrac{1}{6}$ or $\tfrac{1}{3}$.

The addition rule can be extended to cover more than two events. Thus, *if A, B, C, . . . , K are all mutually exclusive,* then

$$P(A \text{ or } B \text{ or } C \cdots \text{ or } K) = P(A) + P(B) + P(C) + \cdots + P(K) \quad (9.2)$$

If we have a population composed of 100 upper-, 200 upper-middle-, 400 lower-middle-, and 300 lower-class persons, for example, the probability of getting an upper-class, *or* an upper-middle-class, *or* a lower-middle-class person in a single draw would be

$$\frac{100}{1,000} + \frac{200}{1,000} + \frac{400}{1,000} = \frac{700}{1,000} = .7$$

if every person had an equal chance of being selected.

Since probabilities are essentially proportions, it follows that if we have all possible simple events, each being mutually exclusive of the others, the sum of these events must be unity. Thus, if we add the probabilities of getting a spade or a heart or a club or a diamond, we must obtain a sum of 1. The probability of event A *not* occurring is equal to the sum of the probabilities of all the remaining (mutually exclusive) events. If we subtract $P(A)$ from unity, we thus have the probability of not getting

A since

if $\qquad 1 = P(A) + P(B) + P(C) + \cdots + P(K)$

then $\qquad 1 - P(A) = P(B) + P(C) + \cdots + P(K)$

The probability of not getting a queen, for example, is $1 - \frac{1}{13}$, or $\frac{12}{13}$.

So far we have been concerned only with mutually exclusive events. A more *general form* of the addition rule can be stated as follows: *If A and B are any events* (not necessarily mutually exclusive)

$$P(A \text{ or } B) = P(A) + P(B) - P(A \ \& \ B) \qquad (9.3)$$

where $P(A \ \& \ B)$ represents the probability of getting *both* A and B.[7] In the general case, the probability of getting either A or B can be obtained

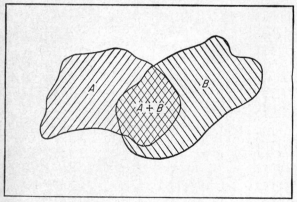

Figure 9.2 Geometric representation of probabilities, with areas proportional to $P(A)$, $P(B)$, and $P(A \ \& \ B)$.

by first adding the probability of getting A to the probability of getting B and then subtracting out the probability of getting both A and B simultaneously. The reason for subtracting out $P(A \ \& \ B)$ is that the probability of this joint occurrence has been figured in twice, once in $P(A)$ and again in $P(B)$. Figure 9.2 may help to indicate why this is the case.

[7] The word *or* as used by the mathematician includes the possibility that both A and B hold. Therefore, the expression "A or B" means "A and/or B." In terms of set-theory notation, "A or B" means the same as $A \cup B$, whereas "A and B" means the same as $A \cap B$.

In Fig. 9.2 the probabilities of A and B have been represented by certain areas which are proportional to their numerical values, the area of the rectangle being taken as unity. In the general case there will ordinarily be some overlap, that is, A and B will not be mutually exclusive. The probability of obtaining either A or B (or both) is represented by the total cross-hatched area. Since the smaller shaded area will have been added in twice, once in A and again in B, we see why it is necessary to subtract out $P(A \& B)$ in order to obtain the total cross-hatched area.[8]

Let us take a numerical example. Suppose A is the event that one obtains a queen in a single draw, and let B be the event that the card is a spade. Then A and B are not mutually exclusive since it is possible to draw both simultaneously (i.e., the queen of spades). Therefore,

$$P(A \text{ or } B) = P(A) + P(B) - P(A \& B)$$

$$= \tfrac{4}{52} + \tfrac{13}{52} - \tfrac{1}{52} = \tfrac{16}{52} = \tfrac{4}{13}$$

This result can be verified intuitively by noting that either A or B could be obtained by drawing any spade or one of the three remaining queens, i.e., any one of 16 cards. Had we simply added $P(A)$ and $P(B)$, the queen of spades would have been considered twice. In the next section we shall take up a general rule for computing $P(A \& B)$ since it will not always be so simple to obtain this quantity. Notice that if the two events are mutually exclusive, there will be no overlap and $P(A \& B) = 0$. Therefore, the general rule reduces to the special case of the addition rule discussed above.

The multiplication rule   A third property of probabilities enables us to obtain the probability of two (or more) events occurring jointly. We can state this property as follows: If A and B are *any two events, the probability of getting both A and B is the product of the probability of getting one of these events times the conditional probability of getting the other given that the first event has occurred.* In symbols,

$$P(A \& B) = P(A)P(B|A) = P(B)P(A|B) \tag{9.4}$$

The symbols $P(A|B)$ and $P(B|A)$ represent what are called conditional

---

[8] You should convince yourself that in order to obtain the probability of A or B *but not both,* we would subtract $2P(A \& B)$ from $P(A) + P(B)$. You should also attempt to extend the general form of the addition rule by drawing a similar diagram for events A, B, and C. (See Exercise 4*b*.)

probabilities. $P(A|B)$ should be read "the probability of A given that B has occurred." The term *conditional probability* means that we recognize that the probability of A may be dependent on whether or not B occurs. In other words, the probability of A given B may differ from the probability of A given that B has *not* occurred. Thus, if B is the event that a man drives recklessly and A the event that he is in a traffic accident, then we would expect $P(A|B)$ to be greater than $P(A)$ since reckless driving is a cause of accidents.

Before illustrating the use of the multiplication rule, let us introduce a new and important concept. Two events A and B are said to be *statistically independent* if and only if $P(A|B) = P(A)$ and $P(B|A) = P(B)$. Thus, if the probability of A occurring remains the same regardless of whether or not B has occurred, and if the same holds true for B, the two events are statistically independent of each other. Practically speaking, this means that knowledge that one of the events has occurred does not help one predict the other. For example, the probability of getting an ace, given that the card is red, is $\frac{2}{26}$ since there are two red aces and a total of 26 red cards. This is numerically the same as the unconditional probability of getting an ace $(\frac{4}{52})$. Therefore color and face value are statistically independent. Knowledge that a card is red does not help one predict whether or not it is an ace. Likewise, knowing the card to be an ace does not help one predict its color. Notice, incidentally, that mutually exclusive events are not independent. If A and B are mutually exclusive, we must always have $P(A|B) = P(B|A) = 0$. Why?

In the case where A and B are statistically independent, we have $P(B|A) = P(B)$ and the multiplication rule takes on the simple form

$$P(A \text{ \& } B) = P(A)P(B) \qquad \text{(if } A \text{ and } B \text{ are independent)}$$

This special case of the multiplication rule is ordinarily much easier to use than the more general rule.

We shall first illustrate the multiplication rule in the special case where A and B are statistically independent. Ordinarily we think of replications of an experiment as being independent of each other. Thus, if we flip a coin once, we expect that the result will have no effect on what happens on the next flip; the probability of heads remains constant from one flip to the next. Knowing that we get a head does not help us predict the outcome of the second toss.[9] Using the multiplication rule, we can therefore

---

[9] We are assuming that the true probability is known and that our task is to predict the outcome of any specific trial. It is of course true that without this knowledge, the probability might be *estimated* by using results of previous trials and this estimate then used to predict future results. This is *not* what we mean when we say that in

calculate the probability of getting successive heads on two flips by multiplying the probabilities of getting a head on any given trial. For an honest coin the probability of two successive heads would be $(\frac{1}{2})(\frac{1}{2})$ or $\frac{1}{4}$. Similarly, if A is the event that we get a red card and B the event that we get an ace, the probability of getting a red ace $[P(A \& B)]$ would be

$$P(A \& B) = P(A)P(B) = \frac{1}{2} \frac{1}{13} = \frac{1}{26}$$

Let us take up two examples in which statistical independence does not hold. The first of these involves a situation in which two variables are related so that knowledge of one helps us predict the other. Suppose we have the following purely hypothetical data:

Trait	Brunettes	Blondes	Redheads	Total
Aggressive	300	600	300	1,200
Nonaggressive	600	100	100	800
Total	900	700	400	2,000

If a girl is drawn at random[10] from this population as a blind date, what is the probability that she will be an aggressive redhead? Since there are 300 aggressive redheads out of 2,000 girls, the probability of getting one of this select group is clearly 300/2,000 or .15. This same probability will now be obtained using the multiplication rule.

Let A be the event that we get a redhead and B be the event that the date is aggressive. Since there are 400 redheads in all, $P(A) = 400/2,000$ or .2. Among the 1,200 aggressive girls, however, there are 300 redheads. Therefore, if we are given knowledge that the date is aggressive, the probability of her being a redhead is 300/1,200 or .25. Similarly the probability of getting an aggressive girl is 1,200/2,000 or .6, but if it is known that the date is a redhead, the probability of her being aggressive

---

the case of statistical independence, knowledge of one event does not help us predict another. Thus, knowledge of 20 successive heads would lead us to predict a biased coin, i.e., that the true probability of getting a head is some value greater than .5. This in turn would lead to a prediction of a head on the twenty-first trial. The assumption is, however, that such a bias, if it exists, is already known. Therefore, if it is known that p actually is .8, knowledge of 20 successive heads will not improve our ability to predict the next outcome.

[10] A random sample will be defined later in this chapter. In a random sample all individuals, and all combinations of individuals, have an equal chance of being selected.

is $^{300}\!/_{400}$ or .75. We thus have

$$P(A) = .2 \qquad P(A|B) = .25$$

$$P(B) = .6 \qquad P(B|A) = .75$$

Using the multiplication rule we obtain the following probability of getting an aggressive redhead:

$$P(A \ \& \ B) = P(A)P(B|A) = (.2)(.75) = .15$$

$$= P(B)P(A|B) = (.6)(.25) = .15$$

For the second example, let us suppose we want to calculate the probability of getting two aces in two draws from an ordinary deck of cards. Let A be the event that we get an ace on the first draw and B the event that an ace turns up on the second draw. Are A and B statistically independent? This depends on whether or not we replace the first card and reshuffle before drawing the second. If we sample *with replacement,* our two draws will be independent since the probability of getting an ace remains constant from one draw to the next and the result of the first cannot possibly affect the second. In this case

$$P(A \ \& \ B) = P(A)P(B) = (\tfrac{1}{13})(\tfrac{1}{13}) = \tfrac{1}{169}$$

Now suppose we sample *without* replacement, i.e., we do not put the first card back in the deck. If we should happen to get an ace on the first draw, the probability of getting a second ace would then be $\tfrac{3}{51}$ since there would be only three aces in the remaining 51 cards. On the other hand, if we did not get an ace on the first draw, the probability of getting one on the second draw would be $\tfrac{4}{51}$. Therefore, we do not have statistical independence in this case and would have to use conditional probabilities to compute $P(A \ \& \ B)$. Thus,

$$P(A \ \& \ B) = P(A)P(B|A) = (\tfrac{4}{52})(\tfrac{3}{51}) = \tfrac{1}{221}$$

It should be mentioned that the multiplication rule we have been discussing can also be extended to cover more than two events. For example, *if A, B, and C are all statistically independent of each other*

$$P(A \ \& \ B \ \& \ C) = P(A)P(B)P(C)$$

The principles for conditional probabilities can likewise be extended very simply. If, for instance, we were to draw four cards without replacement, we could calculate the probability of obtaining four aces as follows:

$$P(4 \text{ aces}) = \frac{4}{52} \frac{3}{51} \frac{2}{50} \frac{1}{49} = \frac{1}{270,725}$$

If there are three events A, B, and C which are not mutually independent, then we may obtain the probability of their joint occurrence through the formula

$$P(A \ \& \ B \ \& \ C) = P(A)P(B|A)P(C|A \ \& \ B)$$

where $P(C|A \ \& \ B)$ refers to the probability of C, given that both A and B have occurred. Of course we can also utilize similar formulas by reordering the positions of A, B, and C. Suppose we have the following population:

Attitude	Whites		Nonwhites		Total
	Republicans	Democrats	Republicans	Democrats	
Favor increased public welfare	50	100	25	225	400
Oppose increased public welfare	350	200	25	25	600
Total	400	300	50	250	1,000

If A is the event that we draw a white, B the event that we get a Republican, and C the event that the person favors increased public welfare, and since there are only 50 white Republicans who favor welfare, then $P(A \ \& \ B \ \& \ C) = 50/1,000 = .05$. From the table we also see that $P(A) = 700/1,000$; $P(B|A) = {}^{400}/_{700}$; and $P(C|A \ \& \ B) = {}^{50}/_{400}$. The last of these figures results from the fact that among the 400 people who are both A and B (white Republicans), there are only 50 who support welfare. Applying the multiplication rule, we thus obtain the result

$$P(A \ \& \ B \ \& \ C) = P(A)P(B|A)P(C|A \ \& \ B)$$
$$= \frac{700}{1,000} \frac{400}{700} \frac{50}{400} = \frac{50}{1,000} = .05$$

As a check we might have used the formula

$$P(A \ \& \ B \ \& \ C) = P(C)P(B|C)P(A|B \ \& \ C)$$
$$= \frac{400}{1,000} \frac{75}{400} \frac{50}{75} = \frac{50}{1,000} = .05$$

The idea of statistically independent events is very closely related to that of independence between two (or more) *variables*, a notion that we shall discuss in some detail in subsequent chapters. We have already used the example of a deck of cards which has the property that face value and suit are independent, meaning that knowledge of the one does not help predict the other. In both the example relating hair color of date to her behavior and that interrelating race, political preference, and welfare attitudes, we found it necessary to use conditional probabilities to obtain the correct results. In these instances we say that the variables concerned are *not* independent or that they are correlated. For simplicity let us consider the example of the dates. Suppose that exactly the same percentage (60 per cent) of blondes, brunettes, and redheads were aggressive, in which case knowledge of hair color would be of no value in predicting behavior. If we retained the same marginal totals, the results would then be as follows:

Trait	Brunettes	Blondes	Redheads	Total
Aggressive	540	420	240	1,200
Nonaggressive	360	280	160	800
Total	900	700	400	2,000

First, you should verify that for these hypothetical data there is no need to use conditional probabilities. Notice, also, that the probability (or proportion) corresponding to each cell in the table is equal to the *product* of the two probabilities in the corresponding margins. For example, if we examine the top left cell, we see that the probability $540/2,000 = .27$ is just the product of the probabilities corresponding to the first column marginal (that is, $900/2,000 = .45$) and the first row marginal (that is, $1200/2,000 = .6$). The same holds for every remaining cell in the table. Whenever the categories of two variables can be arranged in a cross-classification that has this property, we say that the *variables* are statistically independent of each other. In subsequent chapters we shall develop statistical tests for independence as well as measures of dependence that are based on this very simple idea.

*A note on Bayes' theorem   Given that $P(A \ \& \ B) = P(A)P(B|A)$, we may solve for the conditional probability, obtaining

$$P(B|A) = \frac{P(A \ \& \ B)}{P(A)} = \frac{P(B)P(A|B)}{P(A)}$$

But $P(A)$ in the denominator may be decomposed into the two terms $P(B)P(A|B) + P(\bar{B})P(A|\bar{B})$ since B and $\bar{B}$ (not B) are mutually exclusive and exhaustive possibilities. This leads to the equation

$$P(B|A) = \frac{P(B)P(A|B)}{P(B)P(A|B) + P(\bar{B})P(A|\bar{B})}$$

which is known as Bayes' theorem. This theorem can be generalized to several alternatives $B_1, B_2, \ldots , B_k$ as long as these alternatives are mutually exclusive and exhaustive, so that $\sum_{i=1}^{k} P(B_i) = 1$. The probability of any given B_i, given that A has occurred, can be written as follows:

$$P(B_i|A) = \frac{P(B_i)P(A|B_i)}{\sum_{i=1}^{k} P(B_i)P(A|B_i)}$$

Bayes' theorem can of course be applied whenever we are given all the conditional and unconditional probabilities, but these applications are not especially useful. But it can also be applied in instances where "psychological probabilities" have replaced relative-frequency notions. Hays [5] cautions against such usage, and the direct applications of the Bayesian approach to statistics are still relatively untested. Nevertheless it seems advisable to suggest ways in which it may be used. Consider, first, a very simple problem. Suppose an individual selects one of two urns at random and then selects a marble at random from whichever urn has been selected. The first urn contains half white and half black marbles, whereas the second contains two-thirds white and one-third black marbles. We learn that the individual selects a white marble and wish to assign a probability to his having selected, say, the first urn. Notice that this is a kind of "inverse probability" that is peculiarly appropriate to a notion of probabilities as reflecting our own state of knowledge. One might say that he either did or did not select the first urn, and therefore the probability of having selected it is either 1 or 0. But if we were to make a bet

on the basis of our knowledge that he drew a white ball, what odds would
we be willing to give that he selected the first urn? This is certainly a
reasonable way of phrasing the problem.

If we let A be the event that a white ball was selected, B be the event
that the first urn was selected, and $\bar{B}$ be the event that the second urn
was selected, then applying Bayes' theorem we get

$$P(B|A) = \frac{P(B)P(A|B)}{P(B)P(A|B) + P(\bar{B})P(A|\bar{B})}$$

$$= \frac{(\tfrac{1}{2})(\tfrac{1}{2})}{(\tfrac{1}{2})(\tfrac{1}{2}) + (\tfrac{1}{2})(\tfrac{2}{3})} = \frac{\tfrac{1}{4}}{\tfrac{1}{4} + \tfrac{1}{3}} = \frac{3}{7}$$

a result which would not have been predicted using common-sense arguments. Notice that since both urns were selected with equal probability,
we have $P(B) = P(\bar{B}) = .5$, and Bayes' formula could have been
simplified.

Let us next consider a kind of problem that is far removed from statistics, per se, but one which nevertheless is reasonably realistic from the
standpoint of psychological probabilities involving an observer's lack of
knowledge of relative frequencies or other considerations that could be
used to give a priori probabilities. Suppose we know that an action group
has available four alternative means that have different costs and expected chances of succeeding. Assume that an observer, on the basis of
his assessment of the relative costs of the alternative means, labeled
B_1, B_2, B_3, B_4, assigns them the subjective probabilities .4, .3, .2, and .1
respectively. Suppose he estimates the probabilities of success for the
alternative means as .3, .5, 6., and .9, respectively. He learns that the
group succeeds in its action but cannot ascertain which means they have
used. How should he reassess his original estimates of the probabilities
of each means, given the knowledge that success (A) has occurred?
Applying the more general form of Bayes' theorem in the case of the
first means (B_1) we get

$$P(B_1|A) = \frac{P(B_1)P(A|B_1)}{\sum_{i=1}^{k} P(B_i)P(A|B_i)}$$

$$= \frac{(.4)(.3)}{(.4)(.3) + (.3)(.5) + (.2)(.6) + (.1)(.9)} = \frac{.12}{.48} = .25$$

Thus, on the basis of this additional knowledge, the observer might
assign the subjective probability .25 to the first means. Using similar

calculations, he would assign the remaining means subjective probabilities of .3125, .25, and .1875, respectively.

9.3 Permutations

It is now necessary to introduce a further complication. So far we have taken up very simple problems, ones that could easily have been solved almost intuitively. Needless to say, most problems in probability are far more complex than those discussed up to this point. In order to handle problems of somewhat greater complexity, we shall find it necessary to take into consideration the order in which the events may occur. For example, suppose we wish to find the probability of getting an ace, king, and queen in three draws with replacement. We can obtain the probability of getting an ace on the first draw, a king on the second, and a queen on the third. This probability would be $(\frac{1}{13})^3$. But this represents the probability of getting an ace *followed by* a king *followed by* a queen. There are other ways of getting an ace, king, and queen in three draws *if we are not concerned about the order in which they are drawn*. As a matter of fact there are six ways they could be obtained: AKQ, AQK, KAQ, KQA, QAK, and QKA. Each of these possibilities can be seen to have the same probability. Therefore, if we are interested in the probability of getting these cards *in any order*, we can add their separate probabilities (since they are mutually exclusive), obtaining $6(\frac{1}{13})^3$.

Thus, in using the multiplication rule we have let event A refer to the outcome of the first draw, B to that of the second, and so forth. In other words, we have taken order into consideration, whereas usually we are more interested in the probabilities of obtaining a certain set of outcomes. We may want to know the probability of four aces in a bridge hand or the probability of getting a certain percentage of blacks in a sample, regardless of the order in which they were drawn. In computing such probabilities *it will usually be simplest first to determine the probability of any given ordering of outcomes*, and if all other orderings are equally likely, we can simply multiply the number of orderings by the probability of any one of them occurring. In so doing, it will be noticed, we are employing both the multiplication and the addition rules. There are definite formulas which can be used to enable one to count exactly how many orderings there will be in a given problem.

Whenever we have N different events that occur in a particular order, we refer to this as a *permutation* of these events. Whenever the order is irrelevant, we refer to the grouping of events as a *combination*. For example, in the case of the single combination (AKQ) there will be six distinct orderings or permutations, as we have just seen. Let us now see how formulas for counting the numbers of permutations can be obtained in simple instances.

Let us begin with the situation where all N events are distinct. How many ways can they be arranged? Clearly, if we consider N different ordinal *positions* (say, N chairs arranged in a line), the first of these can be filled by any one of the objects or events. Having filled this position, we can fill the second by any of the $N - 1$ remaining events, the third by the remaining $N - 2$, and so forth. When we come to the final position, there will be only one remaining possibility. Thus there will be

$$N(N - 1)(N - 2) \cdots (3)(2)(1) = N!$$

possible orderings, where the symbol $N!$ stands for the long product on the left-hand side and is termed "N factorial." Suppose, for example, that we have thirteen cards, one of each face value. We turn them up one-by-one. How many different possible permutations are there? The first card can have any one of thirteen values. Given that this card has already been turned up, the second card can be any of the twelve remaining values. Thus there are 13×12 possible outcomes for the first two cards. Proceeding through the pile, we conclude that there will be

$$(13)(12)(11)(10) \cdots (3)(2)(1) = 13! = 6{,}227{,}020{,}800$$

possible ways of arranging these thirteen cards.

Suppose next that the events are not all distinct. We may again have thirteen cards, but two of them may be aces, and we may not be distinguishing among the different suits. In this case, the order in which the two aces happen to be selected is irrelevant. Suppose they are selected at the fifth and eleventh drawings. Had they been distinct, and labeled as ace_1 and ace_2, then for every distinct permutation in which ace_1 were drawn before ace_2, there would be an identical permutation in which ace_2 preceded ace_1. We thus see that whenever we cannot distinguish between these two aces, there will be only half as many permutations as when all events are distinct. Therefore the total number of permutations in this case will be $N!/2! = N!/2$.

Suppose there had been three aces instead of two. Had these been labeled as ace_1, ace_2, and ace_3, we see that there would have been $3! = 6$ permutations among these aces that could not be distinguished. Therefore the total number of permutations of the thirteen cards will be $13!/3!$ In general, if there are N objects, three of which cannot be distinguished from the others, there will be $N!/3!$ permutations. This line of reasoning can readily be generalized to more than one set of objects that are not distinct. Suppose our thirteen cards contain three aces and four kings, with the remaining six cards all being distinct. Since the aces, if distinguished, could have been arranged in 3! ways, and the four kings in 4! ways, we must divide 13! by 3!4! to arrive at the number of truly distinguishable permutations.

The general rule should now be obvious. If we have N events that are subdivided in such a way that the first set contains r_1 indistinguishable elements, the second contains r_2 indistinguishable elements, and in general the ith set contains r_i such elements, then if there are k such sets, all distinguishable from each other, the total number of permutations will be $N!/r_1!r_2! \cdots r_k!$ To take another example, if there are 25 children, 6 of whom are three years old, 8 of whom are four years old, 9 of whom are five years old, and 1 each are aged six and seven, then there are $25!/6!8!9!1!1!$ permutations of these children if they are to be distinguished only by age.

The general rule for determining the number of permutations of events, not all of which are distinct, has a very important special case whenever there are only *two* kinds of events (e.g., successes and failures). If there are N events, r of which are successes and $N - r$ failures, and if the successes are indistinguishable among themselves, and similarly for the failures, then the general formula for the number of permutations reduces to $N!/r!(N - r)!$ For example, if we toss a coin 10 times and get 6 heads, then the number of distinct arrangements of heads and tails will be $10!/6!4! = 210$. We shall have occasion to use this important special case extensively in the next chapter when we consider the binomial distribution.

*Working with factorials can become extremely tedious without the benefit of computational shortcuts. Fortunately, when working with ratios of factorials, a considerable amount of cancellation will be possible, as in the case of the above example involving the ratio $10!/6!4!$. The following are the numerical values of factorials from 1 to 20:

$1! = 1$	$11! = 3.992 \times 10^7$
$2! = 2$	$12! = 4.790 \times 10^8$
$3! = 6$	$13! = 6.227 \times 10^9$
$4! = 24$	$14! = 8.718 \times 10^{10}$
$5! = 120$	$15! = 1.308 \times 10^{12}$
$6! = 720$	$16! = 2.092 \times 10^{13}$
$7! = 5,040$	$17! = 3.557 \times 10^{14}$
$8! = 40,320$	$18! = 6.402 \times 10^{15}$
$9! = 362,880$	$19! = 1.216 \times 10^{17}$
$10! = 3,628,800$	$20! = 2.433 \times 10^{18}$

For larger values of N it is possible to pin down the limits between which $N!$ must lie by using Stirling's approximation

$$\sqrt{2N\pi}\left(\frac{N}{e}\right)^N < N! < \sqrt{2N\pi}\left(\frac{N}{e}\right)^N\left(1 + \frac{1}{12N - 1}\right)$$

where $\pi \simeq 3.14159$ and $e \simeq 2.71828$. Those students familiar with the use of logarithms will find it most convenient to work with logs of factorials, thereby converting products to sums and ratios to differences. For example,

$$\log\left(\frac{8!}{3!}\right) = \log\frac{8\ 7\ 6\ 5\ 4\ 3\ 2\ 1}{3\ 2\ 1}$$
$$= \{\log 8 + \log 7 + \log 6 + \log 5 + \log 4 + \log 3 + \log 2 + \log 1\}$$
$$- \{\log 3 + \log 2 + \log 1\} = \log 8 + \log 7 + \log 6 + \log 5 + \log 4$$

Some examples Let us next consider some applications of these principles to probability problems of a somewhat more complex nature than we have discussed up to this point. As implied in the introduction to this section, an important general strategy in many problems in which order of selection is irrelevant is to calculate the probability of one particular permutation and then to multiply this by the number of permutations involved. For example, suppose we wish to obtain the probability of getting exactly one ace and at least two kings in four draws, assuming replacement. We note that this can be accomplished by getting either an ace and three kings *or* an ace, two kings, and some other card which cannot be either an ace or a king. Representing these possibilities symbolically as AKKK and AKKO ("O" standing for "other"), we note that there are $4!/3! = 4$ ways of arranging the ace and three kings, whereas there are $4!/2! = 12$ ways of arranging the AKKO combination. It is because the number of permutations differs in these two situations that we must keep them distinct. If we sample with replacement, the probability of getting an ace on a single draw is $\frac{1}{13}$, as is the probability of getting a king, whereas the probability of getting an O is $\frac{11}{13}$. Therefore the probability of getting exactly one ace and two or more kings is

$$4(\tfrac{1}{13})^4 + 12(\tfrac{1}{13})^3(\tfrac{11}{13}) = \frac{136}{28,561} = .0048$$

Now suppose we wish to obtain the probability of getting exactly one ace and at least two *hearts* in four draws, with replacement. An additional complication has entered the picture since the ace may be one of the hearts. It will be convenient to distinguish among four kinds of cards: the ace of hearts (AH) which has the probability $\frac{1}{52}$ of being selected; aces of nonhearts ($A\overline{H}$) which have the probability of $\frac{3}{52}$ of selection; non-ace hearts ($\overline{A}H$) with the probability of $\frac{12}{52}$ of selection; and nonace nonhearts ($\overline{A}\,\overline{H}$) with the probability of $\frac{36}{52}$ of being drawn. Naturally the sum of these probabilities is unity since these types are mutually exclusive and exhaustive.

We next display the combinations that can produce exactly one ace and two or more hearts and calculate the number of permutations of each. These are as follows:

(a) Exactly two hearts:

$AH, \overline{A}H, \overline{A}\,\overline{H}, \overline{A}\,\overline{H}$ $(4!/2!)[\frac{1}{52} \cdot \frac{12}{52} \cdot \frac{36}{52} \cdot \frac{36}{52}] = .02552$

$A\overline{H}, \overline{A}H, \overline{A}H, \overline{A}\,\overline{H}$ $(4!/2!)[\frac{3}{52} \cdot \frac{12}{52} \cdot \frac{12}{52} \cdot \frac{36}{52}] = .02552$

(b) Exactly three hearts:

$AH, \overline{A}H, \overline{A}H, \overline{A}\,\overline{H}$ $(4!/2!)[\frac{1}{52} \cdot \frac{12}{52} \cdot \frac{12}{52} \cdot \frac{36}{52}] = .00851$

$A\overline{H}, \overline{A}H, \overline{A}H, \overline{A}H$ $(4!/3!)[\frac{3}{52} \cdot \frac{12}{52} \cdot \frac{12}{52} \cdot \frac{12}{52}] = .00284$

(c) Exactly four hearts:

$AH, \overline{A}H, \overline{A}H, \overline{A}H$ $(4!/3!)[\frac{1}{52} \cdot \frac{12}{52} \cdot \frac{12}{52} \cdot \frac{12}{52}] = \underline{.00094}$

                                                           .06333

Adding these probabilities of mutually exclusive events we get a total probability of .063.

As our final example let us consider a situation where it is more convenient to draw what is referred to as a tree diagram to represent the various possibilities. It is sometimes the case that a sequence of events terminates at different points depending on the outcomes of previous events. The most familiar illustration of this occurs in the case of athletic events, where a team will be declared the winner if it wins two out of three games, or perhaps four out of seven, and where there is no point in proceeding further once the fixed number of games has been won. Suppose there are two teams A and B playing in a "two-out-of-three" series. Suppose also that A is the better team and that on the basis of its past performance is given a probability of .6 of winning any given game. In a more realistic example, the probabilities of winning each game may change according to results of the previous games, and this can be handled by

the method we shall propose. But for simplicity let us take the probability
of team A winning each game as $p = .6$, letting $q = .4$ represent team
B's chances of winning any given game. The successive trials are therefore assumed to be independent. What is team A's probability of winning the series? What are the individual probabilities of each possible
sequence of wins and losses?

We may diagram the possible sequences as follows:

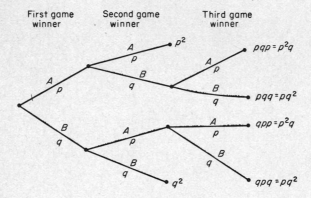

The upper branch of the tree represents the possibilities assuming A has
won the first game, whereas the lower branch represents those involving
an initial win by B. If A wins the second game after having won the first,
the series stops and A wins with a probability of p^2. However, if A wins
the first game and B the second, then a third game must be played. If
this is won by A, the series stops and A wins with probability p^2q. But if
B wins the third game, it wins the series with probability pq^2. This kind
of series produces a perfectly symmetrical tree, though one could obviously invent contests involving handicaps that produced asymmetrical
trees. For example, team A might have to win four games whereas B
must win only three.

Probabilities of the various possibilities can now be computed as
follows:

Sequences for which team A wins ($p = .6$)		Sequences for which team B wins ($q = .4$)	
$p^2 =$	.360	q^2	160
$pqp =$	.144	$pqq =$	.096
$qpp =$	.144	$qpq =$	.096
Probability of winning series	.648		.352

The fact that the probabilities sum to unity can also be shown algebraically as follows:

$$p^2 + 2p^2q + 2pq^2 + q^2 = p^2 + 2pq(p + q) + q^2$$

$$= p^2 + 2pq + q^2 = (p + q)^2 = 1$$

9.4 Expected Values

An idea that probably originated in the gambling casino has important statistical applications. This is the notion that if one repeats an experiment a large number of times, making various bets on the outcomes, it should be possible to calculate his expected winnings (or losses), under different assumptions about the nature of the game being played. To take a very simple example, suppose you are flipping coins, betting each time on heads, and that each time a head turns up you win $1 but that each time a tail is obtained you lose $2. Under the assumption that the coin is honest, you would obviously not want to remain in the game very long. But how can your expected gains or losses be calculated in more complex examples?

In this very simple example common sense would suggest multiplying the probability of each outcome by the gain or loss if this outcome occurs, and then adding. We would thus obtain for our expected "gain" the quantity $(1)(\frac{1}{2}) + (-2)(\frac{1}{2}) = -.5$. This means that, on the average, one would expect to lose 50 cents per trial. Of course one's actual gains or losses might differ from this expected value, but if we were to rely on probability theory and play the game a very large number of times, our expected total loss should be approximately $.5N$, where N is the number of trials.

As a second example, suppose we were to toss a single die, receiving a dollar if the face is even, losing two dollars if we get a one or three, but winning three dollars if a five appears. Assuming that each face has an equal chance of turning up, our expected winnings would be

$$(-2)(\tfrac{1}{6}) + (1)(\tfrac{1}{6}) + (-2)(\tfrac{1}{6}) + (1)(\tfrac{1}{6}) + (3)(\tfrac{1}{6}) + (1)(\tfrac{1}{6})$$
$$= \tfrac{1}{3} = \$.333$$

per game. In general, if there are k possible outcomes $X_1, X_2, \ldots, X_k$, and if the probability of X_i is given by $p(X_i)$, then we may define the expected value of the variable X, denoted by the symbol $E(X)$, to be $E(X) = \sum_{i=1}^{k} X_i p(X_i)$. In the examples thus far considered the X_i have been payoffs (in dollars) for each of a set of outcomes, but the idea of expected value can be conceived in much more general terms.

Suppose, for example, that we have a population containing N individuals with scores on the X variable. If we select randomly from this population, then each individual will have a probability of $1/N$ of being selected. What is the expected value of X? In this instance we would have

$$E(X) = X_1 p(X_1) + X_2 p(X_2) + \cdots + X_N p(X_N)$$
$$= (X_1 + X_2 + \cdots + X_N)(1/N) = \bar{X}$$

and we obtain the interesting result that the expected value of X is its mean *under the assumption of random sampling.*

Beginning in the next chapter we shall deal extensively with probability distributions, called sampling distributions. Strictly speaking, these distributions are infinite since they refer to probabilities which are here defined only in limiting terms. Nevertheless, we may refer to these probability distributions as having expected values which may be interpreted in the following way. We imagine random samples being repeatedly drawn from some population. If the population has a mean denoted by the Greek letter μ (mu), then $E(X) = \mu$. We shall also wish to find expected values of other quantities, such as the *sample* mean $\bar{X}$ which likewise turns out to have its expected value $E(\bar{X})$ equal to μ in the case of random sampling. Another expression that is of considerable theoretical interest in statistics is $E[X - E(X)]^2$ which, in the case of random samples for which $E(X) = \mu$, is $\displaystyle\sum_{i=1}^{N} (X_i - \mu)^2 p(X_i) = 1/N \sum_{i=1}^{N} (X_i - \mu)^2$, or the variance of X. Although we shall not make major use of expected-value notation, you are likely to find references to it in more advanced texts since it is extensively used in proofs in mathematical statistics.

9.5 Independence and Random Sampling

All the statistical tests to be discussed in this text make use of the assumption that there is independence between events and that therefore conditional probabilities do not have to be used when multiplying probabilities.[11] In other words, it is assumed that there is independence of selection within a sample—the choice of one individual having no bearing on the choice of another individual to be included in the sample.

[11] This will be seen in the case of the binomial discussed in the next chapter. In the case of other tests, however, you will simply have to accept the truth of this statement.

There are many instances in which this important assumption is likely to be violated, however. One should therefore develop the habit of always asking himself whether or not the independence assumption is actually justified in any given problem. It will be helpful at this point to indicate a few examples of situations in which this assumption is likely to be overlooked.

Statisticians often obtain what is called a *random sample* (or *simple random sample*) in order to meet the required assumption of independence as well as to give every individual in the population an equal chance of appearing in the sample. By using a table of random numbers or some equivalent device, one can obtain a sample in essentially the same way that one would draw cards from a well-shuffled deck or numbers in a bingo game. A random sample has the property *not only of giving each individual an equal chance of being selected but also of giving each combination of individuals an equal chance of selection.*[12]

Strictly speaking, since we practically always sample without replacement, the assumption of independence is not quite met. Whenever the population is large relative to the size of the sample, however, one can safely neglect the resulting minor distortion due to the fact that no individual is given a chance to be drawn a second time. For example, if 500 persons are selected out of a population of 100,000, the chances are very slight of any given person being selected again if his name were replaced. Likewise, it makes relatively little practical difference if we replace when drawing only three cards from a deck, but if we were to draw 35 cards, it would make considerable difference. If the sample is relatively large as compared with the population, a correction factor can sometimes be applied to compensate for lack of replacement.[13]

Although the problems introduced by failure to replace are not serious ones, the failure to give every *combination* of individuals an equal chance of appearing in the sample may result in a serious violation of the independence assumption. Suppose, for example, that one were to sort ordinary playing cards into four piles, one for clubs, one for spades, etc. Then suppose he were to select one of these piles randomly. Clearly, every card in the deck would have an equal chance (1 in 4) of being selected, but certainly all combinations would not be possible, let alone equally probable. Knowing that the top card is a spade tells us immediately that all others in the sample are also spades.

Area or cluster samples commonly used in social surveys do not meet

---

[12] In Chap. 21, random sampling will be distinguished from other commonly used types of sampling such as systematic, stratified, and cluster sampling.
[13] See Sec. 21.1.

the assumption of independence for this same reason. If 100 blocks in a city are selected at random and then every third household in these 100 blocks is included in the sample, it is very clear that not all combinations of households have an equal chance of appearing in the sample. For example, two houses in the same block have a much better chance of appearing in the same sample than do two houses in different blocks. Since city blocks are usually relatively homogeneous with respect to characteristics such as the income and education of the head of the household, the result of this type of sampling is to yield less accuracy than a random sample of the same size. This can be seen intuitively if we imagine a situation in which all blocks are completely homogeneous (as was true for the piles of cards). In this case we would need to obtain information about only one household in each block, and the number of "cases" would in effect be the number of *blocks* selected, i.e., a much smaller N. As will be seen in Chap. 21, it is possible to get some extremely misleading results if, having obtained such a cluster sample, a researcher then makes use of statistical tests that assume a random sample.

An analogous problem is likely to be encountered whenever one is interested in behavioral acts of individuals. For example, suppose a social psychologist runs an experiment in which he uses 30 subjects, each of whom makes 50 separate judgments. There would then be 1,500 judgments, and one might be tempted to make use of such an artificially inflated N in a statistical test, assuming in effect that the 1,500 judgments constituted a random sample of judgments from some sort of a population. But, obviously, it would be absurd in most instances to assume that the judgments of a single individual are statistically independent of each other. His first 30 judgments are very likely to affect the remaining ones since, unlike a coin, a person does have a memory.

Suppose a social scientist is primarily interested in pairs of persons rather than in the single individual as a unit. He may have a group of 20 persons, each of whom is in interaction with all of the others. He would therefore have $(20)(19)/2$ or 190 pairs of persons but would not be in a position to consider each pair as being independent of the others. Obviously, knowledge about the Smith-Brown pair is likely to yield some information about the Smith-Jones or Brown-Jones pairs since the same persons appear in several pairs.

Ecologists, anthropologists, and other social scientists interested in generalizing about communities, societies, or other spatially defined units also need to be concerned about lack of independence in much of their work. Here the problem seems to stem from the fact that units selected are often not clearly distinct. The boundaries of a society or community may be difficult to define, and one such unit may shade into another with the division points being more or less arbitrary.[14] For example, if census tracts within a city or counties within a state are used as units, it is often possible to predict from one unit to a contiguous one. If the delinquency rate is high in one tract, it is also likely to be high in an adjacent one since it is even possible that the same gangs of delinquents will be drawn from both tracts. That "something is wrong" in relation to the assumption of independence can be seen intuitively by realizing that whenever units are not clearly distinct, it would be possible to inflate the number of "units" to any desired size by simply slicing the cake into many small pieces. Thus, if there are not enough societies in the world to obtain statistical significance, one might subdivide each society into 10 subregions and obtain 10 times as many "cases."

In a text such as this it is not possible to discuss solutions to problems involving violations of the independence assumption. To the writer's knowledge, many of these problems have not been satisfactorily resolved. It is often rather difficult to assess the seriousness of errors introduced when required assumptions, such as that of independence, are *not* met. We are on safe ground whenever we can be assured that assumptions required for any test *are* met; if they are not met, it is seldom possible to determine just how much we are departing from these assumptions. To be on the safe side, you should develop the habit of examining every assumption carefully. If you have reason to question the validity of a particular assumption, you should consider seriously the possibility of making use of another procedure which does not involve such an assumption. For example, you might decide to make use of a different unit of analysis—the person rather than behavioral acts or pairs of persons, or individual delinquents rather than delinquency rates for a census tract.

Although social scientists and others who use applied statistics have sometimes tended to ignore assumptions, thereby reaching unwarranted conclusions, it is also possible to be overly perfectionistic. Since we never deal with situations as simple as coin flipping or drawing cards from a perfect deck, it is always possible to question every procedure as falling short of the ideal. One can be so much afraid of violating assumptions that he refuses to use any statistical technique at all. Especially in a discipline characterized by exploratory studies and relatively imprecise scientific techniques, it is necessary to make compromises with reality. The most sensible procedure would seem to be to make as few compromises as possible within the limits of practicality.

[14] This situation would be somewhat analogous to a deck of cards each of which gradually shades into the others so that it is difficult to determine where one card ends and another begins. Also, each card would be capable of influencing the face values of its nearest neighbors!

Glossary
Event
Limit
Mutually exclusive events
Probability
Random sample
Statistical independence

Exercises
1. In a single toss of an honest die what is the probability of:
    a. Getting a 6?
    b. Not getting a 6?
    c. Getting a 1 or a 6? (*Ans.* $\frac{1}{3}$)
    d. Getting a 1 and a 6?
    e. Getting either an odd number or a 6?

2. What is the probability of getting each of the following in *three* draws from a well-shuffled deck of cards:
    a. Three jacks, with replacement? (*Ans.* $\frac{1}{2,197}$)
    b. Three jacks, without replacement? (*Ans.* $\frac{1}{5,525}$)
    c. A spade, heart, and diamond (in any order), with replacement?
    d. Exactly two aces, with replacement?
    e. At least one ace, with replacement? (*Hint:* What is the alternative to at least one ace?) (*Ans.* $\frac{469}{2,197}$)
    *f. At least one ace *and* at least one king, with replacement? [*Hint:* In (*f*) and in certain of the exercises that follow, it will be helpful to divide the problem into three steps: (1) determine the various combinations of cards which will yield at least one ace and at least one king (e.g., one ace, one king, and one other card; two aces and one king, etc.); (2) determine the probability of getting these cards in any particular order; and (3) for each of these combinations determine the number of possible orderings.]

3. Suppose 1,000 freshmen are asked about their musical tastes. It is found that 400 of these students are lovers of classical music; the remainder are not. Of these lovers of classical music, only 100 like "rock" music. There are 400 persons who do not like either type of music and the remainder like rock music only.
    a. If a student is selected at random from this population, and if A is the event that he likes classical music and B the event that he likes rock music, what are $P(A)$, $P(B)$, $P(A|B)$, and $P(B|A)$?
    b. Verify numerically that

$$P(A \ \& \ B) = P(A)P(B|A) = P(B)P(A|B)$$

    c. What is the probability of getting a person who likes one of the two types of music but not both?
    *d. Noting that a person can have one of four kinds of tastes (likes both, likes neither, etc.), what is the probability that three persons selected at random as roommates will *all* have the same set of tastes? (Assume replacement.) (*Ans.* .10)

*e. What is the probability that there will be *at least* two rock fans in a corridor of eight persons? (Assume random sampling, with replacement.)

*4. In the data given below, let A be the event of getting a male, B be the event of getting a college-educated person, and C be the event of getting a person with high prejudice.

Degree of prejudice	College-educated		Less than college-educated	
	Male	Female	Male	Female
High	100	50	200	250
Low	150	100	150	200

a. Find $P(A \ \& \ B \ \& \ C)$ on a single draw without using a formula. Verify that the formula for $P(A \ \& \ B \ \& \ C)$ holds for the numerical data of this exercise.

b. Do the same for $P(A \text{ or } B \text{ or } C)$. You will need to *develop* the formula for $P(A \text{ or } B \text{ or } C)$.

c. What is the probability of getting exactly one college-educated male, exactly one female with a college education, and exactly one person with high prejudice in a random sample of three persons? (Assume replacement.)

*5. Students enrolled in introductory sociology at the University of Michigan were classified as to occupational aspirations for self or spouse, depending on the sex of the respondent. The following data were obtained:

Sex	High aspirations	Low aspirations	Total
Male	43	10	53
Female	71	93	164
Total	114	103	217

Suppose you were to draw individuals randomly from this population of 217 students.

a. What is the probability of getting a student with high aspirations? What is the probability of getting a student with high aspirations, given that the student is a male? A female?

b. Suppose you were to select individuals at random (without replacement) from this population, each time guessing whether the individual had high or low aspirations. How often would you guess he had high aspirations? Low aspirations? Why? In 217 trials, how many errors would you expect to make? (*Ans.* 103)

c. Suppose the sex of the student were known. Given that the individual is a male, how many errors would you expect to make in assigning the 53 males to either the high- or low-aspirations category? How many expected errors for females? (*Ans.* 10; 71)

d. How might you construct an index showing the proportional reduction of errors if the respondent's sex is known as compared with errors expected if sex is unknown?

As will be seen in Chap. 15, such an index can be used to measure the strength or degree of relationship between the respondent's sex and his occupational aspirations.

*6. Use a tree diagram to evaluate the probabilities of all possible outcomes of a World Series (the best of seven games), given the assumption that the probability of the National League's team winning each particular game is .6.

References

1. Alder, H. L., and E. B. Roessler: *Introduction to Probability and Statistics*, 4th ed., W. H. Freeman and Company, San Francisco, 1968, chap. 5.

2. Feller, William: *An Introduction to Probability Theory and Its Applications*, 3d ed., John Wiley & Sons, Inc., New York, 1967.

3. Freund, J. E.: *Modern Elementary Statistics*, 3d ed., Prentice-Hall, Inc., Englewood Cliffs, N.J., 1967, chaps. 5 and 6.

4. Gelbaum, B. L., and J. G. March: *Mathematics for the Social and Behavioral Sciences*, W. B. Saunders Company, Philadelphia, 1969, chaps. 2–4.

5. Hays, W. L.: *Statistics*, Holt, Rinehart and Winston, Inc., New York, 1963, chaps. 2 and 4.

6. Kemeny, J. G., J. L. Snell, and G. L. Thompson: *Introduction to Finite Mathematics*, 2d ed., Prentice-Hall, Inc., Englewood Cliffs, N.J., 1966, chaps. 3 and 4.

7. Mueller, J. H., K. Schuessler, and H. L. Costner: *Statistical Reasoning in Sociology*, 2d ed., Houghton Mifflin Company, Boston, 1970, chap. 8.

8. Savage, L. J.: *The Foundations of Statistics*, John Wiley & Sons, Inc., New York, 1954, chaps. 1–3.

Testing Hypotheses: The Binomial Distribution   10

In the social sciences we frequently encounter simple dichotomies such as whether or not an individual possesses a certain attribute or whether an experiment has been a success or failure. Whenever it is possible to hypothesize a certain probability of success in such instances, whenever trials are independent of each other, and whenever the number of trials is relatively small, it is possible to make use of statistical tests involving what is known as the binomial distribution. Although there are numerous statistical tests that are more practical than those which make use of the binomial, it is advisable to devote considerable time to this distribution primarily because of its simplicity. In using the binomial distribution, you can follow relatively easily all of the steps involved and can thereby gain insight into the general procedures used in all statistical tests.

You will probably find this chapter an unusually difficult one because of the fact that a number of new ideas are presented in fairly compact fashion. Many of these same ideas are again taken up in Chap. 11, and you may prefer to treat these two chapters as a single unit, reading Chap. 11 before really mastering materials in the present chapter. In particular, you may wish to postpone reading Sec. 10.3 dealing with various applications of the binomial and Sec. 10.4 on extensions.

10.1 The Binomial Sampling Distribution

Before discussing each of the steps involved in statistical tests, it will be necessary to examine how binomial distributions are obtained. For the time being, it will simplify matters if we confine our attention to the flipping of coins. In this type of problem the number of flips constitutes the sample size, and our interest centers on the number of heads (successes) obtained in N trials.

Assuming that the N trials (coin flips) are statistically independent of one another, we can immediately evaluate the probability of getting r heads and $N - r$ tails in some particular order. For example, we can obtain the probability of getting r successive heads followed by $N - r$ tails. Let p be the probability of obtaining a head; the probability of getting a tail, denoted by q, will then be $1 - p$. Since the trials are independent, we can simply multiply the unconditional probabilities. The probability of getting exactly r heads *in the order described above* will then be

$$\underbrace{p\,p\,p\,\cdots\,p}_{r \text{ terms}}\underbrace{q\,q\,q\,\cdots\,q}_{N - r \text{ terms}} = p^r q^{N-r}$$

Clearly, under the assumptions of statistical independence and a constant probability of success (e.g., the coin does not wear thin unevenly), the probability of getting any other particular ordering of r heads and $N - r$ tails will also be $p^r q^{N-r}$. Therefore, in order to obtain the probability of getting *exactly r heads in any order* it is only necessary to count the number of distinct ways we can get r heads and $N - r$ tails. If N is even moderately large, this task becomes very tedious, however. As we have seen, there is available a mathematical formula that makes such a counting operation unnecessary. The number of possible ways we can order r successes and $N - r$ failures, written symbolically as $\binom{N}{r}$ or sometimes as C_r^N, is

$$\binom{N}{r} = \frac{N!}{r!(N - r)!} \tag{10.1}$$

Formula (10.1) may be simplified for computational purposes by noting that some of the terms in the numerator and denominator cancel each other out.[1] Since $r \leq N$, we can write $N!$ as a product of two terms as follows:

$$N! = [N(N - 1)(N - 2) \cdots (N - r + 1)][(N - r) \cdots (3)(2)(1)]$$

$$= [N(N - 1)(N - 2) \cdots (N - r + 1)][(N - r)!]$$

[1] It should be noted that the symbol $\binom{N}{r}$ is not to be confused with N/r or N divided by r.

and we see immediately that $(N - r)!$ can be taken out of both numerator and denominator. We are then left with

$$\binom{N}{r} = \frac{N(N-1)(N-2)\cdots(N-r+1)}{r!} \qquad (10.2)$$

Thus if we want to find the number of ways of getting four heads in ten flips we have

$$N - r + 1 = 10 - 4 + 1 = 7$$

and therefore
$$\binom{10}{4} = \frac{(10)(9)(8)(7)}{(4)(3)(2)(1)} = 210$$

Notice that in using Eq. (10.2) there are the same number of factors in both numerator and denominator. This will always be the case. This second form is computationally simpler than the first. If $r > N/2$ we begin to get certain terms appearing in both numerator and denominator and therefore canceling each other. For example, if $r = 6$ we have

$$\binom{10}{6} = \frac{(10)(9)(8)(7)}{(1)(2)(3)(4)}\left[\frac{(6)(5)}{(5)(6)}\right] = 210$$

which gives us the same result as obtained in computing $\binom{10}{4}$. In general it can be shown that

$$\binom{N}{r} = \binom{N}{N-r}$$

so that either r or $N - r$ may be used, depending on whichever is the smaller.

If we now wish to obtain the probability of getting *exactly* r successes in N trials and are not interested in the order in which they occur, we can multiply the probability of getting any particular sequence by $\binom{N}{r}$. Denoting the desired probability by $P(r)$, we have

$$P(r) = \binom{N}{r} p^r q^{N-r}$$

or

| Probability of exactly r successes | = | no. of ways of getting r successes | × | probability of any given sequence | (10.3) |

If the coin were an honest one, i.e., if $p = q = \frac{1}{2}$, the probability of getting exactly four heads in ten trials would be

$$P(4) = \binom{10}{4}\left(\frac{1}{2}\right)^4\left(\frac{1}{2}\right)^6$$

$$= 210(\tfrac{1}{2})^{10}$$

$$= 210/1,024 = .205$$

Similarly, we can obtain the probabilities of getting exactly 0, 1, 2, . . . , 10 heads in 10 trials.

No. of heads	Probabilities (with $p = \frac{1}{2}$)
0	$1/1,024 = \ .001$
1	$10/1,024 = \ .010$
2	$45/1,024 = \ .044$
3	$120/1,024 = \ .117$
4	$210/1,024 = \ .205$
5	$252/1,024 = \ .246$
6	$210/1,024 = \ .205$
7	$120/1,024 = \ .117$
8	$45/1,024 = \ .044$
9	$10/1,024 = \ .010$
10	$1/1,024 = \ .001$
	$\overline{1.000}$

Notice that whenever r is zero, the quantity $\binom{N}{r}$ is undefined, and the formula breaks down. We see, however, that there can be only one possible order when $r = 0$ (all tails). In this example, the distribution of probabilities is perfectly symmetrical. Using the fact that $\binom{N}{r} = \binom{N}{N-r}$, you should satisfy yourself that $\binom{N}{r}$ will always be symmetrical but that the factor $p^r q^{N-r}$ will be exactly symmetrical only when $p = q = \frac{1}{2}$.

In the above example, probabilities have been associated with each of the 11 possible outcomes of the experiment. In this simple example there were only a small number of conceivable outcomes, given the assumption that only two outcomes were possible on each flip. In other experiments the number of possible outcomes may be very large or even infinite, and it may be necessary to group certain outcomes together and to associate a probability with the entire set of outcomes. Thus, if the coin had been flipped 1,000 times, we might have obtained the probabilities of getting 400 to 449, 450 to 499, or 500 to 549 heads.

Whenever we associate probabilities with each possible outcome of an experiment, or with sets of outcomes, we refer to the resulting probability distribution as a sampling distribution. Remembering that we are using the concept *probability* to refer to the limit of the ratio of successes to total number of trials, we see that *a sampling distribution refers to the relative number of times we would expect to get certain outcomes in a very large number of experiments.*

In the numerical example under consideration, each experiment consists of flipping a coin ten times and noting the number of heads. Our computations tell us that if we were to perform the experiment 1,024,000 times, we could expect to get approximately (but not exactly) 1,000 occurrences of no heads, 10,000 occurrences of exactly one head, 45,000 of two heads, etc. Furthermore, we would expect that the larger the number of times the experiment is performed, the closer the empirical proportions will be to these theoretical probabilities.

The researcher never actually obtains a sampling distribution by empirical means since he usually only performs an experiment or draws a sample once or at most a very few times. It is important to realize that sampling distributions are hypothetical, theoretical distributions which would be obtained only if one were to repeat an experiment an extremely large number of times. A sampling distribution is obtained by applying *mathematical* or deductive reasoning as was done in the previous example.

Since sampling distributions are not the kind of distributions a researcher actually sees from his data, persons who are not mathematically inclined are likely to have difficulty understanding the role of these hypothetical distributions in statistical inference. Yet, unless the notion of sampling distribution is clearly grasped, you will find it almost impossible to obtain anything more than a "cookbook" understanding of statistics. For this reason, it will be helpful at this point to discuss more systematically the steps made in testing a statistical hypothesis and to see exactly how these sampling distributions are used.

10.2   Steps in Statistical Tests

There are a number of specific steps involved in all statistical tests. It should again be emphasized that each of these steps should be carried out prior to the inspection of one's data. They can be listed as follows:

1.   Making assumptions
2.   Obtaining the sampling distribution
3.   Selecting a significance level and critical region
4.   Computing the test statistic
5.   Making a decision

Each of these steps will be discussed in some detail in this chapter and again in Chap. 11 so that you may become familiar with the general processes involved in all statistical tests.

1. *Making Assumptions* In order to make use of probability theory in obtaining a sampling distribution, the researcher must make certain assumptions about both the *population* to which he is generalizing and the *sampling* procedures used. The assumptions made about the population and sampling procedure usually fall into one of two categories: (1) those of which the researcher is relatively certain or which he is willing to accept, and (2) assumptions which seem most dubious and in which he is therefore most interested. Assumptions in the first category can be lumped together into what we shall call the *model*. The assumptions in the second category are the ones which the researcher wants to test and are called *hypotheses*.

Usually, at least in the simpler tests taken up in the next several chapters, there will be only one hypothesis. It is important to realize that *from the standpoint of the statistical test itself all assumptions have the same logical status*. If the results of the test warrant rejection of the assumptions, all that one can say *on the basis of the test itself* is that at least one (and possibly all) of the assumptions is probably false. Since the test itself can supply no information as to which of the assumptions is erroneous, it is essential if results are to be meaningful that only one of the assumptions be really in doubt. It will then be possible to reject this assumption (the hypothesis) as the faulty one.

Students often ask the following type of question: "On what basis does one choose a particular statistical test in preference to another?" One criterion that can be given at this point is an appropriate model. In other words, the researcher should select a test that involves only a single dubious assumption (his hypothesis). If a certain test requires two or more assumptions that are in doubt, it will be difficult, if not impossible, to decide which should be rejected. In such an instance one should attempt to find an alternative test that does not require as many doubtful assumptions.

To illustrate with our coin example, the binomial test requires the assumption that the 10 flips constitute a random sample of all possible flips with the same coin and that the flips are independent of each other. We are also assuming that the coin is an honest one. The latter assumption would ordinarily be our hypothesis and the former our model, since interest would probably center on whether or not the coin were honest. Conceivably, however, we might be suspicious of the person doing the flipping. If we were relatively sure of the coin, having previously determined that it usually produced about half heads, we could turn the problem around and test an hypothesis concerning the method of flipping (the method of sampling). Suppose we were unwilling to accept as our model the honesty of the coin or the honesty of the flipper. If 50 heads came up in succession, we would decide that at least one of our assumptions was undoubtedly wrong, but we would be unable to choose between them. Usually, of course, we pay careful attention to our sampling methods in order to be reasonably assured that assumptions regarding sampling are actually justified.

Taking a sociological example to illustrate the same point, let us suppose that we are required to make only two assumptions in a particular statistical test: (1) that in the population sampled the proportions of middle- and lower-class persons with high mobility aspirations are the same, and (2) that a random sample of all persons has been obtained. Suppose also that these assumptions lead to certain conclusions which cannot be supported by the data. Perhaps the sample data show a much higher percentage of middle-class persons with high aspirations. We conclude that one or the other of the assumptions is probably erroneous. But which of the two should be rejected? We might like to conclude that the first is wrong, but perhaps we have used biased sampling methods. We must have additional knowledge beyond what can be learned from the test itself.

In this particular example, if we have gone to great lengths to assure the selection of a random sample, we can take assumption (2) as our model and conclude that (1) is probably false. Here, our willingness to accept (2) is based on knowledge about the sampling methods used, i.e., our research methodology. In other instances we may accept certain assumptions on the basis of previous research findings. The important point is that *the test itself cannot be used to enable one to locate the faulty assumption or assumptions*. It is in this sense that the assumptions all have the same logical status. To emphasize this fact and to call your attention to the assumptions in the model, we treat the hypothesis actually being tested as merely one among a number of assumptions required by the test.

As previously mentioned, a researcher is usually interested in setting up an hypothesis which he really would like to reject. The hypothesis that is actually tested is often referred to as a *null hypothesis* (symbolized as H_0) as contrasted with the *research hypothesis* (H_1) that is set up as an alternative to H_0. Usually, although not always, the null hypothesis states that there is no difference between several groups or no relationship between variables, whereas the research hypothesis may predict either a positive or negative relationship. The researcher may actually expect that the null hypothesis is faulty and should be rejected in favor of the alternative H_1. Nevertheless, in order to compute a sampling distribution, he must for the time being proceed as though H_0 is actually correct. He would assume that the coin is an honest one, for example.

Notice that the assumption of an honest coin provides a way of computing exact probabilities using the binomial formula. If one were to hypothesize that the coin is "dishonest," he would find that he could not obtain a sampling distribution until he made his hypothesis more specific. He would have to commit himself on a specific value for p, say .75. Seldom will he be in a position to do this. Likewise, the research hypothesis that there is a larger proportion of persons with high mobility aspirations among the middle class is not as specific as the null hypothesis that there is absolutely no difference between the two classes.

2. *Obtaining the Sampling Distribution* Having made the necessary assumptions, we are in a position to make use of mathematical reasoning to obtain the sampling distribution in which we associate probabilities with outcomes. Such a distribution of probabilities will tell us just how likely each of the possible outcomes is *if the assumptions made are actually correct.* If the above assumptions about the coin and the flips were actually true, we have seen that in the long run only 1 time in 1,024 would we expect to get all heads, only 10 times in 1,024 would we get 9 heads, etc.

Knowledge of the likelihood of any particular outcome occurring by chance if the assumptions were really true can now be used to make a rational decision about the conditions under which we could risk rejecting these assumptions. Suppose, for example, that we were to get all 10 heads. There are two possibilities: (*a*) either the assumptions are correct and this is one of those occasions on which a very rare event occurred, or (*b*) at least one of the assumptions (presumably the null hypothesis) is false. Unfortunately, we can never be positive which alternative is the correct one. If we could, we would have known ahead of time about the assumptions, and there would have been no point in performing the experiment. But we can say that the first alternative is very unlikely.

Let us establish the rule that every time we get 10 heads in 10 trials, we automatically conclude that at least one of the assumptions is false and should be rejected. In the long run, we shall occasionally make erroneous decisions in adhering to this rigid rule since we know that even with an honest coin we can expect to get all 10 heads one time in 1,024 simply by chance. Such a rule will not help us determine the correctness of our decision for any particular experiment, but the laws of probability tell us exactly what proportion of the time we can expect to make correct decisions *in the long run.* In a sense, our faith is in the procedure we are following rather than in the decision we make on any particular occasion. This *procedure* will yield correct decisions most of the time, even though we cannot be absolutely certain of being correct on any specific decision.

3. *Selecting a Significance Level and Critical Region* Ideally, the researcher's decisions should be made prior to the actual experiment or analysis of data. From his knowledge of the sampling distribution he selects a set of alternatives which, should they occur, would require him to reject his assumptions. These unlikely outcomes are referred to as the *critical region.* Thus, he divides the possible outcomes into two categories: (*a*) those for which he will reject (the critical region), and (*b*) those whose occurrence would not permit him to reject. In order to make a choice of critical region, he must make two decisions in addition to his choice of model and hypothesis. First, he must determine the risks he is willing to take of making types I and II errors. Second, he must decide whether or not he wants his critical region to include both tails of the sampling distribution.

As indicated in Chap. 8, one must take into consideration two types of possible errors. The first type of error consists of rejecting a set of assumptions when they are in fact true. A type II error, on the other hand, involves a failure to reject assumptions when they are actually false. From the sampling distribution one can determine the exact probabilities that certain outcomes will occur *if the assumptions are actually true.* If the researcher decides that he will reject whenever a specified set of unlikely outcomes (say either zero or ten heads) occurs, then if the assumptions are true, he will make a type I error whenever he obtains any of these outcomes.

The probability of making a type I error is the sum of the probabilities of each of the outcomes within the critical region. For example, if the critical region consists of zero or ten heads, the probability of a type I error would be 2/1,024 or .002. If a larger critical region were selected, the risk of this type of error would be greater. Suppose it were decided to reject the assumptions if zero, one, nine, or ten heads were obtained. Then the probability of a type I error would be $(1 + 1 + 10 + 10)/1,024$ or .022. The probability of making a type I error is referred to as the *significance level* of the test and can be set at any desired level.

Before discussing possible criteria for deciding upon the significance level to be used, something should be said about type II errors. In view of our earlier discussion of the fallacy of affirming the consequent, it is clearly incorrect to conclude that if certain assumptions cannot be rejected, they therefore must be true. Another set of assumptions might also have led to a sampling distribution for which similar conclusions would have been reached. For example, if the true probability of heads were .51 rather than .50, the correct sampling distribution would be almost identical to the one we calculated. Therefore, exactly the same critical region would probably have been selected, and the decision as to whether or not to reject might have been identical. Yet, strictly speaking, the hypothesis that $p = .5$ would be false and should be rejected. If we were unable to reject it, we would not want to accept it outright as the single correct hypothesis since there are a large number of additional hypotheses which also could not be rejected. We simply decide that we should "not reject" our hypothesis.

Even if we conservatively refuse to accept an hypothesis, we would still like to be able to eliminate as many false hypotheses as possible. In this sense we are making an error whenever we fail to reject a false hypothesis. What can be said about the probability of making a type II error? Unfortunately, it is no simple matter to compute type II errors as was the case with type I errors. We shall have to defer our discussion of type II errors until Chap. 14. One important fact should be noted, however. For any given test the probabilities of types I and II errors are inversely related. In other words, *the smaller the risk of a type I error, the greater the probability of a type II error*. This can be seen in our coin-flipping example. You should convince yourself that if one selects a small critical region (say, zero and ten heads), he will be less likely to reject *any* assumptions than if he were to use a more inclusive region (say, zero, one, nine, and ten heads). In the former case, while he is less likely to reject true assumptions, he is also less likely to reject false assumptions. Therefore he is more likely to make a type II error.

It is thus impossible to minimize the risks of both types of errors simultaneously unless one redesigns his study and selects additional cases or a different statistical test. In practice, we set the probability of a type I error at a fixed level (say .05) and then try to select the statistical test that minimizes the risk of a type II error. In choosing among alternative tests, we select that test which has an appropriate model and which is most powerful in the sense of minimizing the risk of a type II error.[2]

The decision as to the significance level selected depends on the relative costs of making the one or the other type of error and should be evaluated accordingly. Sometimes a practical decision must be made according to the outcome of the experiment. A manufacturer may decide to install expensive equipment; a researcher may decide to draw another sample and replicate his study; or public health authorities may have to decide

[2] For further discussion of this point, see Sec. 14.1.

whether or not to attempt mass innoculations of a new serum. In other instances there is no practical decision required. A sociologist may simply report the results of his study in a journal article and may not have to take the consequences of one or the other type of error.

It is in situations in which a practical decision has to be made that the choice of a significance level is especially difficult. In the coin-flipping example, suppose that the decision involved refusing to continue gambling with a coin the honesty of which were in doubt. If our hypothetical gambler were faced with the prospects of a nagging wife should he return home with empty pockets, he would do well to quit the game if there were even a reasonable doubt about the coin. In such a case he would select a large critical region since the penalty for making a type II error (i.e., staying in the game when the coin is actually dishonest) would be quite large. On the other hand, if he were to run the risk of insulting his boss if he claimed that the coin was dishonest, he would want to be very sure of this fact before he made his decision. In the latter case he should select a very small critical region, thereby minimizing the risk of a type I error. Similarly, if the cost of mass innoculations were considerable or if the serum were potentially harmful, one would want to be very sure of its effectiveness before putting it to use. He would want to make it very difficult to reject the null hypothesis that the serum has no beneficial effect.

If there is no practical decision to be made other than whether or not to publish the results of a study, another rule of thumb should be followed. *The researcher should lean over backwards to prove himself wrong or to obtain results that he actually does not want to obtain.* Usually, but not always, one sets up a null hypothesis that he really wants to reject. Since he would like to be able to reject, he should make it very difficult to achieve the desired result by using a very small critical region.

There are occasions—and you should be alerted to their existence—in which one actually does not wish to reject the null hypothesis. For example, the null hypothesis may take the form of a prediction that there will be no class or religious differences with respect to fertility rates. If one really wishes to establish such differences, he should select a very small critical region, making it difficult for him to reject the null hypothesis. But suppose he actually wishes to show that there are no such differences. Perhaps he is attempting to demonstrate that certain commonly held theories about fertility differentials are incorrect or inadequate. Or he may be hoping that these differences do not exist so that he will not have to control for class or religion in relating fertility rates to other variables.

In the above instances the researcher is in one sense on the wrong end of the hypothesis and should be primarily interested in minimizing the risk of a type II error. In other words, he should be especially concerned that he not retain the null hypothesis of no differences when it is actually false. One is therefore not always being conservative by selecting a small critical region, thus making it difficult to reject a null hypothesis that he may really want to retain. Significance levels commonly used in statistical research are the .05, .01, and .001 levels. It should be realized in view of the above discussion that there is nothing sacred or absolute about these levels. Although a person would usually be conservative in using such levels, if he actually did not want to reject the null hypothesis, he would be on safer ground using perhaps the .10, .20, or even .30 level, thereby reducing his risk of a type II error.

A word of caution is necessary in interpreting the results of significance tests since it is possible to obtain rather misleading results even when the .001 level is used and when rejection is desired. Significance tests tell us how likely a given set of sample results would be if certain assumptions about the population parameters were true. There are several factors that determine the likelihood that we shall be able to reject these assumptions. The first is how inadequate these assumptions really are. If, for example, the true probability of heads is .9, it is very likely that we shall be able to reject the hypothesis that p is .5 because we are actually apt to get a sufficiently high proportion of heads to end up in the critical region. On the other hand, if the true probability is .53, we are less likely to get the extreme results necessary for rejection.

The *number of cases* is another important factor in determining how extreme the results must be before rejection is possible. With only 10 flips or cases it has been seen that very extreme results are required in order to reject. But if N is large, the *proportion* of successes need differ from the hypothesized p by only a small amount in order to reject. If the coin were flipped 10,000 times instead of 10, we would be able to reject the hypothesis if, say, we were to get over 5,200 heads. In other words, under the assumption that p is *exactly* one-half, 5,200 or more heads in 10,000 trials would be even more unlikely than 10 heads in 10 trials, even though the results would not be nearly as extreme. This is, of course, consistent with our greater intuitive faith in large samples and with the realization that in the case of very small samples extreme results could occur quite frequently by chance. Similarly, with a sample of 10,000 persons we could obtain very small differences in the fertility rates between middle- and lower-class women and still be able to reject the null hypothesis that there are *no* differences whatsoever in the population.

With a very large number of cases it is practically always possible to reject any false hypothesis we might set up, regardless of how far our hypothesized value may differ from the true one. This means that if we have 10,000 cases, we should not be very surprised if we are able to reject at the .001 level, and we should be on guard against reporting our finding as though it were a highly important one. Statistical significance should not be confused with practical significance. Statistical significance can tell us only that certain sample differences would not occur very frequently by chance if there were no differences whatsoever in the population. It tells us nothing directly about the magnitude or importance of these differences. A factor that is large enough to produce differences that are statistically significant in a small sample is therefore much more worthy of one's attention than a factor that produces small differences that can only be shown to be statistically significant with a very large sample. If the study involves a large number of cases, we are usually more interested in other kinds of problems than tests of significance. This question will be discussed more thoroughly in Chap. 15 when we take up measures of degree of relationship. For the present it is sufficient to point out that statistical significance does not necessarily imply striking differences or ones that are important to the social scientist.

Another kind of decision must be made before the critical region can be determined. There are a number of outcomes or sets of outcomes the probability of which may be less than the significance level selected. For example, the probability of getting *exactly* eight heads is 45/1,024 or .044. Therefore it would be possible, although not very sensible, to decide to reject the null hypothesis if exactly eight heads were to occur, but otherwise not to reject. The probability of a type I error would then be .044. The choice of such a critical region would seldom make sense theoretically, however, since one would ordinarily be even more hesitant about accepting the null hypothesis if nine or ten heads were to turn up, and yet these alternatives would not belong to the critical region. We are practically always interested in using at least an entire tail of the distribution. We are not interested in the probability of getting *exactly* eight heads but in the probability of getting eight *or more* heads, i.e., the probability of getting eight heads or something *even more unusual*.

But why not also include zero, one, and two heads in the critical region since these alternatives are just as unlikely as eight, nine, and ten heads? Often we are not in a position to predict ahead of time the direction in which the unusual results may occur. In this example we may merely suspect that the coin is dishonest but may have no hint as to whether it is biased in favor of heads or tails. Furthermore, we may not care. In such a case we would want to play safe and make use of both tails of the sampling distribution. For if we were to make use of a critical region consisting only of eight, nine, and ten heads, then were we to obtain exactly one head, we would be in the unfortunate position of not being able to reject the null hypothesis even though it might be incorrect.

There are a number of occasions, however, when we are either able to predict the direction of deviance or are primarily interested in deviations in one direction only. For example, previous information may have led us to predict that the coin is biased in favor of heads. Or we may be betting on tails each time, so that if the coin happens to be biased in favor of tails, we need have no fear of continuing in the game. In more realistic examples it is often possible to predict direction on the basis of theory or previous studies. It may have been predicted that Catholics will have larger families than Protestants, for instance. If one is interested in showing his theory to be correct, he will make significance tests only when results occur in the predicted direction. If they occur in the opposite direction, he need make no test since the data obviously do not support the theory anyway.

Whenever direction has been predicted, one-tailed tests will be preferable to two-tailed tests at the same significance level since it will be possible to obtain a larger tail by concentrating the entire critical region at the proper end of the sampling distribution. This advantage of a one-tailed test is illustrated in Fig. 10.1 in the case of a smooth sampling

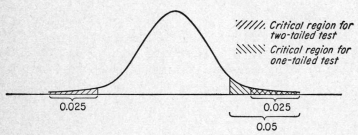

Figure 10.1 Comparison of critical regions for one- and two-tailed tests, using .05 significance level.

distribution having the form of a normal curve. In this figure the probabilities of making a type I error are the same in both instances since the two critical regions are of the same size (as measured in terms of areas). But if the results actually occur in the predicted direction, the researcher will be more likely to reject the hypothesis using a one-tailed test since there is a greater probability of falling into the larger critical region in this direction. The risk of making a type II error if the true probability is in the predicted direction is less than the risk when using a two-tailed test.

At this point you should not really expect to understand intuitively the relationship between type II errors and one- and two-tailed tests. Many of these rather difficult notions will become clear only after several practical examples have been discussed. A more detailed treatment of type II errors must be postponed until Chap. 14.

To be specific in this example, let us select the .05 level and make use of a two-tailed test. The critical region will consist of the alternatives zero, one, nine, and ten since including additional alternatives would increase the probability of a type I error beyond the .05 level. In this example the significance level actually used will be $(1 + 1 + 10 + 10)/1,024$ or .022. In other instances in which the sampling distribution is continuous rather than discrete, it will be possible to use the exact level desired (e.g., .05, .01, or .001).

4. *Computing the Test Statistic* It is always necessary to compute what is referred to as a test statistic, the sampling distribution of which is to be used in the test. Up to now we have dealt only with statistics such as sample proportions, means, and standard deviations, which are directly comparable to the same quantities in the population and which may be used as measures for summarizing the data. A test statistic is a statistic which usually is of no inherent interest descriptively but which is used in testing hypotheses. It is this statistic that has the sampling distribution that is directly used in the test. In other words, we compute a quantity from sample data that varies in a known way according to probability theory. We then compare its value with the sampling distribution and make a decision by evaluating the probability of its occurrence. Of course there are numerous quantities that can be computed from sample data, but only a relatively small number have known sampling distributions that can be used for purposes of testing hypotheses.

In this example of the binomial test the test statistic is so simple that it seems hardly worthwhile calling it to your attention. It is merely the number of successes in N trials and does not require any further computations. In other problems, however, the test statistic will have to be computed. In the case of the binomial test, we have let the number of successes r take on all possible values from zero to N, and we have associated probabilities with each value. Let us suppose in this particular problem involving 10 flips that the number of successes (heads) actually turns out to be 8. We now have all the information necessary in order to make our decision.

5. *Making a Decision* After having selected his critical region and computed the test statistic, the researcher will either reject or fail to reject

the assumptions depending upon the outcome of the experiment. If the outcome falls within the critical region, he will reject with a known probability of a type I error. If it does not fall within the critical region, he will not reject the assumptions and will take the risk of making a type II error. In this example, since the outcome of eight heads does not fall within the critical region, he should not reject the null hypothesis that the coin is an honest one.

Ideally, all decisions prior to steps 4 and 5 should be made before the tabulation of results. Often in exploratory work a person will first examine his data and then make tests of significance. Although this is sometimes necessary, it should be noted that whenever this occurs one is not completely living up to the rules of the game. In such instances it would be better not to put forth the claim that hypotheses were actually being tested. Results could be presented as suggestive, however, and anyone doing a follow-up study would then be in a position to make legitimate statistical tests.

The above comments may sound overrigid and perfectionistic in view of the exploratory nature of much social science research. The writer takes the position, however, that it is preferable to establish a strict "statistical conscience" rather than to leave the impression that anything goes. Unless one makes his decisions prior to the analysis of data, he cannot legitimately make use of probability theory since his analysis is essentially ex post facto. The trouble with ex post facto analyses is that the experiment can be set up so that the researcher cannot possibly lose. Suppose, for example, that he has tentatively decided to use the .05 level of significance. If he finds his results to be significant at the .07 level, he may then decide to reject his hypothesis anyway. But suppose they had been significant at the .09 or .13 or .18 level. Where does one stop? Another way of cheating is to wait until after the experiment to decide whether or not to use a one-tailed test. Then if the results show more heads than tails, one simply decides that he should have used a one-tailed test since he was subconsciously predicting a bias in favor of heads. This way, no matter what the direction of deviance, he can obtain a larger critical region than with a two-tailed test.

10.3 Applications of the Binomial

The sign test   Suppose a social scientist is making use of a simple "before-after" or "after-only" experimental design in which there are a small number of cases and in which he is only able to determine for each case whether or not his experiment has been successful.[3] For example, he may want to find out whether or not an experience in an interracial camp is successful in reducing stereotypes toward minorities. He gives a stereotype test to his subjects both before and after the experiment and is able to determine whether or not this type of prejudice has been reduced. Let us indicate by a + (success) each instance in which prejudice is reduced and by a − (failure) the cases for which prejudice is increased. If there are any persons showing absolutely no change, these persons will be excluded from the analysis. Unless measurement has been extremely crude, there should be relatively few of these persons.[4]

The binomial requires the assumption of independence of trials. The social scientist will therefore want to assume that his experimental group constitutes a random sample drawn from some population about which he wishes to generalize and that there has been little or no mutual influence among participants with respect to prejudice scores. Let us suppose that he wishes to establish that the camping experience is actually effective in reducing prejudice. Since this cannot be done directly, he can set up the null hypothesis that the experiment has no effect. If it actually has no effect, then if the entire population from which the sample has been drawn were to undergo similar experiences, we would expect to find the same number of persons whose prejudice was reduced as persons having increased prejudice. In other words, there would be equal proportions of pluses and minuses.

Since each member of the population has an equal chance of appearing in a random sample, the *probability* of getting a + in any given draw will be .5 under the null hypothesis. An assumption about the *proportion* of pluses in the population, when combined with the assumption of randomness, thus permits us to say something about the *probability* of success in any given trial. Randomness also assures independence of trials. Let us emphasize again that *it is necessary to make assumptions about both the population and the method of sampling*. In this example, interest is centered on the effectiveness of the experiment, i.e., the proportion of successes in the population. Therefore, the social scientist will want to make sure that he uses correct procedures for obtaining a random sample.

If there are eight persons in the sample, the sampling distribution of successes would be as follows:

[3] For a discussion of these and other types of experimental designs, see [6].
[4] The problem of ties or no change is especially troublesome in the case of ordinal variables and will be discussed in Chaps. 14 and 18. For an extended discussion see Bradley [3], chap. 3.

No. of successes	Probability
0	$\frac{1}{256}$ = .004
1	$\frac{8}{256}$ = .031
2	$\frac{28}{256}$ = .109
3	$\frac{56}{256}$ = .219
4	$\frac{70}{256}$ = .274
5	$\frac{56}{256}$ = .219
6	$\frac{28}{256}$ = .109
7	$\frac{8}{256}$ = .031
8	$\frac{1}{256}$ = .004
	1.000

Let us suppose that the social scientist wishes to use the .05 level of significance. Since direction has been predicted, a one-tailed test can be used. The critical region can be determined by cumulating probabilities starting with eight successes, then seven, etc., until the sum becomes greater than the significance level. It will ordinarily not be necessary to obtain the entire sampling distribution since only the tails are actually used in determining the size of the critical region. In this case, the probability of eight successes is .004; the probability of seven or eight successes is .035; and the probability of six, seven, or eight successes is .144. Since the sum of the probabilities of outcomes within the critical region must be less than or equal to the significance level selected, we see that the critical region can consist only of seven or eight successes.

Suppose the social scientist carries out the experiment and finds that in six cases prejudice has been reduced while in the remaining two it has been increased. He therefore will not reject the hypothesis that the experiment has no effect since the probability of getting such a result, or one even more unusual, is greater than .05.

Testing for nonrandomness   In the above example randomness was assumed and interest was centered on the proportion of successes in the population. In other types of problems there may be information about the proportion of persons in a population having a certain characteristic, but there may be a question about selectivity. For example, one may test to see whether or not professionals are overrepresented on boards or blacks underrepresented on jury panels. Suppose a mayor appoints nine persons to a commission, claiming that these persons are representative in the sense that all types of adults have an equal chance of being selected. If it is known that 35 per cent of the labor force is white-collar and yet six out of nine members of the commission are white-collar, a binomial test can be used in order to determine how likely such an occupational distribution would be under the assumption of random sampling.

In this particular problem the probability of success under the null hypothesis would be .35, and the sampling distribution would not be symmetrical. We would look upon each of the nine positions on the commission as a *trial*. The probability of getting a white-collar worker as the first commissioner would be .35, and similarly for each of the eight remaining positions.

Other uses of the binomial The binomial can be used in a number of other types of problems in addition to those mentioned above. Positional measures, such as the median or quartiles, can sometimes be used to enable one to test whether a small subsample of persons is significantly different from what we would expect by chance. From a large survey it may be possible to obtain a very good estimate of the income distribution for a particular city. If data have been obtained for only seven Puerto Ricans, and if six of these persons are in the lowest quartile, we may test to see how likely this would be, provided, of course, that decisions have been made prior to the test.[5] Since by definition one-fourth of a population will be in the lowest quartile, the binomial distribution gives the probability of getting a certain proportion of a subsample below the population quartile, under the assumption that such a subsample essentially constitutes a random sample from the larger population.

For example, since the probability of any given person being in the bottom quartile is .25, the probability of getting exactly six Puerto Ricans in the lowest quartile would be

$$P(6) = \binom{7}{6} \left(\frac{1}{4}\right)^6 \left(\frac{3}{4}\right)^1 = \frac{21}{16,384}$$

Also $$P(7) = \binom{7}{7} \left(\frac{1}{4}\right)^7 \left(\frac{3}{4}\right)^0 = \frac{1}{16,384}$$

Since we need to obtain the probability of getting six *or more* successes, we add these probabilities getting

$$P(6) + P(7) = \frac{21 + 1}{16,384} = .0013$$

[5] We must have a very large number of cases in order to obtain an accurate estimate of the positional measure (e.g., Q_1). Otherwise there will be sufficient sampling error in this estimate to require the use of a two-sample test. The reason for this should become clearer after two-sample tests have been presented in Chap. 13.

Another use of the binomial might involve testing the adequacy of a theory which correctly predicted the direction of certain differences in, say, 11 out of 15 independent trials. In order for such trials to be independent, they would have to involve different samples. For example, one sample might consist of young Protestant males, another of young Protestant females, a third of older Catholic males, etc. Each subsample might be too small to yield statistical significance separately, but if the subsamples were independently selected, a binomial could legitimately be used to test whether or not a sufficient number of subsamples gave results in the predicted direction. Each subsample would constitute a trial, and the probability of the result being in the predicted direction on any given trial would be .5 under the null hypothesis that the theory has absolutely no predictive value, i.e., that it predicts direction wrongly as often as it does so correctly. Notice that such a test could not be used if 15 observations were taken on the *same* sample of persons.

*10.4 Extensions of the Binomial

There are several possible ways of extending the basic approach exemplified by the use of the binomial distribution. Though they are not commonly used in statistical tests in the social sciences, you should at least be familiar with their existence. The first is the *multinomial* distribution, which can be used in situations where there are more than two kinds of events. We have already seen that if there are k distinct kinds of events, and if r_i is the number of events in the ith class, then the number of permutations for these events is given by the expression $N!/r_1!r_2! \cdots r_k!$. If events are statistically independent, and if the probabilities of getting the various kinds of events are given by p_i with $i = 1, 2, \ldots k$, and with $\sum_{i=1}^{k} p_i = 1$, then the probability of getting *exactly* r_1 events of type 1, r_2 events of type 2, $\ldots$ and r_k events of type k in some *particular order* will be

$$\underbrace{(p_1 p_1 p_1 \cdots)}_{r_1 \text{ terms}} \underbrace{(p_2 p_2 p_2 \cdots)}_{r_2 \text{ terms}} \cdots \underbrace{(p_k p_k p_k \cdots)}_{r_k \text{ terms}} = p_1^{r_1} p_2^{r_2} \cdots p_k^{r_k}$$

If we multiply this expression by the number of permutations, we obtain the formula

$$P(r_1, r_2, \ldots, r_k) = \frac{N!}{r_1! r_2! \cdots r_k!} p_1^{r_1} p_2^{r_2} \cdots p_k^{r_k}$$

It is important to recognize that this formula gives us the probability of getting *exactly* the specified numbers of events of each type. For example, suppose we know that a school contains 50 per cent Caucasians, 30 per cent blacks, and 20 per cent Orientals. What is the probability of having the "first string" of the football team contain exactly 3 Caucasians, 7 blacks, and 1 Oriental under the assumption that the racial composition of the team is subject to purely random selection processes? Using the multinomial distribution we would get

$$P(3,7,1) = \frac{11!}{3!7!1!} (.5)^3(.3)^7(.2)^1 = .007$$

We immediately encounter a difficulty that creates complications in the use of the multinomial distribution in statistical tests. In many instances it is not at all obvious how we can specify unambiguously a specific set of outcomes that are more "unusual" than the one obtained. In this example there are many kinds of "unusual" combinations. For example, the team may contain no blacks at all. Or no Orientals. But which outcomes belong in the critical region? If this can be specified, a proper test can be developed. For example, if Caucasians and Orientals were to be lumped together, then we could be concerned about the probability of getting seven *or more* blacks on the team. But in this instance, and in many others as well, we would really be utilizing the binomial rather than the multinomial distribution.

A second kind of modification of the binomial is possible whenever there has been sampling *without* replacement from a relatively small population. If a population of size M contains M_1 elements of type 1, M_2 elements of type 2, and, in general, M_i elements of type i, and if the corresponding sample sizes are N and N_i, then the probability of getting exactly N_1, N_2, . . . , N_k cases of each type is given by what is termed the *hypergeometric distribution* as follows:

$$P(N_1, N_2, \ldots, N_k) = \binom{M_1}{N_1}\binom{M_2}{N_2} \cdots \binom{M_k}{N_k} \Big/ \binom{M}{N}$$

For example, if we wished to determine the probability of getting exactly six spades, six clubs, and one diamond in a bridge hand of thirteen cards (selected randomly but without replacement), this would be

$$P(6,6,1) = \binom{13}{6}\binom{13}{6}\binom{13}{1} \Big/ \binom{52}{13}$$

which is an exceedingly small number. Again, we have the same difficulty of specifying the alternatives that would be considered "more unusual" than this particular combination. In Chap. 15 we shall take up a small-sample test for 2×2 tables, Fisher's exact test, which is based on the hypergeometric distribution involving only two distinct types of events.

Finally, it can be shown that the binomial distribution can be approximated by other distributions when the total sample is so large as to make computations extremely tedious. Whenever N is large and p intermediate in value, so that the product $Np > 5$, the binomial may be approximated by a normal distribution, in which case we may substitute tests based on *proportions* of successes. These tests will be considered in Chaps. 11 and 13.

It is sometimes the case that the sample size is moderately large, whereas p is extremely small (or extremely large). For example, p (or q) may refer to a rare event, such as contracting an unusual disease or committing suicide. If we set the problem up so that p refers to the probability of the rare event, so that $p < q$, and if $Np < 5$, then the binomial may be approximated by the *Poisson distribution* given by the following formula:

$$P(r) = \frac{\lambda^r e^{-\lambda}}{r!}$$

where r refers to the number of successes in N trials, where $\lambda = Np$, and where e is the natural constant approximately equal to 2.718. There are tables of both $r!$ and $e^{-\lambda}$ (see Spiegel [8]) which can be used to ease the computational burden.

To illustrate the use of the Poisson approximation, suppose that it is known that the probability of ever being arrested in a given community is .06, but that in a sample of 50 adult Japanese-Americans only one has been arrested. Then $Np = 3.0$ and

$$P(1) = \frac{3^1 e^{-3}}{1!} = 3e^{-3}$$

Similarly,

$$P(0) = \frac{3^0 e^{-3}}{0!} = e^{-3}$$

where we have used the convention of defining 0! to be unity. In order to obtain the probability of one or fewer Japanese-Americans being arrested, we add $P(1)$ and $P(0)$, getting

$$P(1) + P(0) = 4e^{-3} = 4(.0498) = .199$$

10.5 Summary

This chapter has contained a considerable number of new and very basic ideas, in addition to the mechanics of the binomial distribution itself. Many of these ideas will be discussed again at some length in the following chapter, where they will be illustrated in terms of hypotheses about means and in terms of two other sampling distributions. The important similarities in all tests can be seen in the steps involved in testing hypotheses and in the general concepts that have been introduced in this chapter. Let us again review these very briefly.

It is first necessary to make assumptions about *both* the population under study and the method of sampling from this population. Together, these assumptions plus robability theory enable us to make specific probability statements about outcomes under the null hypothesis. For example, in the case of the binomial these assumptions made it possible to assign a specific numerical value (for example, $p = .5$) to the probability of success on any given trial. In order to make a decision as to the critical region (i.e., the set of outcomes for which we will reject H_0), we need to obtain what is called a sampling distribution, which is a probability distribution that assigns a specific numerical probability to each outcome or to each set of outcomes.

We then decide on the significance level, which is the probability of rejecting the null hypothesis when it is in fact true (a type I error). Ideally, this decision should be made by evaluating the costs of making a type I error, as compared with those of making a type II error, or failing to reject H_0 when it is in fact false. When we also decide whether to use a one- or a two-tailed test, this determines our critical region. This set of outcomes for rejection is found by cumulating the probabilities, beginning with the most extreme outcomes and moving toward the center, until the resulting sum of probabilities is just slightly less than the selected significance level (e.g., .05). We then look at the data, compute the test statistic (e.g., number of successes), and make our decision. If the outcome falls within the critical region, we are committed to rejecting H_0, knowing that we will be making a type I error with a probability equal to that of the significance level selected. If the outcome does not fall within the critical region, we do not reject, thereby running the risk of a type II error. Although (as we shall see in Chap. 14) it is difficult to pin down the probability of a type II error because of the fact that it will depend on just how false our null hypothesis happens to be, we know that for a fixed sample size the smaller we make the risk of a type I error, the greater the risk of a type II error.

Glossary
Binomial distribution
Critical region
Hypergeometric distribution
Model versus hypothesis
Multinomial distribution
One- and two-tailed tests
Poisson distribution
Sampling distribution
Significance level

Exercises
1. In 11 flips of an honest coin, what is the probability of getting exactly four heads? Exactly seven heads? Less than three heads? [*Ans.* $P(4) = 330/2{,}048$]

2. Suppose the coin in Exercise 1 is dishonest and that the probability of getting a head is actually .6. Without doing the computations, indicate how this would affect each of the probabilities obtained above (i.e., raise, lower, or leave them unchanged). [*Ans.* lower $P(4)$]

3. Suppose you wish to test the null hypothesis that the coin is honest by flipping it 11 times. Indicate the critical region you would use:
   a. For a two-tailed test at the .05 level (*Ans.* 0, 1, 10, or 11 heads)
   b. For a two-tailed test at the .10 level
   c. For a two-tailed test at the .01 level
   d. For a one-tailed test at the .05 level, predicting direction that P(head) $> .5$ (*Ans.* 9, 10, or 11 heads)
   e. For a one-tailed test at the .10 level, predicting direction that P(head) $< .5$

4. In a particular community 10 per cent of the population is Jewish. A study of the boards of directors of various service agencies indicates that of a total of seven chairmen of the boards, four are Jewish. How likely is it that this could happen by chance? In this and other exercises involving tests of hypotheses, indicate your reasoning and list the assumptions being made. (*Ans.* $P = .0027$)

5. A social psychologist takes 12 groups which he matches, pair by pair, according to size. He thus has six pairs of groups, with one group in each pair constituting an experimental group and the other the control group. The experiment involves an attempt to increase the cohesiveness of the groups, and the experimenter is able to evaluate whether or not the experimental group is more cohesive than the control group with which it has been paired. How can he make use of the binomial to test the null hypothesis that the experiment has no effect? In this problem you should indicate all assumptions required, compute the sampling distribution, and make a choice of the critical region.

*6. Suppose you are studying a small group of 12 persons and wish to test the hypothesis that the higher the degree of conformity to group norms, the higher one's status in the group. For both variables (conformity and status) you are simply able to evaluate whether the individual is above or below the median. How would you use the binomial to test the null hypothesis that there is no relationship between these two variables? Be sure to indicate your reasoning.

*7. Suppose that it is known that the probability of committing suicide among a certain age group is .003. It is found that among a randomly selected sample of 1,200 Navaho Indians within the same age range, there have been no suicides. How likely is this to have occurred by chance?

References

1. Alder, H. L., and E. B. Roessler: *Introduction to Probability and Statistics*, 4th ed., W. H. Freeman and Company, San Francisco, 1968, chap. 6.
2. Anderson, T. R., and M. Zelditch: *A Basic Course in Statistics*, 2d ed., Holt, Rinehart and Winston, Inc., New York, 1968, chap. 11.
3. Bradley, J. V.: *Distribution-free Statistical Tests*, Prentice-Hall, Inc., Englewood Cliffs, N.J., 1968, chaps. 3 and 7.
4. Hays, W. L.: *Statistics*, Holt, Rinehart and Winston, Inc., New York, 1963, chap. 5.
5. Pierce, Albert: *Fundamentals of Nonparametric Statistics*, Dickenson Publishing Company, Inc., Belmont, Calif., 1970, chaps. 9 and 12.
6. Selltiz, C., M. Jahoda, M. Deutsch, and S. W. Cook: *Research Methods in Social Relations*, Henry Holt and Company, Inc., New York, 1959, chap. 4.
7. Siegel, Sidney: *Nonparametric Statistics for the Behavioral Sciences*, McGraw-Hill Book Company, New York, 1956, pp. 36–42.
8. Spiegel, M. R.: *Theory and Problems of Statistics*, Schaum's Outline Series, McGraw-Hill Book Company, New York, 1961, chap. 7.
9. *Tables of the Binomial Probability Distribution*, National Bureau of Standards, Applied Mathematics Series, no. 6, 1950.

Single-sample Tests Involving Means and Proportions   11

In this chapter we shall be concerned with tests of hypotheses about population means and proportions. A sample mean or proportion, obtained from a single sample, will be compared with the hypothesized parameter and a decision made as to whether or not to reject the hypothesis. You will soon discover that tests of the form discussed in this chapter have much less practical utility than tests involving several samples. At this point, however, it is more important to obtain a good understanding of fundamental ideas than to be overly concerned with practical applications. Unfortunately, the simplest tests are not always the most useful ones.

You will recall that statistical tests involving the binomial made use of the multiplication rule in order to obtain a sampling distribution. We were thus able to see exactly how probability theory was used in obtaining the sampling distribution. From here on, mathematical considerations become far more complex, so much so that in spite of the fact that it would be desirable to understand what is behind every argument, you will have to begin taking more and more statements on faith. Mathematical proofs are available, of course, but most involve more advanced work in probability or even considerably more mathematical background.

11.1 Sampling Distribution of Means

A rather remarkable theorem is based on the same principles and rules of probabilities as is the binomial but cannot be proved in a text such as this. This theorem can be stated as follows: *If repeated random samples of size N are drawn from a normal population, with mean μ and variance σ^2, the sampling distribution of sample means will be normal, with mean μ and variance σ^2/N.* Let us now take a careful look at what this theorem says.

We first start with a normal population, recognizing, of course, that in real life there is no such thing as a perfectly normal population. We then imagine ourselves drawing a very large number of random samples of size N from this population.[1] For each of these samples we obtain a mean $\bar{X}$. These sample means will naturally vary somewhat from sample to sample, but we would expect them to cluster around the true population mean μ. The theorem says that if we plot the distribution of these sample means, the result will be a normal curve. Furthermore, the standard deviation of this normal distribution of sample means will be $\sigma/\sqrt{N}$. Therefore, the larger the sample size selected, the smaller the standard deviation in the sampling distribution, i.e., the more the clustering of sample means (see Fig. 11.1). If we consider the sample means as

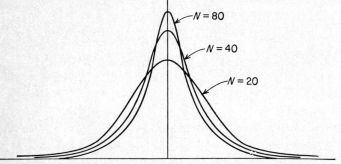

Figure 11.1 Comparison of normal sampling distributions for different-sized samples.

estimates of the population mean, we can say that there is a certain amount of error in our estimation process owing to sampling fluctuations. Therefore, we refer to the standard deviation of a sampling distribution as the *standard error*. In this case the standard error of the *mean*, indicated symbolically as $\sigma_{\bar{x}}$, is $\sigma/\sqrt{N}$.

You should keep clearly in mind that there are three distinct distributions involved, two of which happen to be exactly normal. *First*, there is the population which is assumed to be normal with a mean of μ and a variance of σ^2 [hereafter written in abbreviated form as $Nor(\mu,\sigma^2)$]. *Second*, there is a distribution of scores *within each sample*. If N is large, this distribution will probably be reasonably representative of the population and may therefore be approximately normal. Notice that this is

[1] Be careful not to confuse the *number of samples* (which is infinite) with the *size* (N) *of each sample.*

the only distribution that one actually obtains empirically.[2] *Third*, there is the sampling distribution of a statistic (here, the mean). We have just seen that the sampling distribution for the mean will also be normal but will have a smaller standard deviation than the population (unless the sample size N is one).

The relationship between the population and the sampling distribution is diagramed in Fig. 11.2. The larger the sample size N, the more peaked

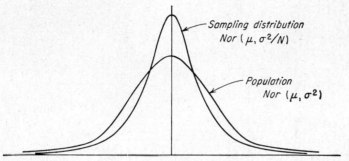

Figure 11.2 Comparison between population and sampling distribution.

will be the sampling distribution, as was shown in Fig. 11.1. It should be kept clearly in mind that although their standard deviations are directly related to each other, they are completely distinct distributions. All of the "cases" in the sampling distribution are *means* of separate samples. As was true in the case of the binomial and as will be true in all other statistical tests, it is the sampling distribution rather than the parent population that is used directly in significance tests. Assumptions about the population may appear in the model. It is through probability theory that statements about the population and about the sampling methods become translated into statements about a sampling distribution.

In summary, the means and standard deviations of the three kinds of distributions are as follows:

	Mean	Standard deviation
Population	μ	σ
Sample	$\bar{X}$	s
Sampling distribution	μ	$\sigma/\sqrt{N}$

[2] Since this is the distribution that the researcher actually sees, there is likely to be a tendency to confuse this second kind of distribution with the sampling distribution.

The theorem is consistent with the common-sense intuition that, assuming biases have been avoided, one can have more faith in estimating the mean from a large sample than a small one.[3] In effect it says that sample means will vary less from sample to sample if N is large. But it is a considerable refinement of common sense in that it gives an indication of how much more faith we should have if N is increased by a given amount. For example, we can see that in order to cut the standard error in half, we need to quadruple N. It also tells us that the more homogeneous the population is to begin with, i.e., the smaller the value of σ, the smaller the standard error $\sigma/\sqrt{N}$ and the greater the clustering of sample means about the population mean.

*A theoretical justification of this important theorem can be given by introducing the notion of linear combinations, an idea that we shall use later at several points. A mean is of course a simple linear function of the scores X_i since $\bar{X} = \dfrac{1}{N}(X_1 + X_2 + \cdots + X_N)$. More generally, it can be shown that if we have a variable Y that is any linear combination of the X_i, and if these X_i have been independently selected, as is the case when we draw a simple random sample, then we may obtain simple expressions for the mean (expected value) of Y and for the variance of Y. Specifically, if

$$Y = c_1X_1 + c_2X_2 + c_3X_3 + \cdots + c_NX_N$$

and if the X_i are independently selected, then

$$E(Y) = c_1E(X_1) + c_2E(X_2) + \cdots + c_NE(X_N)$$

and $\quad\quad \mathrm{Var}\,Y = \sigma_Y{}^2 = c_1{}^2\sigma_{X_1}{}^2 + c_2{}^2\sigma_{X_2}{}^2 + \cdots + c_N{}^2\sigma_{X_N}{}^2$

* In the case of random samples the expected value of each X_i is μ. If we set each $c_i = 1/N$, then Y becomes the sample mean and we have

$$E(\bar{X}) = E(Y) = \left(\frac{1}{N}\right)[\mu + \mu + \cdots + \mu] = \frac{1}{N}(N\mu) = \mu$$

[3] Notice that we have more faith in *estimates* that are based on large samples, but in rejecting a null hypothesis at the .05 level we take the same risk of a type I error regardless of the size of N. As we shall presently see, the size of the critical region used in the test takes the sample size into consideration, thus accounting for the apparent inconsistency.

and

$$\sigma_Y{}^2 = \sigma_{\bar{X}}{}^2 = \frac{1}{N^2}\sigma_{X_1}{}^2 + \frac{1}{N^2}\sigma_{X_2}{}^2 + \cdots + \frac{1}{N^2}\sigma_{X_N}{}^2$$

$$= \frac{1}{N^2}[\sigma^2 + \sigma^2 + \cdots + \sigma^2]$$

$$= \frac{1}{N^2}(N\sigma^2) = \frac{\sigma^2}{N}$$

The latter formula follows from the fact that the variance of each X_i is just σ^2 since we are dealing with individual cases selected with equal probability from a population with variance σ^2. Intuitively, the idea is that if we repeated an experiment of drawing the "first" case a very large number of times, the distribution of these first cases should be approximately $Nor(\mu,\sigma^2)$. The same should also hold for repeated drawings of second cases, and so forth.

The central-limit theorem We are now in a position to state a more general theorem, known as the central-limit theorem, as follows: *If repeated random samples of size N are drawn from any population (of whatever form) having a mean μ and a variance σ², then as N becomes large, the sampling distribution of sample means approaches normality, with mean μ and variance σ²/N.*

This theorem is even more remarkable than the previous theorem. It says that no matter how unusual a distribution we start with, provided N is sufficiently large, we can count on a sampling distribution that is approximately normal. Since it is the sampling distribution, and not the population, which will be used in significance tests, this means that whenever N is large, we can completely relax the assumption about the normality of the population and still make use of the normal curve in our tests.

You should require convincing that the central-limit theorem makes sense empirically. The best way to obtain a good grasp of what the central-limit theorem means, and at the same time to convince yourself that the standard error is really $\sigma/\sqrt{N}$, is to draw a number of samples from a population with known mean and standard deviation, compute the sample means, find the standard deviation of these means, and compare the result obtained with $\sigma/\sqrt{N}$.[4] But why should the sampling distribu-

---

[4] See Exercise 1 at the end of the chapter.

tion become normal if the population distribution is not normal? Let us take a look at a population that is far from normal and see what happens as we take larger and larger samples.

Imagine we are rolling some mathematically ideal dice for which the probabilities of getting each of the six faces are exactly $\frac{1}{6}$. The probability distribution for the toss of a single die is then rectangular, i.e., all numbers (from 1 to 6) have an equal chance of occurring. This type of distribution is in marked contrast to a normal distribution in which extreme values are less likely than those closer to the mean. Such a rectangular distribution can be represented as in Fig. 11.3. Strictly

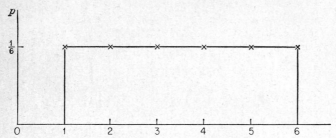

Figure 11.3   Population distribution of probabilities of getting face values of 1, 2, 3, 4, 5, or 6 with a perfect die.

speaking, of course, the distribution would be discrete and not continuous as is implied in the diagram.

Considering such a distribution as our population of all possible dice throws, let us calculate the sampling distribution of the means of samples of size 2. This means that we shall toss two dice, sum up the face values, and divide by 2. As experienced "craps" players are well aware, these sums range from 2 to 12, with 7 being the most likely value. In obtaining the probabilities of occurrence of each of these sums, we first note that there are (6)(6) or 36 possible outcomes if the two dice are distinguished. Thus the first die can come up with any one of six faces showing, and so can the second. To obtain the probability of getting a sum of scores of 7, and hence a mean of 3.5, we need only count the number of ways such a result can occur. Clearly, there are six pairs that will yield a score of 7: (1,6), (2,5), (3,4), (4,3), (5,2), and (6,1). A sum of 6 can be obtained in only five ways: (1,5), (2,4), (3,3), (4,2), and (5,1). Likewise, there is only one way we can get a sum of 12 (6,6) or a sum of 2 (1,1). The probability distribution of *means* can therefore be represented as follows:

Mean	Probability	Mean	Probability
1.0	$\frac{1}{36}$	4.0	$\frac{5}{36}$
1.5	$\frac{2}{36}$	4.5	$\frac{4}{36}$
2.0	$\frac{3}{36}$	5.0	$\frac{3}{36}$
2.5	$\frac{4}{36}$	5.5	$\frac{2}{36}$
3.0	$\frac{5}{36}$	6.0	$\frac{1}{36}$
3.5	$\frac{6}{36}$		$\frac{36}{36}$

When plotted, this sampling distribution takes the form of a triangle (Fig. 11.4).

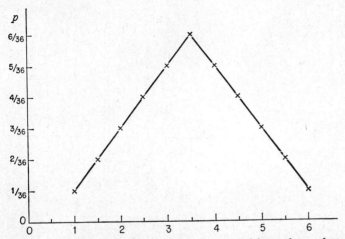

Figure 11.4 Sampling distribution of means of faces, for perfect dice, and samples of size 2.

If three dice are tossed, the faces summed, and means obtained, the sampling distribution of means will be as follows:

Mean	Probability	Mean	Probability
1.00	$\frac{1}{216}$	3.67	$\frac{27}{216}$
1.33	$\frac{3}{216}$	4.00	$\frac{25}{216}$
1.67	$\frac{6}{216}$	4.33	$\frac{21}{216}$
2.00	$\frac{10}{216}$	4.67	$\frac{15}{216}$
2.33	$\frac{15}{216}$	5.00	$\frac{10}{216}$
2.67	$\frac{21}{216}$	5.33	$\frac{6}{216}$
3.00	$\frac{25}{216}$	5.67	$\frac{3}{216}$
3.33	$\frac{27}{216}$	6.00	$\frac{1}{216}$
			$\frac{216}{216}$

This distribution, as can readily be seen in Fig. 11.5, is beginning to approximate the form of a normal curve even though the sample size is only 3. After a careful examination of the above figures, you should be able to grasp intuitively what is happening and why it is that a bell-shaped curve is being approximated as the sample size N becomes larger and larger. Even though on a single toss a 6 is as likely as a 3 or 4, and as a matter of fact two 6's are as likely as two 3's, there is only one way

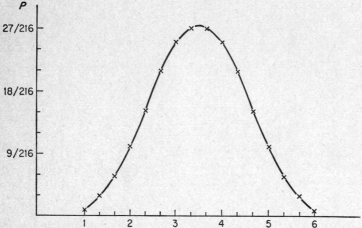

Figure 11.5 Sampling distribution of means of faces, for perfect dice, and samples of size 3.

of getting two 6's whereas there are a number of different ways of getting an *average* of 3.0 on 2 or more tosses. In common language we say that large numbers are likely to be counterbalanced by small ones, especially if N is large.

11.2 Test for Population Mean, σ Known

Let us now see how the central-limit theorem can be used in statistical tests. To begin with, we shall take up the simplest possible model for illustrative purposes. Since some of the assumptions required in this model are impractical, they will be relaxed later on. Each of the five steps discussed in Chap. 10 will again be treated in some detail in order to help you gain greater familiarity with the process of developing statistical tests.

PROBLEM Suppose a researcher is interested in checking on the adequacy of the sampling procedures used in a local survey. The survey has been taken by inexperienced interviewers, and the researcher suspects that middle- and upper-income families may have been oversampled, i.e., given a greater probability of appearing in the sample than lower-income families. Census data, involving a complete enumeration, are available which show the mean family income of the community to be $7,500 and standard deviation $1,500. The smaller survey involves 100 families, supposedly randomly selected, and the mean family income of this sample is found to be $7,900. Does the researcher have reason to suspect a biased sample?

1. *Making Assumptions* In order to make use of the central-limit theorem certain assumptions must be made. As previously indicated, there must always be an assumption about the method of sampling. In this case we assume the sample to be random. It is actually this assumption that we are interested in testing since we are skeptical of the interviewers' ability to give all families an equal chance of being selected. Presumably, we are willing to accept certain assumptions about the population, i.e., that the census data are accurate. If we cannot accept the census figures, there will be at least two assumptions in doubt, and interpretation of the results will be exceedingly difficult. Random sampling, then, will be our hypothesis; the remaining assumptions about the population will constitute the model.

A normal population is required if N is not too large. Here the question arises, "How large does N have to be before we can relax the normality assumption and make use of the central-limit theorem?" There is no simple answer to such a question since an answer depends, among other things, on (1) how precise an estimate of the probability of a type I error is desired, and (2) how good an approximation to a normal population we have. Although you should be wary of simple rules of thumb, it may be suggested that whenever $N \geq 100$, the normality assumption can practically always be relaxed. If $N \geq 50$ *and* if there is empirical evidence to the effect that departure from normality is not serious, then tests of the type discussed in the present section may also be used with a degree of assurance. If $N \leq 30$, however, one should definitely be on guard against the use of such tests unless the approximation to normality is known to be good. Whenever very small samples are used, one usually lacks such information since there are not enough cases in the sample to indicate the form of the population distribution. Therefore, other kinds of tests should ordinarily be used for small samples. Let us suppose in this problem that we can legitimately make use of the central-limit theorem. As we know, income distributions are usually somewhat skewed. On the other hand, we have a reasonably large sample.

In addition to the above assumptions, if we are to use the central-limit theorem, it is also necessary to accept the census figures for μ and σ and to assume an interval scale. Therefore we have the following assumptions:

Level of Measurement: Interval scale
Model: Normal population (can be relaxed)
  $\mu = \$7,500$
  $\sigma = \$1,500$
Hypothesis (null): Random sampling

2. *Obtaining the Sampling Distribution*   Fortunately, the task of getting the sampling distribution has already been done for us. Since we know that the sampling distribution of the sample means is normal or approximately normal, we can go directly to the normal table. From now on, sampling distributions will always be given in the form of tables in Appendix 2. It is important to realize, however, that these tables have been computed using probability theory. It is very easy to get so lost in computational details that one forgets that whenever he is making use of a table in his statistical tests, he is actually making use of a sampling distribution.

3. *Choosing a Significance Level and Critical Region*   The choice of the proper significance level depends, of course, on the relative costs involved in making types I and II errors. If the researcher fails to reject the hypothesis of random sampling when in fact the sample is biased, he runs the risk of reporting misleading findings. On the other hand, if he rejects when the hypothesis is actually true, he may have to repeat the survey at considerable cost. Ideally, he should make a rational decision based on the relative costs of these two kinds of error. Actually, he will probably have a difficult time doing this. Let us assume that he decides on the .05 level. Next, he should decide to use a one-tailed test since the direction of bias is predicted. If the sample mean were to turn out to be less than \$7,500, he would hardly suspect his interviewers of oversampling the middle- and upper-income groups.[5] Given the choice of the .05 level and a one-tailed test, the critical region is determined from the normal table. Since only 5 per cent of the area of the normal curve is to the right of an ordinate 1.65 standard deviations larger than the mean, we know that if the result is more than 1.65 standard deviations larger than μ, the hypothesis should be rejected (see Fig. 11.6).

[5] In this problem, the data for the sample have actually been given, and we *know* the direction of the result. You should imagine, however, that this decision is being made prior to one's knowledge of the outcome.

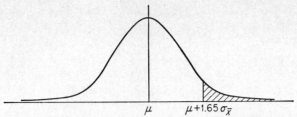

Figure 11.6 Normal sampling distribution with shaded area representing a one-tailed critical region at the .05 significance level.

4. *Computing the Test Statistic* We know that if all assumptions are correct, the sampling distribution of $\bar{X}$'s will be $Nor(\mu, \sigma^2/N)$. In terms of our example,

$$\mu = \$7,500$$

$$\sigma_{\bar{X}} = \frac{\sigma}{\sqrt{N}} = \frac{1,500}{\sqrt{100}} = \$150$$

In order to make use of the normal table it is necessary to convert to standard scores, or in other words to obtain a statistic Z that is $Nor(0,1)$. Previously we used the formula

$$Z = \frac{X - \bar{X}}{s}$$

This formula holds for a sample that is $Nor(\bar{X}, s^2)$ but *not* for the sampling distribution. Let us recall each of the steps in our procedure. We have made a series of assumptions in order to obtain a sampling distribution. This latter distribution tells us how likely a *given* $\bar{X}$ would be if the assumptions were actually true. The investigator has obtained a single $\bar{X}$ from his sample and then will use the theoretical sampling distribution to help him evaluate the likelihood of getting a result as unusual or more unusual than his particular $\bar{X}$. It is the sampling distribution he is dealing with when he uses the normal table. Each "case" in this distribution is an $\bar{X}$, the mean is μ, and the standard deviation is $\sigma/\sqrt{N}$. Therefore, $\bar{X}$ replaces X, μ replaces $\bar{X}$, and $\sigma/\sqrt{N}$ replaces s in the above formula for Z. Hence

$$Z = \frac{\bar{X} - \mu}{\sigma/\sqrt{N}}$$

$$= \frac{7,900 - 7,500}{150} = 2.67$$

In other words, the sample mean is 2.67 standard errors larger than the population mean.

5. *Making a Decision* Since $\bar{X}$ deviates from the assumed μ in the predicted direction by more than 1.65 standard deviations, the hypothesis should be rejected at the .05 level. In fact, having computed Z exactly, we can say more than this. The probability of getting a Z this large or larger is .0038, using a one-tailed test. In practice it is usually advisable to compute the exact significance level whenever possible. In so doing we indicate that the result has fallen within an even smaller critical region than the one originally established. Since the reader may prefer to use a different level of significance than does an author, it is ordinarily helpful to supply exact or more nearly exact probabilities so that the reader may draw his own conclusions about whether or not to accept the findings. In this example, the investigator would reject the null hypothesis that the sampling was random. He would then want to decide whether or not to draw another sample.

11.3   Student's t Distribution

In most instances it is completely unrealistic to treat σ as known. Usually one goes to considerable trouble to assure randomness since he is primarily interested in testing assumptions about the population being studied. In tests of the sort being discussed in this chapter, he is likely to want to test an hypothesis concerning μ. But if this were the case, would he ever be in a position to know the value of σ? Practically never. For if he possessed knowledge of σ, he would undoubtedly also be in a position to know μ unless, of course, someone such as his statistics instructor were deliberately withholding information. Usually he would not know the values of either μ or σ. What can he do in such a situation? Since the central-limit theorem involves σ, he cannot completely ignore its value. One solution would seem to be to replace σ with s, the *sample* standard deviation. As a matter of fact, this was commonly done prior to the development of modern statistics. In the formula for Z, $\sigma/\sqrt{N}$ was simply replaced by $s/\sqrt{N}$, and since s could be computed directly from the sample data, there were no unknowns left in the formula. As it turns out, this procedure yields reasonably good results when N is large. As we shall see presently, however, probabilities obtained in this manner can be misleading whenever N is relatively small. Let us see why this is the case.

We can construct an alternative test statistic

$$t = \frac{\bar{X} - \mu}{s/\sqrt{N - 1}}$$

This statistic was introduced by W. S. Gossett, writing under the name of "Student," and has a sampling distribution known as Student's t distribution. Comparing t with Z, we notice that whereas the numerators are identical, the denominators differ in two respects: (1) there is an $N - 1$ under the radical, and (2) σ has been replaced by s. In order to understand these modifications let us examine each in turn. In so doing we shall have to introduce several new ideas.

The sample standard deviation s can be used as an estimate of σ. Although the problem of estimation will be treated in the next chapter, it is sufficient at this point to mention that we often want an estimate to possess certain properties. One of the properties of a "good" estimate is that it be unbiased. Contrary to what one might expect, it turns out that s is not quite an unbiased estimate of σ. Another quantity, which we can designate as $\hat{\sigma}$ and which is obtained by the formula

$$\hat{\sigma} = \sqrt{\frac{\sum\limits_{i=1}^{N} (X_i - \bar{X})^2}{N - 1}}$$

can be shown mathematically to be an unbiased estimate of σ.[6] The only difference between $\hat{\sigma}$ and s is in the $N - 1$ factor in the denominator. Thus, although you have learned to compute s, you are now faced with the fact that another formula should be used in estimating σ. In the present problem it is $\sigma/\sqrt{N}$ rather than σ which needs to be estimated, since it is the former expression which appears in the denominator of Z. Although it is true that $\hat{\sigma}/\sqrt{N}$ is the best estimator of $\sigma/\sqrt{N}$, it is possible completely to avoid the computation of $\hat{\sigma}$ if s has already been obtained. Notice that

$$\frac{\hat{\sigma}}{\sqrt{N}} = \frac{\sqrt{\left[\sum\limits_{i=1}^{N} (X_i - \bar{X})^2\right] / (N - 1)}}{\sqrt{N}}$$

---

[6] Strictly speaking, $\hat{\sigma}$ is not an unbiased estimate of σ, but $\hat{\sigma}^2$ is an unbiased estimate of σ^2. This subtle distinction need not bother us. In this text we shall commonly use a circumflex (^) over a Greek letter to indicate an estimate of the parameter. Some texts define s with $N - 1$ in the denominator, but we shall keep the two formulas distinct.

Remembering that $\sqrt{a}/\sqrt{b}$ can be written as $\sqrt{a/b}$, we have

$$\frac{\hat{\sigma}}{\sqrt{N}} = \sqrt{\frac{\sum\limits_{i=1}^{N} (X_i - \bar{X})^2}{N(N-1)}}$$

$$= \frac{\sqrt{\left[\sum\limits_{i=1}^{N'} (X_i - \bar{X})^2\right]/N}}{\sqrt{N-1}} = \frac{s}{\sqrt{N-1}}$$

Thus we can take a somewhat biased estimate of σ, divide by a quantity which is slightly smaller than $\sqrt{N}$, and come out with $s/\sqrt{N-1}$ as an unbiased estimate of $\sigma/\sqrt{N}$. It is for this reason that $N-1$ appears in the denominator of t.[7]

In replacing Z with t, the modification introduced by using $N-1$ is relatively slight, but the substitution of s in place of σ may be of considerable significance if N is small. Since s varies from sample to sample, the denominator of t varies as well as the numerator. For a given value of $\bar{X}$, if the s for a particular sample happens to be too small, t will be quite large; if s is large, t will be relatively small. There will thus be greater variability among t scores than among comparable Z values. This means that the sampling distribution of t will be flatter than normal. The t distribution will therefore have larger tails. Just how flat a t distribution is will depend on the size of the sample. If N is very small, the t distribution will be very flat as compared with the normal curve. In other words, it will be necessary to go out a larger number of standard deviations from the mean in order to include 95 per cent of the cases. As N becomes large, the t distribution comes closer and closer to approximating a normal distribution, always being slightly flatter than the normal curve, however. Thus there is a different t distribution for each size sample. The fact that t distributions approach normality makes sense intuitively when we realize that as N becomes large, s becomes a very accurate estimate of σ, and it makes very little difference whether we use s or σ in the denominator.

In order to make use of the t distribution a normal population must be assumed, especially if N is relatively small. The computation of the sampling distribution of t requires that the numerator $(\bar{X} - \mu)$ be

---

[7] Some texts recommend the use of $N - 1$ for small samples and N for large ones. Such a procedure seems to add unnecessary confusion. In the case of large samples, of course, it makes little difference which figure is used.

normally distributed and that it also vary independently of the denominator $s/\sqrt{N-1}$. Ordinarily, we would not expect independence between numerator and denominator since s is actually computed by taking deviations about $\bar{X}$, and therefore it would be surprising to find $\bar{X}$ and s statistically independent of each other. Knowing the sample $\bar{X}$, we would expect to improve our ability to predict s for that same sample. It so happens that for normal populations and random sampling, the sample mean and standard deviation are statistically independent, however. Since this property does not generally hold for all population distributions, and since $\bar{X} - \mu$ will not generally be normally distributed unless N is large, we must assume a normal population when using the t test.

PROBLEM Suppose you are evaluating the programs of a random sample of 25 casework agencies selected from the population of all casework agencies in the state of New York. Each agency keeps a record of the percentage of successful cases, judged according to some standard criterion. A standard has been set that the mean success percentage for all agencies ought to be 60 per cent. In your sample you find the mean percentage to be 52 per cent and the standard deviation to be 12 per cent. Do you have reason to suspect that for the population of agencies as a whole the level of performance is below the standard expected?

1. *Making Assumptions* Necessary assumptions may be listed as follows:

Level of Measurement: Interval scale
Model: Random sampling
        Normal population
Hypothesis: $\mu = 60$ per cent

Notice that no assumption about σ is necessary since s has actually been obtained empirically and can be directly used in the t test. A few comments about the level of measurement are needed. Since each client of an agency is either a success or a failure, and since figures obtained for each agency are percentages of successes, it might be thought that we are dealing with a simple dichotomized nominal scale rather than an interval scale. Indeed, if the units of analysis were *clients* rather than agencies, this would be the case. But remember that the units being studied are *agencies*. A score for each agency (e.g., percentage of successes) has been obtained, and this score represents a legitimate interval scale. For example. a difference between 30 and 40 per cent is the same as a differ-

2. *Obtaining the Sampling Distribution*   Sampling distributions for *t* are given in Table D of Appendix 2. Since these distributions differ for each sample size, the table has been condensed so as to give only the tails of each distribution. In using the table it is first necessary to locate the proper sample size by looking down the left-hand column. These sample sizes are ordinarily given in terms of *degrees of freedom* (df), which in this type of problem is always $N - 1$.[8] Next, locate the proper significance level by reading across the top. Figures within the body of the table indicate the magnitude of *t* necessary to obtain significance at the designated level.

3. *Selecting a Significance Level and Critical Region*   Let us use the .05 level and a one-tailed test. From Table D we see that for 24 degrees of freedom a *t* of 2.064 or larger is needed in order to obtain significance at the .05 level for a two-tailed test. For a one-tailed test and the .05 level, we need a *t* of only 1.711 or larger. In the case of one-tailed tests we simply halve the significance levels required for two-tailed tests. This is because we go out the same number of standard deviations from the mean in order to obtain a one-tailed critical region of .05 as we would to get a two-tailed region of .10.

4. *Computing the Test Statistic*   Although it is true that the sampling distribution of $\bar{X}$ is $Nor(\mu,\sigma^2/N)$ and therefore that the distribution of Z is $Nor(0,1)$, this information is of no real use to us since σ is unknown. Instead, we compute the value of *t*, obtaining

$$t = \frac{\bar{X} - \mu}{s/\sqrt{N - 1}} = \frac{52 - 60}{12/\sqrt{24}} = -3.27$$

5. *Decision*   It has been determined that any *t* the numerical value of which is ≥ 1.711 will be within the critical region. Therefore we reject the hypothesis that $\mu = 60$ and conclude, with a certain risk of error, that the actual level of performance of the agencies is below the standard expected.   Reading across the row corresponding to 24 degrees of freedom in Table D, we see that for a one-tailed test the significance level corresponding to a *t* of 3.27 is somewhere between .005 and .0005.[9]

Several facts about the *t* distribution can be noted at this point. If

---

[8] For a discussion of degrees of freedom, see Sec. 12.1.

[9] Although exact probabilities cannot be obtained from the table, interpolation is always possible. Usually, however, it is sufficient to indicate that *p* is between two values, for example, $.0005 \leq p \leq .005$.

you will examine the column corresponding to $p = .05$ for a two-tailed test, you will notice that as the sample size becomes larger, t values become smaller and converge fairly rapidly to 1.96, the value necessary for significance if the normal table were used. These values should give a reasonably good idea of the degree of approximation to the normal curve for any given sample size. For values of $N - 1$ larger than 30, interpolation will ordinarily be necessary, and for values much larger than 120, the normal table will have to be used since t values are not given. Some texts arbitrarily state that it is only necessary to use the t table when $N \leq 30$. Although such a rule of thumb yields reasonable results, the position taken here is that it is always preferable to *use the t table whenever σ is unknown* and whenever a normal population can be assumed. Since the t table is no more difficult to use, it seems sensible to use exact values in preference to a normal approximation. It should also be emphasized that there is not a unique theory applying to small samples and an entirely different one applying to large samples, as some texts imply.

As can be seen from the t table, only when the sample size is relatively small do the normal and t distributions differ considerably. Also, a normal population must be assumed whenever t is used unless N is quite large, in which case t can be approximated by Z. Therefore, the practical value of the t test is in situations where one has small samples and where a normal population can be assumed. Unfortunately, it is when samples are small that we are ordinarily most in doubt about the exact nature of the population. For example, if a researcher is doing an exploratory study with 17 cases, is he very likely to be in a position to accept the normality assumption? Probably not. As we shall see in Chap. 14, there are tests that can be used as alternatives to the t test and that do not involve the normality assumption.

11.4 Tests Involving Proportions

Up to this point in the chapter we have considered only examples involving an interval scale. Furthermore, normality of the population also had to be assumed for small samples. In this section we shall see how the central-limit theorem can be used to cover tests involving proportions whenever N is fairly large. In fact, proportions will be treated as special cases of means so that our previous discussion will still apply.

Suppose we have a simple dichotomized nominal scale. We might want to test an hypothesis concerning, say, the proportion of males in a population. Let us arbitrarily assign the value one to males and zero to females and treat the scores as an interval scale. Although there is no clearly conceived unit, unless it be the attribute *maleness* which is either possessed or not possessed, we can treat these arbitrary scores as an interval scale because there are only two of them. If a third category were added, this would no longer be possible, however, since it would become necessary to determine the exact position of this category relative to those of the other two. What we are in effect saying, here, is that it is unnecessary to make a distinction between nominal, ordinal, and interval scales in the case of a dichotomy since the problem of comparing distances between scores never arises.

We now have a population made up entirely of ones and zeros. This is a bimodal distribution, having all cases concentrated at one of two points, and is certainly not normal. But if N is sufficiently large, we know that regardless of the form of the population, the sampling distribution of sample means will be approximately $Nor(\mu, \sigma^2/N)$. All that remains to be done is to determine the mean and standard deviation of this population of ones and zeros.

Let p_u represent the proportion of males in the population and q_u the proportion of females, the subscript u indicating that we are dealing with the entire universe. In order to obtain the mean of the ones and zeros in the population, we simply add the values and divide by the total number of cases. The *number* of ones will be the total number of cases multiplied by the proportion of males. Regardless of the number of zeros, their contribution to the sum will be zero. Therefore the population mean will be

$$\mu = \frac{M p_u}{M} = p_u$$

where M represents the size of the population (as distinguished from the sample size N). Therefore, the mean of a number of ones and zeros is exactly the proportion of ones. By similar reasoning $\bar{X} = p_s$, where p_s represents the proportion of males in the sample.

By using the general formula for the standard deviation, we can show that $\sigma = \sqrt{p_u q_u}$. Making use of the symbols for population parameters, the formula for σ becomes

$$\sigma = \sqrt{\frac{\sum_{i=1}^{M} (X_i - \mu)^2}{M}} = \sqrt{\frac{\sum_{i=1}^{M} (X_i - p_u)^2}{M}}$$

Looking at the numerator of the quantity under the radical, we see that there will be only two types of quantities representing the squared deviations from the mean p_u. For each score of one, the squared deviation from the mean will be $(1 - p_u)^2$, and for every zero it will be $(0 - p_u)^2$. Since there will be Mp_u ones and Mq_u zeros in the sum of squares, we get

$$\sigma = \sqrt{\frac{Mp_u(1 - p_u)^2 + Mq_u(0 - p_u)^2}{M}} = \sqrt{\frac{Mp_u q_u^2 + Mq_u p_u^2}{M}}$$

Factoring $Mp_u q_u$ from each term in the numerator, we get

$$\sigma = \sqrt{\frac{Mp_u q_u(q_u + p_u)}{M}} = \sqrt{\frac{Mp_u q_u}{M}}$$
$$= \sqrt{p_u q_u}$$

Notice, incidentally, that since M cancels out in both the formulas for μ and σ, the population mean and standard deviation are independent of the actual size of the population, as we would expect.

Therefore we can use the central-limit theorem to give

$$\sigma_{\bar{X}} = \sigma_{p_s} = \frac{\sigma}{\sqrt{N}} = \sqrt{\frac{p_u q_u}{N}}$$

where the symbol σ_{p_s} indicates that we are dealing with the standard error of sample proportions. In our new terminology, p_s replaces $\bar{X}$, p_u replaces μ, and σ_{p_s} replaces $\sigma_{\bar{X}}$ in the formula for Z. Thus

$$Z = \frac{\bar{X} - \mu}{\sigma_{\bar{X}}} = \frac{p_s - p_u}{\sqrt{p_u q_u/N}}$$

Notice that although it appears as if we have a completely different formula from that previously used, there is really nothing new except a change of symbols. This is true since we have been able to show that proportions can be treated as special cases of means. It should be emphasized, however, that the central-limit theorem requires that N must be large in order to make use of the normal approximation. Whenever N is small, the binomial would be a more appropriate test.

*There is a close connection between this test involving proportions and the binomial distribution. It has been pointed out that if N is large, and if $Np > 5$, where $p < q$, then the binomial distribution may be approximated by a normal distribution. Of course in the case of the binomial distribution we were dealing with *numbers* of successes, rather than proportions. The expected value of the number of successes turns out to be Np, and the standard deviation of the number of successes is $\sqrt{Npq}$. In order to convert each of these into proportions we may divide each by N, getting p for the expected value and

$$\frac{1}{N} \sqrt{Npq} = \sqrt{\frac{Npq}{N^2}} = \sqrt{\frac{pq}{N}}$$

for the standard deviation. Thus in the case of large samples we might have formulated a binomial problem in terms of proportions, changed our symbols to p_u and q_u, and handled the problem according to the procedures of the present chapter. For example, in the case of a sign test we could have used the null hypothesis that $p_u = .5$, comparing this value with the proportion of successes p_s actually found in the sample.

PROBLEM   You are interested in evaluating the program of a particular casework agency and have drawn a random sample of 125 cases from its files. The percentage of successful cases is found to be 55 per cent as compared with the standard of 60 per cent. Can you conclude that the agency is below standard performance?

1. *Making Assumptions*

Level of Measurement: Dichotomized nominal scale
Model: Random sampling
Hypothesis: $p_u = .60$

This example is purposely similar to the previous one in order to emphasize the difference in units of analysis. Here, a single agency is being studied and the sample is of clients who are either successes or failures. In the earlier example agencies, rather than people, were the units being sampled, and the measure for each agency consisted of the *percentage* of successful cases. Notice that no assumption other than the hypothesis is required about the population, as it is implicitly assumed that it is bimodal.

2. *Obtaining the Sampling Distribution*   The sampling distribution will be approximately normal since N is large.

3. *Selecting a Significance Level and Critical Region*   Let us select, for sake of variety, the .02 level and a one-tailed test.

4. *Computing the Test Statistic*   We compute Z as follows:

$$Z = \frac{p_s - p_u}{\sqrt{p_u q_u / N}} = \frac{.55 - .60}{\sqrt{[(.60)(.40)]/125}} = \frac{-.05}{.0438} = -1.14$$

Note that p_u and q_u are used in the denominator rather than p_s and q_s. In case you might be tempted to use t rather than Z, you should notice that given an hypothesized p_u, the value of σ is determined by the formula $\sigma = \sqrt{p_u q_u}$.

5. *Making a Decision*   From the normal table it can be seen that a Z of -1.14 or less would occur approximately 13 per cent of the time by chance if the assumptions were true. We therefore do not reject the hypothesis at the .02 significance level. On the basis of the evidence at hand it cannot be established that the agency is below standard.

Glossary
Central-limit theorem
Rectangular distribution
Standard error
t distribution

Exercises
1. Using the table of random numbers given in Table B of Appendix 2 (see Sec. 21.1 for an explanation of the use of this table), select 10 samples, each of size 4, from the population of 65 cases given in Exercise 1 of Chap. 4. Compute the mean for each of these 10 samples and obtain the standard deviation of these 10 means. You now have a very rough and slightly biased estimate of the standard error of the mean. How does your figure compare with the standard error obtained by using the central-limit theorem, using the population standard deviation you computed in Exercise 2 of Chap. 6?

*2. Verify the sampling distribution of the mean of three dice throws that is diagramed in Fig. 11.5.

3. A sample of size 50 has a mean of 10.5 and a standard deviation s of 2.2. Test the hypothesis that the population mean. is 10.0 using (a) a one-tailed test at the .05 level, and (b) a two-tailed test at the .01 level. Do the same for samples of size 25 and 100 and compare results. [*Ans.* For $N = 50$, $t = 1.59$; fail to reject for (a) and (b)]

4. Suppose it is known that the mean annual income of assembly-line workers in a certain plant is $7,000 with a standard deviation of $900. You suspect that workers with active union interests will have higher than average incomes and take a random sample of 85 of these active members, obtaining a mean of $7,200 and a standard deviation of $1,000. Can you say that active union members have significantly higher incomes? (Use the .01 level.) (*Ans.* $Z = 2.05$; fail to reject)

5. You have taken a poll of 200 community residents of voting age and have found that of two candidates for an office, candidate A received 54 per cent of the votes sampled. Are you justified in concluding that A will win? Use the .05 level. List all of the assumptions you will have to make. (*Ans. Z* = 1.13)

6. Suppose that a test measuring "conformity needs" has been standardized on college students across the country. Fifty per cent of these college students had raw scores of 26 or higher (high scores indicating high conformity needs). Suspecting that conformity needs would generally be higher in the case of adults without a college education, a social scientist selects a random sample of adults 25 or older residing within his community. He finds (1) that 67 per cent of the 257 adults without college education have scores of 26 or higher, and (2) that 59 per cent of the 80 adults with college education have scores in this range.

   a. Can he conclude that the scores of either set of adults within his community are significantly higher than those of the college students on whom the test has been standardized? (Use .001 level.)

   b. Suppose the social scientist knows the entire exact distribution of college students' scores on the test. On the basis of the materials in the present chapter, what are some alternative procedures for testing for the significance of the departures of the two sets of adults' scores from the standardized scores? Do these alternative procedures require any additional assumptions? Explain.

References

1. Freund, J. E.: *Modern Elementary Statistics*, 3d ed., Prentice-Hall, Inc., Englewood Cliffs, N.J., 1967, chaps. 9 and 11.

2. Hagood, M. J., and D. O. Price: *Statistics for Sociologists*, Henry Holt and Company, Inc., New York, 1952, chaps. 15 and 16.

3. Hays, W. L.: *Statistics*, Holt, Rinehart and Winston, Inc., New York, 1963, chap. 10.

4. Wallis, W. A., and H. V. Roberts: *Statistics: A New Approach*, The Free Press of Glencoe, Ill., Chicago, 1956, chaps. 11 and 13.

Point and Interval Estimation

Up to this point the discussion of statistical inference has been concerned solely with the testing of hypotheses. There may also be an interest in *estimating* population parameters, and it is with this topic that the present chapter is concerned. After discussing the principles involved in estimation, we shall proceed to a discussion of the interrelationships between estimation and hypothesis testing. Modifications required for the t distribution and proportions will then be discussed. Finally, we shall take up the general question of the determination of sample size, illustrating the problem with estimation procedures.

In the previous two chapters you may have noted that in a number of practical problems the testing of specific hypotheses is not feasible because we are unable to specify a particular hypothetical value for the parameter, say μ. We shall presently see how estimation procedures can provide a very useful alternative to actual tests in such instances. Also, the social scientist may be directly interested in estimates rather than tests of hypotheses. For example, in survey research the practical aim of the study may be to estimate the proportion of persons using a certain product or voting in an election. Or it may be necessary to estimate the median income in an area or the mean number of children per married couple. Tests of specific hypotheses might be of some use in such instances, but estimation would be the more obvious procedure.

There are basically two kinds of estimation: point estimation and interval estimation. In point estimation we are interested in the best single value that can be used to estimate a parameter. For example, we may estimate that the median income in New York City is $8,500. Usually, however, we also want to obtain some idea of how accurate an estimate we have. We would like to predict that the parameter is somewhere within a given interval on either side of the point estimate.

Thus we may want to make a statement such as, "The median income in New York City is somewhere between $8,000 and $9,000." These two types of estimation are discussed in the sections that immediately follow.

12.1 Point Estimation

The problem of what statistic to use as an estimate of a parameter seems, on the surface, to be completely straightforward and a matter of common sense. If one wants to estimate the *population* mean (or median or standard deviation), why not use the *sample* mean (or median or standard deviation)? Although common sense would not lead us too far astray in these instances, we shall see that the problem is not quite so simple. Obviously, we *could* estimate a population mean in a number of ways. In addition to the sample mean, we could use the sample median or mode, we could use a number half way between the two extreme values, or we could use as our estimate the value of the thirteenth observation. Certain of these procedures would be better than others. We therefore need criteria with which to evaluate each kind of estimate in order to determine just how good it is. The social scientist using statistics as an applied tool seldom has to worry about such criteria. He is usually simply told to use a particular estimate. Nevertheless, it seems to be worthwhile to know just what criteria are used by the mathematical statistician in deciding which estimate to use. Two of the most important criteria are bias and efficiency. We shall discuss each of these in turn. For other criteria, such as sufficiency, consistency, and the principle of maximum likelihood, you should refer to more advanced texts.

Bias An estimate is said to be *unbiased* if the *mean of its sampling distribution is exactly equal to the value of the parameter being estimated*. In other words, the expected value of the estimate in the long run is the parameter itself. Notice that nothing is said, here, about the value of any particular sample result. According to this definition, for random samples $\bar{X}$ is an unbiased estimate of μ since the sampling distribution of $\bar{X}$ has μ as its mean or expected value. This does not mean, however, that we can expect that any particular value of $\bar{X}$ will equal μ, nor will we ever know in any realistic problem whether or not our sample mean does in fact equal the population mean. You should keep clearly in mind that the term *bias*, as used in this sense, refers to long-run results. In practical research you may be accustomed to using the term to refer to peculiar properties of the particular sample you have drawn.

It was mentioned in the previous chapter that the sample standard deviation s is a slightly biased estimate of σ. The statistic s has a sampling distribution as does $\bar{X}$. In other words, sample standard deviations will be distributed about the true population standard deviation just as sample means are distributed about μ. It can be shown mathematically, however, that for random samples the mean of the sampling distribution of s^2 is $[N - 1/N]\sigma^2$ and not σ^2. Therefore, s^2 is a biased estimate of σ^2. To find an unbiased estimate of σ^2 we take the quantity

$$\frac{N}{N-1}s^2 = \frac{N}{N-1}\frac{\displaystyle\sum_{i=1}^{N}(X_i - \bar{X})^2}{N}$$

$$= \frac{\displaystyle\sum_{i=1}^{N}(X_i - \bar{X})^2}{N-1} = \hat{\sigma}^2$$

Since the mean of the sampling distribution of s^2 is $[N - 1/N]\sigma^2$, we see that $\hat{\sigma}^2$ has a sampling distribution with mean exactly equal to

$$\frac{N}{N-1}\left[\left(\frac{N-1}{N}\right)\sigma^2\right] = \sigma^2$$

Although the basic reason why $\hat{\sigma}^2$ (and not s^2) is the unbiased estimate is that the mathematics works out that way, an intuitive explanation is sometimes given in terms of the concept *degrees of freedom*, a notion that will be frequently used in later chapters. The number of degrees of freedom is *equal to the number of quantities that are unknown minus the number of independent equations linking these unknowns.* You will recall that in order to arrive at a unique solution to a set of simultaneous algebraic equations, it was necessary to have the same number of equations as unknowns. Thus in order to solve for X, Y, and Z there must be three equations linking these variables. Had there been only two equations, we could have assigned any value we pleased to one variable, say Z. The values of the other two variables could then be determined by means of the two simultaneous equations. If there were five unknowns and only three equations to be solved simultaneously, we could assign arbitrary values to any two of the unknowns, and then the values of the remaining unknowns would be determined. There would thus be 2 degrees of freedom in this case since we are free to assign values to any two variables.

In computing a standard deviation from sample values, we must make use of an equation linking the N variable X's with the sample mean, i.e., the equation $\sum_{i=1}^{N} X_i/N = \bar{X}$. Given the value of $\bar{X}$, we can assign arbitrary values to $N - 1$ of the X_i's, but the last will then be determined from this equation. Since we have lost 1 degree of freedom in determining the value of the sample mean about which deviations have been taken, we must divide by $N - 1$ rather than N to obtain our unbiased estimate of σ^2. If you prefer to think of it this way, you can consider that we have adjusted the number of cases slightly in order to compensate for the fact that we have taken deviations about the sample mean rather than the true population mean. We have essentially used up one case in computing the mean for the sample. You will find that unbiased estimates are frequently obtained by dividing by the degrees of freedom, rather than by the total number of cases.

Efficiency The *efficiency* of an estimate refers to the *degree to which the sampling distribution is clustered about the true value of the parameter.* If the estimate is unbiased, this clustering can be measured by means of the standard error of the estimate: the smaller the standard error, the greater the efficiency of the estimate. Efficiency is always relative. No estimate can be completely efficient since this would imply no sampling error whatsoever. We can compare two estimates and say that one is more efficient than the other, however. Suppose, for example, that we have a normal population. Then for random sampling the standard error of the mean is $\sigma/\sqrt{N}$. If the sample median were used to estimate the population mean, the standard error of the median turns out to be approximately $1.253\sigma/\sqrt{N}$ for random samples.[1] Therefore, since the standard error of the mean is smaller than that of the median, the mean is the more efficient estimate. This, of course, is the reason the sample mean is usually used in preference to the median even where, as in the case of a normal population, the population mean and median are identical. We say that the mean is less subject to sampling fluctuations, i.e., it is more efficient.[2]

Of the two criteria we have discussed, efficiency is the more important.

[1] Here the population mean and median would be identical.

[2] It is not always true that the mean is the more efficient estimate, although for most populations, especially if departure from normality is not too great, this will be the case. Notice that the question of relative efficiency is an entirely different question from that of which measure is the more appropriate descriptive measure of central tendency. The latter concerns only the problem of finding the best single measure to represent the *sample* data.

If two estimates have the same efficiency, we would of course select the one that is least biased. This is why $\hat{\sigma}$ is used in preference to s. But an efficient estimate that is slightly biased would be preferable to an unbiased estimate that is less efficient. A simple diagram should help you see why this is the case. In Fig. 12.1 the peaked curve with a slight bias would be preferable since, even though in the long run we would tend to underestimate the parameter by a slight amount, we are more likely on any given trial to obtain a sample estimate that is fairly close to the parameter.

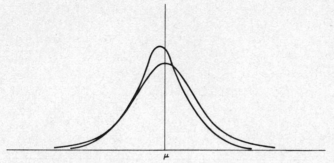

Figure 12.1 Comparison of sampling distribution of a biased estimate with high efficiency and an unbiased estimate with lower efficiency.

Knowledge that estimates will average out to the correct figure in the long run is little consolation if, for any given sample, the estimate is likely to be quite far from the parameter.

12.2 Interval Estimation

You may recall that when you took elementary physics, you were taught to weigh a block of wood several times and then to take the mean value and indicate a possible range of error. Thus you might have indicated that the weight of the wood was 102 ± 2 grams, meaning that you would estimate that the true weight was somewhere between 100 and 104 grams. In so doing, you were admitting the fallibility of the measurement procedure and indicating just about how much faith you had in the accuracy obtained. Although at the time it may not have been brought explicitly to your attention, you would also have admitted that you were not absolutely certain that the true value was actually in the interval obtained. If it were made wider, however, you would have been even more sure that it was in the interval. Thus you might have been almost positive that the true value was between 98 and 106 grams, and willing to bet your last dollar that it was between 2 and 202 grams. In obtaining interval estimates for parameters, we do essentially the same thing as does the physicist, except that we shall be in a position to evaluate the exact probability of error.

The actual procedure used in obtaining an interval estimate, or what is referred to as a *confidence interval*, is very simple and involves no really new basic ideas. We shall first merely state how the interval is obtained, and then we can examine why it is constructed in this manner. One first decides on the risk he is willing to take of making the error of stating that the parameter is somewhere in the interval when in fact it is not. Let us say that he decides he is willing to be wrong .05 of the time or that he uses what is referred to as the 95 per cent confidence interval.[3] The interval is obtained by going out in both directions from the point estimate (e.g., the sample mean) a certain multiple of standard errors corresponding to the confidence level selected. Thus, to estimate the population mean μ we obtain an interval as follows (using the 95 per cent level):

$$\bar{X} \pm 1.96\sigma_{\bar{X}} = \bar{X} \pm 1.96 \frac{\sigma}{\sqrt{N}}$$

where the value 1.96 corresponds to the critical region for the normal curve, using the .05 level and a two-tailed test. If $\bar{X} = 15$, $\sigma = 5$, and $N = 100$, the confidence interval would be

$$15 \pm 1.96 \frac{5}{\sqrt{100}} = 15 \pm 0.98$$

In other words, the interval would run from 14.02 to 15.98.[4]

In order to interpret intervals obtained by this method, we need to return to what we know about sampling distributions, in this case the sampling distribution of the mean. Let us suppose we have a normal sampling distribution with mean μ and standard deviation $\sigma/\sqrt{N}$. For our purposes there are two kinds of sample means: (1) those that do not fall within the critical region, and (2) those that do fall within this

[3] Notice that we refer in the case of confidence intervals to one minus the probability of error. This indicates that we have "confidence" of being correct, say, 95 per cent of the time.

[4] These end points of the interval are referred to as *confidence limits*.

area. Let us first suppose that we have obtained an $\bar{X}$ ($\bar{X}_1$ in Fig. 12.2) that does not fall within the critical region. We know that such an $\bar{X}$ must be within $1.96\sigma_{\bar{X}}$ of μ. If we place an interval on either side of this $\bar{X}$ by going distances equal to $1.96\sigma_{\bar{X}}$, we therefore must cross past μ, the mean of the sampling distribution, no matter whether $\bar{X}$ is to the right or left of μ. Similarly, if the $\bar{X}$ obtained lies within the critical region (see $\bar{X}_2$ in Fig. 12.2), then this $\bar{X}$ will be farther than 1.96 standard errors from μ and the confidence interval will not extend as far as μ.

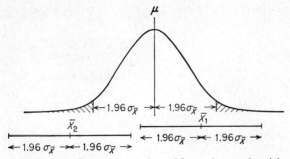

Figure 12.2 Comparison of confidence intervals with the sampling distribution of the mean, showing why 95 per cent confidence intervals include μ 95 per cent of the time.

But we also know that 95 per cent of the time we shall obtain $\bar{X}$'s that do not lie in the critical region whereas only 5 per cent of the time will they fall within this area. In other words, we know that *only 5 per cent of the time would we get intervals by this procedure that would not include the parameter* (for example, μ). The remaining 95 per cent of the time the procedure will yield sample means close enough to the parameter that the confidence intervals obtained will actually include the parameter.

Several words of caution are necessary in interpreting confidence intervals. The beginning student is likely to use vague phrases such as, "I am 95 per cent confident that the interval contains the parameter," or "the probability is .95 that the parameter is in the interval." In so doing he may not clearly recognize that the parameter is a fixed value and that it is the intervals that vary from sample to sample. According to our definition of probability, the probability of the parameter being in any given interval is either zero or one since the parameter either is or is not within the specific interval obtained. A simple diagram indicating the fixed value of the parameter, in this case μ, and the variability of the intervals may help you understand the correct interpretation more clearly. Figure 12.3 emphasizes that one's faith is in the procedure used rather than in any particular interval. We can say that *the procedure is such that in the long run 95 per cent of the intervals obtained will include the true (fixed) parameter.* You should be careful not to imply or assume that the particular interval you have obtained has any special property not possessed by comparable intervals that would be obtained from other samples. Sometimes it is stated that if repeated samples were

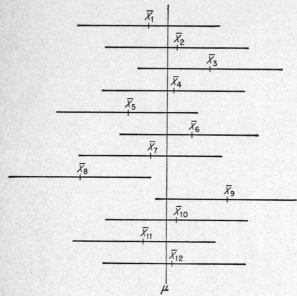

Figure 12.3 Distribution of variable confidence intervals about fixed value of parameter μ.

drawn, 95 per cent of the means from these samples would fall within the confidence interval one has actually computed (for example, 15 ± 0.98). This implies, of course, that the $\bar{X}$ obtained in the researcher's sample exactly equals μ or at least that it is a very close approximation to μ. Actually, the particular interval obtained may be so far out of line that very few $\bar{X}$'s would fall within it. As is always the case in statistical inference, our confidence is not in any particular sample result but in the procedure used.

It is possible to set the risk of error at any desired level by using the proper multiple of the standard error. You should notice, however, that in reducing the risk of error, one necessarily increases the width of the interval unless he simultaneously increases the number of cases. The wider the interval, the less he is really saying about the parameter. To say that the median income of New York families is somewhere between $1,000 and $25,000 is to claim the obvious. Thus, the researcher is faced with a dilemma. He can state that the parameter lies within a very narrow interval, but the probability of error will be large. On the other hand, he can make a very weak statement and be virtually certain of being correct. Exactly what he chooses to do will depend upon the nature of the situation. Although 95 and 99 per cent confidence intervals are conventionally used, it should be emphasized that there is nothing sacred about these levels.

Confidence intervals and tests of hypotheses Although the explicit purpose of placing a confidence interval about an estimate is to indicate the degree of accuracy of the estimate, confidence intervals are also implicit tests of a whole range of hypotheses.[5] They are implicit tests in the sense that specific hypotheses are not actually stated but only implied. In a confidence interval we have an implicit test for every possible value of μ that might be hypothesized. Figure 12.4 indicates how confidence intervals are related to tests of hypotheses.

Focus on the confidence interval drawn about $\bar{X}$. Suppose that, instead of having obtained such an interval, we had hypothesized several alternative values for μ and proceeded to test these hypotheses. Assume, for simplicity, that the value of σ was given and that the .05 significance level and a two-tailed test were used. First suppose that we had hypothesized a value such as μ_1 (Fig. 12.4a) actually falling within this confidence interval. Then the sample mean $\bar{X}$ would clearly not fall within the critical region, and the hypothesis would not have been rejected at the .05 level. On the other hand, had we hypothesized a value outside the interval such as μ_2 (Fig. 12.4b), the distance between the hypothesized μ_2 and $\bar{X}$ would be greater than $1.96\sigma_{\bar{x}}$, and this second hypothesis would have been rejected. Clearly, then, if we were to hypothesize values for μ that fell anywhere within the confidence interval, we would not reject these hypotheses at the appropriate level of significance. Were we to hypothesize values for μ falling outside the interval, we know that these hypotheses would be rejected.

Thus, having obtained a confidence interval, we can tell at a glance what the results would have been had we tested any particular hypothesis. If the nature of our problem were such that no particular hypothesis were

[5] It should be emphasized that although interval estimation and hypothesis testing involve closely interrelated ideas, they are distinct procedures.

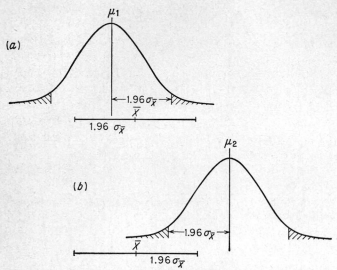

Figure 12.4 Comparison of 95 per cent confidence interval with tests of hypotheses at the .05 level, showing nonrejection of hypothesized mean₁ lying within the interval but rejection of hypothesized mean₂ lying outside the interval.

suggested as being preferable to the others, then obviously the practical alternative to a whole series of tests would be to obtain a single confidence interval.[6] You should satisfy yourself that examples taken up in the previous chapter could just as easily have been handled by the confidence-interval method.

Assumptions made for confidence intervals   The use of confidence intervals does not free us from the necessity of making assumptions about the nature of the population and the sampling method used. Basically, the assumptions for a confidence-interval problem are the same as those required for any tests that are being implicitly made except that it is of course unnecessary to hypothesize a particular value for the parameter being estimated. Random sampling will always be assumed in this text. Also, if a normal sampling distribution is being used, we must either assume a normal population or have a sufficiently large sample. If the t distribution or some other sampling distribution is to be used, the usual assumptions required in comparable tests would have to be made.

[6] However, it should be noted that when we test a given null hypothesis, we obtain a specific probability value, such as $P = .032$, which we would not ordinarily obtain in connection with a confidence interval.

12.3   Confidence Intervals for Other Types of Problems

The discussion of confidence intervals has thus far involved only instances where the parameter being estimated is a population mean and σ is known. If the problem is altered, procedural modifications are completely straightforward, and the basic interpretation of confidence intervals and their relationship to tests of hypotheses remains the same. A confidence interval for a parameter is always obtained by taking the estimate of the parameter and enclosing it in an interval the length of which is a function of the standard error of the estimate.[7]

If the t distribution must be used because σ is unknown, we simply make use of the estimate of the standard error and replace the multiple obtained using the normal table by the comparable figure in the t table. Thus, for a 99 per cent confidence interval for the mean and 24 degrees of freedom we would get

$$\bar{X} \pm 2.797 \hat{\sigma}_{\bar{X}} = \bar{X} \pm 2.797 \frac{s}{\sqrt{N-1}}$$

Had the example in Sec. 11.3 of the previous chapter been worked by means of a 99 per cent confidence interval, the result would have been

$$52 \pm 2.797 \left(\frac{12}{\sqrt{24}} \right) = 52 \pm 6.85$$

Therefore the 99 per cent confidence interval goes from 45.15 to 58.85. We see that the above result is consistent with that previously obtained (that is, $.001 < p < .01$) in that the hypothesized μ of 60 is actually outside the interval computed, and therefore we know that the hypothesis would have been rejected at the .01 level (for a two-tailed test).

Similarly, we can obtain confidence intervals for proportions. Replacing $\bar{X}$ by p_s and $\hat{\sigma}/\sqrt{N}$ by $\sqrt{p_u q_u / N}$, the 95 per cent confidence interval would be

$$p_s \pm 1.96 \sqrt{\frac{p_u q_u}{N}}$$

We encounter a difficulty here that did not occur when a particular value for p_u could be hypothesized. Since p_u will obviously not be known,

---

[7] In some instances, however, such as in the case of confidence intervals for correlation coefficients, the point estimate may not be exactly at the center of the interval.

it becomes necessary to estimate the standard error. Two simple procedures, one of which is more conservative than the other, can be recommended.[8] First, since the sample size must be large in order to justify the use of normal tables, p_s will ordinarily be a reasonably good estimate of p_u. Therefore if we simply substitute p_s for p_u (and q_s for q_u), we can obtain an interval that will usually be a close approximation to the correct one. Thus, in the example in Sec. 11.4 of the previous chapter, we would have obtained the 98 per cent confidence interval as follows:

$$p_s \pm 2.33 \sqrt{\frac{p_s q_s}{N}} = .55 \pm 2.33 \sqrt{\frac{(.55)(.45)}{125}} = .55 \pm 0.1037$$

If one objects to the use of an estimate of the standard error without in some manner correcting for the additional sampling error thereby introduced, he may prefer a more conservative method of obtaining the interval. Since the product pq reaches a maximum value when $p = q = .5$, it follows that the widest possible confidence interval will be obtained when the value .5 is used as an estimate of p_u.[9] Since a narrow interval is ordinarily desired, we are being conservative by obtaining an interval that is as large as it could possibly be regardless of the value of p_u. Using this more conservative method, we obtain a somewhat different interval

$$.55 \pm 2.33 \sqrt{\frac{(.5)(.5)}{125}} = .55 \pm 0.1042$$

Notice that this second interval is only slightly wider than the first. Whenever $.3 \leq p \leq .7$ the two methods will yield approximately the same results.

*If p_s turns out to be either very large or very small, the conservative method may give an interval that is far too wide. If one hesitates to use the first method in which p_u is estimated by p_s, it is possible to combine the two methods in order to obtain a more reasonable interval which is nevertheless a conservative one. We first make use of the more conservative method to obtain an approximate confidence interval. Suppose this interval runs from .10 to .25 with p_s being .175. We would then be reasonably sure that the actual value of p_u is somewhere within this approximate (and conservative) interval. In computing the more

[8] For a third and somewhat more accurate procedure see [2], p. 231.
[9] You should convince yourself that this is true.

exact interval, we now take as our estimate of p_u the value within the approximate interval that is closest to .5. In the above numerical example, we would select the value .25 since the use of this value in the formula for the standard error will result in a wider interval than will any other value within the interval .10 to .25. In other words, instead of using our actual p_s (that is, .175), we select the largest value that we think p_u is likely to have. We therefore compute the 95 per cent confidence interval as follows:

$$.175 \pm 1.96 \sqrt{\frac{(.25)(.75)}{N}}$$

The above interval will be wider and therefore more conservative than that obtained by using p_s under the radical, and yet it does not involve the use of the value .5, which we suspect is far too large.

12.4 Determining the Sample Size

In keeping with the practice of introducing a few new ideas at a time, we have postponed the question of how it is that one's sample size can be determined in advance of data collection. One of the most frequent questions asked of the statistician is, "How many cases do I need?" The answer depends, of course, on what one wishes to do with the sample results. More specifically, there are several facts which must be determined before an adequate answer can be given. Generally, what we must do is to work backwards from the data we expect to get in order to determine the unknown sample size. Thus far we have been taking the sample size as a known quantity. Statistics such as the sample mean and standard deviation can be obtained from the sample results. Once we have decided on the significance level of a test or the desired confidence level, we can then put all of these values into a formula and determine the width of the confidence interval or whether or not to reject a null hypothesis. In the kind of problem we are considering in this section, however, the sample size will be unknown. This means that in order to solve our equation for N we must know each of the other quantities in the formula. Once we have put these values into the equation, solving for N becomes a straightforward problem of algebra. We shall make use of a confidence-interval problem in order to illustrate the process.

Suppose we wish to know how many cases will be required to estimate the mean number of years of schooling completed by persons of foreign-born parentage. Before we can determine the answer to this question, we shall need to have the following pieces of information: (1) the confidence level to be used, (2) the degree of accuracy within which we wish to estimate the parameter, and (3) some reasonable estimate of the values of any parameters that may appear in the formula. For example, we may wish to estimate the mean to within an accuracy of $\pm.1$ year of schooling and to make use of a 95 per cent confidence interval. Notice that both of these quantities must be specified since, for example, we can always obtain accuracy to within $\pm.1$ year if we permit ourselves a high risk of error. We now make use of these values in the formula for a confidence interval

$$\bar{X} \pm \underbrace{1.96 \frac{\sigma}{\sqrt{N}}}_{.1}$$

Knowledge of the level of confidence desired has enabled us to insert the value 1.96. Since we wish accuracy of $\pm.1$, or a total interval width of .2, we know that the quantity $1.96\sigma/\sqrt{N}$ must equal .1. Although the value of $\bar{X}$ will be unknown, we see immediately that it is irrelevant in this problem since we wish to obtain an interval of a certain width regardless of the value of $\bar{X}$.

Suppose we now try to solve the equation

$$.1 = 1.96 \frac{\sigma}{\sqrt{N}}$$

for N. There is still one unknown quantity, the value of σ. But how can we obtain σ before the data have been gathered? Clearly, its value must be estimated by some method which in a sense goes beyond the data we shall collect. Essentially, we have to make an enlightened guess as to its value by using either expert knowledge, the results of previous studies, or conceivably a pilot study of some sort. Usually a pilot study will be too expensive and reliance must be placed on either of the remaining methods. Admittedly, the most satisfactory procedure would be to determine σ exactly, but if this could be done, there would probably be no point in drawing a sample at all. Notice that the type of estimation necessary in this kind of problem is completely different from that used in estimating σ from *sample* data. Therefore there is no point in estimating σ with $\hat{\sigma}$ or in using the t distribution. If we are going to guess anyway, we might as well guess the value of σ rather than that of $\hat{\sigma}$ or s. In this example suppose that on the basis of the best available information we estimate that σ will be approximately 2.5 years. Making use of this value and solving for the required sample size we get

$$.1 = 1.96 \frac{2.5}{\sqrt{N}}$$

or
$$\sqrt{N} = \frac{1.96(2.5)}{.1} = 49$$

and
$$N = 2,401$$

Notice that we solved for N by placing all quantities except $\sqrt{N}$ on one side of the equation and then simplifying. Finally, we squared both sides of the equation in order to get rid of the radical.

To be sure, we can only obtain an approximate value for the desired sample size since parameters will have to be estimated. There would certainly be no point in taking exactly 2,401 cases, for example. Nevertheless, such an approximation will usually give far better results than some intuitive hunch about the number of cases necessary. In practical applications we usually study more than one variable at a time, a fact that may complicate the picture considerably. We are also ordinarily limited by available funds and often have to settle for whatever degree of accuracy we can get. Even so, it will often be helpful to compute the needed sample size as a guide to one's research design.

Although the question of determining sample size will not be discussed in subsequent chapters in connection with other statistical procedures, you will find several exercises that require you to estimate N for other kinds of problems. In all cases, extensions are straightforward, although the algebra may at times become fairly involved.

Glossary
Confidence interval
Degrees of freedom
Efficiency of an estimate
Interval estimate
Point estimate
Unbiased estimate

Exercises
1. Obtain confidence intervals for Exercises 3, 4, and 5 of Chap. 11. Are your results consistent with those of these previous exercises? How do you know? (*Ans.* for Exercise 5, .47–.61)

2. You take a random sample of 200 families within a community and find that in 36 per cent of these families the husband makes over half of the financial decisions.

What is the 99 per cent confidence interval for the percentage of families in which the husband makes over half of such decisions? In what specific sense does your interval give you implicit tests of hypotheses?

3. How many cases will you need to establish a 99.9 per cent confidence interval for the mean, if the total width of the confidence interval is to be no more than $500 and if the standard deviation is estimated to be $1,300? (*Ans. N* = 295)

4. If you suspect that the proportion of homeowners in a certain residential community is approximately .75, how many cases will you need in order to obtain a 95 per cent confidence interval that will be no wider than .03 (total width) when you express the width in terms of proportions? Suppose the proportion of homeowners were estimated to be .5. How many cases would then be needed?

5. Using the fact that for normal populations the sampling distribution of the median has an approximate standard error of $1.253\sigma/\sqrt{N}$, we can place a confidence interval about the sample median. Suppose in Exercise 3 above that you wished to place an interval of the same width about the sample median. Using the same estimate of the standard deviation, how many cases would you need? What does your result show about the relative efficiencies of the mean and median? (*Ans. N* = 463)

*6. It has been argued that a 95 per cent confidence interval represents a series of implicit *two-tailed* tests at the .05 level. Explain why a 95 per cent confidence interval does not represent implicit *one-tailed* tests at the .05 level.

References
1. Freund, J. E.: *Modern Elementary Statistics*, 3d ed., Prentice-Hall, Inc., Englewood Cliffs, N.J., 1967, chaps. 9 and 11.
2. Hagood, M. J., and D. O. Price: *Statistics for Sociologists*, Henry Holt and Company, Inc., New York, 1952, chaps. 15 and 16.
3. Hays, W. L.: *Statistics*, Holt, Rinehart and Winston, Inc., New York, 1963, chaps. 7 and 9.
4. Wallis, W. A., and H. V. Roberts: *Statistics: A New Approach*, The Free Press of Glencoe, Ill., Chicago, 1956, chap. 14.

Bivariate and Multivariate Statistics

Part 4

Two-sample Tests: Difference of Means and Proportions   13

In Chap. 11 tests involving a single sample were discussed. It was found that such tests were not very practical to the social scientist because it is not usually possible to find an hypothesis that is specific enough to predict a value for μ or p_u. When interest is focused on *comparisons* between several categories or samples, however, it becomes unnecessary to specify the absolute levels for either group. Instead, one can simply test the null hypothesis that there are no differences between them. For example, it would be extremely difficult to predict the income level of blacks in Detroit or the prejudice level of whites in that city. But suppose one were interested in testing the hypothesis that the average income of blacks is the same as that of foreign-born whites or that Jews and gentiles have the same amount of prejudice toward blacks. It is this latter type of hypothesis that will be considered in this chapter.

In a social science such as sociology, interest is likely to be focused on establishing *relationships* between variables. This is in contrast to the fact-finding type of survey in which, as we have seen, point and interval estimation of a parameter for a single variable may be of primary concern. Whenever comparisons are made between two samples, we have the most simple kind of problem in which two variables can be related. Up to this point we have been concerned with only one variable at a time. This is perhaps the main reason why the tests discussed so far have not been too useful to sociologists. In this chapter we shall be taking up tests in which a simple dichotomized variable can be related to a second variable. For example, in comparing Jews and gentiles with respect to prejudice, we are relating religion to prejudice. Similarly, one might wish to compare the two sexes with respect to "other directedness" or various other personality characteristics. Comparisons may also be made between a control group and an experimental group into which some variable has been introduced. In subsequent chapters tests involving more than two samples will be discussed.

13.1 Difference-of-means Test

In order to extend the single-sample means test to a test in which a comparison can be made between the means of two samples, we must again make use of the central-limit theorem. An important derived theorem can be stated as follows: *If independent random samples of sizes N_1 and N_2, respectively, are drawn from populations that are $Nor(\mu_1,\sigma_1{}^2)$ and $Nor(\mu_2,\sigma_2{}^2)$, respectively, then the sampling distribution of the difference between the two sample means $(\bar{X}_1 - \bar{X}_2)$ will be $Nor(\mu_1 - \mu_2, \sigma_1{}^2/N_1 + \sigma_2{}^2/N_2)$.* As was true in the case of single samples, this theorem can be generalized in the case of large samples to cover any populations with means of μ_1 and μ_2 and variances $\sigma_1{}^2$ and $\sigma_2{}^2$ respectively. As N_1 and N_2 become large, the sampling distribution of $\bar{X}_1 - \bar{X}_2$ approaches normality as before. Let us now examine the above theorem more closely.

Reference is made to independent random samples. This means that samples must be selected *independently of each other*. The fact that a sample is random assures independence *within* the sample in the sense that knowledge of the score of the first individual selected does not help us predict the score of the second. This is *not* what is meant by the phrase "independent random samples." Not only must there be independence within each sample (assured by randomness), but there must also be independence *between* samples. For example, the samples cannot be matched, as might be the case in comparisons between control and experimental groups. If the two sexes were being compared, one could not use the difference-of-means test on samples composed of husband and wife pairs.

This requirement that samples be independent of each other is extremely important and might be overlooked in applied research, particularly when one is dealing with a cluster sample. If the overall sample is strictly random, and if one is comparing two subsamples selected from a single larger sample, this assumption of independence between samples will automatically be met since all cases in the total sample will have been selected independently of each other. For example, if one compares males and females, then he will also have a random sample of all males, and an independently selected sample of all females. That is, the selection of Bob Jones has no bearing whatsoever on the probability of Susie Smith's being selected. Usually in social research we draw a single larger sample, although for purposes of analysis we may conceptualize the data as having come from several distinct and independent samples. In most such instances the problem of the lack of independence between samples does not arise unless we have deliberately matched the samples. But since there may be circumstances under which the sampling design is not so simple, you should be alerted to the possibility that the assumption of independence between samples may not be met.

In the above theorem we are told that if we were to continue sampling indefinitely, each time selecting two samples and plotting the difference between their means, the sampling distribution of this *difference* between means would be normal or approximately normal. You should attempt to picture exactly what is occurring here. Keep in mind that as a social scientist you will actually obtain only two samples and a single difference, but that here we are dealing with a hypothetical distribution of all possible differences. Since the sampling distribution is for a *difference* between sample means, the mean of the sampling distribution is given by the difference between two population means rather than either of them separately. In the special case where μ_1 and μ_2 are equal, the mean of the sampling distribution will be zero. If $\mu_1 > \mu_2$, we expect that most $\bar{X}_1$'s will be larger than the comparable $\bar{X}_2$'s, and the mean of the sampling distribution will therefore be positive. For example, if $\mu_1 = 60$ and $\mu_2 = 40$, the distribution of $\bar{X}_1 - \bar{X}_2$ will have 20 as its mean or expected value.

It is not quite so easy to see why the variance should be $\sigma_1^2/N_1 + \sigma_2^2/N_2$, or the *sum* of the variances of the sampling distributions for the separate means. Obviously, a difference of variances $\sigma_1^2/N_1 - \sigma_2^2/N_2$ could not be used since it would then be possible to obtain a zero or negative variance for the sampling distribution. But the variance $\sigma_1^2/N_1 + \sigma_2^2/N_2$ is *larger* than either of the variances σ_1^2/N_1 or σ_2^2/N_2. Why should this be the case? Although a complete justification for the formula cannot be given without resorting to mathematical reasoning, some sort of intuitive explanation is possible. In essence, we expect the standard error for the difference of means to be larger than either of the separate standard errors because there are now two sources of error, one in each sample. Thus about half the time the two $\bar{X}$'s will be in error in opposite directions. For simplicity let us assume that $\mu_1 = \mu_2$. Then if $\bar{X}_1$ is larger than μ_1 and $\bar{X}_2$ smaller than μ_2, a large positive number will result on subtraction because the errors are in opposite directions. For example, if $\bar{X}_1$ is 20 greater than μ_1, whereas $\bar{X}_2$ is 15 less than μ_2, the resulting difference $\bar{X}_1 - \bar{X}_2$ will differ from $\mu_1 - \mu_2$ by 35, thus compounding the errors involved. Similarly if $\bar{X}_1$ is small and $\bar{X}_2$ large, a substantial negative difference may occur. In other words, we shall get relatively large differences between sample means quite frequently since each mean will vary independently of the other. Therefore the sampling distribution of a difference will have a larger standard deviation than either of the separate sampling distributions.

*The formulas for the expected value and variance of $\bar{X}_1 - \bar{X}_2$ can be derived by once more using the expressions for linear combinations. You will recall that if $Y = c_1 X_1 + c_2 X_2$ then $E(Y) = c_1 E(X_1) + c_2 E(X_2)$ and $\sigma_Y^2 = c_1^2 \sigma_{X_1}^2 + c_2^2 \sigma_{X_2}^2$, provided that X_1 and X_2 are independent. If we now let Y represent a difference of *means*, replacing X_1 by $\bar{X}_1$ and X_2 by $\bar{X}_2$ and setting $c_1 = 1$ and $c_2 = -1$, then we have as a special case the results

$$E(Y) = E(\bar{X}_1 - \bar{X}_2) = (1)E(\bar{X}_1) + (-1)E(\bar{X}_2) = \mu_1 - \mu_2$$

and
$$\sigma_Y^2 = (1)^2 \sigma_{\bar{X}_1}^2 + (-1)^2 \sigma_{\bar{X}_2}^2 = \frac{\sigma_1^2}{N_1} + \frac{\sigma_2^2}{N_2}$$

Notice that if we had formed the *sum* of $\bar{X}_1$ and $\bar{X}_2$, the expression for the variance of this quantity would have been the same as that for their difference. In Chap. 16 we shall study more complex kinds of comparisons that involve a generalization of this simple comparison of two sample means.

We shall now take up an example illustrating the use of the difference-of-means test. The case in which the σ's are known will not be treated since this problem is straightforward and rather impractical. Instead, it will be presumed that the σ's are unknown. Two special cases will be considered. In model 1 it will be assumed that $\sigma_1 = \sigma_2$; in model 2 the two σ's will be assumed to be unequal. Obviously, these two models cover all possible alternatives.

PROBLEM   A comparison is made between two types of counties, those which are predominantly urban and those which are primarily rural. The counties are compared with respect to the percentage of persons voting Democratic in a presidential election with the following results:

Urban counties	*Rural counties*
$N_1 = 33$	$N_2 = 19$
$\bar{X}_1 = 57\%$	$\bar{X}_2 = 52\%$
$s_1 = 11\%$	$s_2 = 14\%$

Do the data give reasonable grounds for concluding that there is a significant difference in voting preference between these two types of counties? Assume that the counties have been randomly selected from a list of all counties in the Far West and that previous studies have shown that the respective population distributions are approximately normal.

Model 1: $\sigma_1 = \sigma_2$

1. *Assumptions*

Level of Measurement: Percentage of Democratic vote an interval scale
Model: Independent random samples
   Normal populations, $\sigma_1 = \sigma_2 = \sigma$
Hypothesis: $\mu_1 = \mu_2$

The normality assumption can be relaxed whenever the N's are large (e.g., both over 50). The assumption that $\sigma_1 = \sigma_2$ actually can be tested separately by means of the F test, which will be discussed in Chap. 16. This test involves a comparison of the two sample standard deviations. If s_1 and s_2 do not differ markedly, one cannot reject the hypothesis that $\sigma_1 = \sigma_2$. If the assumption of equal standard deviations appears to be reasonable according to the results of the F test, it will be more efficient to take advantage of it in estimating the common value of σ. Given the assumption that both populations are normal, the additional assumptions of equal means and equal standard deviations amount to postulating that the two populations are identical.

Since we are interested in seeing whether or not there is a difference between the two types of counties, our null hypothesis will be that there is no difference. Presumably, we suspect that there will actually be a difference and therefore set up an hypothesis that we wish to reject. In this instance, we can legitimately refer to the hypothesis as being a "null" hypothesis indicating no relationship between the variables "type of county" and "voting preference." It is conceivable that we might have been in a position to specify that the difference between population means could be expected to be some constant other than zero. For example, the hypothesis might have taken the form $\mu_1 - \mu_2 = 10$ if it had been predicted that the Democratic vote would be 10 per cent higher in urban counties. In the social sciences we are seldom in a position to be this specific, however.

2. *Sampling Distribution*
The t distribution will be used since the σ's are unknown and since the total number of cases is much less than 120.

3. *Significance Level and Critical Region*
Let us select the .01 level and a two-tailed test.

4. *Computing the Test Statistic* It will be remembered that t is computed by taking the difference between the obtained sample value and the mean of the sampling distribution and then dividing by an estimate of the standard error of the sampling distribution. We are here concerned with the *difference* between sample means $\bar{X}_1 - \bar{X}_2$. Since the mean of the sampling distribution is $\mu_1 - \mu_2$, we obtain the following expression for t:

$$ t = \frac{(\bar{X}_1 - \bar{X}_2) - (\mu_1 - \mu_2)}{\hat{\sigma}_{\bar{X}_1 - \bar{X}_2}} \tag{13.1} $$

where $\hat{\sigma}_{\bar{X}_1 - \bar{X}_2}$ is an estimate of the standard error of the difference between sample means. Since under the null hypothesis it has been assumed that $\mu_1 = \mu_2$, the expression for t in this *special case* reduces to

$$ t = \frac{\bar{X}_1 - \bar{X}_2}{\hat{\sigma}_{\bar{X}_1 - \bar{X}_2}} $$

The resemblance between the above numerator and the one used in the single-sample test is more or less coincidental, a result of the fact that the μ's dropped out under the null hypothesis. You should *not* draw the conclusion that the μ in the first type of problem has simply been replaced by the *sample* mean of the second sample. Actually, the expression $(\bar{X}_1 - \bar{X}_2)$ has replaced $\bar{X}$; $(\mu_1 - \mu_2)$ has replaced μ; and $\hat{\sigma}_{\bar{X}_1 - \bar{X}_2}$ has replaced $\hat{\sigma}_{\bar{X}}$.

It now remains to evaluate $\hat{\sigma}_{\bar{X}_1 - \bar{X}_2}$. We know, of course, that

$$ \sigma_{\bar{X}_1 - \bar{X}_2} = \sqrt{\frac{\sigma_1^2}{N_1} + \frac{\sigma_2^2}{N_2}} $$

Since in this case $\sigma_1 = \sigma_2$ we can indicate their common value as σ, take it out from under the radical, and simplify the expression for $\sigma_{\bar{X}_1 - \bar{X}_2}$ as follows:

$$ \sigma_{\bar{X}_1 - \bar{X}_2} = \sqrt{\frac{\sigma^2}{N_1} + \frac{\sigma^2}{N_2}} = \sigma \sqrt{\frac{1}{N_1} + \frac{1}{N_2}} = \sigma \sqrt{\frac{N_1 + N_2}{N_1 N_2}} \tag{13.2} $$

The common variance σ^2 can now be estimated by obtaining a *pooled estimate* from both samples. Since the two sample variances will ordinarily be based on different numbers of cases, we can obtain an estimate of σ^2 by taking a weighted average of the sample variances, being careful to divide by the proper degrees of freedom in order to obtain an unbiased estimate. Taking the square root we get an estimate $\hat{\sigma}$ as follows:

$$\hat{\sigma} = \sqrt{\frac{N_1 s_1{}^2 + N_2 s_2{}^2}{N_1 + N_2 - 2}} \qquad (13.3)$$

Since $N_1 s_1{}^2 = \sum_{i=1}^{N_1} (X_{i1} - \bar{X}_1)^2$ we may replace $N_1 s_1{}^2$ by $\sum_{i=1}^{N_1} x_{i1}{}^2$, where $x_{i1} = X_{i1} - \bar{X}_1$. If we do the same for $N_2 s_2{}^2$ we get

$$\hat{\sigma} = \sqrt{\frac{\sum_{i=1}^{N_1} x_{i1}{}^2 + \sum_{i=1}^{N_2} x_{i2}{}^2}{N_1 + N_2 - 2}}$$

Thus if we take the sum of squares about the mean of the first sample and add to this the sum of the squared deviations about the second sample mean, finally dividing by $N_1 + N_2 - 2$, we obtain a pooled estimate of the common variance.

Notice that the symbol $\hat{\sigma}$ is now being used to represent a different estimate from that discussed in the previous two chapters. The symbol "^" is often used in the statistical literature to indicate an unbiased estimate. Since we have lost 2 degrees of freedom, one each in computing s_1 and s_2 from $\bar{X}_1$ and $\bar{X}_2$, the total degrees of freedom becomes $N_1 + N_2 - 2$. We have used both samples in obtaining our estimate, having given greater weight to the variance from the larger of the two samples. Such a pooled estimate will be more efficient than estimates based on either sample alone. As a computational check, the numerical value of $\hat{\sigma}$ will ordinarily be between that of s_1 and s_2.

Finally, we obtain an estimate of $\sigma_{\bar{X}_1 - \bar{X}_2}$ by taking our estimate of σ and multiplying by $\sqrt{\dfrac{N_1 + N_2}{N_1 N_2}}$ as in Eq. (13.2) above. Thus

$$\hat{\sigma}_{\bar{X}_1 - \bar{X}_2} = \sqrt{\frac{N_1 s_1{}^2 + N_2 s_2{}^2}{N_1 + N_2 - 2}} \sqrt{\frac{N_1 + N_2}{N_1 N_2}} \qquad (13.4)$$

Notice that Eq. (13.4) differs from Eq. (13.2) in that the σ of Eq. (13.2) has been replaced by its estimate $\hat{\sigma}$, as defined in Eq. (13.3). At this point the formula seems formidable. You should review the algebraic steps discussed above in order to convince yourself that the formula is not as complicated as it first appears.

In our numerical example we obtain the following results:

$$\hat{\sigma}_{\bar{X}_1-\bar{X}_2} = \sqrt{\frac{33(121) + 19(196)}{33 + 19 - 2}} \sqrt{\frac{33 + 19}{33(19)}} = (12.42)(.288) = 3.58$$

Therefore,
$$t = \frac{(\bar{X}_1 - \bar{X}_2) - 0}{\hat{\sigma}_{\bar{X}_1-\bar{X}_2}} = \frac{57 - 52}{3.58} = 1.40$$

Notice that our estimate $\hat{\sigma} = 12.42$ lies between $s_1 = 11$ and $s_2 = 14$.

5. *Decision*   Since a pooled estimate of the common standard deviation was used, the degrees of freedom associated with t will be $N_1 + N_2 - 2$, or 50. We found that $t = 1.40$, the probability of which would be considerably greater than .01 if all the assumptions were correct. We therefore decide not to reject the null hypothesis at the .01 level, and we conclude that there is no significant difference in voting preferences in Far West urban and rural counties.

Model 2: $\sigma_1 \neq \sigma_2$ Now let us see what modifications are necessary when it is impossible to assume that the two populations have the same standard deviations. Presumably, we have tested and rejected the hypothesis that $\sigma_1 = \sigma_2$. It is now no longer possible to simplify the formula for $\sigma_{\bar{X}_1-\bar{X}_2}$ by introducing a common value for σ, nor is it possible to form a pooled estimate. In this instance we estimate the two (different) standard deviations separately. We estimate σ_1^2/N_1 with $s_1^2/(N_1 - 1)$ and σ_2^2/N_2 with $s_2^2/(N_2 - 1)$, obtaining

$$\hat{\sigma}_{\bar{X}_1-\bar{X}_2} = \sqrt{\frac{s_1^2}{N_1 - 1} + \frac{s_2^2}{N_2 - 1}} \tag{13.5}$$

In the example used above

$$\hat{\sigma}_{\bar{X}_1-\bar{X}_2} = \sqrt{121/32 + 196/18} = \sqrt{3.78 + 10.89} = \sqrt{14.67} = 3.83$$

Therefore,
$$t = \frac{57 - 52}{3.83} = 1.31$$

Thus, the results obtained in the two different models are not greatly different.

Although the procedure used in model 2 is both conceptually and computationally simpler, the estimate of $\sigma_{\bar{X}_1-\bar{X}_2}$ is not quite as efficient as the one previously obtained. Also, even if we assume normal populations, model 2 is somewhat questionable in instances where the N's are not too large or where the sample sizes are very different. The difficulty arises in the selection of the proper degrees of freedom. For example, if the first sample were unusually small, it would be highly misleading to use $N_1 + N_2 - 2$ as the degrees of freedom since s_1 would be a very poor estimate of σ_1 and since the value of $s_1^2/(N_1 - 1)$ would ordinarily be much larger than that of $s_2^2/(N_2 - 1)$. This is true because if the values of s_1^2 and s_2^2 are not too different, the relative sizes of the two fractions will be primarily determined by their denominators. It has been suggested that unless the N's are large the following expression be used to obtain an approximation to the correct degrees of freedom.

$$\text{df} = \frac{\left(\dfrac{s_1^2}{N_1 - 1} + \dfrac{s_2^2}{N_2 - 1}\right)^2}{\left(\dfrac{s_1^2}{N_1 - 1}\right)^2 \left(\dfrac{1}{N_1 + 1}\right) + \left(\dfrac{s_2^2}{N_2 - 1}\right)^2 \left(\dfrac{1}{N_2 + 1}\right)} - 2 \quad (13.6)$$

In the above example, therefore, we obtain

$$\text{df} = \frac{(14.67)^2}{(3.78)^2(\tfrac{1}{34}) + (10.89)^2(\tfrac{1}{20})} - 2 = 33.89 - 2 = 31.89 \simeq 32$$

Notice that all the quantities in the formula for degrees of freedom have already been computed. From the t table, using 32 degrees of freedom, we see that the null hypothesis should not be rejected at the .01 level.

As far as assumptions are concerned, the only difference between model 1 and model 2 is the assumption that $\sigma_1 = \sigma_2$. Notice that there is nothing in the second procedure which *requires* that the standard deviations be unequal. If they happen to be equal (or nearly so), the second model will simply be less efficient. It might seem as though the second procedure is generally preferable since it does not require the assumption that $\sigma_1 = \sigma_2$. But this second model, as we have seen, requires approximations for the degrees of freedom. For large samples, the two methods will usually yield similar results *if the standard deviations are in fact equal*, since both sample standard deviations will ordinarily be good estimates of the common σ.

If the σ's for both populations happen to be known, their values can be placed directly in the formula for $\sigma_{\bar{x}_1 - \bar{x}_2}$ since no estimation will be necessary. Z can then be computed and the normal table used. With known σ's there will of course be no need to distinguish between models

1 and 2. Needless to say, instances in which both σ's are known will be exceedingly rare in practical research.

13.2   Difference of Proportions

As was true in the case of tests involving single-sample proportions, a difference between two proportions can be treated as a special case of a difference between two means. If we are comparing two independent random samples with respect to the proportion of prejudiced persons, we can formulate the null hypothesis that the proportions, p_{u_1} and p_{u_2}, respectively, of prejudiced persons in the two populations are equal. Since it has already been shown in the case of proportions that $\sigma_1 = \sqrt{p_{u_1}q_{u_1}}$ and $\sigma_2 = \sqrt{p_{u_2}q_{u_2}}$, it follows that the two population standard deviations must also be equal. The following example therefore makes use of essentially the same procedures as used for the first model in the case of the difference-of-means test.

PROBLEM Suppose a comparison of recreational habits is made between assembly-line workers and persons whose work is nonrepetitive and not paced by the machine. Let us assume that the researcher suspects that assembly-line workers will be more inclined to select "passive" spectator-type forms of recreation. Among a random sample of 150 assembly-line workers at a given plant it is found that 57 per cent list passive forms of recreation as their favorites. In the second sample, also selected randomly, 46 per cent of the 120 workers list passive forms as favorites. Is there a significant difference at the .05 level?

1. *Assumptions*

   Level of Measurement: Type of recreation is a dichotomy
   Model: Independent random samples
   Hypothesis: $p_{u_1} = p_{u_2}$ (Implies $\sigma_1 = \sigma_2$)

2. *Sampling Distribution* Since the N's are both relatively large, the sampling distribution of the difference between proportions will be approximately normal with mean $p_{u_1} - p_{u_2} = 0$ and a standard deviation of

$$\sigma_{p_{s_1}-p_{s_2}} = \sqrt{\frac{\sigma_1{}^2}{N_1} + \frac{\sigma_2{}^2}{N_2}} = \sqrt{\frac{p_{u_1}q_{u_1}}{N_1} + \frac{p_{u_2}q_{u_2}}{N_2}} \qquad (13.7)$$

where q_{u_1} and q_{u_2} are equal to $1 - p_{u_1}$ and $1 - p_{u_2}$, respectively.[1]

[1] If samples are small, we use Fisher's exact test described in Chap. 15.

3. *Significance Level and Critical Region* The problem specifies that we are to use the .05 level. A one-tailed test is indicated since the direction of the difference is predicted ahead of time. Therefore any positive value of Z greater than 1.65 will indicate that results are so improbable under the assumptions that the null hypothesis may be rejected.

4. *Computing the Test Statistic* Since by hypothesis $p_{u_1} = p_{u_2}$ it follows that $\sigma_1 = \sigma_2 = \sigma$, and the special formula

$$\sigma_{p_{s_1}-p_{s_2}} = \sigma \sqrt{\frac{N_1 + N_2}{N_1 N_2}}$$

can be used. Previously in the single-sample test for proportions it was possible to avoid estimating σ since the actual value of p_u was hypothesized. This time, however, the hypothesis merely states that $p_{u_1} = p_{u_2}$ without specifying what the numerical value of either of these proportions actually is. For this reason we need a pooled estimate of the standard error. Rather than directly obtaining a weighted average of the two sample variances, which is what we did before, we can obtain a slightly better estimate by computing a pooled estimate ($\hat{p}_u$) of p_u. We then obtain $\hat{q}_u$ by subtraction. Since

$$\sigma = \sqrt{p_u q_u}$$
$$\hat{\sigma} = \sqrt{\hat{p}_u \hat{q}_u}$$

we can set

Thus

$$\hat{\sigma}_{p_{s_1}-p_{s_2}} = \hat{\sigma} \sqrt{\frac{N_1 + N_2}{N_1 N_2}} = \sqrt{\hat{p}_u \hat{q}_u} \sqrt{\frac{N_1 + N_2}{N_1 N_2}} \qquad (13.8)$$

In order to obtain $\hat{p}_u$, a weighted average of the sample proportions is taken as follows:

$$\hat{p}_u = \frac{N_1 p_{s_1} + N_2 p_{s_2}}{N_1 + N_2} \qquad (13.9)$$

Notice that the numerator of the above expression is simply the total *number* of individuals in both samples preferring passive forms of recreation. Therefore, in the case of our numerical example we get

$$\hat{p}_u = \frac{150(.57) + 120(.46)}{150 + 120} = .521$$

Therefore

$$\hat{q}_u = 1 - \hat{p}_u = .479$$

and $\qquad \hat{\sigma}_{p_{s_1} - p_{s_2}} = \sqrt{(.521)(.479)} \; \sqrt{\dfrac{150 + 120}{(150)(120)}}$

$$= (.4996)(.1225) = .0612$$

Hence $\qquad Z = \dfrac{(p_{s_1} - p_{s_2}) - 0}{\hat{\sigma}_{p_{s_1} - p_{s_2}}} = \dfrac{.57 - .46}{.0612} = 1.80$

5. *Decision* Since with a one-tailed test the probability of obtaining a value of Z as large or larger than 1.80 is .036 if the null hypothesis is actually correct, we may reject the null hypothesis at the .05 level. We conclude that there is a significant difference with respect to preferences for passive forms of recreation between the two types of employees at this plant.

It should be mentioned at this point that there are several alternative kinds of tests, the most important of which is the chi-square test to be discussed in Chap. 15, which can be used in place of the difference-of-proportions test. Since the use of the difference-of-proportions test is restricted to two samples and to a dichotomized variable, it is not as practical as the chi-square test, which can be employed on three or more samples as well. One advantage of the difference-of-proportions test, however, is that with suitable modifications it can be used when cluster or area samples have been selected. These modifications required for cluster samples are unfortunately beyond the scope of this text.

*Difference of differences of proportions We may easily extend the principle of a test for a difference of proportions (or means) to that of a difference of differences, or even a difference of differences of differences. For example, suppose we had data for both male and female workers and wished to compare the sexes with respect to the *relationships* between work assignments and recreational preferences. Perhaps we would find a difference such as that just illustrated in the case of the men, but none for the women. Or perhaps the direction of the difference might even be reversed for the sexes. Extending this illustration, we might want to add the age factor into the picture. Conceivably we might have one difference of differences (between men and women) for the younger workers but a different result for the older workers. It can be seen that we are anticipating problems that may arise when we are dealing with *more than two variables* and where the several variables may have rather peculiar combined effects. In such instances, we say that there is "interaction" among the variables, or that their joint effects are nonadditive. We shall have occasion to study these kinds of possibilities in greater detail in Chaps. 16 and 20.

In the very simple example where we wish to compare the differences of proportions between men and women, suppose that p_{u_1} and p_{u_2} represent the population proportions for men, as in the previous example. Then we will have two similar proportions p_{u_3} and p_{u_4} to represent the women, and we might make a similar test of the null hypothesis that, for the women, $p_{u_3} = p_{u_4}$. But we can also test the more complex hypothesis that the (population) differences for the two sexes are also identical. Our null hypothesis thus becomes

$$p_{u_1} - p_{u_2} = p_{u_3} - p_{u_4} \quad \text{or} \quad (p_{u_1} - p_{u_2}) - (p_{u_3} - p_{u_4}) = 0$$

Put differently, we are hypothesizing that the *relationship* between work assignment and recreational preferences (as measured by a difference of proportions) is the same for both sexes. An alternative hypothesis might be that the difference is greater for men than for women.

Once more we can use the principle of linear combinations by setting

$$Y = c_1 p_{s_1} + c_2 p_{s_2} + c_3 p_{s_3} + c_4 p_{s_4}$$

For the null hypothesis under consideration we set $c_1 = c_4 = 1$ and $c_2 = c_3 = -1$ and thus (assuming independently selected samples)

$$E(Y) = E(p_{s_1}) - E(p_{s_2}) - E(p_{s_3}) + E(p_{s_4}) = (p_{u_1} - p_{u_2}) - (p_{u_3} - p_{u_4})$$

and
$$\sigma_Y{}^2 = \frac{p_{u_1} q_{u_1}}{N_1} + \frac{p_{u_2} q_{u_2}}{N_2} + \frac{p_{u_3} q_{u_3}}{N_3} + \frac{p_{u_4} q_{u_4}}{N_4}$$

We can then form Z as follows:

$$Z = \frac{(p_{s_1} - p_{s_2}) - (p_{s_3} - p_{s_4})}{\sqrt{\dfrac{p_{u_1} q_{u_1}}{N_1} + \dfrac{p_{u_2} q_{u_2}}{N_2} + \dfrac{p_{u_3} q_{u_3}}{N_3} + \dfrac{p_{u_4} q_{u_4}}{N_4}}}$$

and use the normal table in the straightforward way. Since the denominator contains unknown p_{u_i} and q_{u_i}, we may either estimate these by the corresponding p_{s_i} and q_{s_i}, or we may conservatively set each equal to .5.

It is important to notice that the expression for the variance of Y involves four different N_i which appear as denominators in separate fractions. Since the products $p_{u_i}q_{u_i}$ are generally close to the numerical value .25, we see that the value of each fraction will be primarily a function of the subsample size. In practical terms, if there is one subsample that is very small, this subsample may dominate the expression for the variance of Y, and therefore the denominator of Z. Thus for maximum efficiency we wish to use subsamples of approximately the same size. If one subsample is very small, the above test will probably not prove significant because of the large denominator for Z, and in addition the normal approximation may not be justified.

Exactly the same procedures can be followed in connection with differences among means, for example, $(\bar{X}_1 - \bar{X}_2) - (\bar{X}_3 - \bar{X}_4)$. We shall postpone a consideration of this approach, however, until we take up general comparisons among k means in Chap. 16.

13.3 Confidence Intervals

In the case of single-sample problems we saw that the construction of confidence intervals is often a much more useful procedure than is the testing of hypotheses. In sociological research, confidence intervals are seldom used as alternatives to two-sample tests, however. The reason for this is that we are usually interested in establishing the existence of a relationship between two variables, i.e., a significant difference. There is less concern with estimating the actual magnitude of the difference. Seldom does a social scientist attempt to conclude that the difference between two means is probably between 17 and 28, for example. He is usually satisfied when he finds any significant difference at all. This state of affairs undoubtedly reflects the immaturity of the social sciences and the preponderance of exploratory studies. Perhaps as hypotheses become more precise, there will be a greater need for confidence intervals in two-sample problems.

The procedure used in establishing confidence intervals is a straightforward extension of that previously discussed. One simply takes the sample results, in this case a *difference* between sample means, and places an interval about $\bar{X}_1 - \bar{X}_2$ which is an appropriate multiple of the standard error. For example, if a 95 per cent confidence interval were desired we would obtain an interval as follows:

$$(\bar{X}_1 - \bar{X}_2) \pm 1.96\sigma_{\bar{X}_1 - \bar{X}_2}$$

If an estimate of the standard error and the t distribution were required, the formula would be modified in the usual manner.

13.4 Dependent Samples: Matched Pairs

Sometimes it is advantageous to design a study in which samples are not independent of each other. One of the most common types of problems of this sort is one in which cases in the two samples have been matched pair by pair. There may be experimental and control groups, members of which have been matched on relevant characteristics. Or a simple "before-after" design may be used in which the same persons are compared before and after an experimental variable has been introduced. In this latter instance the "two" samples consist of the very same individuals. Obviously such samples are not independent of each other; knowing the scores of the first members of each pair (first sample) would help to predict the scores of the second members. In fact the whole aim of matching, or of using the same individuals twice, is to control as many variables as possible other than the experimental variable. The attempt is to make the two samples as much alike as possible, much more alike than if they had been selected independently.

A researcher might be tempted to use a difference-of-means test in such problems. But it should be apparent that this procedure would be unjustified since we do not have $2N$ cases (N in each sample) that have been independently selected. Since samples have been deliberately matched, any peculiarities in one sample are most likely to occur in the other as well. In reality there are only N independent cases, each "case" being a *pair* of individuals, one from each sample. Therefore if we treat each pair as a single case, we can legitimately make statistical tests provided other required assumptions are met. Instead of making a difference-of-means test, we can make a direct pair-by-pair comparison by obtaining a difference score for each pair. If we use the null hypothesis that there is no difference between the two populations, thereby assuming that the experimental variable has no effect, we can simply hypothesize that the mean of the pair-by-pair *differences* in the population (μ_D) is zero. The problem then reduces to a single-sample test of the hypothesis that $\mu_D = 0$.

PROBLEM   Suppose an action group is interested in influencing urban voters to vote in favor of public-housing proposals in the coming elections. Cities within the state are carefully matched on variables thought to be relevant, and two different techniques for influencing the voters are attempted. The technique for group A involves an indirect approach through influencing city leaders but making no direct mass appeal. In cities of group B the organization acts as a pressure group and makes a direct appeal to the voter as an outside organization. The following figures indicate the percentages of voters favoring public housing. Is either technique superior?

Pair no.	Group A, %	Group B, %	Difference, (B − A) %
1	63	68	5
2	41	49	8
3	54	53	−1
4	71	75	4
5	39	49	10
6	44	41	−3
7	67	75	8
8	56	58	2
9	46	52	6
10	37	49	12
11	61	55	−6
12	68	69	1
13	51	57	6
			52

1. *Assumptions*

Level of Measurement: Per cent voting favorably is an interval scale
Model: Random sampling
    Population differences distributed normally
Hypothesis: $\mu_D = 0$

It must be assumed that the pairs appearing in the samples have been selected randomly from some population of pairs. As will be discussed below, this assumption sometimes causes a difficult problem of interpretation. Since it is the differences for each pair in which we are directly interested, it must be assumed that the population of all possible differences is normally distributed. If N were large, this assumption could be relaxed.

2. *Sampling Distribution* Since the population standard deviation of differences is not given, it will be necessary to use the t distribution with $N - 1$ or 12 degrees of freedom. Notice that this is half the degrees of freedom that would have been used had a difference-of-means test (with $\sigma_1 = \sigma_2$) been possible.

3. *Significance Level and Critical Region* Let us use the .05 level and a two-tailed test. Therefore, for 12 degrees of freedom, if $|t| \geq 2.179$ we shall reject the null hypothesis.

4. *Computing the Test Statistic* First we find the mean of the sample differences by adding the difference column and dividing by N ($= 13$). The sample standard deviation of differences is also obtained.

$$\bar{X}_D = {}^{52}\!/_{13} = 4.0$$

$$s_D = \sqrt{\frac{\Sigma(X_D - \bar{X}_D)^2}{N}} = \sqrt{\frac{328}{13}} = 5.023$$

Therefore $$t = \frac{\bar{X}_D - \mu_D}{s_D/\sqrt{N-1}} = \frac{4.0 - 0}{5.023/\sqrt{12}} = 2.76$$

Notice that, once having obtained the difference column, we pay no further attention to the remaining columns. This principle applies to more complex situations where, for example, we may have a difference of differences for each pair. (See Exercise 5.)

5. *Decision* With 12 degrees of freedom, a probability of .02 corresponds to a t of 2.681. We therefore decide to reject the null hypothesis, and noting the direction of the difference, we conclude that method B is superior to method A.

13.5 Comments on Experimental Designs and Significance Tests

Although it is not possible in a general text such as this to go very deeply into questions of design of experiments, a few comments are in order.[2] You may have asked yourself why it is that one would ever want to make use of matched samples in preference to independent samples. Clearly, degrees of freedom are sacrificed. Since the use of matched samples requires cutting the number of cases in half (as far as the test is concerned), do we not lose more than we gain? This all depends on how successful we have been in matching cases. The purpose of matching is, of course, to minimize differences due to extraneous variables. What this means is that careful matching should reduce considerably each of the pair-by-pair differences. In other words, the better the matching, the smaller the standard deviation of differences. Thus although the number of cases is reduced, s_D should also be reduced. If there is a large reduction in the standard deviation of differences relative to the loss in cases, then we gain by matching. Since cases will ordinarily be lost in the matching

[2] For further discussion of experimental designs, see any standard text on research methods. In particular, see [8], chap. 4.

process (see below), the moral is this: do not match unless you are reasonably sure you have located the important relevant variables. If you are studying delinquency and match according to hair color, you will probably be worse off than if you did not match at all.

Methods texts usually mention the fact that there is likely to be considerable loss of cases due to the matching process. That is, a large number of cases will have to be eliminated because there are no similar cases with which they can be paired. Such an attenuation of cases can play havoc with the randomness assumption. A social scientist may start with a random sample of 1,000 cases, ending up with 200 that are matchable. In so doing, he is probably biasing his final sample considerably since he is undoubtedly eliminating many of the more extreme or unusual cases that cannot be matched. Thus it is often difficult to determine the nature of the population to which one is generalizing. For this reason, extreme caution should be used in generalizing results. This type of design is therefore likely to be most useful in studies in which there is minimal interest in generalizing to a specific finite population, such as native whites in Chicago.

In connection with such an attenuation of cases and the resulting difficulties in generalizing to a specific population, it is sometimes argued that there is no real interest in the population itself since the researcher's concern is primarily with establishing relationships between variables. For example, a social psychologist may begin by using only those white male college freshmen taking introductory psychology who volunteer as subjects for study. Still further selectivity may occur as some subjects are eliminated in the matching process. Suppose a relationship is then found between the experimental variable and some dependent variable. It would be tempting to conclude that the same relationship would hold regardless of the population studied, i.e., that it is a universal relationship. If this should indeed prove to be the case, the social scientist may very well assert that he has no interest in generalizing to any particular finite population. But on what basis can he assume that the relationship found for such a very restricted population will hold for other populations as well? Obviously, the experiment must be performed on a large number of very different populations before such an assertion can legitimately be made. Although one may gain control over a number of variables in a carefully designed experiment, there is practically always a comparable loss in the degree to which results can be generalized to more inclusive populations.

In pair-by-pair matching it is desirable to randomize within each pair by flipping a coin to decide which member of the pair should be assigned to the experimental group and which to the control group. This procedure makes interpretation of results more meaningful in the sense that possible self-selection can be ruled out. For example, in the attempt to influence voters on the public-housing issue, suppose community leaders were permitted to choose which of the two types of influences they preferred or which they thought would be most effective in their particular community. It is possible that all or most communities with a certain type of leadership might receive the indirect approach, whereas those with another type of leadership would receive the direct method. We would then have an uncontrolled variable (leadership type), the effects of which would be hopelessly confounded with those of the experimental variable. Specifically, suppose group B turned out to have the higher percentage of favorable votes but also the more democratic leadership because of the fact that democratic leaders tended to favor this approach for their communities. How would we know whether or not the difference in vote was actually due to the superiority of method B or to the leadership differences between the two communities?

It might be argued that leadership type should have been controlled in the matching process so that two communities in any given pair would have the same type of leadership. But it is obviously impossible to control for all relevant variables in the matching process, not only because of practical difficulties but also because of our limited knowledge of which variables are actually the most important. At some point we must admit that there may be important variables, many of which are unknown to the researcher, which have not been controlled in the matching process. It is precisely at this point that we rely on randomization, i.e., on the laws of probability, in the expectation that the effects of uncontrolled variables will be neutralized. For example, we expect with a fairly large N that roughly half of the communities with the more democratic leadership will fall into group A and half into group B. The same will also be true for other uncontrolled variables.

In ex post facto experimental designs, in which the researcher comes on the scene *after* the experiment has taken place and consequently has had no opportunity to make such random assignments, the possibility of self-selection can never be ruled out. Nor can the laws of probability be used to help one evaluate the effects of the experimental variable as contrasted with possible effects of variables for which the groups have not been matched. One of the major advantages of laboratory experiments over so-called "natural" or ex post facto experiments is this randomization control over possible self-selection.

Other methods of matching samples are often suggested as alternatives to pair-by-pair matching. These alternative methods usually have the advantage of reducing the attenuation of cases but lead to difficulties when it comes to statistical analysis. One such method involves matching by frequency distributions. For example, care may be taken that the two groups are similar with respect to mean income, mean age, general income distribution, and so forth. The groups are thus comparable with respect to these summarizing measures even though a given individual may have no exact counterpart in the other group with whom he can be paired. In this type of design we are again clearly violating the independence assumption, but to the writer's knowledge there is no simple way to make use of a statistical test which is at the same time efficient and also does not involve dubious assumptions. One could pair cases as well as possible and proceed as above, but pairing would undoubtedly lead to an inefficient design. Certainly it would not be legitimate to use a difference-of-means test with $N_1 + N_2 - 2$ degrees of freedom.

Significance tests and generalizations to populations   There has been an extensive debate in the sociological literature concerning the appropriateness of significance tests in instances where one is dealing with the entire population. (See especially [3], [7], [9], and [10].) For example, one may have data available for all of the counties or states within the United States or a particular region. If so, there is no larger population to which one wishes to generalize, and it may be difficult to conceptualize the generalization process as involving an extrapolation to some larger universe of possibilities, or to these same cases under similar circumstances. If so, then tests of significance would seem inappropriate since no sampling error would be involved.

The position one takes on this question seems to depend primarily on whether he is satisfied with generalizations to fixed populations, or whether he wishes to make inferences about the causal processes that may have generated the population data. In this text we have conceptualized the problem as though our only objective is that of inferring to fixed populations, but obviously whenever we wish to link our findings with *theoretical* analyses, our objectives are never this simple. The problem of making causal inferences from nonexperimental data, based either on samples or complete populations, is far too complex to be considered in this elementary text. Nevertheless, there is a way of looking at tests of significance that is much more compatible with theoretical explanations as to *why* a particular relationship has been found.

Suppose, for example, that we have used all 50 states and that we have found a difference between Northern and Southern states, or between those that have Republican and Democratic governors. Ordinarily we would not want to stop with a simple description of such a difference, but we would want to provide an explanation, say, in terms of regional or political differences. Let us say we have found that Southern states spend relatively larger proportions of their budgets on highways but less on higher education. Before we can make any claims that our explanation should involve looking for causal factors producing this regional difference, we might think of the hypothetical skeptic who could pose a very simple alternative explanation for our finding, namely, "chance" processes. In effect, he might say: "You claim to have found a difference due to regional characteristics. But I could have used a table of random numbers to divide up your 50 states. Or perhaps they might have been divided up alphabetically according to the third letter of their names. If I could show that such a random or nearly random process could have produced a difference as great or greater than yours, then your explanation is no more plausible than mine."

Notice that there is no question, here, of generalizing to a population larger than the total of 50 states. The argument revolves around the processes that could have generated differences among subpopulations delineated in various ways. Certainly, if one could have frequently obtained differences as great as the regional differences by using a table of random numbers, then since the skeptic's "theory" is much simpler than our own, there is not much point in looking further into the data. If we take this viewpoint of the generalization process, then clearly it makes sense to make significance tests even when one has data for the entire population. It would seem as though most social scientists in fact have this more inclusive objective of saying something about causal processes, and therefore they should always make tests in order to rule out the simple "chance-processes" alternative. However, it should be emphasized that a significance test will *not* rule out many other kinds of alternative explanations, as for example those that introduce additional variables as common causes of the two variables under consideration. We shall return to this more difficult issue in Chap. 19.

Exercises

1. Fifty census tracts in a city are selected at random. It is found that twenty of these are serviced by community centers, the remainder are not. You compare the delinquency rates for these two types of tracts and obtain the following delinquency rates (given in terms of the number of delinquents per 1,000 adolescents):

Measure	With center	Without center
Sample size	20	30
Mean	27	31
Standard deviation (s)	6	8

Test for the significance of the difference between these two types of tracts (.01 level) using (a) model 1, and (b) model 2. How do the two results compare? [*Ans.* (a) $t = 1.87$; fail to reject]

2. A random sample of married women still living with their husbands is selected and the women are classified as being either "satisfied" or "unsatisfied" with their marital lives. The two groups of women are then compared with respect to the duration of their marriages with the following results:

Duration of marriage (nearest whole year)	Satisfied f_1	Unsatisfied f_2
0–2	34	10
3–4	41	16
5–9	50	23
10–14	39	25
15–19	18	14
20–39	15	16
Total	197	104

Is there a significant difference between the two groups at the .01 level?

3. Suppose you expect to find the difference between the mean annual incomes of samples of doctors and dentists to be about $500 (that is, $\bar{X}_1 - \bar{X}_2 = 500$). You estimate the standard deviations to be $1,900 and $1,600 respectively. You plan to take the same number of doctors and dentists in the total sample. How many cases would you need in order to establish a significant difference between the mean incomes of doctors and dentists, using the .05 level? Suppose you intended to take twice as many doctors as dentists. How many cases would you then need? (*Ans.* 95 of each)

4. You have classified a random sample of college students as either "other-directed" or "inner-directed." You find that 58 per cent of the seniors are other-directed, whereas 73 per cent of the freshmen belong to this category. There are 117 seniors and 171 freshmen in the total sample. Is this difference significant at the .001 level?

*5. Suppose you have designed a before-after experiment with a control group. In other words, you have matched two groups, pair by pair, and have before-and-after measures on both groups. Make use of the *t* test to test for the effectiveness of your experimental variable (a) using only the "after" scores, ignoring the "before" scores; (b) using the "before" and "after" scores of the experimental group only; and (c) using all four sets of scores. (*Hint:* How can you use all four scores to separate out the effects of the experimental variable from extraneous factors that might have affected both groups?) Contrast the advantages and disadvantages of methods (a) and (b). What are the advantages of (c) over both (a) and (b)? [*Ans.* (a) $t = 1.25$; fail to reject]

Pair	Control group		Experimental group	
	Before	After	Before	After
A	72	75	66	77
B	61	60	61	65
C	48	37	43	49
D	55	64	55	53
E	81	76	76	91
F	50	59	52	68
G	42	49	40	51
H	64	55	65	74
I	77	75	67	79
J	69	78	64	63

*6. In Table 15.4 of Chap. 15 you will find some data relating grades children receive to their ability, effort, and social class.

a. Looking at the middle class only, make a test to see if the relationship between effort and grades varies according to the ability level of the student.

b. Extend this test to see if the "interaction" tested for in (a) differs according to the social class of the student.

Note: In effect, in (b) you will be testing for an interaction of an interaction, or what is called a second-order interaction.

References

1. Alder, H. L., and E. B. Roessler: *Introduction to Probability and Statistics*, 4th ed., W. H. Freeman and Company, San Francisco, 1968, chaps. 8 and 10.

2. Downie, N. M., and R. W. Heath: *Basic Statistical Methods*, 2d ed., Harper and Row, Publishers, Incorporated, New York, 1965, chaps. 11 and 12.

3. Gold, David: "Statistical Tests and Substantive Significance," *American Sociologist*, vol. 4, pp. 42–46, 1969.

4. Goodman, L. A.: "Modifications of the Dorn-Stouffer-Tibbetts Methods for 'Testing the Significance of Comparisons in Sociological Data,'" *American Journal of Sociology*, vol. 66, pp. 355–359, 1961.

5. Hagood, M. J., and D. O. Price: *Statistics for Sociologists*, Henry Holt and Company, Inc., New York, 1952, chap. 19.

6. Hays, W. L.: *Statistics*, Holt, Rinehart and Winston, Inc., New York, 1963, chap. 10.

7. Kish, Leslie: "Some Statistical Problems in Research Design," *American Sociological Review*, vol. 24, pp. 328–338, 1959.

8. Selltiz, C., M. Jahoda, M. Deutsch, and S. W. Cook: *Research Methods in Social Relations*, Henry Holt and Company, Inc., New York, 1959, chap. 4.

9. Selvin, H. C.: "A Critique of Tests of Significance in Survey Research," *American Sociological Review*, vol. 22, pp. 519–527, 1957.

10. Winch, R. F., and D. T. Campbell: "Proof? No. Evidence? Yes. The Significance of Tests of Significance," *American Sociologist*, vol. 4, pp. 140–143, 1969.

Ordinal Scales: Two-sample Nonparametric Tests   **14**

As yet we have not had occasion to discuss significance tests involving ordinal scales, although it was pointed out in Chap. 2 that ordinal scales are very frequently used in social science research. In the present chapter we shall take up certain two-sample tests that can be used with ordinal scales, tests which are directly comparable to tests involving differences of means and proportions discussed in the previous chapter. Tests taken up in this chapter can therefore be used in relating ordinal-scale variables to variables involving a dichotomized nominal scale. In subsequent chapters we shall take up tests which permit one to relate an ordinal scale to a nominal scale with any number of categories or to another ordinal scale.

The tests discussed in this chapter are often referred to as *nonparametric* or *distribution-free* tests in that they do not require the assumption of a normal population. Actually, both the terms *nonparametric* and *distribution-free* are somewhat misleading. We do not imply that such tests involve distributions that do not have parameters. Nor can a population be "distribution-free." Both terms are actually used to refer to a large category of tests which do not require the normality assumption or any assumption that specifies the exact form of the population. Some assumptions about the nature of the population are required in all nonparametric tests, but these assumptions are generally weaker and less restrictive than those required in parametric tests. We have already come across certain nonparametric tests. The binomial, sign, and difference-of-proportions tests, for example, do not require the assumption of normality since they all refer to dichotomized nominal scales. In contrast with these particular nonparametric tests, the tests taken up in this chapter all involve ordinal scales, thus making it possible to use a somewhat higher level of measurement. In the following chapter two additional nonparametric tests will be considered, both involving only nominal scales.

What is the advantage of nonparametric tests as compared with a test such as the difference-of-means test? In using the t test for a difference of means, we saw that it was necessary to assume not only an interval scale but normal populations as well. The assumption of normality could of course be relaxed in the case of large samples, but it was argued that it is precisely when samples are small that the normality assumption is most questionable. Therefore we would expect to find that nonparametric alternatives to the difference-of-means test will be most useful whenever either of two conditions is met: (1) we cannot legitimately use an interval scale but ordering of scores is justified, or (2) the sample is small and normality cannot be assumed. Since these nonparametric tests involve weaker assumptions than the difference-of-means test, they may not take advantage of all the available information. Thus if an interval scale can legitimately be used and if the normality assumption can be either made in the case of small samples or relaxed in the case of larger ones, a difference-of-means test will ordinarily be preferable to one of the nonparametric tests.

In what sense can we speak of one test being preferable to another? What criteria are used in making such a decision? First, as we have already implied above, if a test requires us to make certain dubious assumptions that cannot themselves be tested, it will not be as satisfactory as one that does not require such assumptions. If everything else were equal, which is practically never the case, we would always select the test that required the weakest assumptions. If the results of the test called for rejection, we could more readily turn to the null hypothesis as the single faulty assumption. Unfortunately, however, the problem is not quite so simple. If it were, we would always make use of nonparametric procedures. It usually turns out that a test which requires stronger assumptions is a more powerful test in the sense that its use will involve a lower risk of a type II error. We thus have two criteria which work in opposite directions and which must be evaluated accordingly. Nonparametric tests require weaker assumptions, but they are less powerful. We shall get a better idea of what is meant by "strong" and "weak" assumptions when we come to the specific nonparametric tests that can be used as alternatives to the difference-of-means test. First, however, we must consider the question of how the relative power of a test is evaluated.

*14.1 Power and Power Efficiency

The *power* of a test is defined as $1 -$ (probability of a type II error) or as $1 - \beta$. Thus the power of a test is inversely related to the risk of failing to reject a false hypothesis. Assuming a fixed sample size, the greater the ability of a test to eliminate false hypotheses, the greater is its relative power. As we have already indicated, it is much more diffi- cult to evaluate the risk of type II than type I errors. Not only must we know the exact form of the population, but we must also know the degree to which the hypothesized parameter differs from the true value. In other words, the probability of a type II error, and therefore the power of the test, depends upon which alternative hypothesis is actually correct. For these reasons, we seldom actually compute the probabilities of type II errors in applied research. As implied earlier, however, the power of a test must be used in evaluating its relative efficiency. Several alternative tests may be made to involve the same risk of a type I error. We then use the relative risks of making type II errors to select the test which will be most appropriate under a given set of conditions. Although the problem of determining the power of a test is fairly complex and beyond the scope of this text, we can indicate in a general way what is involved in making such comparisons. In order to do so, we must introduce the notion of a power function.

The general form of a power function for a two-tailed test is given in Fig. 14.1. Such a function gives us the power of a test for the various

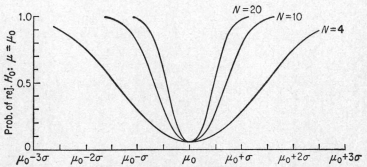

Figure 14.1 Power functions for two-tailed tests, with $\alpha = .05$, for samples of varying size. (*By permission from W. J. Dixon and F. J. Massey,* Introduction to Statistical Analysis, *McGraw-Hill Book Com- pany, New York, 1957, fig. 14.6, p. 252.*)

possible correct alternatives to H_0. To be specific, let us assume that we have hypothesized a particular value μ_0 for the population mean. Sup- pose, however, that the true population mean is actually two standard errors away from the hypothesized mean. Clearly, then, H_0 is false and should be rejected. Since the power of a test is $1 - \beta$, the power of the test actually gives us the probability of *rejecting* H_0 when it is false. This latter probability, rather than the probability of error, is given by the height of the curve. If the true mean is two standard errors from μ_0, the probability of rejecting H_0 can be determined by finding the height of the curve at this point on the X axis. Thus the values along the X axis indicate the possible *correct* values of μ whereas those on the Y axis indicate the probabilities of rejecting H_0.

Notice that when the correct value of the mean is actually μ_0 (and therefore we would be making an error in rejecting H_0), the height of the power function is given by the significance level of the test. Why? Notice also that if the correct value of μ is not too distant from μ_0, the power of the test, as indicated by the height of the curve, will be less than will be the case if the true value is quite different from μ_0. This tells us that our risk of a type II error is relatively large whenever the hypothesized value is not too far from the correct value, but that if we have missed the mark by a considerable amount, we shall have a much higher probability of rejecting our false hypothesis. This is consistent with the intuitive argument we developed earlier in connection with the binomial. It is also in line with our practical interests. If the null hypothesis is almost correct, we are not too bothered if we fail to reject it, even though technically we are in error in so doing. It is when H_0 is substantially incorrect that we are really interested in rejecting it.

*To generate the height of the power function at any given point along the horizontal axis, we must be in a position to assume the form of the sampling distribution. In this particular example we assume the sampling distribution of $\bar{X}$ to be $Nor(\mu, \sigma^2/N)$. If the true mean μ is to the right of the hypothesized mean μ_0, as diagramed in Fig. 14.2, then the true sampling distribution (about μ) will be to the right of the hypothesized sampling distribution (about μ_0). Of course we use the hypothesized sampling distribution to determine the critical region since the true μ

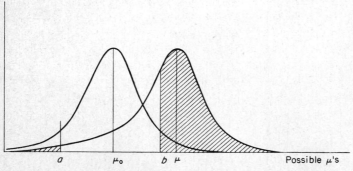

Figure 14.2   Derivation of power as a function of $(\mu - \mu_0)$.

will be unknown. Let us say that the critical region turns out to be the set of $\bar{X}$'s less than a but greater than b. In determining the power of the test, we must now evaluate the true probability of landing within the critical region, given that the true mean is μ rather than μ_0. This is obtained by calculating the shaded area *under the true sampling distribution* which is to the left of a and to the right of b in the diagram. We see that when μ and μ_0 are far apart, this shaded area is almost unity, but when μ and μ_0 are very close together, the shaded area approaches α (say .05) as its lower limit.

In order to give a better indication of how power functions are actually used, we can compare the power function of a two-tailed test (Fig. 14.1) with those of some one-tailed tests. Again let us suppose that H_0 predicts that the true mean is μ_0. Consider the one-tailed test in which we have used the upper or positive tail as our critical region. If the true value of μ is actually greater than μ_0, most of the sample means drawn from the population will also be greater than μ_0 and we have a better chance of ending up in this one-tailed critical region than would have been the case had we used a two-tailed test at the same significance level. In other words, if μ is actually to the right of μ_0, we have a better chance of rejecting H_0 with a one-tailed test in this direction. This means, of course, that the power of this particular one-tailed test will be greater for values of μ in the positive direction. But suppose the true value of μ is actually to the left of μ_0. Then most of the $\bar{X}$'s will fall to the left of μ_0 and very few will fall into the critical region at the opposite (or positive) end of the continuum. In this case, therefore, we shall practically never be able to reject H_0, and the power of the one-tailed test will be very weak indeed. The opposite kind of pattern will obviously hold in the case of one-tailed tests with critical regions in the lower or negative tails.

The power functions of one- and two-tailed tests can be compared as in Fig. 14.3. In summary, we see that a one-tailed test will be more powerful than its two-tailed counterpart (using the same significance level) for alternatives which are in the direction of the critical region, but it will be much less powerful than a two-tailed test if the parameter actually lies in the direction opposite to that predicted. The risk of a type II error is therefore considerable if one makes a one-tailed test and happens to predict the wrong direction. In such an event the data cannot be used to support the theory anyway. Therefore one would probably have no interest in proceeding with the test unless, for exploratory purposes, he wanted to ascertain whether or not a completely opposing theory would have any merits.

We have seen in comparing one- and two-tailed tests that one test may be more powerful for certain alternatives but less powerful for others. In

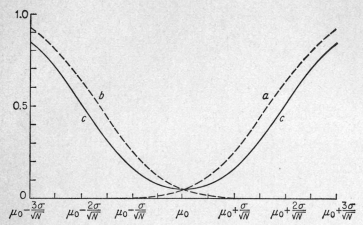

Figure 14.3 Comparison of power functions for one- and two-tailed tests, with $\alpha = .05$. (a) Reject if $Z > 1.645$. (b) Reject if $Z < -1.645$. (c) Reject if $Z > 1.96$ or if $Z < -1.96$. (*By permission from W. J. Dixon and F. J. Massey,* Introduction to Statistical Analysis, *Mc-Graw-Hill Book Company, New York, 1957, fig. 14.5, p. 249.*)

general, this may also occur in comparing two very different kinds of tests. For example, we shall soon see that one particular nonparametric test may be more powerful than a second under one set of circumstances but less so under another. It is this fact which makes it difficult to develop relatively simple generalizations concerning the superiority of one test over another. The situation is further complicated by the fact that one test may be powerful for large samples but relatively less so for smaller samples. The power of any given test will of course increase with the sample size since for any given significance level an increase in sample size essentially makes it possible to reject the null hypothesis with smaller deviations from hypothesized values. We have seen, for example, that the standard error of the mean decreases as N increases and that therefore as N increases, the sample mean must be closer to the hypothesized value if we are to retain H_0. What we are saying, then, is that we can more easily reject a false hypothesis when N is large. But although the power of a test may increase with N, the *rate* of increase in power may not be the same for all tests. A test which has relatively weak power for small N's may possibly "catch up" with another test so that the former test is actually more powerful for large samples.

In order to compare the relative powers of two tests we can ask ourselves how many cases would be needed with the first test in order to get the same power as with a given number of cases using the second test.

Usually we compare the power of a given test with that of the most powerful alternative. In the case of the first three nonparametric tests considered in this chapter, the most powerful alternative will be the t test for the difference of means. The term *power efficiency* is usually used to refer to the power of a certain test relative to that of its most powerful alternative. If we refer to the power efficiency of one of these nonparametric tests as 95 per cent, we mean that the power of the nonparametric test using 100 cases is approximately the same as that of the t test using 95 cases *if the model used in the t test is correct.*

Since it is necessary to assume a particular form for the population in order to evaluate the power of a test, we imagine in the above illustration that we actually have an interval-scale level of measurement and that the populations are both normal in form. In determining the power efficiency of the nonparametric test, we are essentially asking ourselves how much our failure to accept the normality assumption will cost us if in fact such an assumption were actually legitimate. Here we see that the failure to accept this assumption and our consequent use of the nonparametric test would cost us an extra five cases above the 95 used in the difference-of-means test. With such a small loss in efficiency we would probably go ahead with the nonparametric test if we were at all in doubt about the assumptions required by the difference-of-means test. On the other hand, if the power efficiency were only 60 per cent and if departures from normality were not too great (or if N were large), we would probably use the difference-of-means test.

As indicated in the previous chapter, it is when samples are small that we need to be most concerned about the normality assumption. For small N's it will not in general be possible to translate power-efficiency statements into comparisons of exact sample sizes since the latter quantities must always be integers. Thus with 95 per cent efficiency, a sample of size 10 using the nonparametric test would be approximately equivalent to one of 9.5 using the t test. Although such a statement is operationally meaningless, it is at least helpful for comparative purposes.

Before closing this section, it is again necessary to remind you that the power efficiency of a given test may depend upon the sample size selected; it may be highly efficient for small samples but much less so for larger ones.

14.2 The Wald-Wolfowitz Runs Test

In the runs test and also in the following two tests to be considered in this chapter, we assume that we have two independent random samples and that the level of measurement is at least an ordinal scale. In all three tests our null hypothesis will be that the two samples have been drawn from the same continuous population (or identical populations). The underlying dimension will be assumed to be continuous rather than discrete, although we recognize that tied scores may result because of the crudity of the measuring instrument. The hypothesis that the two samples have been drawn from the same population is actually very similar in nature to our assumptions in the difference-of-means test. As previously indicated, when we put together the assumptions of normality, equal variances, and equal means, we are in effect assuming the two populations to be identical. In the case of the runs test we are hypothesizing that the two populations have exactly the same form and hence can be thought of as identical. We do not have to specify the nature of this form, however. It might be normal, or it might not. We are therefore making a weaker set of assumptions than required in the difference-of-means test, weaker in the sense that the difference-of-means test (with equal σ's) requires all of the assumptions of the runs test *plus* the assumption of normality and the use of an interval scale.

In the difference-of-means test our interest centered on differences in central tendency rather than differences in dispersion or differences in form. The runs test essentially tests for all of these possible differences simultaneously. As we shall see presently, its main use is in testing for differences in dispersion or form, since there are more powerful non-parametric tests available for testing for differences in central tendency. Notice, incidentally, that the null hypothesis has not been stated in terms of means or standard deviations but rather in terms of any differences whatsoever. This will also be the case for other nonparametric tests discussed in this chapter. With ordinal scales, it is of course meaningless to think in terms of means and standard deviations.

The basic principle involved in the runs test is very simple, as are the computations. We first take the data from both samples and rank the scores from high to low, ignoring the fact that they come from two different samples. If the null hypothesis is correct, we would expect that the two samples will be well mixed. In other words, we would not expect a long run of cases from the first sample followed by a run of cases from the second. For example, if we refer to the samples as A and B, we expect that the rank ordering will be more or less as follows:

ABBABAAABABBABBAAABAAB

rather than

AAAAAAAAABABBBBBBBBBBB

In order to test to see how well the two samples are mixed when ranked, we can simply count the number of runs that occur. A *run* is defined as any sequence of scores from the same sample. In the first example above we have a run of a single A, followed by a run of two B's, then a single A, a single B, a run of three A's, and so on. The total number of runs is therefore 14. In the second example, however, the A's are bunched toward the lower end of the continuum, and we have only four runs. The procedure of counting runs will usually be simplified and errors reduced by drawing a line under all scores in the first sample and a line above those in the second. We then only need to count the number of separate lines. If the number of runs turns out to be quite large, as in the first case, the two samples will be well enough mixed that we shall not be able to reject the null hypothesis. On the other hand, a small number of runs probably means that the hypothesis is incorrect and therefore should be rejected. The sampling distribution of runs can be used to establish the critical region used in rejecting the null hypothesis.

PROBLEM   Suppose that judges have ranked 19 social organizations according to their prestige, giving a score of 1 to the group with the highest standing and 19 to that with the lowest. Ten of these groups restrict their membership to gentiles, whereas the remaining nine admit Jewish persons. Assuming that these social organizations have been selected randomly from a list of all such organizations in the community, can we conclude that in the population there is a significant difference in the prestige of restrictive and nonrestrictive social organizations?

Restrictive membership: Ranks 1, 2, 4, 5, 6, 7, 9, 11, 14, 17 ($N_1 = 10$)
Nonrestrictive membership: Ranks 3, 8, 10, 12, 13, 15, 16, 18, 19 ($N_2 = 9$)

1.   *Assumptions*

   Level of Measurement: Prestige is an ordinal scale
   Model: Independent random samples
   Hypothesis: Samples have been drawn from populations having the same continuous distributions

2.   *Sampling Distribution*   If both N_1 and N_2 are less than or equal to 20, the exact sampling distribution of the number of runs r is given in Table E of Appendix 2. For larger N's the sampling distribution of r is approximately normal with

$$\text{Mean} = \mu_r = \frac{2N_1N_2}{N_1 + N_2} + 1 \qquad (14.1)$$

and   Standard deviation $= \sigma_r = \sqrt{\dfrac{2N_1N_2(2N_1N_2 - N_1 - N_2)}{(N_1 + N_2)^2(N_1 + N_2 - 1)}}$ (14.2)

Notice that although normality of the population is not assumed, the sampling distribution of r will be approximately normal even with small N's. As we shall presently see, a number of other nonparametric test statistics also have this property. Note also that the formulas for the mean and standard error involve only the sample sizes and therefore do not require us to estimate population parameters as was necessary for the difference-of-means test. The comparative simplicity of formulas for the sampling distributions of nonparametric statistics is in part due to the fact that since scores have been ranked and therefore must always take on the numerical values 1, 2, 3, . . . , N, such quantities as the sum and standard deviation of ranks depend only on the number of cases used.

3. *Significance Level and Critical Region*   Since Table E, Appendix 2, gives only the number of runs necessary for rejection at the .05 level, for small samples we are restricted to this level of significance, although more complete tables can be found in [9]. Notice that the runs test does not take into consideration the direction of the relationship between prestige and restriction of membership. On the other hand, when we make use of the sampling distribution of r, we are interested in only one tail since we can reject the null hypothesis only when there is a small number of runs (regardless of the direction of the difference).[1] Strictly speaking, therefore, we are using the runs test as a one-tailed test even though direction of relationship has not been predicted. The same sort of situation will occur in the Mann-Whitney test discussed in the next section and in several important tests to be taken up in subsequent chapters. In order to avoid ambiguities we shall therefore distinguish between one-tailed tests and situations in which direction has been predicted. Up to this point, such a distinction has been unnecessary since all one-tailed tests have involved predictions as to direction.

In the case of normal sampling distributions we have seen that whenever direction has been predicted in advance, we may in effect cut a significance level in half by utilizing a single tail of the sampling distribution. In the case of the runs test and many other kinds of applications, we may rely on a somewhat different kind of justification for halving significance levels when direction has been predicted. In the context of the present example let us suppose that there is absolutely no difference in the population of social organizations with respect to the prestige of restrictive

---

[1] There are other applications of the runs test, however, in which both tails may be used. For example, there may be too many runs if the samples have been artificially rather than randomly mixed, and this fact may be used in a test for randomness.

and nonrestrictive organizations. Let A be the event that we achieve sample results that are significant, say, at the .05 level without having predicted direction. Then clearly $P(A) = .05$. Now let B be the event that the direction of the difference for the samples is as predicted, assuming absolutely no population differences. Then $P(B) = .5$ if we rule out the possibility that the difference will be exactly zero.

Since A and B will ordinarily be independent events, then the probability of getting significance at the .05 level without predicting direction *and* the probability of predicting direction correctly will be given by $P(A \& B) = P(A)P(B) = (.05)(.5) = .025$. The same principle can always be utilized whenever the sampling distribution of a test statistic is either symmetrical or insensitive to the direction of a difference. For example, if we had been interested in comparing three samples (as will be done in the next two chapters), and if we had been able to predict the exact ordering of these differences (for example, $\bar{X}_1 > \bar{X}_2 > \bar{X}_3$), then the probability of getting differences in exactly this order would be $\frac{1}{6}$, under the assumption that $\mu_1 = \mu_2 = \mu_3$, and we could thus justify taking the significance level as $\frac{1}{6}$ of the level obtained without predicting direction. Of course this procedure is wide open to ex post facto reasoning and can only be applied when predictions have been made in advance of looking at the data.

Numbers in the body of the table give us the number of runs which will yield significance at the .05 level, assuming direction has not been predicted. Therefore any value of r which is equal to or less than the figure in the table will indicate that we have so few runs that we may reject the null hypothesis at this level. Since the numbers of cases in the two samples are ten and nine respectively, we see that we may reject if we get six or fewer runs.

4. *Computing the Test Statistic* Arranging the organizations in order of prestige, drawing lines under the scores of the first sample and above those of the second, we see that there are 12 runs.

$$\underline{1\ 2}\ \overline{3}\ \underline{4\ 5\ 6\ 7}\ \overline{8}\ \underline{9}\ \overline{10}\ \underline{11}\ \overline{12\ 13}\ \underline{14}\ \overline{15\ 16}\ \underline{17}\ \overline{18\ 19}$$

Although the number of cases is somewhat too small for the normal approximation to hold, we can go ahead with the computations using this approximation in order to illustrate its use and to compare results with those obtained using Table E, Appendix 2. As usual, we compute the value of Z, which will tell us how many standard deviations the obtained number of runs is from the mean or expected number of runs under the null hypothesis. Thus

$$\mu_r = \frac{2(10)(9)}{10 + 9} + 1 = 10.47$$

and
$$\sigma_r = \sqrt{\frac{2(10)(9)[2(10)(9) - 10 - 9]}{(19)^2(18)}} = 2.11$$

We therefore get
$$Z = \frac{r - \mu_r}{\sigma_r} = \frac{12 - 10.47}{2.11} = .725$$

Since the number of runs obtained actually exceeds the mean or expected number, we need go no further since small numbers of runs are needed for rejection. Had the number of runs been less than the expected number, we would have looked up the Z value in the normal table, using the table as though we were making a two-tailed test (e.g., rejecting at .05 level if $Z \leq -1.96$).

5. *Decision*   Since the number of runs turned out to be greater than six, the figure given in Table E, we decide not to reject the null hypothesis at the .05 level. As we have just seen, the use of the normal approximation also leads us to this same conclusion. On the basis of our data we therefore do not conclude that there is a significant difference between the two types of organizations with respect to prestige.

Ties   In the above data there were no two organizations that received tied scores. The assumption of underlying continuity theoretically rules out the possibility of ties since no two scores would ever be exactly the same. But owing to crudities of measurement, and such crudities will almost certainly exist in most social research, ties do arise in practice. Notice that if two organizations within the same sample had been tied with respect to prestige scores, the runs test would have been unaffected. But suppose ties occurred across samples. Then the number of runs can be considerably affected depending on how the ties are broken. Suppose, for example, that two organizations (from different samples) had been tied for eighth and ninth positions. Had the positions of these two groups been switched from the order previously used, we would have obtained 10 runs instead of 12. In other words, we would get either 10 or 12 runs depending on the order used. Since this order would be completely arbitrary, we might find ourselves sometimes rejecting and other times failing to reject the null hypothesis. The safest procedure we can use in the case of ties is to compute the number of runs using all possible ways of breaking the ties. If all orderings lead to the same decision (rejection or nonrejection), we may safely adhere to this decision. If they lead to different decisions, it will be possible to resolve the problem by flipping a coin, but perhaps the safest procedure is to withhold judgment. Bradley [1] recommends the sensible procedure of providing the reader with the range of probabilities obtained by breaking ties in all possible ways. Obviously if there are a large number of tied ranks, the runs test should not be used.

14.3 The Mann-Whitney or Wilcoxon Test

Another nonparametric test that can be used in situations where the runs test is appropriate is a test which seems to have been invented independently by several persons and which is commonly known as either the Mann-Whitney or Wilcoxon test. This test requires exactly the same assumptions as the runs test and, like the latter test, involves a very simple procedure. We again combine the scores of both samples and rank them from 1 to 19. We then focus on the scores in the second sample (or whichever sample is the smaller). Taking each score in the second sample, we count the number of scores in the first sample that have larger ranks. Having done this for each of the scores in the second sample, we then add the results to give us the statistic U. The sampling distribution of U can then be obtained exactly if the N's are small, or it can be approximated by a normal curve in the case of larger samples. If U is either unusually small or unusually large, we can reject the assumption that the two samples have been drawn from the same population.

An alternative form of exactly the same test may be used with the normal approximation. Instead of obtaining U directly, we can compute the sum of the ranks for each of the samples. We then go through a procedure which is analogous to that used in the difference-of-means test. We take a difference of the sums of the ranks for each sample, subtracting from this difference a quantity representing the expected difference under the null hypothesis. This difference of differences, which is analogous to $(\bar{X}_1 - \bar{X}_2) - (\mu_1 - \mu_2)$, is then divided by the standard error in order to obtain Z. The analogy is not perfect since we are dealing with sums of ranks rather than means of ranks, but the parallel with the difference-of-means test is obvious. Again, a large numerical value of Z will lead to rejection. We shall now proceed to illustrate the Mann-Whitney test by making use of the same example as used above. We shall then compare the power efficiency of this test with that of the runs test.

PROBLEM   Same as used for runs test.

Restrictive membership: ranks 1, 2, 4, 5, 6, 7, 9, 11, 14, 17 ($N_1 = 10$)
Nonrestrictive membership: ranks 3, 8, 10, 12, 13, 15, 16, 18, 19 ($N_2 = 9$)

1.   *Assumptions*   Same as those required in runs test.

2.   *Sampling Distribution*   The sampling distribution of U will be found
in Table F of Appendix 2 if neither N_1 nor N_2 is larger than eight, and
in Table G if one of the N's is between 9 and 20 and the other between 1
and 20. Notice that these two tables have different formats. Table F
is set up with different combinations of N_1 and N_2 across the top, with
U values down the left margin, and with probability values in the body
of the table. Thus if $N_2 = 6$ and $N_1 = 4$, with N_2 always being the larger
of the two sample sizes, and if $U = 5$, we see that the probability of
obtaining $U \leq 5$ is .086 with direction predicted. The separate tables
under Table G, on the other hand, each correspond to a different significance level, with critical values of U being given in the body of the
table. Thus for $\alpha = .001$ with direction predicted, for $N_1 = 13$ and
$N_2 = 10$ (where N_2 is not necessarily larger than N_1), we find that a
value of U of 17 or *smaller* will imply significance. For larger N's the
sampling distribution of U will be approximately normal with

$$\text{Mean} = \mu_U = \frac{N_1 N_2}{2} \tag{14.3}$$

and $$\text{Standard deviation} = \sigma_U = \sqrt{\frac{N_1 N_2 (N_1 + N_2 + 1)}{12}} \tag{14.4}$$

3.   *Significance Level and Critical Region*   For purposes of comparability we shall continue to use the .05 level without predicting the direction
of relationship.

4.   *Computing the Test Statistic*   The statistic U can be computed by
either of two methods. With small N's it will be relatively simple to
compute U by carrying out the procedure implied in the definitional
formula. Focusing on each of the nine groups in the second sample, let
us count the number of cases in the first sample which have lower prestige
and therefore larger rank scores. Since the first organization in the second
sample has been ranked third in prestige, there are eight groups in the
first sample having lower prestige scores. Similarly, the second group
in sample 2 is ranked eighth, and therefore there are four groups in the

other sample with lower prestige scores. Continuing the process for each of the remaining organizations in sample 2, and then summing, we get

$$U = 8 + 4 + 3 + 2 + 2 + 1 + 1 + 0 + 0 = 21$$

Notice that had we carried out the same procedure but focusing on the groups within the first sample, we would have gotten

$$U' = 9 + 9 + 8 + 8 + 8 + 8 + 7 + 6 + 4 + 2 = 69$$

Either of these quantities could be used to test for the significance of the relationship, but since the tables have been set up in terms of the *smaller* U value, we always make use of the lesser of these two quantities. It will not be necessary to compute both U and U' since once either value has been obtained, the other can be computed from the formula

$$U = N_1 N_2 - U' \qquad \text{or} \qquad U' = N_1 N_2 - U \qquad (14.5)$$

In this case we would use the value 21 as our test statistic.

If the number of cases is relatively large or if ties occur, it will probably be more convenient to obtain U by summing the ranks of the separate samples, calling these sums of ranks R_1 and R_2, and then using the formulas

$$U = N_1 N_2 + \frac{N_2(N_2 + 1)}{2} - R_2 \qquad (14.6)$$

or $$U' = N_1 N_2 + \frac{N_1(N_1 + 1)}{2} - R_1 \qquad (14.7)$$

whichever is the more convenient. Summing the ranks we thus get

1	3
2	8
4	10
5	12
6	13
7	15
9	16
11	18
14	19
17	
$R_1 = 76$	$R_2 = 114$

As a check, we should have

$$R_1 + R_2 = \frac{N(N + 1)}{2}$$

or
$$76 + 114 = \frac{19(20)}{2} = 190$$

where N represents the total number of cases in *both* samples. Therefore

$$U = 10(9) + \frac{9(10)}{2} - 114 = 90 + 45 - 114 = 21$$

*The sums of ranks R_1 and R_2 could have been used directly in making the test, making it unnecessary to compute U. Since exact tables for small N's are usually given in terms of U, you will ordinarily find it advantageous to think in terms of the U statistic. The use of the sums of ranks can be used heuristically to point up the similarity of the Mann-Whitney test to the difference-of-means test, however. The use of a little algebra will convince you that we can take Eqs. (14.3) to (14.7) and obtain the result that for the normal approximation the statistic

$$Z = \frac{R_1 - R_2 - (N_1 - N_2)(N + 1)/2}{\sqrt{N_1 N_2 (N + 1)/3}} \qquad (14.8)$$

will be approximately $Nor(0,1)$. Expressing Z in this form, we notice that the numerator consists of the difference $R_1 - R_2$ together with a term which turns out to be the expected or long-run value of this difference under the null hypothesis. This correction factor is of course necessary since we are dealing with a difference of *sums* rather than means, requiring us to take into consideration the fact that the two N's will not ordinarily be equal. If N_1 and N_2 are the same, we note that this second factor becomes zero, and we are left with a numerator of $R_1 - R_2$. Thus we see the similarity with the difference-of-means test in which the numerator reduced to $\bar{X}_1 - \bar{X}_2$ in the case of the null hypothesis of no differences. The Mann-Whitney test may therefore be thought of as a *difference-of-summed-ranks* test.

5. *Decision* Making use of Table G, Appendix 2, we see that at the .05 level we need a U of 20 or *smaller* in order to reject the null hypothesis

if direction has not been predicted. Therefore we barely fail to reject the hypothesis that there is no difference between the two types of organizations. Notice, however, that had direction been predicted in advance, we would have needed a U of 24 or less at the .05 level. We see, incidentally, that although the same decision was reached in using both the runs and Mann-Whitney tests, we came much closer to rejecting when the latter test was used. If H_0 were really false, we therefore in this instance would have less of a risk of a type II error than with the runs test.

Had our N's been larger, we could have made use of the normal approximation. In order to illustrate the procedure used, we can compute Z for the above data. We get

$$Z = \frac{U - N_1 N_2/2}{\sqrt{N_1 N_2 (N_1 + N_2 + 1)/12}} = \frac{21 - 45}{\sqrt{10(9)(20)/12}} = -1.96$$

If we had replaced U by U' $(= 69)$, we would have obtained

$$Z = +1.96$$

*Had we used Eq. (14.8) we would also have gotten

$$Z = \frac{76 - 114 - (10 - 9)(20)/2}{\sqrt{10(9)(20)/3}} = -1.96$$

The use of the normal approximation thus yields the conclusion that, with direction not predicted, we may just barely reject at the .05 level. Of course the exact tables are always preferable to the normal approximation whenever they are available.

Ties If ties occur, we must again assume that they are due to crudities of measurement and that the underlying distributions are really continuous. If ties occur within classes, there will of course be no effect on U, and we can proceed as before. If ties occur across classes, we give each of the cases the average of the scores they would have had if no ties existed. Thus if two organizations are tied for eighth and ninth, each receives a score of $(8 + 9)/2$ or 8.5. Had the tenth organization also been tied with these two groups, each would have received the rank $(8 + 9 + 10)/3$ or 9.0. In computing U, it will now probably be less confusing to use the sum-of-ranks method. The correction factor involves the standard error of U and therefore appears in the denominator of Z.

The revised formula becomes

$$Z = \frac{U - N_1 N_2 / 2}{\sqrt{[N_1 N_2 / N(N-1)][(N^3 - N)/12 - \Sigma T_i]}} \qquad (14.9)$$

where $N = N_1 + N_2$ and $T_i = (t_i^3 - t_i)/12$, where t is the *number* of observations tied for a given rank.

In computing ΣT_i we first note all the instances in which ties occur. Perhaps two groups are tied for eighth and ninth, and three for the lowest honors. Thus we have one t of two, and another of three. Therefore

$$\sum T_i = T_1 + T_2 = \frac{t_1^3 - t_1}{12} + \frac{t_2^3 - t_2}{12}$$

$$= \frac{2^3 - 2}{12} + \frac{3^3 - 3}{12} = \frac{6}{12} + \frac{24}{12} = 2.5$$

and

$$Z = \frac{21 - 45}{\sqrt{\frac{10(9)}{19(18)} \left(\frac{19^3 - 19}{12} - 2.5 \right)}} = -1.964$$

This correction for ties can be used only with the normal approximation since the exact tables have not been computed allowing for ties. The correction factor will ordinarily have negligible effect unless the number of ties is quite large.[2] If the number of ties is extremely large, the Smirnov test (see below) should probably be used as an alternative to the Mann-Whitney test.

Comparison of the Mann-Whitney and Runs tests The null hypothesis for both tests is that the two samples have been drawn from identical populations. Usually our interest centers on differences in central tendency, as was true in the difference-of-means test. Occasionally, however, we may be more interested in differences in dispersion or form. As a general proposition, we can say that the Mann-Whitney test will be more powerful than the runs test whenever the major differences between the two populations are with respect to central tendency, whereas the runs test will be more powerful in situations in which the populations differ only slightly in central tendency but substantially in dispersion or form.

[2] It might seem as though the correction for ties always reduces the denominator without changing the numerator, but we must keep in mind that such ties will ordinarily bring U and U' closer together and therefore reduce the numerator as well.

A simple example can be used to illustrate this point. Suppose we had two populations with identical medians but in the one case with a very homogeneous distribution and in the other a very heterogeneous one. We then might expect results such as the following:

Sample 1	Sample 2
5	1
6	2
7	3
8	4
9	13
10	14
11	15
12	16
$R_1 = 68$	$R_2 = 68$

In this extreme example, the Mann-Whitney test would not lead to rejection of the null hypothesis (which is obviously false) because R_1 is exactly equal to R_2. Using the runs test, on the other hand, we would be able to reject since there would be only three runs. Since failing to reject would mean making a type II error, we see that the power of the runs test is greater than that of the Mann-Whitney test. In most instances, however, we are more likely to find differences in central tendency with relatively minor differences in dispersion. You should convince yourself that for populations of this sort, we are likely to obtain a fairly large number of runs toward the center of the distribution. For such data the runs test will turn out to be much less powerful than the Mann-Whitney test. For most sociological applications the Mann-Whitney test seems to be the more useful of the two tests.

*If an interval-scale level of measurement had been attained and normal populations legitimately assumed, the *t* test for the difference between means could have been made. Under such conditions how much would we lose by using the Mann-Whitney test, thereby dropping back in level of measurement and using a weaker model? The evidence is that for moderate and large samples, the power efficiency of the Mann-Whitney test is approximately 95 per cent as compared with the *t* test. The power efficiency for small samples is also very high, although exact numerical values are not easily obtainable. Bradley [1] notes that, in general, the efficiency of many nonparametric tests, including the Mann-Whitney test, is relatively greater for small samples than for large ones. The Mann-Whitney test is thus a very powerful alternative to the *t* test. In view of the fact that it requires much weaker assumptions, it should therefore be used in instances where there is reasonable doubt of the legitimacy of either the interval scale or normality. Less is known about the power efficiency of the runs test. Smith [8] has found efficiencies of approximately 75 per cent in several empirical examples where sample sizes were about 20 and where the normal populations had equal standard deviations. Bradley [1] notes that the large-sample efficiency of the runs test, as compared with the t test, is approximately one-third under these same conditions.

14.4 The Kolmogorov-Smirnov Test

The Kolmogorov-Smirnov test, which we shall refer to simply as the Smirnov test, is another two-sample nonparametric test requiring the same assumptions as the runs and Mann-Whitney tests. The power of the Smirnov test is, in general, difficult to evaluate, but in situations where the populations differ only with respect to central tendency, its power appears to be intermediate between those of the runs and Mann-Whitney tests. (Bradley [1], pp. 291–292.) Strictly speaking, the Smirnov test also assumes no ties, but as we shall see, the procedure is extremely convenient to use in situations where there are large numbers of ties resulting from the grouping of data into ordered categories.

Quite frequently in sociological research we make use of variables which are actually ordinal scales but for which data have been grouped into three or more large categories. If there are four or more such ordered categories, the Smirnov test will be especially useful, whereas the number of ties involved would prohibit the use of the Mann-Whitney test. A sociologist may have divided community residents into six social classes, treating all persons within one class as tied with the other members of the class with respect to overall status. Or occupations may have been ranked according to status, with all persons within the same occupational class receiving tie scores. Perhaps an attitudinal variable has been found to yield a Guttman scale with seven response types. In all of these examples we may wish to conceptualize the variable as being continuous in nature but the measuring instrument as being exceedingly crude so as to yield data that are grouped into a relatively small number of ordered categories. As was true in the case of interval scales, the finer the distinctions made and the larger the number of categories used, the less information lost.

The principle behind the Smirnov test is also a very simple one. If the null hypothesis that independent random samples have been drawn from identical populations is correct, then we would expect the cumulative frequency distributions for the two samples to be essentially similar. The test statistic used in the Smirnov test is the maximum difference between the two cumulative distributions. If the maximum difference is larger than would be expected by chance under the null hypothesis, this means that the gap between the distributions has become so large that we decide to reject the hypothesis. We can take either the maximum difference in one direction only (if direction has been predicted) or the maximum difference in both directions.

PROBLEM   Suppose we have divided a random sample of adult males in a community into six social classes and have also classified them as having either low or high mobility aspirations. The latter two categories can be considered to be independent random samples from the larger populations of adult males with low and high aspirations respectively, since a completely random total sample assures independence between any subsamples we might select. Let us suppose that we have predicted that those with high mobility aspirations will tend to have higher class standing than those with low aspirations. Can we conclude that the results are significant at the .01 level?

Class	Low aspirations	High aspirations
Lower-lower	58	31
Upper-lower	51	46
Lower-middle	47	53
Upper-middle	44	73
Lower-upper	22	51
Upper-upper	14	20
Total	236	274

1.   *Assumptions*   Same as required for Mann-Whitney and runs tests.

2.   *Sampling Distribution*   The sampling distribution of D, the maximum difference between the two cumulative distributions, can be given exactly in the case of small N's (≤ 40) when $N_1 = N_2$ ([7], page 129). This case will not be treated, since with relatively small N's the Mann-Whitney test may be used instead of the Smirnov test, and since in most sociological examples we ordinarily do not obtain samples of exactly the same size.   If both samples are larger than 40 and if direction has not been predicted, we shall need a value of D at least as large as

$$1.36 \sqrt{\frac{N_1 + N_2}{N_1 N_2}}$$

in order to reject at the .05 level. For the .01 and .001 levels, the coefficient 1.36 can be replaced by 1.63 and 1.95 respectively. In the case of the .10 level, the comparable coefficient is 1.22.

If direction has been predicted, we may use a chi-square approximation. The chi-square test statistic (χ^2) will be considered in the following chapter, and the chi-square table will become more familiar at that time.[3] The approximation formula, however, is as follows:

$$\chi^2 = 4D^2 \frac{N_1 N_2}{N_1 + N_2} \qquad (14.10)$$

where the degrees of freedom associated with chi square are always two in this particular application. Although continuous population distributions are assumed in using the chi-square approximation, if data are actually discrete and therefore result in large numbers of ties, the probabilities obtained will be in the conservative direction if rejection is desired. In other words, the true probabilities will be less than those computed.

3. *Significance Level and Critical Region* The problem calls for the .01 level of significance. Since direction has been predicted, we shall use the chi-square approximation.

4. *Computing the Test Statistic* We first obtain the cumulative frequency distributions for each of the samples (see Table 14.1), expressing the F values as *proportions* of the total sample sizes. Thus the first value entered in the F column for sample 1 will be $58/236$ or .246; the second will be $109/236$ or .462, and so on. The last entries in both columns will of course be unity. We now form a difference column $F_1 - F_2$ and locate the largest difference with a positive sign, since we predicted higher percentages of lower-class persons with low aspirations, i.e., higher F_1 values. This value of D turns out to be .187, as indicated by the arrow. Next, we compute the value of chi-square, using Eq. (14.10).

5. *Decision* Note that the larger the value of D, the larger will be the chi square. Therefore, we need to determine just how large chi square must be in order to reject the null hypothesis. Referring to the chi-square table (Table I, Appendix 2), locating the degrees of freedom down the left-hand margin and the significance level across the top, we see that for 2 degrees of freedom, the value 9.210 corresponds to the .01 level. This

[3] For this reason, you may prefer to postpone consideration of the Smirnov test until after you have read Chap. 15.

means that if the null hypothesis were actually true, we would get a chi square this large or larger by chance less than 1 per cent of the time. Since we obtained a chi square of 17.74, we see that we can reject the null hypothesis. This same chi-square test can be used for small samples when direction has been predicted, and if one is interested in rejecting the null hypothesis, the chi-square approximation will actually be conservative. In other words, the probabilities obtained by this method will be larger than the true probabilities.

Table 14.1   Computations for Smirnov two-sample test

Class	Mobility aspirations				Difference $F_1 - F_2$
	Low F_1		High F_2		
Below upper-lower	58	.246	31	.113	.133
Below lower-middle	109	.462	77	.281	.181
Below upper-middle	156	.661	130	.474	.187 ←
Below lower-upper	200	.847	203	.741	.106
Below upper-upper	222	.941	254	.927	.014
Total	236	1.000	274	1.000	

$$\chi^2 = 4D^2 \frac{N_1 N_2}{N_1 + N_2} = 4(.187)^2 \frac{236(274)}{236 + 274} = 17.74$$

Had direction not been predicted, we would have needed a value of D at least as large as

$$1.63 \sqrt{\frac{N_1 + N_2}{N_1 N_2}} = 1.63 \sqrt{\frac{236 + 274}{236(274)}} = 1.63(.0888) = .145$$

in order to reach significance at the .01 level. In this case we obtain D by taking the largest difference regardless of sign. Since this value is the same as the D previously used (.187), we see that we may again reject the null hypothesis.

14.5   The Wilcoxon Matched-pairs Signed-ranks Test

All three nonparametric tests discussed up to this point in the chapter have required that the two samples be selected independently of each other. It will be remembered that when pairs were matched, we could not use the ordinary difference-of-means test. Instead, we treated each matched pair as a single case and obtained a difference score for each pair. We then went ahead as though we had a single sample and tested the null hypothesis that $\mu_D = 0$. In addition you will recall that in using the sign test, we also could have made use of matched pairs, considering only the sign of the difference score and testing the null hypothesis by using the binomial distribution. In the sign test we had to throw away any information we might have had about the magnitude of the differences involved. On the other hand, the most powerful test, the t test, required not only an interval scale but also the assumption of a normal population of difference scores. The Wilcoxon matched-pairs signed-ranks test combines some of the features of both these tests and lies between them with respect to power efficiency.

As we shall presently see, the Wilcoxon test requires slightly higher than an ordinal-scale level of measurement. It will be necessary to assume an ordered-metric scale in which it is possible not only to rank the original scores themselves but to rank the *differences* between such scores. Since ordered-metric scales are seldom found in sociological research, this requirement essentially amounts to our needing an interval scale. Since the Wilcoxon test does not assume a normal population, however, it will be considered along with the other two-sample nonparametric tests in this chapter. The power efficiency of this test is substantially higher than that of the sign test, a fact which is not surprising since the sign test takes advantage of so little information. When the assumptions of the t test are actually true, the power efficiency of the Wilcoxon test is approximately 95 per cent for both small and large samples. It is therefore especially useful in situations where we have an interval-scale level of measurement but where the sample size is too small to justify the normality assumption.

The Wilcoxon test involves essentially the same null hypothesis as used in the sign test and also the t test for paired samples. The null hypothesis states that there are no differences between the scores of the two populations. In making use of this test we first obtain the difference scores for each pair. These differences are then ranked, ignoring the sign of the difference. Thus a difference of -6 would receive a higher rank than a difference of $+3$. Having ranked the absolute values of the differences, *always assigning rank 1 to the smallest numerical difference*, we then go back and record the signs. Finally, we obtain the sums of the ranks of both the positive and negative differences. If the null hypothesis is correct, we expect that the sum of the ranks of the positive differences will be approximately the same as the sum of the ranks for negative differences. If these sums are quite different in magnitude, the null hypothesis may be rejected. We form the statistic T which is the *smaller* of these two sums. We can then make use of exact tables for the sampling distribution of T when N is small, and a normal approximation when N is large.

For comparative purposes let us use the same data as were used in the case of the comparable t test. Table 14.2 repeats these data and also gives the necessary computations for the Wilcoxon test. Notice that when we ignore sign, several of the difference scores are tied as to magnitude. In such an instance we again give the tied scores the average of

Table 14.2 Computations for Wilcoxon Matched-pairs test

Pair no.	Group A	Group B	Difference	Rank of difference	Negative ranks
1	63	68	5	(+) 6	
2	41	49	8	(+) 10.5	
3	54	53	−1	(−) 1.5	1.5
4	71	75	4	(+) 5	
5	39	49	10	(+) 12	
6	44	41	−3	(−) 4	4
7	67	75	8	(+) 10.5	
8	56	58	2	(+) 3	
9	46	52	6	(+) 8	
10	37	49	12	(+) 13	
11	61	55	−6	(−) 8	8
12	68	69	1	(+) 1.5	
13	51	57	6	(+) 8	
Total					13.5

the ranks they would have received had the scores not been tied.[4] Thus there are two differences of size 1. Since we are here giving the smallest differences the lowest ranks, each of these differences has received a rank score of 1.5. In the fifth column we have indicated the sign associated with each rank in parentheses to the left of the rank. By inspection we see that the sum of the negative ranks will be less than that of the positive ranks. Therefore we obtain T by adding these negative ranks. It is not necessary to keep the negative signs in looking up the value of T in the

[4] A somewhat more conservative procedure would be to break ties in such a way as to yield the largest possible T value. Pairs having difference scores of exactly zero (e.g., no change whatsoever) should be dropped from the analysis.

table, since T values will always be given as positive. Thus

$$T = 1.5 + 4 + 8 = 13.5$$

Let us now formalize what we have done by listing the steps in the usual manner.

1. *Assumptions*

   Level of Measurement: Ordered-metric scale (difference scores can be
                           ranked)
   Model: Random sampling
   Hypothesis: Sum of positive ranks = sum of negative ranks in popu-
                 lation

2. *Sampling Distribution*   The sampling distribution of T for $N \leq 25$ is given in Table H, Appendix 2. For larger samples, T is approximately normally distributed with

$$\text{Mean} = \mu_T = \frac{N(N + 1)}{4} \qquad (14.11)$$

and $$\text{Standard deviation} = \sigma_T = \sqrt{\frac{N(N + 1)(2N + 1)}{24}} \qquad (14.12)$$

3. *Significance Level and Critical Region*   As in the case of the t test, we shall use the .05 level without predicting the direction of the outcome.

4. *Computing the Test Statistic*   The value of T has already been computed from Table 14.2. We obtained a T of 13.5.

5. *Decision*   Table H of Appendix 2 gives critical values of T for $N \leq 25$. Since T represents the smaller of the two sums of ranks, we need small numerical values of T in order to reject the null hypothesis. Thus we may reject H_0 whenever T is equal to or less than the values given in the body of the table. We see that for an N of 13, we need a T of 17 or smaller in order to reject at the .05 level. We also see that a T of 13 or less would be required for rejection at the .02 level. In using the t test in the previous chapter you will note that we just barely rejected at the .02 level; here we are slightly above the .02 level, but the results of the two tests are very similar.

Although our N is quite small, we can make use of the normal approximation for illustrative purposes. We get

$$Z = \frac{T - N(N+1)/4}{\sqrt{N(N+1)(2N+1)/24}}$$
$$= \frac{13.5 - 13(14)/4}{\sqrt{13(14)(27)/24}} = \frac{13.5 - 45.5}{\sqrt{204.75}} = -2.24$$

Since a Z of -2.24 corresponds to $p = .025$ for a two-tailed test, we again reach the same conclusion. The value of T is much smaller than we would expect by chance, and we may reject the null hypothesis. It should be noted that the above normal approximation does not involve an explicit correction for ties and therefore should not be used if the relative number of ties is extremely large.

14.6 Concluding Remarks

In the present chapter we have discussed four different nonparametric tests. Others will be taken up in later chapters. You have undoubtedly noted that all the nonparametric tests we have discussed thus far involve very simple ideas and considerably fewer computations than, say, the difference-of-means test. This is a further reason why we may predict that in the future sociologists will make much more frequent use of these nonparametric tests. It is unfortunately impossible in a general text to do much more than discuss a few such tests very briefly. Some of the tests taken up in this chapter have other kinds of applications which have not been discussed. For example, the runs test can be used as a test for randomness. The Smirnov test can be used as a one-sample test to compare observed frequencies with those predicted theoretically. In some cases, confidence intervals can be obtained using nonparametric procedures. Once you have gained a certain familiarity with the tests covered in this text, you may therefore want to consult more specialized sources. Fortunately, a good many of these nonparametric procedures can be easily understood by the reader who is not mathematically trained. It is also fortunate that a number of these procedures have been summarized in texts by Siegel [7], Bradley [1], and Pierce [5]. You may also want to refer to an extensive bibliography of nonparametric methods compiled by Savage [6].

In this chapter, as well as in the previous chapter, we have seen the necessity of distinguishing between samples that have been independently selected and those that have been matched or that involve comparisons of scores for the same individuals. Thus the independence or lack of independence between samples is one of the considerations involved in selecting among statistical procedures. In the case of matched samples we form a single score for each pair, then treat the data as though we were dealing with a single sample. Where samples have been independently sampled, and where the sizes of the samples may not be the same, we have formulated null hypotheses to the effect that there has been independent sampling from the same populations, and the sampling distributions of our test statistics (Z, t, r, U, or D) have been based on this assumption. These principles readily generalize to more than two samples. In Chaps. 15 and 16 we shall be concerned with comparisons among three or more independently selected samples, where the second variable may be a nominal, ordinal, or interval scale. Although we shall not explicitly focus on more complex examples involving more than two matched samples, it can be seen from Exercise 5 of the previous chapter and Exercise 5 of this chapter that the extensions are straightforward. The basic idea is that one obtains a single score for each pair (which may be a difference of differences or some more complex function) and then proceeds as though there had been a single sample of size N, where N represents the number of pairs (or triplets, etc.).

A general problem we are confronting for the first time in the present chapter is that of the criteria one uses in choosing among alternative statistical procedures. We have focused primarily on the notion of the relative power efficiencies of the tests versus the fact that some tests require stronger assumptions than others. You should not be left with the impression that the issue is as simple as this distinction implies, however. It has already been pointed out that in most practical instances, one will not know enough about true parameter values to make definitive decisions. In addition, there is another more technical question that we have not discussed. This involves the relative sensitivities of tests to violations of required assumptions. For example, what harm does it do to use a difference-of-means test if the populations have a specified nonnormal form? What if the assumption of interval scales is violated? Statisticians use the term *robustness of a test* to refer to its sensitivity to distortions of various kinds. Robustness is particularly difficult to evaluate whenever there are *several* distortions or unmet assumptions that apply simultaneously. Although parametric procedures such as the difference-of-means test appear to be reasonably robust under many conditions, there are differences of opinions as to the advisability of using such tests when reasonably satisfactory nonparametric alternatives seem available.

Our own position is that whenever clearcut criteria are difficult to apply, it is always wise to use several different tests, both parametric and nonparametric, and to report both sets of results so that a reader may make his own decision. Usually this can be done by simply reporting results of a second test in footnotes, with comments suggesting why conclusions may not have been identical. In situations where there is a nonparametric test (or estimate) available whose power is almost as high as the comparable parametric procedure, as for example the Mann-Whitney test as an alternative to the t test, then it would seem preferable to rely primarily on the nonparametric procedure. But we shall encounter many multivariate parametric procedures for which there is no really satisfactory nonparametric alternative. Rather than using a weak or theoretically unsatisfactory nonparametric alternative, it would seem preferable in these instances to rely most heavily on the parametric procedures with the full recognition that definitive results may not be attainable. In short, one cannot give a simple dogmatic answer to the question of which kind of test or measure is most appropriate.

Glossary
Nonparametric test
*Power efficiency
*Power function
*Power of a test

Exercises
1. A number of Protestant churches in a community have been classified as being (1) predominantly upper or upper-middle class, or (2) predominantly lower-middle or lower class. They are ranked according to how formal their services are, with the following results:

Upper and upper-middle: Ranks 1, 2, 3, 6, 7, 8, 11, 13, 14, 15, 17, 21, 25
Lower-middle and lower: Ranks 4, 5, 9, 10, 12, 16, 18, 19, 20, 22, 23, 24, 26, 27

Using the .05 level, can you establish a significant difference (a) with the runs test, and (b) with the Mann-Whitney test? Which test would you prefer? Why? [*Ans.* (a) $r = 14$, fail to reject; (b) $U = 52$, fail to reject]

2. In Table 18.3, p. 417, data are given on the popularity rankings of the members of a summer work-camp group. Consider persons with participation ranks 1 to 8 as being "active" in group discussions, with the remainder of the group being placed in the "inactive" category. Is there a significant difference at the .05 level between the "actives" and "inactives" with respect to popularity? Use both the runs and Mann-Whitney tests.

3. Suppose you have been able to rank urban occupations from high to low, using the general categories of professional and managerial, clerical, skilled, semiskilled, and unskilled. You have asked every head of the household whether or not he favors

increasing social security benefits at the taxpayer's expense. The results are as follows:

Occupational level	Favors	Opposes
Professional and managerial	46	97
Clerical	81	143
Skilled	93	88
Semiskilled	241	136
Unskilled	131	38
Total	592	502

Is there a significant relationship between occupation and attitude at the .001 level? (*Ans. D* = .282, *P* < .001)

4. Work Exercise 2, Chap. 13, using the Smirnov test. Compare results with those of the *t* test.

5. Work all parts of Exercise 5, Chap. 13, using the Wilcoxon matched-pairs signed-ranks test. How do the results of the two tests compare? [*Ans.* (*a*) *T* = 14.5, fail to reject; (*c*) *T* = 11, fail to reject]

*6. Verify that Eq. (14.8) is algebraically equivalent to the other formula for *Z* given on p. 259.

References

1. Bradley, J. V.: *Distribution-free Statistical Tests*, Prentice-Hall, Inc., Englewood Cliffs, N.J., 1968, chaps. 1–3, 5, 11, and 13.
2. Dixon, W. J., and F. J. Massey: *Introduction to Statistical Analysis*, 3d ed., McGraw-Hill Book Company, New York, 1969, chap. 17.
3. Freund, J. E.: *Modern Elementary Statistics*, 3d ed., Prentice-Hall, Inc., Englewood Cliffs, N.J., 1967, chap. 13.
4. Hays, W. L.: *Statistics*, Holt, Rinehart and Winston, Inc., New York, 1963, chap. 18.
5. Pierce, Albert: *Fundamentals of Nonparametric Statistics*, Dickenson Publishing Company, Inc., Belmont, Calif., 1970, chap. 14.
6. Savage, I. R.: "Bibliography of Nonparametric Statistics and Related Topics," *Journal of the American Statistical Association*, vol. 48, pp. 844–906, 1953.
7. Siegel, S.: *Nonparametric Statistics for the Behavioral Sciences*, McGraw-Hill Book Company, Inc., New York, 1956, chaps. 5 and 6.
8. Smith, K.: "Distribution-free Statistical Methods and the Concept of Power Efficiency," in L. Festinger and D. Katz (eds.), *Research Methods in the Behavioral Sciences*, The Dryden Press, Inc., New York, 1953, pp. 536–577.
9. Swed, F. S., and C. Eisenhart: "Tables for Testing Randomness of Grouping in a Sequence of Alternatives," *Annals of Mathematical Statistics*, vol. 14, pp. 66–87, 1943.
10. Walker, H. M., and J. Lev: *Statistical Inference*, Henry Holt and Company, Inc., New York, 1953, chap. 18.

Nominal Scales: 15
Contingency
Problems

In this chapter we shall study the relationships between two or more nominal scales. We have already seen that the case of two dichotomized nominal scales can be handled as a problem involving a difference of proportions. It is often desirable to make use of a more general test procedure which enables us to test for differences among three or more samples or to compare two (or more) samples with respect to a variable that has more than two categories. The chi-square test discussed in the next section enables us to interrelate nominal scales with any number of categories. Several new ideas will also be introduced. So far we have only been concerned with tests for the *existence* of a relationship between two variables. Some measures indicating the *strength* or degree of relationship will be presented in this chapter. Procedures used in controlling for one or more variables will also be discussed.

15.1 The Chi-square Test

The chi-square test is a very general test that can be used whenever we wish to evaluate whether or not frequencies which have been empirically obtained differ significantly from those which would be expected under a certain set of theoretical assumptions. The test has many applications, the most common of which in the social sciences are "contingency" problems in which two nominal-scale variables have been cross-classified.[1] For example, suppose religious affiliation and voting preference have been interrelated and the data summarized in the following 3×3

[1] For another use of chi square, see Exercise 3 at the end of the chapter.

contingency table:

Party	Protestants	Catholics	Jews	Total
Republicans	126	61	38	225
Democrats	71	93	69	233
Independents	19	14	27	60
Total	216	168	134	518

Notice that if frequencies were converted to percentages, we could say that whereas 58.3 per cent of the Protestants are Republican, only 36.3 per cent of the Catholics and 28.4 per cent of the Jews prefer this political party. We would then want to ask whether or not these differences were statistically significant. Since there are three religious denominations and three categories of political preference, a single difference-of-proportions test cannot be used. In using the chi-square test we can make essentially the same kind of null hypothesis as before, however. We can assume that there are no differences among the three religious populations. This amounts to saying that the proportions of Republicans, Democrats, and Independents should be the same in each of the three groups. Proceeding under the assumption that the null hypothesis is correct and that the samples are random and independently selected, we can then compute a set of frequencies that would be expected given these marginal totals. In other words, we can compute the number of Protestants whom we would expect to be Republicans and compare this figure with that actually obtained. If these and comparable differences for other cells are quite large, we are likely to be suspicious of the null hypothesis.

Some measure of the difference between observed and expected frequencies must be obtained. There are, of course, a large number of possible measures, but we need one for which the sampling distribution is known and tabulated. For this reason, we make use of a measure referred to as chi square (χ^2) which is defined as follows:

$$\chi^2 = \sum \frac{(f_o - f_e)^2}{f_e} \tag{15.1}$$

where f_o and f_e refer respectively to the observed and expected frequencies for each cell.[2] In words, chi square is obtained by first taking the square of the difference between the observed and expected frequencies in each cell. We divide this figure by the *expected* number of cases in each cell

[2] In order to reduce confusion we have dropped the subscript i, it being assumed that we are summing over all the cells.

in order to standardize it so that the biggest contributions do not always come from the largest cells. The sum of these nonnegative quantities for all cells is the value of chi square.

Notice that the larger the differences between observed and expected frequencies, the larger the value of chi square. Chi square will be zero only when all observed and expected frequencies are identical. We can now perform a test of the null hypothesis by looking at the sampling distribution of chi square. We would hardly expect observed and expected frequencies to be exactly the same. If the value of chi square turns out to be larger than that expected by chance, however, we shall be in a position to reject the null hypothesis under the usual procedures.

PROBLEM   We can make use of the example given above but simplified so as to give a 2 × 2 table. The extension to the general case will turn out to be straightforward. Let us assume that Catholics and Jews have been combined and the Independents omitted. We then have the following table:

Party	Protestants	Catholics and Jews	Total
Republicans	126	99	225
Democrats	71	162	233
Total	197	261	458

It is important to note that the figures in each cell are actual frequencies rather than percentages. If figures given are percentages, they must be converted into frequencies, since the chi-square test statistic involves a comparison of frequencies rather than percentages.

1. *Assumptions*

Level of Measurement: Two nominal scales
Model: Independent random samples
Hypothesis: No differences among religious populations with respect
           to political preference

The level of measurement *can*, of course, be higher. Chi-square tests are frequently used with ordinal scales and sometimes even interval scales. As we have seen in previous chapters, however, more powerful tests are available in such instances and would ordinarily be used in preference to chi square. Again, it is necessary to assume independence between samples in order to make use of the chi-square test. The sample size must be relatively large because chi square, as defined by the formula, has a sampling distribution that approximates the distribution given in the table only when N is large.[3]

The null hypothesis can be stated in a number of equivalent ways. Saying that there is no difference among religious groups with respect to political preference is essentially saying that there is no relationship between religious affiliation and voting preference. It must be realized, however, that such a statement would apply only to these variables as they have been operationally defined: in this case political preference and religion would be defined as dichotomized variables. One could also state the null hypothesis by listing the various proportions that are assumed equal. Although this last method is perhaps the most precise, it can become quite cumbersome in the general case.

2. *Significance Level* Let us suppose that we want to demonstrate a difference and that we wish to be extremely conservative. We shall therefore use the .001 level. Suppose, also, that the direction of the difference is not predicted.

3. *Sampling Distribution* The sampling distributions for chi square are given in Table I, Appendix 2. Notice that distributions differ according to the degrees of freedom involved. The determination of degrees of freedom will be discussed below. Since regardless of the direction of the relationship between religion and political preference our interest is in whether or not the obtained chi square is *larger* than would be expected by chance, we are concerned only with the upper tail of the distribution. The lower tail, consisting of very small values of chi square, is not ordinarily used in contingency problems.

4. *Computation of Test Statistic* Our first task in the computation of chi square is to obtain the expected frequencies. The null hypothesis states that there are no population differences as to voting preference. Therefore, regardless of what the true percentage of Republicans in each religious *population* may be, we would expect that in the long run there would be the same proportion of Republicans in both of the *samples*. Since the proportion of Republicans in the combined sample is $225/458$ or .4913, we would expect this same figure in each of the two religious samples. Each sample would thus be expected to have the same percentage of Republicans and the same percentage of Democrats. We can then obtain the expected *number* of Republicans among the Protes-

---

[3] See pages 285–286 for a more complete discussion of this problem.

tants by multiplying .4913 by the total number of Protestants in the sample. Thus the expected number of Republican Protestants would be $(.4913)(197) = 96.8$. The other expected frequencies can be computed in a similar manner. It is generally advisable to keep at least one decimal place in computing expected frequencies. Thus we would not round up to 97.

Before going on, it should be noted that expected frequencies can also be obtained by reasoning the other way around, i.e., in terms of the proportion of Republicans we would expect to be Protestant. Since the proportion of Protestants in the combined sample is $197\!/\!458$ or .4301, we can get the expected frequency of Protestant Republicans as follows: $(.4301)(225) = 96.8$. You should learn to obtain expected frequencies both ways as a check on your computations.

After the procedure has become a familiar one, you will probably find it more convenient to make use of a simple formula described below. If we label the cells and marginal totals as

$$
\begin{array}{cc|c}
a & b & a+b \\
c & d & c+d \\
\hline
a+c \quad b+d & & N
\end{array}
$$

the expected frequency for any cell can be obtained by multiplying the two marginals corresponding to the cell in question and dividing by N. For example, the expected figure for cell a would be

$$(a + b)(a + c)/N = (225)(197)/458 = 96.8$$

The use of this last procedure reduces any rounding errors which may be introduced by first dividing (to obtain a proportion) and then multiplying.

It will be noted that this procedure of multiplying marginals and dividing by the total number of cases is basically the same as that discussed in Chap. 9 in connection with the independence of two variables. This emphasizes the fact that expected frequencies are computed on the basis of the *assumption* that the variables are unrelated, whereas the observed frequencies show us the degree to which this assumption is violated. Recall that if events (or variables) A and B are statistically independent, then knowing the value of one will not aid one in predicting the other. If the observed and expected frequencies were exactly equal, this would mean in our example that knowing a person's religious preference would not enable one to predict his political preference.

By convention, we usually place the expected frequencies in parentheses beneath the frequencies actually obtained for each cell as indicated below.

Party	Protestants	Catholics and Jews	Total
Republicans	126 (96.8)	99 (128.2)	225
Democrats	71 (100.2)	162 (132.8)	233
Total	197	261	458

Computations for chi square can then be summarized in a table such as Table 15.1. Notice that the quantity $f_o - f_e$ has the same numerical

Table 15.1 Computations for chi square

Cell	f_o	f_e	$f_o - f_e$	$(f_o - f_e)^2$	$(f_o - f_e)^2/f_e$
a	126	96.8	29.2	852.64	8.808
b	99	128.2	−29.2	852.64	6.651
c	71	100.2	−29.2	852.64	8.509
d	162	132.8	29.2	852.64	6.420
Total	458	458.0			30.388

value for each cell. You should convince yourself that this will always be the case in 2 × 2 tables but that it will not hold generally. Squaring this value has the effect of getting rid of the negative quantities. It is important that the *expected* frequencies, rather than the observed ones, be used in the denominators. The observed frequencies will vary from sample to sample, and some might even be equal to zero.

It is often more convenient to make use of a computing formula that does not actually require subtraction of each expected frequency from each observed frequency. Expanding the numerator in the expression for chi square and then collecting terms we get

$$\chi^2 = \sum \frac{(f_o - f_e)^2}{f_e} = \sum \frac{f_o^2 - 2f_o f_e + f_e^2}{f_e}$$
$$= \sum \frac{f_o^2}{f_e} - 2 \sum f_o + \sum f_e$$

But since both Σf_o and Σf_e are equal to N we obtain

$$\chi^2 = \sum \frac{f_o^2}{f_e} - N \tag{15.2}$$

Using this formula, which involves only a single subtraction, we get the same result as before (see Table 15.2).

Table 15.2 Computations for chi square, using computing formula

Cell	$f_o{}^2$	$f_o{}^2/f_e$
a	15,876	164.008
b	9,801	76.451
c	5,041	50.309
d	26,244	197.620
Total		488.388

$$\chi^2 = 488.388 - 458$$
$$= 30.388$$

In the case of the 2×2 table only, it is possible to express chi square as a simple function of the cell frequencies and marginal totals. If the table is labeled as before, we get

$$\chi^2 = \frac{N(ad - bc)^2}{(a + b)(c + d)(a + c)(b + d)} \tag{15.3}$$

Although this computing form requires the multiplication of large numbers, the use of logarithms may simplify considerably the computations involved. Incidentally, we see from Eq. (15.3) that chi square will be zero when the diagonal product ad is exactly equal to the product bc. This fact can be used as a quick method to determine whether or not it will be necessary to go ahead with a test for significance. If the diagonal products are almost equal, chi square will be too small to yield significance. These diagonal products·can also be used to determine the direction of the relationship without bothering to compute percentages. The larger of the two products indicates which diagonal contains the bulk of the cases.

*The above formulas for chi square, as well as the procedure for calculating expected frequencies, are sufficient for most purposes, but a somewhat different version applicable to the general $r \times c$ case may be useful for those who wish to follow discussions of chi square in more advanced texts. This alternative formulation will later be used to obtain the upper limit of chi square in the general $r \times c$ case. Additionally, the alternative form for the computing formula does not require the explicit calculation of the expected frequencies.

Let N_{ij} = number observed in (i, j)th cell of the table, and
         e_{ij} = number expected (under H_0) in the (i, j)th cell

for $i = 1, 2, \ldots, r$; and $j = 1, 2, \ldots, c$

Let $N_{i.} = \sum_{j=1}^{c} N_{ij}$ for $i = 1, 2, \ldots, r$ (row totals), and

         $N_{.j} = \sum_{i=1}^{r} N_{ij}$ for $j = 1, 2, \ldots, c$ (column totals)

Then we may express chi square as follows

$$\chi^2 = \sum_{i=1}^{r} \sum_{j=1}^{c} \frac{(N_{ij} - e_{ij})^2}{e_{ij}}$$

But since $e_{ij} = N \frac{N_{i.}}{N} \frac{N_{.j}}{N} = \frac{N_{i.}N_{.j}}{N}$

the computing formula (15.2) becomes

$$\chi^2 = N \left[\sum_{i=1}^{r} \sum_{j=1}^{c} \frac{N_{ij}^2}{N_{i.}N_{.j}} - 1 \right]$$

and we see that there is no need to compute the expected frequencies explicitly.

5. *Decision*   Before using the chi-square table, we must first determine the degrees of freedom associated with this test statistic. In previous problems the degrees of freedom have always depended on the number of cases sampled. For contingency problems, however, the degrees of freedom depend only on the number of *cells* in the table. In computing expected frequencies you may have noticed that it is not necessary to compute values for each cell since most could have been obtained by subtraction. In fact in the 2 × 2 table we need to compute only one expected frequency, and the others will all be determined. This is true because we make use of our sample marginal totals to compute expected frequencies. In other words, if we fill in a value for any one cell, the other values are completely determined since expected frequencies must have the same marginal totals as observed frequencies. We therefore have only 1 degree of freedom.

Having determined that there is only 1 degree of freedom in the 2×2 table, we look across the row corresponding to 1 degree of freedom in the chi-square table until we come to the desired level of significance. Corresponding to the .001 level we find a chi square of 10.827. This means that if all assumptions were in fact correct, we would obtain a value for chi square this large or larger only one time in a thousand. In other words, only very seldom will observed and expected frequencies differ by such an amount as to yield a chi square ≥ 10.827 *if* there were no relationship between religious affiliation and voting preference (as operationally defined in this problem). Since we actually obtained a value for chi square equal to 30.388, we conclude that the null hypothesis can be rejected at the .001 level. We see incidentally that when N is large, it is not at all difficult to obtain significance at the .001 level.

Even though we were concerned only with large values of chi square, the direction of the relationship was not predicted in the above example. Regardless of whether Protestants were more likely to be Republicans or Democrats, the result would have been a large chi square if differences in percentages were also large. In other words, the test statistic is insensitive to the direction of relationship since it involves the squares of deviations and cannot be negative. As noted previously, we can take advantage of predictions as to direction simply by halving the significance level obtained. If chi square is large enough to yield significance at the .10 level without predicting direction, the result will be significant at the .05 level provided, of course, that the direction of the relationship was predicted beforehand.

If the desired significance level cannot be obtained exactly from the chi-square table, a good approximation can be made by taking the square root of chi square and entering the normal table. For example, we know that a chi square of 3.841 with 1 degree of freedom corresponds to the .05 level when direction has not been predicted. The square root of this figure is 1.96 which is the Z value necessary for significance with the normal table. The normal table can be used only in the case of 2×2 contingency problems.

General case   In the general case of the contingency table with r rows and c columns, the assumptions and computations for chi square require only slight modification. The null hypothesis of "no difference" or "no relationship" now implies that each population will have the same proportions for each of the categories of the second variable. The expected frequencies can be obtained in exactly the same way as before, but there will now be rc cells, and the degrees of freedom will be different.

Suppose we make use of the same example as before but in its original

form, that of a 3 × 3 table. Incidentally, this table supplies us with
more information than in the 2 × 2 case in which Catholics and Jews
were combined into a single category. We therefore can expect results
that may differ somewhat from those obtained above. Computing expected frequencies by any of the methods previously suggested, we obtain

Party	Protestants	Catholics	Jews	Total
Republicans	126 (93.8)	61 (73.0)	38 (58.2)	225
Democrats	71 (97.2)	93 (75.6)	69 (60.2)	233
Independents	19 (25.0)	14 (19.4)	27 (15.6)	60
Total	216	168	134	518

A computing table can be constructed as before (see Table 15.3).

Table 15.3   Computations for chi square for 3 × 3 contingency table

Cell	f_o	f_e	$f_o{}^2$	$f_o{}^2/f_e$
a	126	93.8	15,876	169.254
b	61	73.0	3,721	50.973
c	38	58.2	1,444	24.811
d	71	97.2	5,041	51.862
e	93	75.6	8,649	114.405
f	69	60.2	4,761	79.086
g	19	25.0	361	14.440
h	14	19.4	196	10.103
i	27	15.6	729	46.731
Total	518	518.0		561.665

$$\chi^2 = 561.665 - 518 = 43.665$$

To determine the proper degrees of freedom, we notice that once the
first two expected frequencies have been filled in for the first column, the
third is determined by subtraction. The same will be true for the second
column. All the expected frequencies in the third column will then be
determined from the row totals. Generally, for each of the first $c - 1$
columns, it will be possible to fill in all but one, or $r - 1$, of the cells. The
final column will then always be completely determined. Therefore the
number of degrees of freedom for the $r \times c$ contingency table can be given

by the formula

$$df = (r - 1)(c - 1)$$

Notice that this formula gives 1 degree of freedom in the special case where $r = c = 2$.

Since there are 4 degrees of freedom associated with our 3×3 table, we see that a chi square of 18.465 is required for rejection at the .001 level. We therefore reject the null hypothesis. Notice that although a larger value of chi square is required for rejection, there are many more cells contributing to its value. Since chi square represents a sum rather than an average, we would expect that, other things being equal, the greater the number of cells, the larger the chi square. The fact that the value of chi square needed for significance increases with the degrees of freedom should not surprise us.[4]

Correction for continuity   It has been indicated that the chi-square test requires a relatively large N because of the fact that the sampling distribution of the test statistic approximates the sampling distribution given in the chi-square table only when N is large. The question naturally arises, then, as to how large N has to be in order to make use of this test. The answer depends on the number of cells and the marginal totals. Generally, the smaller the number of cells and the more nearly equal are all marginal totals, the smaller the total N can be. The criteria usually used for deciding whether or not the number of cases is sufficient involve the *expected* frequencies in each cell. Whenever any of the expected frequencies are in the neighborhood of 5 or smaller, it is advisable to make some kind of modification as indicated below.

The chi-square distribution is assumed to be a continuous one. Actually, however, when the number of cases is relatively small, it is impossible for the computed value of chi square to take on very many different values. This is true because observed frequencies must always be integers. In correcting for continuity we imagine that observed frequencies actually can take on all possible values, and we make use of those values within a distance of half a unit on either side of the integer obtained which will give the most conservative results. In the case of the 2×2 table a correction for continuity can very easily be made. This correction consists of either adding or subtracting .5 from the observed frequencies in order to reduce the magnitude of chi square. The corrected version of

[4] Note that the opposite was true in the case of the t distribution. Why?

Eq. (15.3) becomes

$$\chi^2 = \frac{N\left(|ad - bc| - \dfrac{N}{2}\right)^2}{(a + b)(c + d)(a + c)(b + d)}$$

To see the effect of correcting for continuity we can take the following tables:

(A)	7	13	20		(B)	7.5	12.5	20
	(10)	(10)				(10)	(10)	
	8	2	10			7.5	2.5	10
	(5)	(5)				(5)	(5)	
	15	15	30			15	15	30
	$\chi^2 = 5.40$					$\chi^2 = 3.75$		

In Table *B* we have corrected for continuity by reducing the differences between observed and expected frequencies in each cell by .5. We have imagined that there are between 6.5 and 7.5 cases in the top left-hand cell and have used the number 7.5 since it is the closest value within this interval to the expected frequency of 10.0. In this example, correcting for continuity reduces the significance level from approximately .02 to somewhat greater than .05. Obviously, corrections for continuity will have less effect when expected frequencies are larger. Since making such a correction actually involves very little additional effort, and since one is on the conservative side in so doing, it is recommended that the correction be made whenever the expected frequency in any cell falls below 10. With very small samples even this correction produces misleading results. An alternative test discussed in the next section is available for 2 × 2 tables.

Corrections for continuity cannot easily be made in the case of the general contingency table. If the number of cells is relatively large and if only one or two cells have expected frequencies of 5 or less, then it is generally advisable to go ahead with chi-square tests without worrying about such corrections. If there are a large number of small cells, however, the only practical alternative may be to combine categories in such a manner as to eliminate these cells. Of course categories should only be combined if it makes sense to do so theoretically. Thus if there were an "other religions" category consisting of such a wide range of religious groups as to make the category theoretically meaningless, it would perhaps be better to exclude these persons from the analysis altogether, although as a general rule it is not good practice to exclude data from one's analysis.

*15.2 Fisher's Exact Test

In the case of 2 × 2 tables where N is small, it is possible to make use of a test developed by R. A. Fisher which gives us exact rather than approximate probabilities. If we label the cells and marginals of a 2 × 2 table as

$$
\begin{array}{cc|c}
a & b & a+b \\
c & d & c+d \\
\hline
a+c & b+d & N
\end{array}
$$

we can obtain the probability of getting *exactly* these frequencies under the null hypothesis that there are no differences in the population proportions. This probability is given by the formula

$$
P = \frac{(a+b)!(c+d)!(a+c)!(b+d)!}{N!\,a!\,b!\,c!\,d!}
$$

This probability formula can be derived by using the hypergeometric distribution for calculating probabilities on the basis of sampling *without replacement*. In this test, as well as certain other nonparametric tests, we may conceptualize the problem as involving repeated samples from a "population" of size N. In other words, we treat our obtained sample as though it were an actual population, and we imagine, in this instance, categorizing our cases as falling into one of the four cells. Given that there are $a + c$ individuals in the first column, $a + b$ in the first row, and so forth, what is the probability that of the $a + b$ individuals in the first row, exactly a will be in the first column and b in the second column? We imagine that we have sampled $a + b$ individuals at random, but without replacement, and placed them in the first row, with the remainder necessarily falling in the second row. In effect, then, we imagine that we fill the cells by essentially a random process and ask how likely our given results would have been had such a process actually been carried out.

Applying the formula for the hypergeometric distribution given in Sec. 10.4, we see that the probability of getting exactly a and b cases in the two cells of the top row is given by

$$
P(a,b) = \frac{\binom{a+c}{a}\binom{b+d}{b}}{\binom{N}{a+b}}
$$

Writing each of the terms in terms of factorials and simplifying we get

$$P(a,b) = \frac{\dfrac{(a + c)!}{a!(a + c - a)!}\dfrac{(b + d)!}{b!(b + d - b)!}}{\dfrac{N!}{(a + b)!(N - a - b)!}} = \frac{\dfrac{(a + c)!}{a!c!}\dfrac{(b + d)!}{b!d!}}{\dfrac{N!}{(a + b)!(c + d)!}}$$

$$= \frac{(a + c)!(b + d)!(a + b)!(c + d)!}{N!a!b!c!d!}$$

It can easily be verified that the same result would have been obtained had we conceived of the problem in terms of selecting a sample of $a + c$ individuals and assigning them to the first *column*.

Notice that there are nine factorials in this formula for P. The task of directly evaluating the formula would therefore be quite formidable. Furthermore, since one is ordinarily interested in the entire tail of the sampling distribution rather than the probability of getting exactly the results obtained, he will then have to add to this first probability the probabilities of getting even more unusual outcomes in the same direction.

A simple numerical example can be used to illustrate what is involved. Suppose we have obtained the following 2 × 2 table:

3	9	12
12	5	17
15	14	29

If we assume that the marginals remain fixed, we immediately see that there are three outcomes (in the same direction) which are even more unlikely than the one obtained. They are as follows:

2	10	12		1	11	12		0	12	12
13	4	17		14	3	17		15	2	17
15	14	29		15	14	29		15	14	29

Notice that we can arrive at the successive tables by each time reducing by one the cells a and d, and increasing by one the cells b and c, until we reach the final table in which cell a is empty.

We shall suppose that it is always cell a which contains the smallest number of cases, since it will always be possible to arrange the tables in this fashion.[5] Let us use the symbol P_0 to denote the probability of getting exactly no cases in cell a (given these marginals) under the null hypothesis; let P_1 represent the probability of getting exactly one case in cell a, P_2 the probability of getting exactly two, etc. Then in this particular problem we must obtain the sum of the probabilities

$$P_0 + P_1 + P_2 + P_3$$

in order to compute the probability of getting three or fewer cases in cell a. Since we are making use of a one-tailed test, we shall have to double the significance level obtained if we were unable to predict direction beforehand.[6]

Rather than compute each of the P_i from the above formula involving products of factorials, it will be much more convenient to obtain P_0 directly and then to obtain the remaining probabilities as simple functions of P_0. In order to distinguish between the various possible combinations of numerical values of a, b, c, and d for fixed marginals, let us make use of the subscript k to denote the magnitude of the smallest cell a. Thus if there are k individuals in cell a, we shall refer to the quantities in the various cells as $a_k\ (= k)$, b_k, c_k, and d_k. Since the marginals are assumed to remain fixed, if we decrease a_k and d_k by 1, we must increase b_k and c_k by 1. We can now simplify the formula for P_0 since $a_0 = 0$ and therefore $a_0! = 1$ (by definition), $(a_0 + b_0)! = b_0!$, and $(a_0 + c_0)! = c_0!$. A number of the factorials therefore cancel out, leaving us with

$$P_0 = \frac{(c_0 + d_0)!(b_0 + d_0)!}{N!d_0!}$$

The numerator now consists of the factorials of just two of the marginals rather than all four, and the denominator involves only $N!$ and $d_0!$. The value of d_0 can be obtained from the last of the four tables given above. In this example, therefore, $(c_0 + d_0) = 17$, $(b_0 + d_0) = 14$, $N = 29$, and $d_0 = 2$. P_0 can now be evaluated by using a table of logarithms of factorials or by writing out the factorials and simplifying.

[5] In rare instances the direction of the relationship will change if one always follows the rule that cell a must always be the smallest cell. For example, if both marginal distributions are grossly unequal, the rule may not work. Thus if a, b, c, and d are 1, 2, 3, and 7 respectively, the product $ad\ (= 7)$ is greater than the product $bc\ (= 6)$. If one then reduces a to 0, the resulting cells will be 0, 3, 4, and 6, and there will be a reversal of direction since $bc > ad$. One should therefore check for such reversals and, if they occur, he should label as a the smallest cell in the lesser of the two diagonals.
[6] In a strict sense, Fisher's test should probably only be used when direction has been predicted in advance, since the two tails will practically never be perfectly symmetrical.

In order to compute the values of P_1, P_2, and P_3, we now need a general formula for P_{k+1} in terms of P_k. Since marginals are assumed fixed, we have

$$P_{k+1} = \frac{(a+b)!(c+d)!(a+c)!(b+d)!}{N!(a_k+1)!(b_k-1)!(c_k-1)!(d_k+1)!}$$

because of the fact that when we add 1 to the a cell we also add 1 to d and subtract 1 from both b and c. If we now divide P_{k+1} by P_k, practically all terms will vanish. The numerators for both probabilities are identical since they involve the same marginals. The factorial N's cancel. We are then left with

$$\frac{P_{k+1}}{P_k} = \frac{a_k!b_k!c_k!d_k!}{(a_k+1)!(b_k-1)!(c_k-1)!(d_k+1)!}$$

But $a_k!/(a_k+1)!$ is just $1/(a_k+1)$ and similarly for $d_k!/(d_k+1)!$. Also, $b_k!/(b_k-1)! = b_k$ and $c_k!/(c_k-1)! = c_k$.

Therefore,
$$\frac{P_{k+1}}{P_k} = \frac{b_k c_k}{(a_k+1)(d_k+1)}$$

or
$$P_{k+1} = \frac{b_k c_k}{(a_k+1)(d_k+1)} P_k$$

and the troublesome factorials have disappeared. Therefore, we can make use of this formula to obtain P_1 from P_0. Having computed P_1, we can then get P_2 and so on.

Returning to our numerical example, we obtain P_0 as follows:

$$P_0 = \frac{14!17!}{29!2!} = .17535 \times 10^{-5}$$

Therefore,

$$P_1 = \frac{b_0 c_0}{(a_0+1)(d_0+1)} P_0 = \frac{12(15)}{1(3)} (.17535 \times 10^{-5}) = 10.521 \times 10^{-5}$$

In computing P_2 we must be careful to make use of a_1, b_1, c_1, and d_1 rather than the figures used in obtaining P_1. We get

$$P_2 = \frac{b_1 c_1}{(a_1+1)(d_1+1)} P_1 = \frac{11(14)}{2(4)} (10.521 \times 10^{-5}) = 202.529 \times 10^{-5}$$

Similarly,

$$P_3 = \frac{b_2 c_2}{(a_2+1)(d_2+1)} P_2 = \frac{10(13)}{3(5)} (202.529 \times 10^{-5}) = 1,755.252 \times 10^{-5}$$

Notice that each of the factors in the numerator is decreased by 1 as we compute P_{k+1} from P_k, whereas the factors in the denominators are each increased by unity. Adding the probabilities we thus get

$$P_0 + P_1 + P_2 + P_3 = [.175 + 10.521 + 202.529 + 1,755.252] \times 10^{-5}$$

$$= 1,968.48 \times 10^{-5} = .0197$$

Therefore the probability of getting three or fewer individuals in cell a under the null hypothesis is .02, and we would make our decision whether or not to reject the null hypothesis accordingly.

Because the Fisher test is an exact test, it is to be preferred over the chi-square test corrected for continuity. Since the chi-square test will ordinarily yield somewhat lower probabilities than the Fisher test, one will be on the conservative side in using this exact test if he actually wants to reject the null hypothesis. In other words, in using the chi-square test we may arrive at probabilities which are actually too small, possibly leading us to the conclusion that the null hypothesis should be rejected when actually it should not. If the smallest expected frequency is quite a bit larger than 5 and if the correction for continuity is used, the two tests will give approximately the same results. Even though we may avoid the use of factorials in the case of the Fisher test, it can be seen that if the smallest cell frequency is greater than 5, the computations involved will become quite tedious. Therefore, the Fisher test will be found to be most useful in the case of very small N's or whenever the total sample size is moderate but one or more of the marginals very small. In cases where both $(a + b)$ and $(c + d)$ are ≤ 30, tables are available in [3] which make this exact test extremely simple to use.

15.3 Measures of Strength of Relationship

Up to this point we have only been concerned with the question of whether or not a relationship between two variables exists. We have set up null hypotheses to the effect that there is no relationship and have then tried to reject these hypotheses. But just how much have we accomplished when we are able to reject? We refer to a relationship as being statistically significant when we have established, subject to the risk of a type I error, that *there is* a relationship between the two variables. But does this mean that the relationship is significant in the sense of being a strong relationship or an important one? Not necessarily. The question of the strength of a relationship is a completely different question from that of whether or not a relationship exists. In this section we shall take up several measures of degree of association which can be used to help answer this second kind of question.

It would seem reasonable, on first thought, to attempt to assess the strength of a relationship by simply noting the significance level attained in a test. For example, it might be reasoned that if one test were significant at the .001 level and another at the .05 level, the former relationship would be the stronger of the two. But is this necessarily the case? Looking at the two significance levels can tell us in which case we can have more faith that a relationship exists. Thus, we would be almost certain of the *existence* of a relationship in the first case but not so sure in the second. We must remember, however, that the significance level attained depends on the sizes of the samples used. As indicated previously, if the samples are very large, it is generally easy to establish significance for even a very slight relationship. This means, in effect, that when samples are large, we are saying very little when we have established a "significant" relationship. For large samples, a much more important question is, "Given that a relationship exists, how strong is it?"

In order to illustrate the above argument, let us take a closer look at a property of chi square. In doing so, you should keep in mind that exactly the same principles apply for other kinds of significance tests. Let us ask ourselves what happens to chi square when the number of cases is increased. For illustrative purposes we can take the following 2 × 2 table:

$$
\begin{array}{cc|c}
30 & 20 & 50 \\
20 & 30 & 50 \\
\hline
50 & 50 & 100
\end{array}
$$

Chi square for this table turns out to be exactly 4.0. Now suppose we were to double the sample sizes, keeping the same proportions in each cell. We would then obtain

$$
\begin{array}{cc|c}
60 & 40 & 100 \\
40 & 60 & 100 \\
\hline
100 & 100 & 200
\end{array}
$$

and chi square would be 8.0, a figure which is exactly double the previous one. By examining the formula for chi square it is very easy to prove that if proportions in the cells remain unchanged, chi square varies directly with the number of cases. If we double the number of cases, we double chi square; if we triple them, we triple chi square. Suppose the original number of cases is multiplied by a factor k. Then since proportions in the cells remain unchanged, each new observed frequency will be exactly k times the old one, and similarly for the expected frequencies. The new chi square can thus be expressed as

$$(\chi^2)' = \sum \frac{(kf_o - kf_e)^2}{kf_e} = \sum \frac{k^2(f_o - f_e)^2}{kf_e} = k \sum \frac{(f_o - f_e)^2}{f_e}$$

Thus the value of the new chi square is exactly k times that of the original one.

The implications of this fact can be brought out by means of another illustration. Suppose we obtain the following results when we relate sex differences to tolerance of deviant behavior:

Tolerance	Males	Females
High	26	24
Low	24	26

In this case chi square is 0.16, and we would rightly report that the relationship is not a significant one. Suppose, however, that the survey had been a very ambitious one and that data had been collected on 10,000 cases with the following results:

Tolerance	Males	Females
High	2,600	2,400
Low	2,400	2,600

Chi square is now 16.0, a value that is highly significant statistically. Had we expressed the results in terms of percentages, however, they would have looked much less interesting. If we said that 52 per cent of the males were highly tolerant whereas *only* 48 per cent of the females were in this category, we would rightly be criticized for emphasizing differences that seemed trivial from the standpoint of theoretical or practical significance. This example illustrates a very important point. A difference may be statistically significant without being significant in any other sense. In the instance where 10,000 cases were selected, we can be very sure that there is some slight relationship that will produce a statistically significant relationship.

We can see that when a sample is small, it requires a much more striking relationship in order to obtain significance. Therefore, with small samples significance tests are far more important. In such cases we may be saying quite a bit when we can establish significance. The significance level depends on two factors: the strength or degree of relationship and the size of the samples. Significance can be obtained with a very strong relationship and very small samples, or with a very weak relationship and large samples. In most social research our primary interest is not so much in finding variables that are interrelated but in locating the important relationships. Although it should be emphasized that not all strong relationships are important (e.g., the relationship between age of husband and age of wife), in order for a relationship to be of some practical importance it must be at least moderately strong. Having first established the existence of a relationship, the researcher should always ask himself, "How strong is it?"

How is strength of relationship measured, then? We are seeking a descriptive measure that can help us summarize the relationship in such a manner that we can compare several relationships and reach a conclusion as to which is the strongest. Ideally, we would also like to have some kind of operational interpretation for the measure that has intuitive appeal. By convention, statisticians have adopted the custom of designing measures that have unity as an upper limit and either zero or -1.0 as a lower limit. Most measures can attain their limit of 1.0 (or -1.0) only when the relationship is a perfect one, and they take on the value zero when there is no relationship at all between the variables, i.e., they are independent. Several measures that can be used with contingency tables will presently be discussed and their properties evaluated.

Before taking up various measures of association that can be used with contingency tables, we should at least mention the rather simple and obvious procedure of reporting differences in terms of percentages. It is certainly possible to get a very good indication of the degree of relationship between two dichotomized variables by comparing percentages. For example, if 60 per cent of the males sampled are classed as highly tolerant whereas only 30 per cent of the females are so categorized, there is a 30 per cent difference between the two groups. Why not use such a figure as a measure of strength of relationship? If, for example, we compare middle- and lower-class individuals with respect to tolerance, obtaining only a 20 per cent difference, we can then claim a stronger relationship between sex and tolerance than between class and tolerance.

In the special case of the 2 × 2 table, percentages can easily be compared in such a manner, and the widespread familiarity with percentages as contrasted with other types of measures would certainly argue for such comparisons of percentages.[7] But what about the general $r \times c$ table?

---

[7] Another advantage of percentages will be seen when we consider slopes in Chap. 17. As noted in connection with tests for differences of differences in proportions, a difference of proportions may be considered a special case of a slope.

Here the use of percentages may make it difficult for the reader to see at a glance how strong the relationship may be. For example, suppose three social classes were used with the following results: upper class, 70 per cent highly tolerant; middle class, 50 per cent highly tolerant; and lower class, 30 per cent highly tolerant. We now have a spread of 40 per cent between the upper and lower classes, a difference which is numerically larger than that between males and females. On the other hand, we would ordinarily expect a larger difference when only the extremes are used. Suppose there had been five classes. What kind of percentage differences would we now expect, and how would we compare the results with those of a 2 × 2 table? To introduce a further complication, suppose we had used four categories of tolerance. Quite obviously, it becomes difficult to make comparisons from one table to the next. We need a single summarizing measure which will have the same upper and lower limits regardless of the number of cells.

Traditional measures based on chi square It has already been noted that chi square is directly proportional to N. We can make use of this fact to construct several measures of association. In the case of the two contingency tables

30	20	50		60	40	100
20	30	50	and	40	60	100
50	50	100		100	100	200

we desire a measure which would have the same value for each table since when we express results in terms of percentages they are the same in both cases. In other words, we would probably say that the degrees or strengths of relationship in the two sets of data are identical, and that the only difference is in the size of the samples. Although the value of chi square for the second table is double that of the first, we notice that if chi square is in each instance divided by the total number of cases, the results are identical. This suggests that χ^2/N or some multiple of this expression would give us one of the properties we desire in our measure, that it yield the same result when the proportions in comparable cells are identical.

Notice that the value of χ^2/N, or ϕ^2 as it is commonly denoted, is 0 when there is absolutely no relationship between the two variables. It turns out that in the case of 2 × 2 (or 2 × k) tables, ϕ^2 also has an upper limit of unity when the relationship between the two variables is perfect.

Suppose we had obtained the following table:

50	0	50
0	50	50
50	50	100

You can easily verify that in this case chi square is 100 and therefore ϕ^2 is $100/100$ or 1.0. It will always be the case that when two diagonally opposite cells are *both* empty, the value of chi square in a 2 $\times$ 2 table will be N, and therefore ϕ^2 will be unity. Obviously, the relationship in the above example is as perfect as it could possibly be. If sex were being related to tolerance, we could then say that all males are highly tolerant and all females intolerant. In terminology with which we shall shortly become more familiar, we can say that all of the variation in tolerance is explained by or associated with sex.[8]

In the general $r \times c$ table, ϕ^2 can attain a value considerably larger than unity. Therefore, several other measures have been developed which are also simple functions of χ^2/N but which also have unity as their upper limits. The first of these measures, referred to as Tschuprow's T, is defined as

$$T^2 = \frac{\chi^2}{N \sqrt{(r-1)(c-1)}} = \frac{\phi^2}{\sqrt{(r-1)(c-1)}}$$

Although the upper limit of T is unity, this limit can be attained only when the numbers of rows and columns are equal. In other words, T must always be less than one in a 2 $\times$ 3 or 3 $\times$ 5 table. If there are considerably more rows than columns (or vice versa), the upper limit of T may be well below unity. To correct for this fact, we can always divide the obtained value of T by the maximum T possible for given numbers of rows and columns. Since more satisfactory measures are available, however, there is no need to discuss such a correction procedure.

*We can show that the upper limit of ϕ^2 is Min $(r-1, c-1)$ by utilizing the formula

$$\chi^2 = N \left[\sum_{i=1}^{r} \sum_{j=1}^{c} \frac{N_{ij}^2}{N_{i.}N_{.j}} - 1 \right]$$

Note that

$$\frac{N_{ij}^2}{N_{i.}N_{.j}} \leq \frac{N_{ij}}{N_{i.}} \qquad \text{for } i = 1, 2, \ldots, r$$

[8] This assumes, of course, that tolerance is taken as a dichotomized variable.

and
$$\frac{N_{ij}{}^2}{N_{i.}N_{.j}} \leq \frac{N_{ij}}{N_{.j}} \qquad \text{for } j = 1, 2, \ldots, c$$

Therefore
$$\sum_{i=1}^{r} \sum_{j=1}^{c} \frac{N_{ij}{}^2}{N_{i.}N_{.j}} \leq \sum_{i=1}^{r} \sum_{j=1}^{c} \frac{N_{ij}}{N_{i.}} = \sum_{i=1}^{r} 1 = r$$

and
$$\sum_{i=1}^{r} \sum_{j=1}^{c} \frac{N_{ij}{}^2}{N_{i.}N_{.j}} \leq \sum_{j=1}^{c} \sum_{i=1}^{r} \frac{N_{ij}}{N_{.j}} = \sum_{j=1}^{c} 1 = c$$

Thus
$$\sum_{i=1}^{r} \sum_{j=1}^{c} \frac{N_{ij}{}^2}{N_{i.}N_{.j}} \leq \text{Min } (r, c)$$

and hence
$$\chi^2 \leq N[\text{Min } (r, c) - 1] = N[\text{Min } (r - 1, c - 1)]$$

Therefore
$$\phi^2 \leq \text{Min } (r - 1, c - 1)$$

There is another measure, introduced by Cramér and which we shall denote by V, defined as follows

$$V^2 = \frac{\chi^2}{N \text{ Min } (r - 1, c - 1)} = \frac{\phi^2}{\text{Min } (r - 1, c - 1)}$$

where $\text{Min}(r - 1, c - 1)$ refers to either $r - 1$ or $c - 1$, whichever is the smaller (minimum value of $r - 1$ and $c - 1$). Although V is not commonly used in the social science literature, it seems to be preferable to T in that it can attain unity even when the numbers of rows and columns are not equal. As can easily be verified, V and T are equivalent whenever $r = c$. Otherwise, V will always be somewhat larger than T. Of course both measures become equivalent to ϕ in the 2×2 case. Also, in the $2 \times k$ case we see that V and ϕ will be identical.

Still another measure of association based on chi square is Pearson's contingency coefficient C which is given by

$$C = \sqrt{\frac{\chi^2}{\chi^2 + N}}$$

Like the other measures, C becomes 0 when the variables are independent. The upper limit of C, however, depends on the number of rows and columns. In the 2×2 case the upper limit of C^2 becomes $N/(N + N)$ since χ^2 can reach a maximum value of N. Therefore, the upper limit of

C is .707. Although the upper limit increases as the number of rows and columns increases, this upper limit is always less than one. For this reason, C is somewhat more difficult to interpret than the other measures unless a correction is introduced by dividing by the maximum value of C for the particular numbers of rows and columns. In the case of the 2×2 table, for example, the obtained C should be divided by .707.

The above measures of strength of relationship are all based on chi square. Since the value of chi square would ordinarily have been previously calculated in order to test for significance, all of these measures require very little additional computation. On the other hand, there is no particular reason why a measure of association has to be based on the comparable test statistic. In fact it can be shown that all measures based on chi square are somewhat arbitrary in nature, and their interpretations leave a lot to be desired. For example, they all give greater weight to those columns or rows having the smallest marginals rather than to those with the largest marginals [2]. Since both T and C are sometimes found in the literature, however, you should be familiar with their properties.

Yule's Q Another commonly used measure is Yule's Q, which is also a special case of a measure γ (gamma) to be discussed in Chap. 18 in connection with ordinal scales. This measure can only be used in a 2×2 table and is defined as

$$Q = \frac{ad - bc}{ad + bc}$$

where a, b, c, and d refer to the cell frequencies. Notice that the numerator, when squared and multiplied by N, is the numerator in the expression for chi square. Like the other measures, Q is zero when the variables are independent, i.e., when diagonal products ad and bc are equal. Unlike ϕ^2, however, Q attains its limits of ± 1.0 whenever any *one* of the cells is zero. In order to understand the nature of the circumstances under which Q can be unity whereas ϕ^2 is less than this value, let us take the following examples:

30	0	30		40	0	40
20	50	70		10	50	60
50	50	100		50	50	100

Although Q takes on the value of unity in both of the above tables, the values of ϕ^2 are .429 and .667 respectively. In both cases it would be impossible to have *two* diagonally opposite cells vanish because of the nature of the marginal totals. Therefore ϕ^2 can take on the value 1 only when certain conditions hold for the marginals. In the 2×2 table the marginals for the first variable have to be identical with those of the second.[9] The greater the discrepancy between row and column marginals, the less the upper limit of ϕ^2.

The question now arises as to whether or not we wish to consider a relationship "perfect" when only one of the cells vanishes. The answer to this question would seem to depend, among other things, upon how the categories of the two variables were formed. Usually it is possible to conceptualize a problem in terms of an independent and a dependent variable. It would then seem reasonable to argue that in order for a relationship to be perfect, the marginals for the dependent variable should naturally "fit" those of the independent variable. Suppose, for example, that there were 60 Protestants and only 40 Catholics and Jews. Then for a perfect relationship we would expect all 60 Protestants to vote Republican and all 40 of the others to vote Democratic. The marginals would then be the same for both variables and both ϕ^2 and Q would be unity. On the other hand, if half of the sample voted Republican and half Democratic, then even though all of the Republican votes came from the Protestants, we would not call the relationship a perfect one since 10 of the Protestants must have voted Democratic. In this case, the marginals for the dependent variable would not coincide with those of the independent variable, and ϕ^2 would be less than unity. Therefore ϕ^2 would seem the more appropriate measure in such an instance since Q would take on the value unity in spite of the imperfect relationship between the two variables.

Occasionally it so happens that the marginals of the dependent variable are fixed by the method used in categorizing. If, for example, the dependent variable were actually continuous but had been dichotomized at the median, then the two sets of marginals could not possibly be identical unless the marginals for the independent variables were also split 50-50. For example, had religious preference been related to political conservatism scores dichotomized at the median, then ϕ^2 could not attain unity (assuming the same religious split as above). In such a case Q might be a more appropriate measure since it can take into consideration the fact that the marginals for the dependent variable have been completely fixed by the method of research.

[9] This does not mean that marginals have to involve a 50-50 split. It means that if one marginal is split 70-30, the other must also be split 70-30. Corrections for unequal marginals are also possible, but, as implied in the discussion that follows, one should be cautious in using such corrections.

Goodman and Kruskal's tau   A number of other measures of association that can be used with contingency tables have been presented by Goodman and Kruskal [5], [6], and [7]. Most of these measures involve what have been referred to as probabilistic interpretations. Since they have an intuitive meaning enabling one to interpret values intermediate between zero and one, these measures would seem to be superior to those based on chi square.

In order to illustrate one of these measures τ_b, let us take a numerical example. We shall refer to the nominal scales being related as A and B, and we shall take B as the dependent variable.

	A_1	A_2	Total
B_1	300	600	900
B_2	600	100	700
B_3	300	100	400
Total	1,200	800	2,000

Now let us suppose that we are given a sample (or population) of 2,000 persons and are asked to place them in one of the three categories B_1, B_2, or B_3 in such a manner that we are to end up with exactly 900 cases in B_1, 700 in B_2, and 400 in B_3. Suppose, first, that we know nothing about the individuals that will aid us in this task. If the individuals are given to us in a completely random order, we can very easily calculate the number of errors we can expect to make in assigning individuals to one of the three categories.

Since we shall be assigning 900 individuals to B_1, whereas 1,100 out of every 2,000 actually do not belong in this class, we can expect to make $900(1,100/2,000)$ or 495 errors in the long run. Similarly, we must assign 700 individuals to B_2, whereas 1,300 out of 2,000 do not belong there. Therefore we can expect to make $700(1,300/2,000)$ or 455 errors in putting individuals into B_2. In other words, of the 700 we place in this category, we can expect that only $700 - 455$ or 245 will be placed correctly. Of course we do not expect to make exactly 455 errors, but this is the figure we would get if we averaged out our errors in the long run. Finally, we would expect to make $400(1,600/2,000)$ or 320 errors in assigning individuals to B_3. Notice that although we make fewer assignments to this smaller category, our risk of error is greater than for the other two categories since only 20 per cent of the individuals actually belong in this class. Therefore, in total, we would expect to make

$$495 + 455 + 320 = 1,270$$

errors in placing these 2,000 individuals. Our batting average would not be very good.

Now suppose we are given some additional information about each individual; we are told whether he is in A_1 or A_2. We now ask ourselves whether knowing the A class will help us reduce the number of errors made in assigning the individuals to B categories. If the variables A and B are statistically independent, we know that knowledge of A will not help us predict B. In this case, then, we would expect to make just as many errors as when we did not use the information about the A class. On the other hand, if the relationship between A and B were perfect, we would be able to predict B with complete accuracy if we knew A. The measure we shall develop tells us the proportional reduction in errors when A is known.

Let us see how we calculate the number of expected errors when A is known. If we are given the fact that the individual belongs in A_1, we can use the first column of figures. We now must place exactly 300 of these 1,200 individuals in B_1, the remaining 600 in B_1 coming from A_2. Since 900 out of the 1,200 individuals in A_1 actually do not belong in B_1, we can expect to make 300(900/1,200) or 225 errors. Similarly, out of the 600 individuals in A_1 we place in B_2, we can expect to make 300 errors, and the expected number of errors for B_3 will be 225. We now take the 800 individuals in A_2 and assign 600 to B_1 and 100 each to the remaining two categories. In so doing, we can expect to make 150, 87.5, and 87.5 errors respectively. Adding these quantities for both A_1 and A_2, we see that we can expect to make a total of 1,075 errors when A is known.

We define the measure τ_b to be the proportional reduction of errors. Thus

$$\tau_b = \frac{\text{no. of errors not knowing } A - \text{no. of errors knowing } A}{\text{no. of errors not knowing } A}$$

$$\tau_b = \frac{1,270 - 1,075}{1,270} = \frac{195}{1,270} = .154$$

In other words, we have saved ourselves 195 errors out of an expected number of 1,270 and have reduced our errors by 15.4 per cent. If τ_b had turned out to be .50 we could thus give it the very simple interpretation that knowledge of A would cut the number of errors in half; a value of .75 would mean reducing the number of errors to one-fourth of the original number, and so forth. No such simple interpretation is possible in the case of ϕ^2 (see [2]). Had we wished to predict A classes from B classes, we would denote the comparable measure as τ_a. τ_a and τ_b will not in general have the same numerical values. Why?

In the case of the 2 $\times$ 2 table it can be demonstrated that $\tau_a = \tau_b = \phi^2$.

This points up two kinds of notational difficulties. Notice that some of our measures (C, Q, T, and V) are denoted by Latin letters, whereas others (ϕ and τ) are denoted by Greek letters. If we were entirely consistent, Greek letters should be reserved for population parameters that are being estimated by sample statistics. Unfortunately, once symbols are in common use, it is difficult to standardize usage, and the best the reader can do is to be aware of the inconsistency. Also, certain measures appear as squared, whereas others do not. In particular, we see that in the 2×2 case the unsquared symbol τ is equivalent to ϕ^2, which we have seen is equal to T^2 and V^2 in this instance. Thus in the case of the more general table it would seem to make sense to compare τ with the other squared coefficients, though noting that they will not be identical. In general, we can expect numerical values of τ to be *smaller than* the unsquared coefficients ϕ, T, and V. If one were to think in terms of certain absolute magnitudes as small, medium, or large (e.g., a value less than .3 is "small"), he could easily be misled unless he clearly recognized the differences among the measures.

Lambda   There is another measure, lambda (λ), that is very similar to τ and that is also asymmetric with respect to A and B. Taking B as the dependent variable to which predictions are being made, notice that the expected number of errors will actually be minimized if we are allowed to place *all* the individuals in the largest of the B_i categories. (See Exercise 5, Chap. 9.) In the previous example this would involve putting all 2,000 cases in B_1, rather than restricting ourselves to 900 cases in this category. If we did so, we would make 1,100 errors since there are a total of 1,100 cases in B_2 and B_3. Notice that this is fewer errors than we made in the case of the denominator of τ_b. Now suppose that we know the category of A to which the individual belongs. If we are allowed to assign all of the 1,200 individuals in A_1 to B_2, the row containing the largest number of A_1 individuals, then we will make only $300 + 300 = 600$ errors. Similarly, if we place all 800 A_2's in the B_1 category, we will make only 200 errors. Thus, knowing the A category and being allowed to make these less restrictive assignments, we would expect to make 800 errors. We form a "proportional reduction in error" measure λ_b as follows:

$$\lambda_b = \frac{1,100 - 800}{1,100} = .273$$

We see that lambda is easier to compute than tau, that it involves an unrestrictive minimizing of errors, and that in this instance it has a considerably larger numerical value than tau. However, it has the undesirable property that it may take on a numerical value of zero in instances where all of the other measures we have considered will not be zero, and where we would not want to refer to the variables as being uncorrelated or statistically independent. This may occur merely because one of the B marginals is much larger than the rest, so that no matter what the A category, the decision is always to place all of the individuals (for all A_i) in the same B category. For example, had the categories B_1 and B_2 been combined in the above hypothetical example, the decision would always be to place all individuals in the B_1 and B_2 category, rather than in B_3, so that the resulting λ_b would have been zero. By the same token, even though a single marginal total (for example, B_1) does not dominate the rest, it is likely that some of the less numerous categories will not enter into the computation of lambda at all. In the above illustration, the decision rule never results in the assignment of individuals to B_3. Had there been an additional row B_4, also with a relatively small number of cases, the measure lambda might have been insensitive to the distribution of cases between B_3 and B_4. For these reasons, tau is to be preferred over lambda in instances where the marginal totals are not of approximately the same magnitudes.

15.4 Controlling for Other Variables

Discussions of tests of significance and measures of association have to this point involved only two variables at a time. In most practical problems, however, it is necessary to control for one or more additional variables which may be either obscuring a relationship or creating a spurious one. Although it is often true that generalizations in the social sciences are stated in terms of only two variables, it is practically always implicitly recognized that relevant variables are assumed to be controlled. Frequently the phrase "other things being equal" is used to emphasize this fact. Ideally, an hypothesis should be stated in such a manner that it is clearly understood what variables are to be controlled. As a discipline advances toward maturity, generalizations become qualified so as to indicate the exact conditions under which they can be expected to hold true. In the initial stages of its development, however, it is frequently impossible to know what the relevant variables are which need to be controlled. For this reason, propositions are not usually stated in the social sciences in such a manner as to suggest what variables are to be controlled. Nevertheless, you should develop the habit of always looking for possible control variables even though you may not have been explicitly told to do so.

As we shall see later, there are several possible methods of controlling statistically. The method discussed in this chapter is perhaps the most straightforward and the one most directly analogous to the laboratory experiment in which control variables are actually physically held constant. In laboratory experiments a control variable is held at a constant value while other variables are interrelated. Thus, the temperature may be held at 70°F while the relationship between pressure and volume is investigated. If a relationship is then found between the latter variables, it may be possible to state much more precisely its nature than if temperature were not controlled. But the scientist would not be justified in stating a generalization as if it always applied unless he were to find that exactly the same relationship actually held for all temperatures. He would undoubtedly perform a series of experiments, each at a different temperature. It is quite likely that he would find the relationship to hold only within a certain range of temperatures. He would then have to qualify his generalization to read, "The relationship between pressure and volume is such and such *provided* the temperature is between −100 and 600°F." Hopefully, he might find a correction factor which would enable him to restate his proposition in a more general form that would apply to a greater range of temperatures. Exactly the same kind of argument would apply to additional control variables. Simultaneous controls could be made for several variables by holding each at a fixed value. Further experiments could then be performed with different *combinations* of values for control variables. With several controls acting simultaneously, a much larger number of replications would be required.

There is a certain resemblance between the kind of statistical controlling procedure we shall presently discuss and the laboratory experiment in which control variables are actually physically manipulated and held constant at different levels. But there is a fundamental difference that is crucial in connection with one's *interpretations* of the results. When we control statistically, we carry out paper and pencil manipulations in which we adjust scores or move individuals around from one table to another. But we do not actually manipulate their true scores. For example, if we "control" statistically for IQ, we do not literally hold someone's intelligence constant. We may adjust IQ scores, adding to some and subtracting from others, so that we can pretend they are equal. But we cannot manipulate a person's real intelligence in a way that is comparable to controls placed on temperature or pressure in a laboratory experiment.

This kind of hypothetical controlling and adjusting is very convenient and will not lead us astray *if* the real world coincides with what we are doing. If a real change in intelligence would, in fact, affect our relationship in a given way, and if holding it constant in an experiment would enable us to infer the true relationship between two other variables "with intelligence held constant," then our paper and pencil operations will make good sense. But it must be clearly recognized that such paper and pencil "controls" can be carried out on *any* variable for which we have measurements (including categories), even those that are causally *dependent* on the variables we are studying, and also those that are spuriously related to either variable for extraneous reasons.

Statistical controls are basically much easier to make than true controls, and we thus have a much greater degree of flexibility in applying them even in instances where it makes no sense to do so. Basically, one needs a *theory* in order to justify the application of controls, and such a theory will involve assumptions as to the causal structure of the system of variables. Although this question is beyond the scope of a general text on statistics, it is necessary to introduce a major note of caution at this point since many misunderstandings of statistical controlling operations have resulted from a blind application of control variables without benefit of an underlying theory.

Returning to the example of the relationship between religious preference and political party, one may control statistically for variables such as sex and social class. In order to hold sex constant, for example, we can look only at the male voter. If the relationship is found to hold for males and also separately for females, then we can say that it generally holds controlling for sex since we have examined both categories of the sex variable. Quite possibly, however, the relationship might hold for males but not for females. In this case the generalization would have to be qualified, and attention could be turned to the question of why it should hold for one sex but not the other. It can be seen that controlling for relevant variables not only provides a more rigorous test of an hypothesis, but it may also lead to additional insights if the relationship is found to differ from one category of the control variable to the next.

Sometimes it may be desirable to control for several variables simultaneously. Because of a shortage of cases it is frequently necessary to control for relevant variables one at a time, but a certain amount of information may be lost in this manner. For example, suppose that sex were ignored and a control introduced for the voter's social class. We would then look separately at each social class to see whether or not the relationship always held. In contrast to this procedure, we might have controlled simultaneously for class and sex by taking all possible combinations of the control variables (e.g., male lower class, female lower class, male middle class, etc.) and studying the relationship within each combination of control categories. Conceivably, the relationship might hold for every combination except that of the female lower-class voter. If this were the case, we would be led to look for any peculiarities of this particular subgroup.

Let us take another concrete example in order to illustrate the process. Suppose we have the following data on school children: their class background, IQ's, school grades, and how hard each child works. It will be helpful to summarize the data in terms of a master table as in Table 15.4.

Table 15.4   Master table for interrelating four variables

Intelligence	Grades	Middle class		Lower class		Totals
		High effort	Low effort	High effort	Low effort	
High	High	60	40	40	18	158
	Low	20	24	16	38	98
Low	High	40	24	6	2	72
	Low	24	12	32	54	122
Totals		144	100	94	112	450

Notice that such a table contains enough cells that all four pieces of information (class, IQ, grades, and effort) can, if necessary, be reconstructed for each individual. That is, we know how many persons have each combination of traits (e.g., lower class, high IQ, low effort, high grades). If we wish less detailed knowledge, we may always combine into larger groupings. For example, we might combine lower- and middle-class students and retain only the distinction between IQ, effort, and grades. But if we are given only the more inclusive and less detailed breakdowns, we cannot recover the complete information without redoing the analysis. Therefore, a master table such as Table 15.4 should be used as a working table from which a series of separate tables can be reconstructed.

Ordinarily it will be most convenient to construct such a master table so that the dependent variable appears as the innermost column on the left-hand side and so that the independent variable of greatest interest appears as the lowest row in the top heading. This will result in subtables with those frequencies that are being directly compared. For example, in Table 15.4 we have four subtables, each of which relates effort to grades. All individuals in the top-left subtable are middle class and high IQ, and so forth. The exact arrangement of rows and columns is not of crucial importance, however, since they can obviously be re-arranged according to the relationships of interest (as has been done in Table 15.5).

Table 15.5 Series of contingency tables relating two variables, with two simultaneous controls

Grades	High effort		Low effort	
	High IQ	Low IQ	High IQ	Low IQ
	Middle class			
High	60	40	40	24
Low	20	24	24	12
	Lower class			
High	40	6	18	2
Low	16	32	38	54

Suppose we suspect a middle-class bias on the part of teachers resulting in a tendency to give high grades to middle-class children regardless of ability or effort but to give high grades to lower-class children only when there is evidence of both high ability and high effort. We would then predict that grades should generally be higher for middle-class children, controlling for both ability and effort, except possibly in the case of children with high ability and effort. We would also predict that the relationships between grades and both ability and effort would be stronger in the lower class than the middle. In other words, if middle-class children always receive high grades, there should be no relationship (or a very slight one) within this class between grades and either effort or ability. Let us concentrate on the relationship between grades and ability and investigate whether or not this relationship is stronger for the lower class. In this case, we shall want to control for effort. In each class there will be both high- and low-effort students. Therefore we can construct four contingency tables as in Table 15.5.

We now compare the two classes with respect to the existence and strength of relationship, looking separately at high- and low-effort students. The direction of relationship can also be noted in each case by simply computing percentages or by comparing the diagonal products. Computing chi square and ϕ for each table, we get the results shown in

Table 15.6. Thus we see that relationships for middle-class children are nonsignificant, but for lower-class children in both effort categories there is a moderately strong positive relationship between ability and grades. We also notice that the relationship is somewhat stronger in the case of the hard-working students.

Table 15.6

Class	Effort	Chi square	Significance level	ϕ
Middle	High	2.565	Not significant	.133
	Low	.188	Not significant	.043
Lower	High	28.064	$p < .001$	.546
	Low	15.582	$p < .001$	.373

You have undoubtedly noticed the marked effect that controlling has on the number of cases appearing in each cell. Instead of having only four cells, we have four times this number when we use two dichotomized control variables. Had a third simultaneous control, say sex, been added, there would have been 32 cells instead of 16. Had any of the variables involved more than two categories, the number of cells would have been further multiplied. Therefore, although simultaneous controls can theoretically be added indefinitely, the number of cases must be very large in order to control by this method. An alternative would simply be to restrict the nature of the population and to generalize only to middle-class males with a college education or some comparable subgroup. A much larger sample of this particular subgroup could then be selected. Usually if simultaneous controlling is to be used, it becomes necessary to select those two or three controls which look most promising. It is, of course, possible to make use of Fisher's exact test when the number of cases in each cell becomes very small, but it should be remembered that it will then be necessary to have a very high degree of relationship in order to obtain significance. Because of this attenuation of cases, the mere fact that a relationship becomes nonsignificant when controls are introduced is not sufficient evidence that the control variable is having an effect. Measures of degree of relationship should always be computed and compared.

Whenever *relationships* differ from one category of a control variable to the next, we have an example of what is termed *nonadditivity* or *statistical interaction.* This possibility has already been discussed in connection with the test for a difference of differences in proportions and will be discussed again in more detail in Chaps. 16 and 20. Whenever one suspects the possibility of interaction, he should *always* make a statistical test for interaction before proceeding further. Since there will inevitably be some slight differences in relationships from one sample to the next, the basic question being asked in such tests is whether or not the sample interactions are sufficiently large that they could have readily occurred by chance even if there were no population interaction. In this example, since all variables have been dichotomized one may make a simple test of a difference of differences in proportions, as suggested in Chap. 13. Since there are two control variables being considered simultaneously, it is even possible that there may be what is called a second-order interaction, or a difference of differences of differences. For example, the difference between the high- and low-effort relationship may be greater for lower-class than for middle-class children.

If interaction is found to be statistically significant and also large enough to be substantively meaningful, then generalizations will have to be qualified by specific reference to the control category. For example, one would have to say: "For lower-class children a relationship was found between grades and ability, but none was found for middle-class children." From that point on, other relationships should be studied separately for the two class levels. But if interaction is statistically insignificant or small enough to be ignored even if statistically significant, then it is reasonable to assume that the relationships are basically similar from one control category to the next. In such instances we will be in a position to simplify the analysis considerably by pooling the separate results. Let us next see what specific kinds of simplifications become possible in the case of categorized data.

First, we may pool the separate chi-square tests into a single overall test provided that they are based on independently selected random samples. The procedure is extremely simple and involves adding the separate values of chi square and also the degrees of freedom, evaluating the result in the familiar manner. For example, suppose that for four 2×2 tables the resulting chi squares were 2.1, 3.3, 2.7, and 2.9. The sum of these values is 11.0, and the sum of the degrees of freedom is 4. From the table we see that a chi square of 11.0 with 4 degrees of freedom is significant at the .05 level. Thus, although none of the separate chi squares were significant, we were able to take advantage of the fact that it makes theoretical sense to pool the results. In effect, we are saying that if a relationship comes out roughly the same each time but the probabilities of the separate results are each greater than .05, we still may ask ourselves how likely such a combination of outcomes would be if there were no relationships in any of the four tables.

Notice that the results of such a pooling operation may very well differ from the total relationship between the two variables *without any controls*. We are essentially getting an *average* relationship *within* categories of the control variable(s) when we pool results. Had we simply ignored the control variable(s), the effects of such controls would have been completely obscured. By pooling we are making a single chi-square test of the overall relationship between two variables, controlling for the additional variables.

Similarly, we might want to obtain a single measure of association by computing a weighted average of the measures based on the four separate tables. One method which has been suggested for doing this is to use weights which are proportional to the number of cases in each table. Thus we might multiply each τ_b by the number of cases in the table, add the results, and finally divide by the total number of cases in all four tables. We would then end up with a single test of significance and a single measure of association representing an average of the results of all four tables.

Another very simple way of obtaining a weighted average can be described briefly. (For further details, see Rosenberg [12].) Basically the procedure involves standardizing across all control categories by obtaining a weighted average of proportions (or percentages). Suppose we have obtained the following results for males and females separately:

	Males				Females			
	Prot-estants	Catho-lics	Jews	Total	Prot-estants	Catho-lics	Jews	Total
Republicans	180	80	20	280	100	50	10	160
Democrats	90	80	50	220	60	30	70	160
Independents	30	40	30	100	40	20	20	80
Total	300	200	100	600	200	100	100	400

First, we convert these figures to proportions adding down to 1.00, since the independent variable is given at the top of each table. The results are as follows:

	Males			Females		
	Protestants	Catholics	Jews	Protestants	Catholics	Jews
Republicans	.60	.40	.20	.50	.50	.10
Democrats	.30	.40	.50	.30	.30	.70
Independents	.10	.20	.30	.20	.20	.20
Total	1.00	1.00	1.00	1.00	1.00	1.00

Provided we are willing to obscure the differences between these two tables by an averaging procedure, we may form a weighted average by multiplying each proportion in the table for the males by .6, since there are 600 males in the total sample of 1,000. Similarly we can weight each figure in the female table by .4. The result is as follows:

	Protestants	Catholics	Jews
Republicans	.56	.44	.16
	(.36 + .20)	(.24 + .20)	(.12 + .04)
Democrats	.30	.36	.58
	(.18 + .12)	(.24 + .12)	(.30 + .28)
Independents	.14	.20	.26
	(.06 + .08)	(.12 + .08)	(.18 + .08)
Total	1.00	1.00	1.00

where each proportion in the derived table is the sum of the two weighted proportions (as indicated in the parentheses) for the separate tables above. Since the sum of the weights is 1.0, the sum of the proportions in the derived table will also be unity for each column. These results could of course also be expressed in terms of percentages.

This procedure of controlling by obtaining weighted averages will be seen to be a very general one. In effect, we have standardized the number of Protestants, Catholics, and Jews so that their relative sizes in the male and female samples have become irrelevant. Had there been simultaneous controls for additional variables, we could have extended this standardizing procedure in a straightforward way. Thus if we had also wanted to control for social class, using three levels, we would have had six original tables, one for each sex-class category. Having previously looked for interaction and decided that no important differences would be obscured by the weighting procedure, we could again assign weights W_i to each of the control tables, making $\Sigma W_i = 1.0$, obtaining a single combined table as above.

In substituting a single measure and test for several separate measures and tests, we run into the usual problems involved whenever summarizing statistics are used. We boil down our data so as to obtain fewer statistics. On the other hand, we run the risk of distorting our results. For example, if one of the tables produced a large chi square and a very high degree of relationship as compared with the rest, then combining the results, thereby obscuring this fact, may prove highly misleading. As always, statistical manipulations can never be a substitute for common sense.

Some of the ideas discussed in the present section, and especially the notion of pooling the results of separate tables, are undoubtedly new and somewhat confusing at this point. It will therefore be helpful to review this section after you have been exposed to some of the material in Chaps. 16 to 20. By that time, several different types of control procedures will have been discussed and compared.

Exercises

1. Compute chi square for the data of Exercise 5, Chap. 9. Taking occupational aspirations as the dependent variable B, what is the value of τ_b? How does the value of τ_b compare with that of the measure you computed in part (d) of Exercise 5?

2. In Exercise 3 of Chap. 14 you made use of the Smirnov test. Taking these same data, what conclusion do you reach using the chi-square test? For these particular data, which test would you prefer? Why? Compute ϕ, T, V, C, τ_b, and λ_b.

*3. The chi-square test can generally be used to compare observed and theoretical frequencies. In particular, it can be used to test the null hypothesis that sample data have been drawn randomly from a normal population. The observed frequencies are compared with those which would be expected if the distribution were actually normal, with the same mean and standard deviation as computed from the sample data. After obtaining the values of $\bar{X}$ and s, you may use the true limits and the normal table to give the expected frequencies within each interval. The degrees of freedom will be $k - 3$, where k represents the number of intervals. One degree of freedom will be lost since the total of the expected frequencies must be N; the remaining two degrees of freedom which have been lost are due to the necessity of using $\bar{X}$ and s as estimates of the actual parameters μ and σ. With these facts in mind, test to see whether or not the following data depart significantly from normality. (*Ans.* $\chi^2 = 2.53$, fail to reject.)

Interval	Frequency
0.0– 9.9	7
10.0–19.9	24
20.0–29.9	43
30.0–39.9	56
40.0–49.9	38
50.0–59.9	27
60.0–69.9	13
	208

4. In a recent study, H. L. Wilensky [14] found a general relationship between union activity and both political orientation and voting intentions, controlling for socioeconomic status. Data for 15 black members with respect to voting intentions tended to support these general findings. Seven of the eight blacks who were inactive members of the union did not follow the "union line" in their 1948 voting behavior. Of the seven members who were really active in the union, five voted as suggested by the union. Test for the significance of a relationship using (a) Fisher's exact test with direction predicted, and (b) chi square corrected for continuity with direction predicted. [*Ans.* (a) $p = .035$; (b) $\chi^2 = 3.22$, $p < .05$.]

5. Make use of the data given below (rearranging the table, if necessary) to obtain information as to the accuracy of statements (*a*), (*b*), and (*c*). Where appropriate, compute measures of the degree of relationship and control for relevant variables.

   a. Females are less prejudiced than males, regardless of religion and social class.
   b. The degree of relationship between religion and prejudice toward blacks will depend on the social class of the "prejudiced" person.
   c. The reason that Jewish persons appear in the table to be less prejudiced than non-Jews is due to the high percentage of women and upper-class persons in the Jewish sample.

Religion	Sex	Amount of prejudice toward blacks				Totals
		High		Low		
		Upper class	Lower class	Upper class	Lower class	
Non-Jew	Male	14	30	15	16	75
	Female	8	13	9	7	37
Jewish	Male	13	7	22	15	57
	Female	18	9	33	21	81
Total						250

6. Using the data for Exercise 5 above, construct tables relating religion to prejudice, with simultaneous controls for sex and social class. Assuming that possible interactions can be neglected, standardize these results so that the relationship between religion and prejudice, with controls, can be presented in a single 2 × 2 table.

*7. Suppose you expect to do a chi-square test on a 2 × 2 table relating religious preference (Protestant-Catholic) to political preference (Republican-Democrat). You plan to take equal sized random samples of Protestants and Catholics, and you predict direction in advance, expecting that the proportion of Protestants who are Republicans will be about .60, whereas the proportion of Catholics who are Republicans will be about .40. About how many cases will you need to establish significance at the .05 level?

References

1. Anderson, T. R., and M. Zelditch: *A Basic Course in Statistics*, 2d ed., Holt, Rinehart and Winston, Inc., New York, 1968, chap. 9.
2. Blalock, H. M.: "Probabilistic Interpretations for the Mean Square Contingency," *Journal of the American Statistical Association*, vol. 53, pp. 102–105, 1958.
3. Bradley, J. V.: *Distribution-free Statistical Tests*, Prentice-Hall, Inc., Englewood Cliffs, N.J., 1968, chap. 8.
4. Downie, N. M., and R. W. Heath: *Basic Statistical Methods*, 2d ed., Harper and Row, Publishers, Incorporated, New York, 1965, chap. 14.

5. Goodman, L. A., and W. H. Kruskal: "Measures of Association for Cross Classifications," *Journal of the American Statistical Association,* vol. 49, pp. 732–764, 1954.

6. Goodman, L. A., and W. H. Kruskal: "Measures of Association for Cross Classifications. II: Further Discussion and References," *Journal of the American Statistical Association,* vol. 54, pp. 123–163, 1959.

7. Goodman, L. A., and W. H. Kruskal: "Measures of Association for Cross Classifications. III: Approximate Sampling Theory," *Journal of the American Statistical Association,* vol. 58, pp. 310–364, 1963.

8. Hagood, M. J., and D. O. Price: *Statistics for Sociologists,* Henry Holt and Company, Inc., New York, 1952, chap. 21.

9. Hays, W. L.: *Statistics,* Holt, Rinehart and Winston, Inc., New York, 1963, chap. 17.

10. McCarthy, P. J.: *Introduction to Statistical Reasoning,* McGraw-Hill Book Company, New York, 1957, chap. 11.

11. Mueller, J. H., K. Schuessler, and H. L. Costner: *Statistical Reasoning in Sociology,* 2d ed., Houghton Mifflin Company, Boston, 1970, chap. 9.

12. Rosenberg, Morris: "Test Factor Standardization as a Method of Interpretation," *Social Forces,* vol. 41, pp. 53–61, 1962.

13. Siegel, Sidney: *Nonparametric Statistics for the Behavioral Sciences,* McGraw-Hill Book Company, New York, 1956, pp. 96–111.

14. Wilensky, H. L.: "The Labor Vote: A Local Union's Impact on the Political Conduct of its Members," *Social Forces,* vol. 35, pp. 111–120, 1956.

Analysis of Variance   16

In Chap. 13 we compared two samples by testing for the significance of the difference between means or proportions. Such tests were capable of handling situations in which one of the two variables being interrelated was a dichotomized nominal scale. In the last chapter we saw how more than two samples could be compared by means of the chi-square test. In the present chapter we shall take up a very important kind of procedure, analysis of variance, which can be used to test for differences among the means of more than two samples. Analysis of variance thus represents an extension of the difference-of-means test and can generally be used whenever we are testing for a relationship between a nominal (or higher order) scale and an interval scale. We shall also see that under certain circumstances analysis-of-variance tests can be extended to situations in which there is a single interval scale and two or more nominal scales. Analogous nonparametric tests will also be considered, as will several measures of degree of association.

16.1   Simple Analysis of Variance

Although analysis of variance can be considered as an extension or generalization of the difference-of-means test, it involves some basically new principles which will require a fairly lengthy explanation. For this reason a brief overview may be helpful so that you will not find yourself becoming lost in the details. The assumptions for analysis of variance are basically the same as required for the difference-of-means test, but the test itself is very different. We shall have to assume normality, independent random samples, and equal population standard deviations, and the null hypothesis will be that population means are equal. The test itself, however, involves working directly with variances rather than means and standard errors.

Suppose the data in Table 16.1 represent murder rates for each of three types of cities: those that are primarily industrial, trade, or political centers. We can compute separate means for each of the three categories or samples, and we can also obtain a grand mean by ignoring the classes and averaging all scores. In this example all samples are of the same size, but this need not always be the case.

Since it is being assumed that all populations have the same standard deviation, we can form two independent estimates of the common variance σ^2. One of these estimates will be directly analogous to the pooled estimate used in the difference-of-means test. This estimate will

Table 16.1 Data for analysis of variance

	Murder rates			Total
	Industrial community	Trade community	Political community	
	4.3	5.1	12.5	
	2.8	6.2	3.1	
	12.3	1.8	1.6	
	16.3	9.5	6.2	
	5.9	4 1	3.8	
	7.7	3.6	7.1	
	9.1	11.2	11.4	
	10.2	3.3	1.9	
Sums	68.6	44.8	47.6	161.0
Means	8.58	5.60	5.95	6.71
No. of cases	8	8	8	24

be a weighted average of the variances *within* each of the separate samples and will always be unbiased even if the sample means differ considerably among themselves. This is true because each sample variance will be computed separately and will involve only the deviations from the mean of that particular sample.

The second estimate of the common variance involves the variance of the separate sample means treated as individual scores. In this case, the deviations of the sample means about the grand mean will be used in estimating σ^2. For the data of Table 16.1 we would obtain the variation of the three sample means 8.58, 5.60, and 5.95 about the total mean of

6.71. This estimate of σ^2 will be unbiased only if the population means are in fact equal. If the population means are equal, the sample means can be expected to differ from one another according to the central-limit theorem, that is, to approach a normal distribution as the size of the sample increases, and we can make use of this law and the actual differences among sample means to estimate the true variance. On the other hand, if the population means are actually different, we expect that the sample means will differ from one another more than would be the case if the population means were the same. Therefore, if the null hypothesis is false, the second estimate of σ^2 will ordinarily be too large, and it will be a biased estimate.

The test used in analysis of variance involves a comparison of the two separate estimates of the population variance. Instead of taking the difference between the two estimates, however, we take the *ratio* of the second estimate to the first. If the null hypothesis is correct, then both estimates will be unbiased, and the ratio should be approximately unity. If the population means actually differ, however, the second estimate will ordinarily be larger than the first and the ratio greater than unity. Since sampling fluctuations are always a factor, we have to ask ourselves how large a ratio we are willing to tolerate before we become suspicious of the null hypothesis. Fortunately, the ratio of the two estimates F has a known sampling distribution, provided the two estimates of variance are actually independent of each other, and therefore a fairly simple test can be made. This, in essence, is what we do in an analysis-of-variance test. Now let us take a more detailed look at the procedure involved.

Breaking total variation into component parts Although our ultimate goal is the formation of two independent estimates of the variance, it will be necessary to introduce a new concept in order to explain how these estimates are obtained. Let us use the term *variation* (as distinct from variance) to refer to the sum of the squared deviations from the mean. The total variation about the grand mean for all samples would be $\sum\limits_{i=1}^{N} (X_i - \bar{X})^2$. The term variation thus will refer to a sum of squares, without dividing by the number of cases involved. We shall now proceed to break this total variation into two component parts, each of which will be used in the computation of the two estimates.

Let us represent our data symbolically as in Table 16.2. The individual scores are represented by X_{11}, X_{21}, . . . , X_{ij}, the sample means by $\bar{X}_{.1}$, $\bar{X}_{.2}$, . . . , $\bar{X}_{.k}$, and the grand mean by $\bar{X}_{..}$. The dots are used

Table 16.2 Symbolic representation of data for analysis of variance

	Categories				Total
	A_1	A_2	$\cdots$	A_k	
Scores	X_{11} X_{21} X_{31} . . . $X_{N_1 1}$	X_{12} X_{22} X_{32} . . . $X_{N_2 2}$	$\cdots$ $\cdots$ $\cdots$ $\cdots$	X_{1k} X_{2k} X_{3k} . . . $X_{N_k k}$	
Sums	$\displaystyle\sum_{i=1}^{N_1} X_{i1}$	$\displaystyle\sum_{i=1}^{N_2} X_{i2}$	$\cdots$	$\displaystyle\sum_{i=1}^{N_k} X_{ik}$	$\displaystyle\sum_i \sum_j X_{ij}$
Means	$\bar{X}_{\cdot 1}$	$\bar{X}_{\cdot 2}$	$\cdots$	$\bar{X}_{\cdot k}$	$\bar{X}_{\cdot\cdot}$
No. of cases	N_1	N_2	$\cdots$	N_k	N

in the subscripts in order to distinguish column means from row means which will be used when we add a second nominal scale. The general symbol X_{ij}, represents the score of the ith individual in the jth column. The sum $\displaystyle\sum_{i=1}^{N_1} X_{i1}$ indicates that the N_1 scores in the first column have been summed, and similarly for the remaining columns.[1]

Now we perform some simple algebra. We can write

$$X_{ij} - \bar{X}_{\cdot\cdot} = (X_{ij} - \bar{X}_{\cdot j}) + (\bar{X}_{\cdot j} - \bar{X}_{\cdot\cdot})$$

or

$$\begin{pmatrix} \text{Individual} \\ \text{score} \end{pmatrix} - \begin{pmatrix} \text{grand} \\ \text{mean} \end{pmatrix} = \begin{pmatrix} \text{individual} \\ \text{score} \end{pmatrix} - \begin{pmatrix} \text{category} \\ \text{mean} \end{pmatrix}$$
$$+ \begin{pmatrix} \text{category} \\ \text{mean} \end{pmatrix} - \begin{pmatrix} \text{grand} \\ \text{mean} \end{pmatrix}$$

having subtracted $\bar{X}_{\cdot j}$ (the mean of the jth column) from X_{ij} and then added it back in immediately. We have therefore expressed the differ-

[1] Since there are two subscripts, i and j, it is important to distinguish between the symbols $\displaystyle\sum_i$ and $\displaystyle\sum_j$. In the latter case, the j values would be summed for any particular (fixed) i, and we would thus obtain the sum of scores in the ith *row*.

ence between a single individual's score and the grand mean as a sum of two quantities: (1) the difference between his score and the mean of the category to which he belongs, and (2) the difference between the category mean and the grand mean. In the above numerical example we can express the difference between the score of the first individual in the first category and the grand mean as

$$4.3 - 6.71 = (4.3 - 8.58) + (8.58 - 6.71)$$

or $$-2.41 = -4.28 + 1.87$$

Squaring both sides of the equation we obtain

$$(X_{ij} - \bar{X}_{..})^2 = (X_{ij} - \bar{X}_{.j})^2 + 2(X_{ij} - \bar{X}_{.j})(\bar{X}_{.j} - \bar{X}_{..}) + (\bar{X}_{.j} - \bar{X}_{..})^2$$

By summing both sides we obtain the sum of the squared deviations for all individuals. We can first sum down each column and then add the resulting figures for each category. When we do this, the middle term becomes zero. In order to see why this is the case, notice that in summing down any particular column the value of j will be constant. Therefore for the jth column the factor $(\bar{X}_{.j} - \bar{X}_{..})$ will be constant and can be taken outside of the summation. Thus for the sum of the scores in the jth column the middle term becomes

$$2(\bar{X}_{.j} - \bar{X}_{..}) \sum_i (X_{ij} - \bar{X}_{.j})$$

But since for every column the deviations about the column mean must be zero, we immediately see that the middle term must vanish for each and every column. We thus get

$$\sum_i \sum_j (X_{ij} - \bar{X}_{..})^2 = \sum_i \sum_j (X_{ij} - \bar{X}_{.j})^2 + \sum_i \sum_j (\bar{X}_{.j} - \bar{X}_{..})^2 \quad (16.1)$$

$$\text{Total sum of squares} = \frac{\text{within sum}}{\text{of squares}} + \frac{\text{between sum}}{\text{of squares}}$$

In so doing, we obtain a double summation which we write as $\sum_i \sum_j$, indicating that we have summed over both rows and columns.

We have divided the total variation into two parts. The first of these is a sum of squared deviations of the individual scores from their own category means. This is referred to as a *within* sum of squares and will be used to obtain our first estimate of the common variance σ^2. Notice that this sum of squares is obtained essentially the same way that the pooled estimate was formed in the difference-of-means test. If we write the within sum of squares as

$$\sum_{i=1}^{N_1} (X_{i1} - \bar{X}_{\cdot 1})^2 + \sum_{i=1}^{N_2} (X_{i2} - \bar{X}_{\cdot 2})^2 + \cdots + \sum_{i=1}^{N_k} (X_{ik} - \bar{X}_{\cdot k})^2$$

we see that the first term is just $N_1 s_1^2$, where deviations have been taken about the category mean, and similarly for the other terms. Therefore

$$\text{Within SS} = N_1 s_1^2 + N_2 s_2^2 + \cdots + N_k s_k^2$$

When we divide by the proper degrees of freedom, which will turn out to be $N - k$, we obtain a pooled estimate based on all k categories. The second or *between* sum of squares involves the deviations of category means from the grand mean and is therefore a measure of the variability between samples. The second estimate of the variance will be based on this between sum of squares.

The between and within sums of squares are often referred to as *explained* and *unexplained* variations respectively. It is perhaps easier to see why the within variation is referred to as being unexplained since this refers to variation which is left unaccounted for by the categorized variable. If within category A there is still some variability about the category mean, then this variability can certainly not be explained by the category. On the other hand, if the category means differ considerably among themselves, a relatively large fraction of the total variation may be attributed to differences among the several categories. Thus, it is the amount of variability within the categories as compared with differences between them which determines how closely the two variables are associated. Homogeneous categories which differ considerably among themselves will explain a high proportion of the variation.[2] In the extreme case, if we had completely homogeneous categories, the within sum of squares would be zero and all of the variability could be attributed to the categorized variable. Thus if all industrial cities had exactly the

[2] This does not imply causality, of course. The word "explained," as used in the statistical literature, is best translated as "associated with" and should in no sense be interpreted as necessarily implying that one has located an explanatory variable in the causal or theoretical sense.

same murder rates that differed from those of all trade centers which were also completely homogeneous, etc., then city type could be said to explain all of the variation in murder rates. Knowing the city type would enable us to predict the murder rate exactly.

In order to obtain estimates from these two separate sums of squares it is only necessary to divide by the appropriate degrees of freedom. The degrees of freedom associated with the total sum of squares is $N - 1$ since we have seen that $\hat{\sigma}^2$ is the unbiased estimate of σ^2, 1 degree of freedom having been lost through the computation of the grand mean $\bar{X}_{..}$. Let us now look at the between sum of squares. This quantity represents the sum of squared deviations of the k sample means from the grand mean. In effect, each category mean is being treated as a single case. Therefore there are $k - 1$ degrees of freedom involved, one having been lost because of the fact that the weighted average of the $\bar{X}_{.j}$ must be $\bar{X}_{..}$. In the case of the within-class estimate, 1 degree of freedom will be lost in each column through the computation of the $\bar{X}_{.j}$. Therefore, in all, there will be $N - k$ degrees of freedom associated with the within estimate. Notice that the degrees of freedom add as do the sums of squares. Thus

$$N - 1 = (N - k) + (k - 1)$$

Total df = within df + between df

Our two estimates of the common variance thus become

$$\text{Within estimate} = \frac{\sum_i \sum_j (X_{ij} - \bar{X}_{.j})^2}{N - k} \tag{16.2}$$

$$\text{Between estimate} = \frac{\sum_i \sum_j (\bar{X}_{.j} - \bar{X}_{..})^2}{k - 1} \tag{16.3}$$

At this point it may have occurred to you that we actually have three separate estimates of the common variance if we include the usual estimate based on the total sum of squares. Why, then, not compare the latter estimate with either of the others since this total estimate may well be a better estimate than either of the other two? It will be recalled that the F test requires that the estimates compared be independent of each other. The estimate based on the total sum of squares is not independent of the others, however, and this is the reason this estimate cannot be used in the F test. Ordinarily, the within and between sums of squares are not independent of each other either. It so happens that the normal distribution has the property that these two quantities are independent in spite of the fact that the same $\bar{X}_{.j}$'s appear in both expressions. It is for this reason that we must assume all populations sampled to be normal. It will be recalled that normality was also required in the case of the t distribution because of the necessity of having the numerator independent of the denominator. As will be seen presently, the t distribution is a special case of the F distribution.

PROBLEM   Let us make use of the above hypothetical data representing murder rates for three types of cities. We wish to find out whether there are significant differences among the means of the three types of cities.

1.   *Assumptions*

    Level of Measurement: Murder rates an interval scale
                                City type a nominal scale
    Model: Independent random sampling
                Normal populations for each city type
                Population variances are equal
$$(\sigma_1{}^2 = \sigma_2{}^2 = \cdots = \sigma_k{}^2 = \sigma^2)$$
    Hypothesis: Population means are equal
$$(\mu_{.1} = \mu_{.2} = \cdots = \mu_{.k})$$

As was the case with difference-of-means tests, it must be assumed that samples are selected independently of each other. In other words, cities are not matched in any way. Since all three populations of city types are assumed to be normal with equal means and variances, we are in effect assuming they are identical. The three samples can therefore be conceived as being randomly drawn from the same population. The researcher is usually interested in the assumption of equal means. In this example, he probably would expect differences in murder rates among the three types of cities and would set up the null hypothesis of no differences. It should be noted that large samples are not required because of the normality assumption. Obviously if there were only one case in each category, however, there could be no variability within the categories, and therefore no test is possible.

The F test itself does not test the assumption of equal variances or *homoscedasticity* (as this assumption is referred to in technical language). In instances where sample variances seem to differ considerably among themselves, an independent test for the equality of variances may have to be made (see [1], pp. 141 to 144). If the results of such a test indicate that there are rather extreme departures from homogeneity of variance, then analysis of variance should not be used. Moderate departures from homogeneity can be tolerated, however. Such departures can often be reduced considerably by the transformation of variables.[3] If a single category is either considerably more or less homogeneous than the others, it may be advisable to omit this category from the analysis-of-variance test. Generally speaking, moderate departures from normality and equality of variances can be tolerated without necessitating the use of nonparametric alternatives (see [1], pp. 220 to 223).

2. *Significance Level and Critical Region* Let us use the .05 level. If the null hypothesis is actually incorrect, then if we always take the ratio of the between to the within estimate, we can expect to find the value of F to be larger than unity. We shall therefore use the upper tail of the F distribution as the critical region. If F turns out to be less than unity, there will be no point in looking up the probability value in the table since F values of greater than unity will be needed to reject the null hypothesis. An F of less than unity would indicate a greater degree of heterogeneity within categories than would be expected by chance. You should again keep in mind that although we shall use only one tail of the F distribution, this does not imply that we are predicting in advance which of the categories' means will be the largest.

3. *Sampling Distribution* The sampling distribution of F is given in Table J, Appendix 2. The use of the table will be described below.

4. *Computation of Test Statistic* In order to obtain a value for F, the ratio of the between and within estimates, it will first be necessary to calculate the total, between, and within sums of squares. Since the total variation is equal to the sum of the other two, we shall only have to compute two of the values, the third being obtained from the other two.

---

[3] For example, it sometimes happens that categories having the highest means are also the least homogeneous. In such instances if one takes as his interval scale the logarithm of the original variable, the effect will be to equalize the variances. For an additional discussion of the use of logarithmic transformations, see Sec. 18.2.

The within sum of squares, it will be remembered, involves a pooling operation. This is considerably more work than that required for the other two sums of squares, and therefore we get the within sum of squares by subtracting the between from the total sum of squares.

The computing formula for the total sum of squares is obtained in the same way as the formula for the variance [see Eq. (6.6)]. Thus,

$$\text{Total sum of squares} = \sum_i \sum_j (X_{ij} - \bar{X}..)^2 = \sum_i \sum_j X_{ij}^2 - \frac{\left(\sum_i \sum_j X_{ij}\right)^2}{N}$$

$$(16.4)$$

This is basically the same formula that we have used in computing standard deviations, but it is now necessary to make use of double summation notation.

The computing formula for the between variation looks considerably more formidable but on inspection is found to involve a relatively simple procedure. It is as follows:

$$\text{Between sum of squares} = \sum_j \frac{\left(\sum_i X_{ij}\right)^2}{N_j} - \frac{\left(\sum_i \sum_j X_{ij}\right)^2}{N} \qquad (16.5)$$

$$= \left[\frac{\left(\sum_i X_{i1}\right)^2}{N_1} + \frac{\left(\sum_i X_{i2}\right)^2}{N_2} + \cdots + \frac{\left(\sum_i X_{ik}\right)^2}{N_k}\right] - \frac{\left(\sum_i \sum_j X_{ij}\right)^2}{N}$$

Notice that the final term in the above expression is the same factor that was subtracted from $\sum_i \sum_j X_{ij}^2$ to obtain the total sum of squares. The first term may confuse some readers, however. Taking this expression apart, we see that we first compute each column sum and then square it to obtain $\left(\sum_i X_{ij}\right)^2$. We then divide this expression by the number of cases in the column, which need not always be the same. We then have for the jth column $\left(\sum_i X_{ij}\right)^2/N_j$. Finally, we do the same thing for each column and add the results.

Computations for our numerical problem, given below, should help to make the procedures more clear. The total and between sums of squares are computed as follows:

$$\sum_i \sum_j X_{ij}^2 = (4.3)^2 + (2.8)^2 + \cdots + (1.9)^2 = 1{,}453.58$$

$$\frac{\left(\sum_i \sum_j X_{ij}\right)^2}{N} = \frac{(161.0)^2}{24} = 1{,}080.042$$

$$\text{Total SS} = 1{,}453.58 - 1{,}080.042 = 373.538$$

$$\text{Between SS} = \frac{(68.6)^2}{8} + \frac{(44.8)^2}{8} + \frac{(47.6)^2}{8} - 1{,}080.042$$

$$= 1{,}122.345 - 1{,}080.042 = 42.303$$

To obtain the within sum of squares we simply subtract the second expression from the first, getting

$$\text{Within SS} = \text{total SS} - \text{between SS}$$

or
$$331.235 = 373.538 - 42.303$$

The estimates of the common variance can now be computed by dividing by the proper degrees of freedom. Finally, F is computed by dividing the between estimate by the within estimate. Computations are summarized in Table 16.3.

Table 16.3 Computations for analysis of variance

	Sums of squares	Degrees of freedom	Estimate of variance	F
Total	373.538	$N - 1 = 23$		
Between	42.303	$k - 1 = 2$	21.152	1.34
Within	331.235	$N - k = 21$	15.773	

5. *Decision* In order to decide whether or not to reject the null hypothesis, we must determine whether the value of F falls within the critical region. It will be noticed that three separate tables for F are given corresponding to the .05, .01, and .001 significance levels respec-

tively. This information cannot be condensed into a single table because two degrees of freedom must be associated with each F, one for the numerator and one for the denominator. The degrees of freedom associated with the numerator, the between estimate, are found across the top of the table. Those for the denominator, the within estimate, are obtained by reading down the table. Notice that all values of F given in the table are ≥ 1.0, indicating that the table is set up directly for one-tailed tests. In other words, the numerator is always the larger of the two estimates. In our problem we obtained an F with 2 and 21 degrees of freedom (written $F_{2,21}$) equal to 1.34. Using the table for the .05 significance level and locating the proper degrees of freedom, we find the figure 3.47. We therefore know that if the assumptions were correct, we would get a value of F equal to or larger than this value less than 5 per cent of the time. Since the value we actually obtained for F is less than 3.47, we fail to reject the null hypothesis at the .05 level. We decide that there is insufficient evidence for concluding that city types actually differ with respect to murder rates.

16.2 Comparisons of Specific Means

It will be noticed that the above problem could have been handled by using difference-of-means tests involving the t distribution. Three separate pair-by-pair comparisons might have been made between industrial and trade, industrial and governmental, and trade and governmental cities. In contrast to this method, analysis of variance offers a *single test* of whether or not all three types differ significantly among themselves or, in other words, whether they all could have come from the same population. The advantage of analysis of variance is that a single test may be used in place of many. Had there been four categories, 4(3)/2 or 6 difference-of-means tests would have been necessary. With 6 categories 15 tests would be required, whereas with 10 categories 45 would be needed. Suppose 15 tests were required and only 4 turned out to be significant. What would we conclude? It might be difficult to say.

There is an easy way out which, on the surface, appears to be a reasonable procedure. Why not simply perform a difference-of-means test on the two categories having the smallest and largest means respectively? Then if these are significantly different, we can conclude that the categories do in fact differ among themselves. But we must remember that (assuming equal sample sizes) we would be selecting the single test most likely to yield significance and ignoring the rest. Since we can expect one test in twenty to yield significance at the .05 level even if all population means are identical, it is apparent that we are loading the dice in favor of rejection. In other words, the significance level actually being used is not the .05 level but perhaps the .5 or .7 level since we are essentially obtaining the probability of getting *at least one* success (significance at the .05 level) in a large number of trials.

It should not be concluded that analysis of variance is always preferable to a series of difference-of-means tests, however. The latter tests, when used cautiously, may yield considerably more information. For example, analysis of variance may lead to significant results primarily because one category is far out of line with the rest. Had this category been excluded, the conclusion might have been quite different. A series of difference-of-means tests might indicate this fact more clearly. Especially if one suspects before making the test that one or more categories will differ considerably from the others, a number of one-tailed difference-of-means tests might be appropriate. It is also sometimes possible to predict the order in which category means will fall. Suppose, for example, that it had been predicted that murder rates would be highest in industrial cities and lowest in governmental. In such a case 2 one-tailed difference-of-means tests could have been used—one predicting a difference between industrial and trade cities, and the other a difference between trade and governmental. Generally, it would seem that the more knowledge we have for predicting the relative magnitudes and/or directions of differences, the more likely it is that separate difference-of-means tests will be appropriate. Analysis of variance, on the other hand, seems to be more useful on the exploratory level.

Finally, the relationship between the t and F distributions can be noted. Had there been only two types of cities, an analysis-of-variance test might also have been made and the results compared with those of a difference-of-means t test. In this case, the degrees of freedom associated with the numerator of F would have been $2 - 1$ or 1. The degrees of freedom for the denominator would have been $N - 2$, the same as for t in a difference-of-means test. Also, it will be recalled that when we assume $\sigma_1 = \sigma_2$ the denominators of both t and F involve pooled estimates of the variance. It turns out that the t distribution can be considered as a special case of the F distribution. If we were to compute the values of t^2 with $N - 2$ degrees of freedom, we would find that they are exactly the same as those for an F with 1 and $N - 2$ degrees of freedom, as can be checked by comparing the F and t tables. In other words, t is the square root of an F having 1 degree of freedom associated with its numerator. This means, of course, that exactly the same conclusions will be reached in the two-sample case regardless of whether we use analysis of variance or the difference-of-means test. In this sense analysis of variance is actually an extension of the difference-of-means test.

*Orthogonal comparisons   In many situations where there are more than
two categories being compared, it is desirable to make a number of
specific comparisons that have been designated in advance of the testing
procedure and that are based on theoretical interests. For example, suppose an experiment involves five groups, one of which is a control group
and the remainder are subject to different sets of experimental manipulations. Perhaps the second and third groups have authoritarian leaders
and have been subjected to moderate and high degrees of frustration,
respectively. The fourth and fifth groups might also have been subjected
to moderate and extreme frustration but might have involved democratic
leadership experiences. We might want to compare the control group with
all four experimental groups, but we might also want to compare the two
authoritarian groups with the two democratic, or the two moderate-
frustration groups with the two that have received extreme frustration.
Are all these comparisons legitimate in the sense that they are not supply-
ing us with redundant information? Put another way, if we know the
results of one comparison, will this automatically shed light on some of
the others? We need a systematic way of deciding whether the comparisons are *orthogonal* or really independent.[4]

   We can again make use of the idea of linear functions in a procedure
that is a straightforward extension of the difference-of-means test. If
we wanted to compare the control group (group I) with the experimental
groups, it would naturally occur to us to subtract the mean of the four
experimental-group means from the mean of the control group. Similarly,
if we wanted to compare the democratic and authoritarian groups, we
would naturally subtract the mean of groups IV and V (democratic)
from that of groups II and III. If we were giving each group equal
weight (regardless of the relative sample sizes), this would involve com-
paring the means of the two means or $(\frac{1}{2})(\bar{X}_2 + \bar{X}_3) - (\frac{1}{2})(\bar{X}_4 + \bar{X}_5)$,
with the null hypothesis being that $(\frac{1}{2})(\mu_2 + \mu_3) - (\frac{1}{2})(\mu_4 + \mu_5) = 0$.

   More generally, let us define a function ψ_i for the ith comparison we
wish to make as follows:

$$\psi_i = c_{i1}\mu_1 + c_{i2}\mu_2 + \cdots + c_{ik}\mu_k = \sum_{j=1}^{k} c_{ij}\mu_j$$

---

[4] The notion of orthogonality derives from a geometrical interpretation of statistical
associations and refers to situations in which the relationship can be represented by
means of perpendicular or orthogonal axes. For our purposes the important point is
that if we also assume homoscedasticity and normality for the distribution of the
dependent variable, then it can be shown that orthogonality implies statistical
independence.

where the c_{ij} are very simple weights that will depend on the comparison being made. If we impose the restriction that the sum of the weights must be zero, i.e., $\sum_{j=1}^{k} c_{ij} = 0$, this will greatly simplify the analysis without in any way restricting the comparisons we can make. Thus if our first comparison involves the control group versus the mean of the four remaining groups, we may take $c_{11} = 1$, with the remaining c_{1j} all being $-\frac{1}{4}$. If a particular comparison simply leaves out one of the categories (e.g., the control group) we merely let the c_{ij} for that category be zero. Thus for the three comparisons under consideration we would have:

	I	II	III	IV	V
ψ_1: control versus others (I versus II, III, IV & V)	1	$-\frac{1}{4}$	$-\frac{1}{4}$	$-\frac{1}{4}$	$-\frac{1}{4}$
ψ_2: authoritarian versus democratic (II & III versus IV & V)	0	$\frac{1}{2}$	$\frac{1}{2}$	$-\frac{1}{2}$	$-\frac{1}{2}$
ψ_3: moderate versus extreme frustration (II & IV versus III & V)	0	$\frac{1}{2}$	$-\frac{1}{2}$	$\frac{1}{2}$	$-\frac{1}{2}$

If the population variances σ_i^2 are approximately equal, the populations approximately normal, and all samples of the same size, the separate comparisons will be mutually independent (as well as orthogonal) if the following relationship among the coefficients holds:

$$\sum_{j=1}^{k} c_{hj}c_{ij} = 0 \qquad \text{for all } h \neq i$$

In particular, we begin by examining the first pair of comparisons ($h = 1$, $i = 2$). In our case we have the following:

$c_{11}c_{21} + c_{12}c_{22} + c_{13}c_{23} + c_{14}c_{24} + c_{15}c_{25}$
$= 1(0) + (-\frac{1}{4})(\frac{1}{2}) + (-\frac{1}{4})(\frac{1}{2}) + (-\frac{1}{4})(-\frac{1}{2}) + (-\frac{1}{4})(-\frac{1}{2}) = 0$

and we see that the condition holds. We next move to the first and third comparisons and finally to the second and third, again noting that the required sum of products is zero. Thus

$$1(0) + (-\frac{1}{4})(\frac{1}{2}) + (-\frac{1}{4})(-\frac{1}{2}) + (-\frac{1}{4})(\frac{1}{2}) + (-\frac{1}{4})(-\frac{1}{2}) = 0$$

and

$$0(0) + (\frac{1}{2})(\frac{1}{2}) + (\frac{1}{2})(-\frac{1}{2}) + (-\frac{1}{2})(\frac{1}{2}) + (-\frac{1}{2})(-\frac{1}{2}) = 0$$

In general, it can be shown that if there are k categories, there can be at most $k - 1$ comparisons that are mutually orthogonal. Also, if the sample sizes are unequal, it becomes necessary to weight by the sample category sizes N_j, with the test criterion for orthogonality becoming

$$\sum_{j=1}^{k} \frac{c_{hj}c_{ij}}{N_j} = 0$$

In our example we have utilized only three mutually orthogonal comparisons, whereas $k - 1$ or four will be possible. Of course in most instances it may not make theoretical sense to utilize all possible orthogonal comparisons, but it is nevertheless instructive to see what the fourth would be. Notice that we have already compared the control group with all the experimental groups, and therefore we would not expect a comparison of the control with any subset (e.g., the authoritarian groups) to be orthogonal with the first comparison. This can easily be seen by applying the test criterion. Notice that we have compared group II (along with either group III or IV) with group V (in combination with groups III and IV). We might therefore expect that if groups II and V were paired together against III and IV, the resulting comparison might be orthogonal to the remaining comparisons, as in fact turns out to be the case. Unless one were specifically looking for interaction, this particular comparison would probably not make sense theoretically since it would involve averaging scores of the medium-frustration authoritarian group with that of the high-frustration democratic group.

Notice that in testing for mutual orthogonality or independence among comparisons, we have said nothing about the actual sample data, except for the sample sizes N_j. The test criterion involves only the weights c_{ij} and not the sample means or variances. In effect, decisions concerning one's comparisons should be made in advance of the data collection. One can then test each comparison for statistical significance, as indicated below. This test will involve the t distribution in a manner exactly analogous to the difference of means test, which is of course the simplest possible comparison where $c_{11} = 1$ and $c_{12} = -1$. The numerator of t will be an estimate of the linear function ψ_i obtained by substituting sample means for the population counterparts. Thus if we form

$$\hat{\psi}_i = c_{i1}\bar{X}_1 + c_{i2}\bar{X}_2 + \cdots + c_{ik}\bar{X}_k$$

we will have the numerator for the ith comparison. For example, in the case of our first comparison involving the control group versus all others, we would have

$$\hat{\psi}_1 = \bar{X}_1 - (\tfrac{1}{4})(\bar{X}_2 + \bar{X}_3 + \bar{X}_4 + \bar{X}_5)$$

as would have been suggested by common sense.

For our denominator of t we wish to use a pooled estimator based on all samples, even in instances where the comparison may not involve all such samples. Recalling our result for the variance of a linear combination we know that

$$\text{var } \hat{\psi}_i = c_{i1}{}^2 \text{ var } \bar{X}_1 + c_{i2}{}^2 \text{ var } \bar{X}_2 + \cdots + c_{ik}{}^2 \text{ var } \bar{X}_k$$

If we are assuming normality and equal variances $\sigma_i{}^2 = \sigma^2$ this expression becomes

$$\text{var } \hat{\psi}_i = c_{i1}{}^2 \frac{\sigma^2}{N_1} + c_{i2}{}^2 \frac{\sigma^2}{N_2} + \cdots + c_{ik}{}^2 \frac{\sigma^2}{N_k} = \sigma^2 \sum_{j=1}^{k} \frac{c_{ij}{}^2}{N_j}$$

which, when we substitute an estimate for σ^2 and take the positive square root, becomes the desired denominator for t, which will have $N - k$ degrees of freedom. This same expression would have been used in the denominator of t in the cases of our second and third comparisons, which deleted the control group. For example, in the case of the second comparison we would have

$$\hat{\sigma}^2 = \frac{N_1 s_1{}^2 + N_2 s_2{}^2 + \cdots + N_5 s_5{}^2}{N - 5} \text{ (within group estimate of variance)}$$

$$\sum_{j=1}^{5} \frac{c_{ij}{}^2}{N_j} = \frac{0}{N_1} + \frac{(\tfrac{1}{2})^2}{N_2} + \frac{(\tfrac{1}{2})^2}{N_3} + \frac{(-\tfrac{1}{2})^2}{N_4} + \frac{(-\tfrac{1}{2})^2}{N_5}$$

$$= \left(\frac{1}{4}\right)\left(\frac{1}{N_2} + \frac{1}{N_3} + \frac{1}{N_4} + \frac{1}{N_5}\right)$$

and therefore $t = \dfrac{(\tfrac{1}{2})(\bar{X}_2 + \bar{X}_3) - (\tfrac{1}{2})(\bar{X}_4 + \bar{X}_5)}{\hat{\sigma}(\tfrac{1}{2}) \sqrt{1/N_2 + 1/N_3 + 1/N_4 + 1/N_5}}$

$$= \frac{(\bar{X}_2 + \bar{X}_3) - (\bar{X}_4 + \bar{X}_5)}{\hat{\sigma} \sqrt{1/N_2 + 1/N_3 + 1/N_4 + 1/N_5}}$$

which is an obvious extension of the difference-of-means test. Notice that the factor $(\tfrac{1}{2})$ canceled from the numerator and denominator, this being a reflection of the fact that the absolute magnitudes of the c_{ij} do not matter as long as $\sum_j c_{ij} = 0$. Again it should be emphasized that the pooled estimator $\hat{\sigma}$ will be just the estimator based on the within sum of squares (as calculated for the F test), and will involve *all* categories, whereas the numerator of t and the expression under the radical in the denominator will not both necessarily involve all categories.

16.3 Two-way Analysis of Variance

Under certain circumstances it is possible to extend analysis of variance by adding further nominal-scale variables. Such a procedure is mainly feasible in controlled experiments in which the researcher can assign individuals randomly to several groups, thereby controlling the number of cases in each category. In natural situations, where no such control can be exercised, the extension described in this section will be less useful. Some of the basic ideas involved in what has been referred to as two-way analysis of variance will be useful in helping you understand certain material presented in Chaps. 19 and 20, however.

If it is possible to introduce another nominal-scale variable in such a manner that all combinations of subcategories of the two nominal scales have the same number of cases, the extension of analysis of variance is a straightforward one.[5] Suppose the categories of the second nominal scale are represented by rows. We now obtain a number of subcells, each with an equal number of cases. In order for this condition to be met, we must of course restrict ourselves to column categories of the same size. To the numerical data given in Table 16.1, let us add the nominal scale "region," using only the two regions Northeast and Southeast. Let us suppose that there are the same number of cities in each of the six cells. If this were not the case, an approximate method would have to be used (see below). The numerical data are now given in Table 16.4, with subcategory sums and means indicated in each box.

If there are the same number of cases in each subcell, it is possible to break the within-columns or unexplained sum of squares into several components. We can, of course, do an analysis of variance across the rows, completely ignoring the columns. Between- and within-rows sums of squares would then be obtained in exactly the same manner that

---

[5] If we place the same number of cases in each category, then if we were to construct a contingency table relating the two nominal scales, we would see that there is no relationship between them in the *sample*. It is this lack of relationship between the nominal-scale variables that permits us to separate the row and column sums of squares in an unambiguous manner.

Table 16.4 Data for two-way analysis of variance

Regions	City type			Total
	Industrial	Trade	Governmental	
Northeast	4.3 5.9 2.8 7.7 $\Sigma X = 20.7$ $\bar{X} = 5.18$	5.1 3.6 1.8 3.3 $\Sigma X = 13.8$ $\bar{X} = 3.45$	3.1 3.8 1.6 1.9 $\Sigma X = 10.4$ $\bar{X} = 2.60$	$\sum_{j} X_{1j} = 44.9$ $\bar{X}_{1.} = 3.74$
Southeast	12.3 9.1 16.3 10.2 $\Sigma X = 47.9$ $\bar{X} = 11.98$	6.2 4.1 9.5 11.2 $\Sigma X = 31.0$ $\bar{X} = 7.75$	6.2 11.4 7.1 12.5 $\Sigma X = 37.2$ $\bar{X} = 9.30$	$\sum_{j} X_{2j} = 116.1$ $\bar{X}_{2.} = 9.68$
Total	$\sum_{i} X_{i1} = 68.6$ $\bar{X}_{.1} = 8.58$	$\sum_{i} X_{i2} = 44.8$ $\bar{X}_{.2} = 5.60$	$\sum_{i} X_{i3} = 47.6$ $\bar{X}_{.3} = 5.95$	$\sum_{i} \sum_{j} X_{ij} = 161.0$ $\bar{X}_{..} = 6.71$

between- and within-column figures were computed. It turns out mathematically that if there are the same number of cases in each subcell, the between-rows sum of squares can be considered to come completely from the within-column or unexplained (by columns) sum of squares. Thus the total variation can now be divided into three portions as follows.

$$\text{Total SS} = \text{between-column SS} + \text{between-row SS} + \text{unexplained SS} \tag{16.6}$$

We have taken the total variation and explained all we could by means of the first nominal scale (city type). Of that which is left unexplained (the within-column sum of squares), a certain portion can then be explained by the second nominal scale (region). The remainder, often called an error term, is the proportion of the total variation left unexplained by both variables. There are now three estimates of the common variance in addition to the estimate based on the total sum of squares, and these can be used to make two separate F tests. The error term can be used in the denominators of both F tests since the estimate based on the unexplained sum of squares will always be unbiased and independent of the other two. The numerators of the F's will be the estimates based on the between-columns and between-rows sums of squares. Each test will be a test for the existence of a relationship between the interval-scale variable and one of the nominal-scale variables, controlling for the other nominal scale.

Although we shall discuss this type of controlling operation in greater detail in Chap. 19, a few words on the subject are necessary at this point since controlling by using two-way analysis of variance involves a somewhat different principle from that discussed in connection with contingency problems. You will recall that up to this point our controlling procedure has involved holding the control variable constant by examining what happens within each category of the control variable. For example, we made a series of chi-square tests, one for each of these categories. Here we make a *single F* test instead of many, as was also done in the case of the pooled chi-square test. In effect, we take the control variable's presence into consideration by *adjusting* values of the interval scale according to the category of the control variable.

You will note from Table 16.4, for example, that the mean murder rate for all Northeastern cities is 3.74, whereas the mean for Southeastern cities is 9.68. Suppose we were to pretend that all cities were in the same region, performing a statistical adjustment of the murder rates by adding a fixed quantity (i.e., 2.97) to all cities in the Northeast and in this case subtracting the same quantity from those in the Southeast so that both categories had the same mean (i.e., the grand mean 6.71). Such a controlling operation involves raising the hypothetical question of what the murder rates would be like *if* all cities were actually exposed to the same regional influences. Rather than actually treating the regions separately, thereby sacrificing cases, we make use of a poor man's substitute by adjusting the murder-rate scores to take this control variable into consideration. What we lose in scientific rigor we gain in efficiency of design since we can then make use of a single test involving the total number of cases.

In adjusting the murder rates in such a fashion we are actually reducing the total variation in the scores. In effect, we are subtracting out that portion of the variation due to region. Taking the *adjusted scores*, we could then compare the between- and within-column estimates in the usual manner. Fortunately, it is unnecessary actually to obtain the adjusted scores. Were we to do so we would find that the results would be identical with those obtained using two-way analysis of variance. In other words, the type of analysis we shall describe below is equivalent to an adjusting operation such as the one we have been discussing. In effect what we do, first, is let the control variable operate on the dependent variable, taking out that portion of the total variation explained by this control variable. We then take the remainder as a "new total" variation and determine how much of this remainder can be explained by the other independent variable. This "new total" is equivalent to the total variation in the *adjusted scores*. In general, we can control for additional variables in the same manner. By making adjustments for each of the control variables, we take out all of the variation that can be explained by these variables. We then look at the remainder to see how much can be explained by the other independent variable. We shall make considerable use of this same type of controlling operation in subsequent chapters.

Interaction We are not yet ready for a numerical example since there is one further complication introduced by the addition of a second nominal scale. Whenever there are at least two cases in each subcell, an additional test should always be made. This is a test for "interaction," or a possible effect due to the peculiar combinations of the two nominal-scale variables. In order to make the two-way analysis-of-variance test previously described, it is necessary to assume the property of additivity. Formally stated, this property requires that mean population differences between columns be the same for each row and, conversely, that differences between rows be the same for each column. Additivity can be illustrated by the following figures representing hypothetical population means:

	A_1	A_2	A_3
B_1	5	10	20
B_2	10	15	25
B_3	25	30	40

Notice that the differences between the first and second columns are 5 for each row; between the second and third, differences are 10 for each row. Also, differences between the first and second rows are all 5, and between the second and third they are 15. Suppose, however, that the mean for the middle cell were 35 instead of 15. Then additivity would not hold. Although A_3 usually produces higher scores than A_2, and B_3 higher than B_2, something peculiar happens when A_2 and B_2 are put together in that a very high mean results. The process is somewhat analogous to what happens when hydrogen and oxygen are combined and water produced. The result is not what would be expected if each element were examined separately.

We have already encountered this possibility of interaction in connection with contingency tables, where we saw that the relationship between two variables may differ according to the level of a third variable. Let us illustrate the idea with several additional examples. Suppose that ordinarily industrial cities have higher murder rates than governmental cities and that Southeastern cities have higher murder rates than those in the Northeast. Conceivably, we might find industrial cities in the Southeast having an unexpectedly low average murder rate. We might then look for some kind of interaction between industry and regional factors that produces a low rate. A second type of example is perhaps more plausible. Suppose a choice between three types of teaching methods is to be made. Four teachers are each asked to use all three methods. Teacher A may be generally more effective than B. Similarly, the first method may generally be superior to the second. But conceivably teacher A may not work well with the first method, producing poorer results than would be expected. In this case there is interaction between teacher and method.

*Before turning to the calculation of the various quantities, it will be instructive to write down a general linear model that will turn out to be analogous to models developed in connection with regression analysis, where we will express an interval-scale variable as a function of several other variables which can be taken as either interval scales or attributes. Letting the score of the kth individual in the ith row and the jth column be represented by X_{ijk}, we may conceive of this score as being composed of the following components: (1) a component "due to" the overall population mean μ; (2) a second due to the effects of being in a particular row i, which we shall label the row effect α_i; (3) a similar effect β_j owing to being in column j; (4) an interaction effect γ_{ij} due to the peculiar combination of the ith row and the jth column; and (5) a unique effect, or error term ϵ_{ijk} produced by factors not explicitly considered in the equation. The equation then becomes

$$X_{ijk} = \mu + \alpha_i + \beta_j + \gamma_{ij} + \epsilon_{ijk}$$

which of course refers to population parameters that must be estimated from the sample data. It turns out that if all of the required assumptions for two-way analysis of variance are met (see below), we may obtain unbiased estimators of the parameters in the above equations as follows:

$$\hat{\mu} = \bar{X}_{..} \qquad\qquad \hat{\gamma}_{ij} = \bar{X}_{ij} - \bar{X}_{i.} - \bar{X}_{.j} + \bar{X}_{..}$$

$$\hat{\alpha}_i = \bar{X}_{i.} - \bar{X}_{..} \qquad\qquad = \bar{X}_{ij} - (\hat{\alpha}_i + \hat{\beta}_j + \hat{\mu})$$

$$\hat{\beta}_j = \bar{X}_{.j} - \bar{X}_{..} \qquad \hat{\epsilon}_{ijk} = X_{ijk} - \bar{X}_{ij}$$

*Each of these estimates makes intuitive sense except, perhaps, for the estimate of the interaction effect γ_{ij}. We use the sample grand mean $\bar{X}_{..}$ to estimate μ and the deviations of the row and column sample means from $\bar{X}_{..}$ to estimate the row and column effects, α_i and β_j, respectively.

The deviation of X_{ijk} from the subcategory sample mean $\bar{X}_{ij}$ of course represents unexplained variation in the sample that estimates the comparable residual term ϵ_{ijk}. The estimate of the interaction component γ_{ij} can then be obtained by subtraction. In effect, then, we have expressed each individual X_{ijk} in terms of the following components:

$$X_{ijk} = \bar{X}.. \quad + (\bar{X}_{i.} - \bar{X}..) + (\bar{X}_{.j} - \bar{X}..)$$

(Grand $+$ (row $+$ (column
mean) effect) effect)

$$+ (\bar{X}_{ij} - \bar{X}_{i.} - \bar{X}_{.j} + \bar{X}..) + (X_{ijk} - \bar{X}_{ij})$$

$+$ (interaction $+$ (error
effect) term)

For example, in the case of the second governmental city in the Northeast we would have

$$1.60 = 6.71 + (3.74 - 6.71) + (5.95 - 6.71) + (2.60 - 3.74$$
$$- 5.95 + 6.71) + (1.60 - 2.60)$$

*The basic procedure in the case of this model, as well as more generally, is to make separate tests for each of the component effects α_i, β_j, and γ_{ij} by evaluating the contributions of each term in relation to the size of the error term. Since, in addition, we always wish to utilize as simple a model as is necessary, we begin by seeing if it makes sense to eliminate the interaction component γ_{ij}. Let us now return to the computational procedure we shall use.

The test for interaction can be made independently of the two tests previously described and involves the same basic procedure. The unexplained sum of squares, or error term, is further broken down by subtracting out that portion which can be accounted for by interaction. The total sum of squares is thus decomposed into

Total SS $=$ between-column SS $+$ between-row SS
$+$ interaction SS $+$ error SS   (16.7)

This can be accomplished by taking each combination of the categories of A and B and treating it as a category of a combined (single) variable. In other words, we treat the problem as though there were a single nominal scale with categories $A_1B_1, A_2B_1, \ldots, A_kB_l$. Obviously, if there were only one case within each subcell, there could be no within-subclass variation. If there is absolutely no interaction, we should get exactly the same error term as obtained by adding the separate effects of the rows and columns [as in Eq. (16.6)]. On the other hand, if there is significant interaction, the error term will be smaller using this second approach. For example, you should convince yourself that if cell ij produced effects out of line with the rest, cell ij would be relatively homogeneous as compared with either column j or row i, and the within-subclass sum of squares would be smaller than the residual obtained by subtracting the sum of the between-column and between-row sums of squares from the total sum of squares.

The difference between the amount of variation explained using these subcells and the amount explained assuming additivity can then be attributed to interaction. Thus we have

$$\text{Total SS} = \text{between-subclass SS} + \text{within-subclass SS}$$

where the between-subclass sum of squares has been broken into three components:

$$\text{Between-subclass SS} = \text{between-column SS} + \text{between-row SS} + \text{interaction SS}$$

Computations   Returning once more to the numerical problem involving city types, region, and murder rates, we can begin by listing the required assumptions.

1. *Assumptions*

   Level of Measurement: Two nominal scales, one interval scale
   Model: Independent random samples
            All subcell, row, and column populations normal
            Subcell population variances equal
   Hypotheses: 1. Population column means equal
               2. Population row means equal
               3. Additivity in population (no interaction)

There are now three separate hypotheses that can be tested independently. The test for interaction should be made first, with subsequent tests depending on the outcome of this test. If hypothesis (3) is not rejected, the usual procedure is to assume additivity in the model, throwing the sums of squares due to interaction (in the sample) back into the error term and using this larger error term to test hypotheses (1) and (2). If the hypothesis of no interaction is rejected, the procedure to be used in the other two tests will depend upon the nature of the data (see below). Notice that we now must assume normality and equality of variances for each of the *subcells* in order to test for interaction. Cases in the different subcells must be selected independently and cannot be matched.

2. *Level of Significance* .05 level

3. *Sampling Distribution* F

4. *Computation of Test Statistics* We have already obtained the total and between-column sums of squares. The between-row sum of squares is computed exactly the same way as the between-column sum of squares. Thus

$$\text{Between-row SS} = \frac{44.9^2}{12} + \frac{116.1^2}{12} - 1{,}080.042$$
$$= 1{,}291.268 - 1{,}080.042 = 211.226$$

In order to obtain the interaction sum of squares, we make use of the sums of each subclass. The between-subclass sum of squares is

$$\text{Between-subclass SS} = \frac{20.7^2}{4} + \frac{47.9^2}{4} + \cdots + \frac{37.2^2}{4} - 1{,}080.042$$
$$= 1{,}341.585 - 1{,}080.042 = 261.543$$

We obtain the error term used in testing for interaction by subtracting the between-subclass sum of squares from the total. Thus

$$\text{Error SS} = 373.538 - 261.543 = 111.995$$

The amount actually due to interaction is the between-subclass sum of squares less the sum of the amounts due to the rows and columns separately. Therefore,

$$\text{Interaction SS} = 261.543 - (42.303 + 211.226) = 8.014$$

The results can be summarized as in Table 16.5.

The degrees of freedom are determined by the usual procedures. With l rows and k columns there will be $l - 1$ degrees of freedom associated with the between-row sum of squares. To obtain the degrees of

Table 16.5   Computations for two-way analysis of variance, with test for interaction

	Sums of squares	Degrees of freedom	Estimate of variance	F
Total	373.538	$N - 1 = 23$		
Between subclass	261.543	$kl - 1 = 5$		
Between columns	42.303	$k - 1 = 2$	21.152	
Between rows	211.226	$l - 1 = 1$	211.226	
Interaction	8.014	$(k - 1)(l - 1) = 2$	4.007	0.644
Error (within subclass)	111.995	$N - kl = 18$	6.222	

freedom for the interaction term, we take the number of subcells less one $(kl - 1)$ and subtract from this quantity the degrees of freedom associated with the between-row $(l - 1)$ and between-column $(k - 1)$ sums of squares. A simpler rule of thumb is to take the *product* of the degrees of freedom associated with the between-columns and between-rows sums of squares. Thus, if we multiply the between-columns and between-rows degrees of freedom, we get $(k - 1)(l - 1) = 2$ degrees of freedom. This is the same result as we would obtain by taking the between-subclass degrees of freedom $(= 5)$ and subtracting from this the sum of the degrees of freedom for the between-rows-and-columns sums of squares $(= 1 + 2)$. This can be expressed algebraically with the following identity.

$$(kl - 1) - (k - 1 + l - 1) = (k - 1)(l - 1)$$

The remaining degrees of freedom, which should equal the total number of cases less 1 degree of freedom for each subclass, can then be associated with the error term.

5. *Decision*   The test for interaction produces an F that is less than unity. We therefore have no reason to reject the null hypothesis of no interaction. This means that the small additional amount explained by interaction within these *samples* can easily be attributed to sampling fluctuations. In this case, we would probably be willing to accept the assumption of additivity even though we are on the wrong end of the test and should therefore be primarily concerned with the risk of a type II error. Incidentally, had tables been available, we might have used a level of significance such as .3 if we had really been interested in keeping the additivity assumption. Having decided that there is no interaction, we can now throw the sum of squares due to (sample) interaction back into the error term, using this larger error term as the basis

Table 16.6 Computations for two-way analysis of variance with interaction thrown into error term

	Sums of squares	Degrees of freedom	Estimate of variance	F	Significance level
Total	373.538	23			
Between columns	42.303	2	21.152	3.525	$p < .05$
Between rows	211.226	1	211.226	35.204	$p < .001$
Error	120.009	20	6.000		

for the error estimate of variance. Doing this, we get Table 16.6 in which the error term of 120.009 is the sum of the interaction and error terms in Table 16.5.

From the table we see that for an F with 2 and 20 degrees of freedom, we need an F of 3.49 or larger to attain significance at the .05 level. We also see that an F with 1 and 20 degrees of freedom of 35.204 is highly significant since an F of only 14.82 would be required for significance at the .001 level. Thus there is little doubt that a relationship exists between region and murder rates. Notice that when we control for region by letting this factor explain all that it can of the variation in murder rates and then letting city type explain what it can of the remainder, we now get a significant relationship between city type and murder rates. It will be recalled that the relationship without a control for region was not significant.

It should be noted that if interaction is not significant, we practically always gain more than we lose by throwing the interaction back into the error term, using this combined error term in the denominator of F. Although the error sum of squares is thereby slightly increased, there will also be more degrees of freedom associated with the larger error term. Since the interaction term will be relatively small, the net effect will usually be to obtain a smaller denominator for F. Also, of course, there will be a larger number of degrees of freedom associated with F, and therefore a smaller value of F will be required for significance.

We must now ask what we would have done had interaction been significant. The answer to this question is not at all simple, but we can at least give a few suggestions. The reader who is interested in a more complete treatment should consult a text such as Hays [7], Kirk [9], or Anderson and Bancroft [1].

If the interaction is significant, it is sometimes possible to locate one or two rows or columns, or even a few subcells, that are producing the interaction. For example, if we had used five regions, we might have noted that the Southeast differed in fundamental ways from the remaining regions. If so, we might have excluded this single region from the analysis and tested for interaction among the remaining categories, recognizing the ex post facto nature of such a procedure. In many other applications it will not be so simple to locate single rows, columns, or cells responsible for the interaction, in which case it may be a theoretical challenge to formulate a reasonably general explanation for the pattern found. In fact, the location of a major interaction effect may be the most important single finding of the study. Although the conceptualization of mathematical models involving interaction is beyond the scope of this text, it should be noted that certain relatively simple alternatives to linear additive models, such as multiplicative models, can be formulated. (See Blalock [2].)

In addition to focusing attention on interaction itself, there may also be an interest in whether or not one or the other of the nominal-scale variables is related to the interval scale. What tests of these relationships can be made? The question boils down to this: "Which estimate of the variance should be used in the denominator of F, the error estimate or the estimate based on the interaction term?" The answer to this question seems to depend on the nature of the two nominal-scale variables, and in particular on whether or not the categories used represent all categories in the population or merely a sample of categories. It is frequently the case in sociological problems, in which we do not usually assign individuals randomly to categories, that these categories represent all of the categories of interest in the classification scheme. Thus, if we divide all cities into three types and exclude no cities in so doing, we expect to include at least some cities from all types. Similarly, if we classify persons as male or female, or Protestant, Catholic, or Jewish, we ordinarily expect to include some representatives from all (or nearly all) categories. On the other hand, our categories might themselves involve a sampling of all types. For example, we might have selected Methodists, Quakers, and Jehovah's Witnesses as three religious groups representing a much larger number. Perhaps each of these denominations is a representative of a certain type of religion. Let us look at each of these situations in turn.

In the first situation our categories of *both* variables represent all or nearly all types that are of interest. There is certainly no sampling error involved in selecting categories as there might be if only three religious denominations were used for comparison. In most of these problems, our interest is likely to be centered on the degree of homogeneity of each type relative to the magnitude of differences between types. The second nominal-scale variable may be considered primarily as a disturbing variable that needs to be controlled. Interaction may be only an interesting by-product of the analysis. In this case, it would make sense to compare the estimate based on the between sum of squares with the estimate based on the unexplained sum of squares. This latter estimate is a within-subclass estimate and involves the variation which still remains unexplained by the joint operation of the major independent variable (say, city type) and the control variable. We are letting the control variable operate first, and then letting the major independent variable explain what it can of the remainder. A certain additional amount is also explained by the interaction of the two variables. Each of these "explained" sums of squares can be compared with the "unexplained" sum of squares or error term. We would then take this error estimate as our denominator in each of the separate F tests. In testing for the significance of a difference among columns we would therefore take the between-column estimate divided by the error term, and similarly for rows. In our numerical problem, had interaction been significant, these F ratios would have been 21.152/6.222 and 211.226/6.222 respectively.

Other considerations arise when the categories of either (or both) nominal-scale variable involve only a small sample of possible categories. If interaction turns out to be significant and therefore is larger than the error estimate, there is always the added question of whether or not this would have occurred had the categories been different. If *both* the row and the column variable involve a sampling of categories, we refer to the model as a *random effects* model, as contrasted with the *fixed effects* model for which neither variable involves a sampling of categories. I have never seen a reasonable illustration of such a random effects model, although *mixed* models involving one (or more) unsampled factors and a sampled factor are reasonably common. The most usual mixed model in the social science applications occurs in instances where persons (teachers, experimenters, interviewers, equipment operators, etc.) are taken as one of the factors. For example, in classroom experiments one may need to consider the "teacher effect" among a set of perhaps five teachers. In a laboratory setting the investigator may have used three different experimenters. Even though instructed to conduct themselves in similar ways, these persons would inevitably introduce idiosyncratic factors into the situation. Or in a survey, the analyst may need to separate out "interviewer effects" from those of other variables. In all such illustrations it would be recognized that the persons involved constitute a very small fraction of the potential number to which one wishes to generalize, and that interactions between persons and the factor of major interest will be especially troublesome.

These intuitive ideas can be given a more rigorous underpinning (see Hays [7], Chap. 13). It suffices here to indicate the preferred procedure.

Let us first assume that one is interested in testing for the effects of the *nonsampled* or fixed factor. If the interaction had been significant, this of course means that the estimate of the variance based on the interaction term must have been greater than the "error" estimate (so as to produce an $F > 1.0$). Given the fact that the second factor has been sampled, and that a second sample might have produced a very different interaction estimate, the more conservative procedure would be to use the interaction estimate (the larger of the two quantities) as the denominator for the F ratio in one's test for the significance of the fixed or nonsampled factor. In effect, the interaction is considered as an error. In our numerical example, suppose one were to consider that region were a sampled factor, since we have selected only two regions out of perhaps five or six. Had our interaction effect been significant and therefore not thrown back into the error term, we would have used the ratio 21.152/4.007 in testing for the significance of the effects of city type on murder rates.

If we are also interested in testing for the effects of the sampled factor (e.g., persons or region), however, we continue to use the error estimate rather than the interaction estimate in the denominator of F. The intuitive justification is that the other factor is *not* being sampled, and therefore sampling error in this factor cannot be a source of error in our estimate of the effects of the sampled factor on the dependent variable. Thus, had the interaction been significant in our example, we would have used the ratio 211.226/6.222 in testing for the effects of region on murder rates. (The fact that the denominator of 6.222 is greater than 4.007 used in connection with the effects of city type is a reflection of the fact that in this example the F used in testing for interaction came out less than unity, whereas a significant interaction would of course require an F of greater than unity.) A fuller justification of this procedure is given in Hays [7].

A further caution with respect to the interpretation of significant interactions is necessary. In the statistical literature one often finds references to the "main effects" of the row or column variable plus "interaction effects." It is possible to interpret these main effects as the average effects of one of the independent variables over the range of variation in the other variable(s). However, if the interaction component is relatively large, this very simple distinction between main and interaction effects will be difficult to translate into substantive or theoretical terms. For when interaction is large, it does not make theoretical sense to obscure real differences by talking about the average effects of, say, city type. Therefore, one should understand that this distinction between main and interaction effects is confined to statistical usage, as is the distinction between "explained" and "unexplained" sums of squares.

It is sometimes easy to fall into the trap of substituting statistical terminology for the terminology of one's own substantive discipline and to believe that there are distinct kinds of "effects" that have a simple counterpart in one's substantive theory. Perhaps the most cautious approach is to realize that whenever one finds statistical interactions of substantial magnitude, this means that two or more variables have joint effects on some dependent variable that are too complex to be adequately described by a simple additive model. The presence of statistical interaction therefore constitutes a clue that relationships are more complex than might have been thought, but by itself interaction should not be treated as though it were something apart from the "main" effects of the variables under consideration.

Extension to three or more nominal scales In theory there is nothing that prevents us from extending analysis of variance to additional variables. In practice, however, we are likely to be restricted by the requirement of equal numbers of cases in each subcell unless we are in a position experimentally to control this factor. If we add a third nominal scale, we can divide the total sum of squares into between A, between B, between C, interaction, and error terms, and we can make a number of F tests of separate hypotheses. This time, however, there will be more than one kind of interaction. There can be interaction between variables A and B, A and C, B and C, and among all three variables acting together. We first make a test for the three-factor interaction $(A \times B \times C)$. If this is not significant, we can throw it back into the error term and test for the three two-factor interactions. Tests of the significance of A, B, and C can then be made. The extension to four or more nominal scales would be made in a similar way. In the event that the researcher is able to control the number of cases in each category by random assignment, there are numerous other experimental designs available, and a textbook on experimental design should be consulted. Many of these alternative designs make it possible to gain greater efficiency (through a reduction of sample size) at a cost of simplifying assumptions about some of the interaction terms. For example, if one is willing to assume that certain interactions are negligible, he may deliberately "confound" these with main effects in a more efficient "incomplete" design.

*Two-way analysis of variance with unequal subclasses When the number of cases in each subclass is unequal, as will usually be the case in sociological research, two-way analysis of variance is no longer straightforward. If the number of cases is sufficiently large, it will of course always be possible to control for a second nominal scale by running separate analyses within each category of the control variable as we did in the case of contingency problems. If the number of cases is relatively small to begin with, certain approximate methods may be used. One of these methods involves the use of logarithms, but in other respects it is straightforward (see [8], pp. 260 to 266).

Another procedure, described in Walker and Lev [11], is conceptually much simpler. This latter method involves treating the means of the various subcells as though they were single cases. Sums of squares and estimates of variance can be obtained for the between-rows, between-columns, and interaction terms by essentially assuming that there is only one case (the mean) in each subcell. The error sum of squares is then obtained as in ordinary two-way analysis of variance by subtracting the between-subclass sum of squares from the total sum of squares using the total number of cases rather than the means of each subcell. The error estimate is obtained by dividing by the error degrees of freedom, as before, and then dividing the latter figure by the harmonic mean of the number of cases in each subcell. This latter operation is necessary in order to make the error estimate, based on the total number of cases, comparable to the estimates based only on the subcell means treated as single cases. *F* tests can then be made in the usual manner.

If the subclasses contain disproportional numbers of cases, as is the usual situation in nonexperimental research, this means that the row and column variables will be intercorrelated. In effect, some of the variation that the column variable "explains" will also be "explained" by the row variable, and there will be ambiguities as to which variable should be assigned the credit for the common explained variance. We will encounter this same difficulty in connection with multiple-regression analysis and, implicitly, in analysis of covariance.

After considering both multiple regression and analysis of covariance, we shall briefly discuss (in Chap. 20) a so-called "dummy-variable" approach that can be used to handle a good many kinds of situations, including that where we have two (or more) intercorrelated nominal-scale independent variables. We shall see, however, that this very general statistical approach does not enable one to overcome *theoretical* difficulties that arise in instances where the independent variables are intercorrelated. Such problems can only be handled by means of simultaneous-equation estimation procedures that are beyond the scope of this text. It should be noted that one of the major advantages of experimental designs is that they permit the manipulation of independent variables in such a way that their effects can be separated unambiguously, thus making it possible to assess the main effects of each variable provided interactions are not too large.

16.4 Nonparametric Alternatives to Analysis of Variance

In case the assumptions for analysis of variance are not met, there are nonparametric tests available that can be used as alternatives to one- and two-way analysis of variance. We shall first consider the Kruskal- Wallis one-way analysis of variance with ranks, then take up the Fried- man test for matched samples that can be used in situations where the row variable constitutes a set of matching variables, and where there is one "case" to each row.

The Kruskal-Wallis test The test considered in this section was developed by Kruskal and Wallis and is appropriate whenever we have a number of independent random samples and an ordinal-scale level of measurement. Its power efficiency for large samples is approximately 95 per cent. The test is basically a very simple one, involving a comparison of the sums of the rankings for each of the categories of the nominal-scale variable. A statistic H is computed in order to measure the degree to which the

Table 16.7   Data and computations for Kruskal-Wallis analysis of variance with ranks

	Industrial city		Trade city		Political city	
	Rate	Rank	Rate	Rank	Rate	Rank
	4.3	10	5.1	11	3.1	5
	2.8	4	1.8	2	1.6	1
	5.9	12	3.6	7	3.8	8
	7.7	16	3.3	6	1.9	3
	12.3	22	6.2	13.5	6.2	13.5
	16.3	24	9.5	18	7.1	15
	9.1	17	4.1	9	11.4	21
	10.2	19	11.2	20	12.5	23
Sums		$R_1 = 124$		$R_2 = 86.5$		$R_3 = 89.5$

various sums of ranks differ from what would be expected under the null hypothesis. If there are more than five cases in each class, the sampling distribution of H is approximately chi square.[6]

For comparative purposes let us illustrate the use of the Kruskal- Wallis test with the same data. In Table 16.7 the murder rates for the three types of cities have been ranked from high to low (low ranks indicating low murder rates).

[6] In the case of three classes and very small N's see [10], pp. 185 to 187.

1. *Assumptions*

  Level of Measurement: Ordinal and nominal scale
  Model: Independent random sampling
  Hypothesis: Samples drawn from same continuous population

2. *Significance Level and Critical Region*   Let us again use the .05 level.

3. *Sampling Distribution*   The sampling distribution of H will be approximately chi square with $k - 1$ degrees of freedom, where k is the number of categories used.

4. *Computation of Test Statistic*   We compute H by means of the formula

$$H = \frac{\left(\dfrac{12}{N(N+1)} \displaystyle\sum_{i=1}^{k} \dfrac{R_i{}^2}{N_i}\right) - 3(N+1)}{1 - \Sigma T_i/(N^3 - N)} \tag{16.8}$$

where N_i and N represent the number of cases in the ith category and in the total sample respectively. The denominator in the above formula represents a correction for ties, where

$$T_i = t_i{}^3 - t_i$$

t_i being the number of observations tied for a given rank.

  In this particular example there is only one pair of scores which is tied. Hence $T_i = 2^3 - 2 = 6$. We thus get

$$H = \frac{[12/24(25)](124^2/8 + 86.5^2/8 + 89.5^2/8) - 3(25)}{1 - 6/(24^3 - 24)} = 2.17$$

5. *Decision*   Referring to the chi-square table, we see that for 2 degrees of freedom we need a chi square of 5.991 or larger in order to obtain significance at the .05 level. Since we obtain an H of 2.17, we decide not to reject the null hypothesis at this level of significance. We see, incidentally, that we have reached the same conclusion as before.

The Friedman two-way analysis with ranks   Ordinal data apparently do not permit one to handle the notion of interaction, except in a very crude and unsatisfactory way. Nevertheless, if one is willing to assume that interaction is irrelevant and if one wishes to control for one or more variables by what amounts to a matching procedure, he may proceed as follows. Match the individuals (in this case cities) according to whatever criteria one wishes to use. One such criterion might be region; another might be size; a third, age of city, and so forth. Then assign one member of each set to an experimental condition, taking as the number of "cases" the number of sets of matched individuals. Obviously, this procedure represents an extension of matched-pairs tests we have already considered. In some instances there may be repeated observations on each individual; in others it may not have been possible to randomize the assignment to treatment or experimental groups, in which case one's interpretations must be much more cautious. In our own case it would obviously be impossible to assign cities randomly to the industrial, trade, and political categories.

We now consider each set of matched individuals as an independent replication. *Within* each of the sets we assign ranks 1, 2, 3, . . . , k according to the scores on the dependent variable. We do the same for each case and add the ranks down, obtaining a sum of ranks T_j for the jth column. If the experimental (column) variable has absolutely no effects, we would expect that the various T_j would all be approximately equal. In effect, we would be assigning the ranks within each row completely randomly and would not expect the total of the rank scores in any one column to be unusually large or small. But there will ordinarily be minor sample differences among the T_j, and we therefore wish to obtain a measure of the differences among the T_j that has a known sampling distribution.

If we compute the statistic

$$S = \sum_{j=1}^{k} (T_j - \bar{T})^2$$

where k is the number of categories and $\bar{T}$ is the mean of the T_j, it turns out that the sampling distribution of S can be calculated exactly for small samples and approximately for larger ones. Tables for the exact distributions are given in Bradley [3] and Siegel [10]. For $k \geq 4$ and $N \geq 10$, where N represents the number of *sets* of matched individuals, we may use a chi-square approximation as follows:

$$\chi^2 = \frac{12S}{Nk(k+1)} = \frac{12}{Nk(k+1)} \sum_{j=1}^{k} T_j^2 - 3N(k+1)$$

where the degree of freedom for chi square is $k - 1$, and where the far right-hand side will be more convenient for computing purposes. We again assume continuous underlying distributions of true scores so that ties are considered to arise solely as a result of crudity of measurement. We may either assign tied scores the means of the ranks they would have received had there been no ties, or more conservatively, we may break ties so as to minimize the value of the obtained chi square. Let us now proceed with an example.

Continuing with the same illustration for the sake of comparability, we retain the assumption of at least an ordinal level of measurement for murder rates and the null hypothesis that samples have been drawn from the same continuous population. In effect, this hypothesis amounts to the assertion that, in the context of an actual experiment, the experimental variable has no effect. We now assume, however, that the samples are matched, in this case into triplets of cities, one industrial, one trade, and one political. There are eight replications, so that $k = 3$ and $N = 8$. Previously in doing a two-way analysis of variance, we utilized only two rows, one for the Southeast and the other for the Northeast. Here we allow for eight separate rows, thus permitting individual matching where this is possible. Of course we could assign each Southeastern city arbitrarily to any one of the bottom four rows, but this would ordinarily be a less efficient design than using more refined controls in the matching process. To be specific, let us assume that we have utilized four different size classes from each of the two regions, so that the cities have been matched simultaneously on size and region. Assume that the cities have been arranged as in Table 16.8.

Table 16.8 Data and computations for Friedman test

Set	Industrial city		Trade city		Political city	
	Rate	Rank	Rate	Rank	Rate	Rank
A	4.3	2	5.1	3	3.1	1
B	2.8	3	1.8	2	1.6	1
C	5.9	3	3.6	1	3.8	2
D	7.7	3	3.3	2	1.9	1
E	12.3	3	6.2	1.5 (1)	6.2	1.5 (2)
F	16.3	3	9.5	2	7.1	1
G	9.1	2	4.1	1	11.4	3
H	10.2	1	11.2	2	12.5	3
T_j		20		14.5 (14)		13.5 (14) $\bar{T} = 16$

The ranks of course do not coincide with those in Table 16.7 since we have treated each set as a separate replication, with ranks in each case going only up to $k = 3$. Notice that there is only one tie within set E, with the average rank 1.5 having been assigned. The more conservative procedure would have been to assign rank 1 to the trade city and rank 2 to the political city since, for the remaining sets, $T_2 > T_3$. Results for the more conservative procedure are given in parentheses. Although we are dealing with a very small number of cases and columns, we shall use the chi-square approximation for illustrative purposes. We obtain

$$\chi^2 = \frac{12}{8(3)(4)} [20^2 + 14.5^2 + 13.5^2] - 3(8)(4) = 3.06$$

which, for $d.f. = k - 1 = 2$, is not significant at even the .10 level. Had we used the more conservative procedure, resulting in $T_2 = T_3 = 14$, we would have obtained a chi square of 3.00.

Bradley [3] notes that the power efficiency of the Friedman test depends not only on the sample size but on the number of categories used. For large samples the efficiency of the test, relative to that of the F test (assuming all the assumptions of the latter are justified), is approximately equal to

$$\frac{3}{\pi} \left(\frac{k}{k + 1} \right)$$

Thus for $k = 2$ the large-sample efficiency would be approximately $2/\pi = .64$, and for $k = 5$ it would be approximately $5/2\pi = .80$. Bradley notes that as k becomes small, the advantages of utilizing rankings across columns diminishes. In the limiting case where $k = 2$, we can assign only the two ranks 1 and 2, and the test becomes equivalent to the sign test and of course has the same relatively low power efficiency.

If the dependent variable (here murder rates) is so crudely measured that only two values *success* and *failure* can be assigned, it is possible to make use of a very similar nonparametric test, referred to as Cochran's Q test. The procedure utilized in Cochran's test involves assigning 1's and 0's to the X's (perhaps according to whether they are above or below the overall median) and utilizing either an exact distribution or a chi-square approximation much the same as in the case of the Friedman test. The Cochran test is discussed in Hays [7] and Bradley [3] and is also appropriate for matched samples.

16.5 Measures of Association: Intraclass Correlation

Analysis-of-variance tests only enable us to determine whether or not a relationship exists between two variables. As we have already seen, it is fairly easy to obtain statistical significance with even a very slight relationship provided there is a sufficiently large number of cases. Having determined that a relationship exists, subject to the risk of a type I error, we next need to measure the strength or degree of relationship. Some indication of the magnitude of the relationship can be obtained by simply comparing the means of the various categories. If these means are very different, the relationship is likely to be a strong one; if the differences are slight, we may not be able to attach much practical significance to them even though we may have obtained statistical significance. A mere comparison of category means may be misleading, however, unless one also notes the degree of homogeneity within each group. Usually, although perhaps not always, our interest centers on the relative magnitude of differences among means *as compared with* differences within categories. In other words, we wish to obtain a measure of the degree to which categories are homogeneous as compared with the total variability in the interval-scale variable. If the categories are perfectly homogeneous, the association between the two variables will be complete; knowing the category to which an individual belongs enables one to predict his score exactly.

Several closely related measures of association have been developed which make use of the total, between, and within sums of squares or of the estimates of variance based on these sums of squares. The *correlation ratio E^2*, perhaps the simplest of these measures, involves our taking the ratio of the explained to the total sum of squares. Thus

$$E^2 = \frac{\text{explained SS}}{\text{total SS}} = \frac{\text{between SS}}{\text{total SS}} \qquad (16.9)$$

As we shall see in the next chapter, the interpretation for the correlation ratio is directly analogous, except for its lack of sign, to that for ordinary product-moment correlation, and we shall make use of this measure in testing for the nonlinearity of the relationship between two interval scales.

The sample correlation ratio is slightly biased, however. You will remember that the sample standard deviation or variance tends to underestimate the population standard deviation or variance, the degree of bias becoming fairly serious for small samples. For this reason we used $N - 1$ rather than N in the denominator in order to obtain an unbiased estimate. Likewise, when the number of cases in each category becomes rather small, the expected value of the variability within each sample, as measured by the standard deviation, s, will tend to be less than that within the population. In order to correct for a comparable bias in the correlation ratio, we can obtain what is referred to as the *unbiased correlation ratio* by making use of the correct degrees of freedom and working directly with the estimates of variance rather than the sums of squares.

The formula for the unbiased correlation ratio ϵ^2 turns out to be

$$\epsilon^2 = 1 - \frac{V_w}{V_t} \qquad (16.10)$$

where V_w and V_t stand for the within and total *estimates* respectively. Although we have not as yet found it necessary to compute the total estimate, its value can easily be obtained by dividing the total sum of squares by $N - 1$. In the numerical example we have been using (see Table 16.3, p. 327), the values of E and ϵ are

$$E^2 = \frac{42.303}{373.538} = .113 \qquad E = .34$$

$$\epsilon^2 = 1 - \frac{15.773}{373.538/23} = .029 \qquad \epsilon = .17$$

Notice that the value of ϵ is smaller than that of E.

A somewhat more commonly used measure of association is the *intraclass correlation coefficient*. This measure derives its name from the fact that basically it involves a product-moment correlation between all possible pairs of cases *within* categories of the nominal-scale variable.[7] Like the other measures discussed in this section, the intraclass correlation coefficient r_i can also be considered as a measure of the degree of homogeneity of the classes relative to the total variability in the interval scale. Its formula is as follows

$$r_i = \frac{V_b - V_w}{V_b + (\bar{n} - 1)V_w} \qquad (16.11)$$

where V_b and V_w are the between and within estimates respectively and $\bar{n}$ represents an average number of cases in each class. An alternative

---

[7] After reading Chap. 17 you may wish to consult [5] to see the precise nature of the relationship between these two measures.

formula for r_i in terms of F is as follows:

$$r_i = \frac{F - 1}{F + (\bar{n} - 1)}$$

If the number of cases in each class is the same, there will of course be no question as to the value of $\bar{n}$. For unequal classes a simple arithmetic mean may be used to give the value of $\bar{n}$. Haggard [5] recommends a somewhat different kind of average value which should be used whenever the number of cases differs considerably from one category to the next. This formula for computing $\bar{n}$ is

$$\bar{n} = \frac{1}{k - 1}\left(\sum_{i=1}^{k} N_i - \frac{\sum_{i=1}^{k} N_i^2}{\sum_{i=1}^{k} N_i}\right) \tag{16.12}$$

where N_i represents the number of cases in the ith category and k the number of categories. In our numerical example all categories are of the same size and therefore $\bar{n} = 8$.

Thus
$$r_i = \frac{21.152 - 15.773}{21.152 + 7(15.773)} = \frac{5.379}{131.563} = .041$$

Several properties of the intraclass correlation coefficient can be noted. When the categories are all perfectly homogeneous, there will be no within-class variation (that is, $V_w = 0$), and the value of r_i will be $+1.0$. At the other extreme, suppose that all of the variation occurred within classes and that the category means were exactly equal. In this case V_b would vanish, and the lower limit would be

$$\frac{-V_w}{(\bar{n} - 1)V_w} = -\frac{1}{\bar{n} - 1}$$

Thus, the lower limit is not -1.0 except in the special case where there is an average of 2 cases within each class. Ordinarily, therefore, the lower limit will be less than unity in absolute value. Actually this seldom bothers us since we rarely find categories that are substantially less homogeneous than would be expected by chance. When the between and within estimates are exactly equal, i.e., when the value of F is exactly unity, r_i will be zero. Thus $r_i = 0$ whenever categories are exactly as homogeneous as would be expected by chance if there were no relationship between the two variables. Ordinarily the values of r_i will be between 0 and 1.0. Unfortunately, there seems to be no simple probabilistic interpretation for values of r_i between these limits.

*The notion of intraclass correlation can easily be generalized to cover two-way analysis of variance. In those situations in which we would use the error term in the denominator of F, we may obtain a measure of the degree of relationship between the column variable and the interval scale, controlling for the row variable, by taking as V_b the between-column estimate and replacing V_w by the error term. Similarly, we might also take the between-row estimate as V_b, thus obtaining a measure of the degree of association between the interval scale and the row variable after having subtracted out the variation due to the column variable. As we shall see in Chap. 19, this procedure is directly analogous to what we do in obtaining "partial" correlations between two interval scales, controlling for a third interval scale.

Glossary
Correlation ratio
Explained and unexplained variation
Homoscedasticity
Interaction
Intraclass correlation
Orthogonal comparisons

Exercises
1. Since the F test can be used to test the null hypothesis that we have two independent estimates of the same variance, we can use this test to test the assumption that $\sigma_1 = \sigma_2$ in difference-of-means problems. Since it will ordinarily not be possible to predict which value of s^2 will be the larger, we take the ratio of the larger to the smaller and then double the probability value given in the F table. With these facts in mind, take the data of Exercise 1, Chap. 13, and test the hypothesis that $\sigma_1 = \sigma_2$. (*Ans.* $F = 1.75$; fail to reject at .10 level)

2. Suppose that the data given below represent the incomes of the chairmen of the boards of various types of community organizations. You have randomly selected five organizations of each type for both large and small communities and therefore have obtained equal numbers of cases in each subclass.
    a. Make use of one-way analysis of variance to test for a relationship between organization type and income of board chairmen, ignoring the size of the community. What are the values of E and ϵ? (*Ans.* $F = 4.97$; $E = .52$; $\epsilon = .47$)
    b. Using two-way analysis of variance, what can you say about the relationship between organization type and income, controlling for community size? How do results compare with those in (*a*)? (*Ans.* For interaction, $F = 3.52$; reject at .05 level)

c. Compute the intraclass correlation for both (a) and *(b) above.

Size of community	Organization type		
	Religious	Social welfare	Civic
Large	$13,000	$15,000	$20,800
	11,500	10,600	18,100
	17,300	12,300	14,600
	19,100	11,400	22,300
	16,700	10,800	16,500
Small	15,000	9,300	14,400
	12,300	10,400	10,800
	13,900	12,900	9,700
	14,300	11,000	12,300
	11,700	9,100	13,100

3. Convert the income data for Exercise 2 to ranks and, using the Kruskal-Wallis test, test for a relationship between organization type and income:
   a. Ignoring community size. (*Ans. H* = 9.2; reject at .05 level)
   b. Controlling for community size.

4. Use the data for Exercise 2 and assume that the communities have been matched in triplets according to size, with there being ten sizes. The organizations in the top row (with incomes of $13,000, $15,000, and $20,800) represent the largest communities; those in the second row the next largest, and so forth. Use a Friedman test to test for the relationship between organization type and income of chairman (.05 level).

*5. Analysis of variance can be performed on grouped as well as ungrouped data. To reduce confusion it will be simplest to make use of Eqs. (16.4) and (16.5) in unmodified form, remembering, however, that in the case of grouped data we treat scores as though they were concentrated at the midpoints of intervals. With these facts in mind, perform a one-way analysis of variance on the data of Exercise 2, Chap. 13. As a check on your computations, how do the values of F and t compare?

6. Using the data of Exercise 2 above:
   a. Ignoring size of community, test for the significance of the difference between the mean income of the chairmen of the religious organizations and that of the welfare and civic organizations combined.
   b. What comparison would be orthogonal to that made in (a)?
   c. Suppose you consider that you have six types of organizations (religious in large community, religious in small community, welfare in large community, etc.). How many mutually orthogonal comparisons would be possible? Find a specific set of the above number of comparisons that are mutually orthogonal and verify that this is the case.

References

1. Anderson, R. L., and T. A. Bancroft: *Statistical Theory in Research,* McGraw-Hill Book Company, New York, 1952, chaps. 17 and 18.
2. Blalock, H. M.: "Theory Building and the Statistical Concept of Interaction," *American Sociological Review,* vol. 30, pp. 374–380, 1965.
3. Bradley, J. V.: *Distribution-free Statistical Tests,* Prentice-Hall, Inc., Englewood Cliffs, N.J., 1968, chap. 5.
4. Dixon, W. J., and F. J. Massey: *Introduction to Statistical Analysis,* 2d ed., McGraw-Hill Book Company, New York, 1957, chap. 10.
5. Haggard, E. A.: *Intraclass Correlation and the Analysis of Variance,* The Dryden Press, Inc., New York, 1958, chaps. 1–5.
6. Hagood, M. J., and D. O. Price: *Statistics for Sociologists,* Henry Holt and Company, Inc., New York, chap. 22.
7. Hays, W. L.: *Statistics,* Holt, Rinehart and Winston, Inc., New York, 1963, chaps. 11–14.
8. Johnson, P. O.: *Statistical Methods in Research,* Prentice-Hall, Inc., Englewood Cliffs, N.J., 1949, chaps. 10 and 11.
9. Kirk, R. E.: *Experimental Design: Procedures for the Behavioral Sciences,* Brooks/Cole Publishing Company, Belmont, Calif., 1968, chap. 3.
10. Siegel, S.: *Nonparametric Statistics for the Behavioral Sciences,* McGraw-Hill Book Company, New York, 1956, pp. 166–172, 184–193.
11. Walker, H. M., and J. Lev: *Statistical Inference,* Henry Holt and Company, Inc., New York, 1953, chap. 14.

Correlation and Regression

17

In the present chapter and in Chap. 18 we shall consider the relationship between two interval scales. An extension to three or more interval-scale variables will be made in Chap. 19 when we discuss multiple and partial correlation. For the time being, our concern is with situations in which we have two interval-scale measures on each individual. For example, we may know both the number of years of education completed and the annual income of adult males in a given community. Or we may be interested in relating the percentage of the labor force engaged in manufacturing to a city's population growth.

In certain problems of this sort we are often not only interested in significance tests and measures of degree of relationship but also may want to describe the *nature* of the relationship between the two variables so that we can predict the value of one variable if we know the other. For example, we may want to predict a person's future income from his education, or a city's rate of growth from the percentage of its labor force engaged in manufacturing. When interest is focused primarily on the exploratory task of finding out *which* variables are related to a given variable, we are likely to be mainly interested in measures of degree of relationship such as correlation coefficients. But once we have found the significant variables, we are more likely to turn our attention to regression analysis in which we attempt to predict the exact value of one variable from the other.

Although you are already familiar with significance tests and measures of association, it will be advisable to begin our discussion by studying the prediction problem. This is because the notion of regression is both logically prior to and theoretically more important than that of correlation. The reason for this will be more apparent as we proceed. After discussing the prediction problem, we shall turn our attention to measuring the strength of relationship. In Chap. 18, which actually represents a continuation of the present chapter, we shall discuss various significance tests as well as rank-order correlation which can be used in relating two ordinal scales.

17.1   Linear Regression and Least Squares

In one sense the ultimate goal of all sciences is that of prediction. This does not imply, of course, that one is only secondarily interested in understanding or providing causal explanations of why two or more variables are interrelated as they are. Perhaps it is correct to say that such explanation is the ultimate goal and that to the degree that understanding becomes perfected, prediction becomes more and more accurate. Presumably, if understanding were complete, perfect prediction would also be possible, provided that certain necessary factual information were also known. For example, if one knows the laws of planetary motion *and* the gravitational field within the solar system *and* the position and velocity of Venus at a particular time, one can then predict its future motion. Regardless of the philosophical implications of such a deterministic point of view, it remains that prediction is the goal of every science.

Predictive statements in sociology and other social sciences are often of necessity rather crudely worded. Usually this is because we have not reached the interval-scale level of measurement. Thus we might predict that the higher a person's status in the group, the greater his conformity to group norms. Such a statement need not imply one-way causality; it merely says that status and conformity are positively related. Making an analogy with mathematical terminology which is not strictly correct, we say that status is a *function* of conformity or conformity is a function of status, begging the question of causality. Notice, however, that we have said very little about the *form* of this relationship other than that it is positive. Unless we have an interval-scale level of measurement for both variables, it becomes very difficult to say much more.

Suppose, however, that we do have two interval scales. It then becomes possible to describe more exactly how the one variable varies with the other. For example, we might be able to say that for every year of schooling completed, the expected income will increase by $1,000. If this were indeed the case, we would have a very simple relationship, a linear or straight-line relationship. But most relationships are not nearly this simple although, as we shall see, it is frequently possible to obtain a very good approximation to the true relationship by assuming lin-

$Y = 1000 \, Edu + A$

earity. The most elegant and simple way to express a relationship between two (or more) variables is by means of a mathematical equation. Thus, you are acquainted with certain laws of physics which state a relationship between pressure, volume, and temperature $(PV/T = k)$ or which indicate a relationship between the rate of acceleration of a falling body, the distance it has fallen, and the length of time it has been falling. We can also represent each of these mathematical equations as some sort of a geometrical curve. Fortunately, in sociology we usually deal with very simple equations and the simplest possible curves (straight lines). When we add more variables, we cannot so easily represent equations as geometrical figures since we run out of dimensions, but we need not worry about this for the present.

Suppose there is a dependent variable Y which is to be predicted from an independent variable X. In some problems X will clearly precede Y in time. For example, a person usually completes his education before earning his income. In these instances such a conceptualization makes very good sense, although we want to be careful not to imply a necessary or causal relationship or that X is the only variable influencing the value of Y. If the direction of causation is ambiguous, or if each variable is thought to cause the other, then if we wish to provide a theoretical *explanation* of the relationship, we shall need to use a simultaneous-equation approach that cannot be discussed in the present text. (See [1], [2], and [6].) If our goal is simple estimation or short-run prediction of Y from X, then there will be no such ambiguities, though it should once more be pointed out that there is nothing in the statistical operations that can prevent us from carrying out mathematical operations that are theoretically meaningless. In this and in the following chapters we shall assume that the variable Y selected as dependent in the mathematical sense is also causally dependent, so that theoretical interpretations can be relatively straightforward.

We have seen that if X and Y are statistically independent, we cannot predict Y from X, or more exactly, knowledge of X does not improve our prediction of Y. Presumably, then, when the variables are not statistically independent, knowledge of X does help us predict Y. The stronger the dependence, the more accurate our prediction will be. Later we shall measure the strength of this relationship by means of correlation coefficients. We now focus on *how* we predict Y from X. For example, we may wish to estimate a man's future income, given that he has completed three years of high school. Without this knowledge of education, our best guess (assuming no inflation) would be the mean income of all adult males. Knowing his education, however, ought to enable us to obtain a better prediction.

The regression equation   Let us conceptualize the problem in the following way. We imagine that for every fixed value of the independent variable X (education), we have a distribution of Y's (incomes). In other words, for each educational level there will be a certain income distribution in the population. Not all persons who have completed high school will have exactly the same income, but these incomes will be distributed about some mean. There will be similar income distributions for grammar school graduates, college graduates, postgraduates, etc. Each of these separate income distributions (for fixed X's) will have a mean, and we can plot the position of these means in the familiar rectangular coordinates. We refer to the resulting *path of these means* of the Y's for fixed X's as a *regression equation of Y on X*. Such a regression equation can be illustrated in Fig. 17.1.

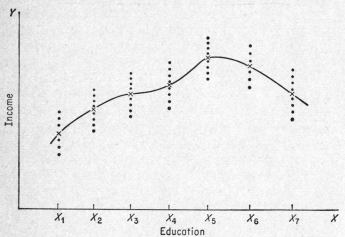

Figure 17.1   General form of regression of Y on X or the path of the means of Y values for fixed values of X.

These regression equations are the "laws" of a science. In some instances there is very little dispersion about the regression equation. In these instances very accurate predictions can be made, and deviations from the law **are** often thought of in terms of measurement error or as being due to minor uncontrolled forces. The "law" can then be stated as though a perfect relationship between Y and X existed. In the ideal, all points would be thought of as falling exactly on the curve, and the relationship would be abstracted as a perfect mathematical function in which there is a unique Y for each X. In the social sciences we cannot be nearly this pretentious. We expect considerable variability about the regression equation and prefer to think in terms of means and variances of a Y distribution for each X. Nevertheless, in principle the procedure is the same in all sciences, although the laws of social science are nowhere near as precise as those of physics.

In Fig. 17.1 we have indicated the general nature of regression equations as involving the paths of the means of Y values for given values of X. We shall now have to make some simplifying assumptions in order to make the problem manageable statistically. Although the idea of regression is perfectly general, most statistical work has been carried out on only the simplest of models. In particular, we shall for the present assume (1) that the form of the regression equation is linear, (2) that the distributions of the Y values for each X are normal, and (3) that the variances of the Y distributions are the same for each value of X. We can look at each of these assumptions in turn, giving most of our attention to the first.

If the regression of Y on X is linear, or a straight-line relationship, we can write an equation as follows

$$Y = \alpha + \beta X \tag{17.1}$$

where both α and β are constants. Equation (17.1) implies that the relationship between X and Y is exact, but we shall shortly introduce an error term into the equation. An alternative way of writing the equation is $E(Y|X) = \alpha + \beta X$, where the notation $E(Y|X)$ emphasizes that we are concerned with the expected value of Y, which depends on X. Greek letters have been used since for the present we are dealing with the total population. In such an equation both α and β have definite geometrical interpretations. If we set X equal to zero, we see that $Y = \alpha$. Therefore, α represents the point where the regression line crosses the Y axis (i.e., where $X = 0$).

The slope of the regression line is given by β since this constant indicates the magnitude of the change in Y for a unit change in X. The fact that the relationship is linear means that any given change in X, say 5 units, always produces the same change in Y (that is, 5β units) regardless of the position on the X axis. (See Fig. 17.2.) You should convince yourself that if $\beta = 1$ and if the units of X and Y are indicated by equal distances along the respective axes, the regression line will be at a 45-degree angle with the X axis. A β larger than unity indicates a steeper slope. The steeper the slope, the larger the change in Y for a given change in X. Similarly, if β is less than unity but greater than zero, it will take a larger change in X to produce a given change in Y. In the

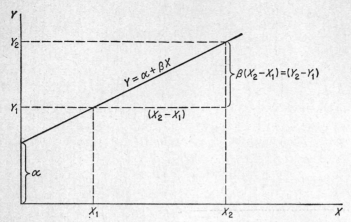

Figure 17.2 The linear regression equation, showing geometrical interpretations for α and β.

limiting case where the line is horizontal, β becomes zero and changes in X produce no change in Y. In other words, if $\beta = 0$, there is no linear relationship between X and Y. Knowledge of X will not help one predict Y if a linear model is assumed.[1] If β is negative, we know that there is a negative relationship between the two variables: as X increases, Y decreases.

A straight line can always be completely determined if we know either two points on the line or a single point and the slope. Therefore, there is a unique line with the equation $Y = \alpha + \beta X$ provided, of course, that α and β are assumed to be fixed (but general) quantities. If the α and β are given, the line can be drawn by simply taking two points on the line. We know that when $X = 0$, $Y = \alpha$. Therefore the point $(0,\alpha)$ lies on the line. Also, when $Y = 0$, we have $0 = \alpha + \beta X$ or $X = -\alpha/\beta$. This point $(-\alpha/\beta, 0)$ is, of course, the point where the line crosses the X axis. If it is inconvenient to use these two points, any other points can be determined by the same procedure.[2]

Assumptions about X and the disturbance term Thus far we have not dealt explicitly with the fact that since there will be scatter about the regression equation, we should represent the actual value of Y for each individual by an equation that contains a disturbance or error term that

[1] As we shall see later, statistical independence assures us that β will be zero, but it does not follow that if β is zero, we necessarily have independence.
[2] For a numerical example see p. 376.

is unique to each individual. If we let Y_i and X_i refer to the scores for the ith individual, then we may represent the (linear) relationship as follows

$$Y_i = \alpha + \beta X_i + \epsilon_i$$

where ϵ_i represents the disturbance term, the behavior of which we need to examine. We may conceive of this term as both involving measurement error in Y (but not X) and being a resultant of all the various causes of Y that have not been explicitly brought into the equation. If most of these omitted causes have rather minor impacts individually, and if they are operating almost independently of each other, then it will be reasonable to assume that the expected value of the disturbance term $E(\epsilon_i)$ will be zero and that ϵ_i will be approximately normally distributed. Most important, the disturbance term will be statistically independent of X. It turns out that in using least squares to estimate the regression coefficients α and β, it is necessary to assume that $E(\epsilon) = 0$ and that X_i and ϵ_i are uncorrelated. The normality assumption, plus the assumption of homoscedasticity that σ_ϵ^2 is constant across all levels of X, will be necessary in tests of significance and for the establishment of confidence limits.

The really crucial assumption that underlies the use of regression analysis is that X is independent of the error term. In experimental applications we often are in a position to select fixed levels of X (as, for example, when we hold temperature constant at intervals of 50 degrees). In such instances, since the level of X is under our control and, presumably, not manipulated in such a way that it varies systematically with the disturbance term, we seldom worry about this particular assumption. A little thought should convince you, however, that in many experimental situations even this assumption is naïve, since in manipulating X one may inadvertently affect other factors left out of the equation and therefore contained in the disturbance term.

In nonexperimental research we commonly take the X's as well as the Y's as observed rather than manipulated, and therefore X as well as Y is a random variable, or what is called a *stochastic* variable that has a probability distribution. Sometimes the distribution of X will be approximately normal, although this is not necessary in the case of regression analysis. What *is* essential, however, is that we make some assumptions about the joint distribution of the X_i and the disturbance term ϵ_i. If we had strong a priori grounds for specifying some particular distribution, this would be sufficient, but we practically always lack this information. Most commonly we assume that X_i and ϵ_i are statistically independent, an assumption that will be justified if the omitted causes of Y are either: (1) numerous, singly unimportant, and not highly interrelated; or (2) unrelated to X in situations where one or two omitted factors predominate. If one is unwilling to make such an assumption in any particular instance, he should attempt to identify the major disturbing factors that have been omitted and to bring these into the equation explicitly as additional variables. In Chap. 19 we shall discuss multiple regression where such additional causal factors have been included.

One of the advantages of the statistical theory of regression analysis is that it is sufficiently developed that such assumptions about the behavior of disturbance terms are explicit. It should be apparent that what we have said about the behavior of omitted variables applies equally well to *all* procedures we have discussed thus far. For example, if one finds a statistically significant difference of means or proportions, and if one wishes to attribute a causal explanation to the independent variable (e.g., sex) in this relationship, he will also need to assume that omitted factors are not systematically related to the dichotomized nominal scale (e.g., sex). One cannot get around unmet assumptions about omitted variables by merely shifting the kind of analysis and hoping that the problem will disappear.

It was indicated above that for significance tests we shall assume that the Y's are distributed normally about each value of X. For stochastic X's it will also be convenient to assume that for each fixed value of Y the X's are also distributed normally. We say that the joint distribution of X and Y is a *bivariate normal* distribution, meaning that there are two variables, each of which is distributed about the other normally. Such a bivariate normal distribution has a definite mathematical equation and can be represented as a three-dimensional surface as in Fig. 17.3. The height of the surface at any given point (X,Y) is proportional to the number of cases at that point. Thus it takes a three-dimensional diagram to represent the joint distribution between X and Y, just as we required two dimensions to represent the frequency distribution of X alone. The exact shape of this figure, which looks very much like a fireman's hat, will depend on how closely the variables are related.

If both variables have been expressed in terms of standard-deviation units, then the more highly related the two variables, the narrower the hat. In the extreme case in which Y can be exactly predicted from X, and therefore all points are exactly on the regression equation, the standard deviations of Y's for each X would be zero, and there would be no thickness to the hat at all. On the other hand, if there were no relationship between X and Y, the base of the hat would be more nearly circular. Any plane perpendicular to the XY plane will cut the surface in a normal curve. A plane parallel to the XY plane will cut it in an ellipse. The bivariate normal distribution has the property that the regression of

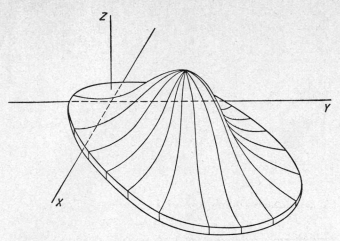

Figure 17.3 The bivariate normal distribution. (*By permission from A. M. Mood*, Introduction to the Theory of Statistics, *McGraw-Hill Book Company, New York, 1950, fig. 41, p. 165.*)

Y on X is linear. Therefore if we have a bivariate normal distribution, we know that if we trace the means of Y's for each X, the result will be a straight line. It does not follow, however, that if the regression is linear, the joint distribution is necessarily bivariate normal.

For tests of significance, we shall also need to assume that the standard deviations of the Y's for each X are the same regardless of the value of X. This assumption will be discussed in connection with the topic of correlation since correlation is essentially a measure of spread about the regression line. For the present it is sufficient to point out that if the joint distribution is bivariate normal, the standard deviations of the Y's for each X will in fact all be identical. This property of equal variances is referred to as *homoscedasticity* and is analogous to the assumption made in analysis of variance that $\sigma_1 = \sigma_2 = \cdots = \sigma_k$.

Linear least squares The regression model we have been discussing is conceptually rather simple, but unfortunately it is not directly useful in its theoretical form. Seldom do we have enough cases to examine the distribution of Y's for successive fixed values of X. More commonly, we find that there are relatively few cases in which the X's are identical or nearly so. If we plot the distribution of cases on the X and Y axes in the conventional manner, we usually find a scattering of dots such as indicated in Fig. 17.4. When we plot the distribution of points in this manner, we obtain what is referred to as a *scattergram* or scatter diagram.

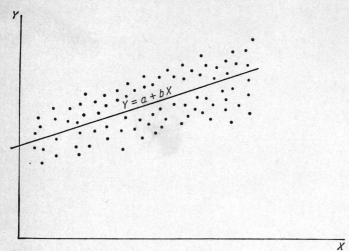

Figure 17.4 Scattergram and least-squares line.

You should develop the habit of always drawing a scattergram before proceeding with further analysis. Mere inspection of the scattergram may indicate that there is no point in going any further. For example, if dots appear to be randomly distributed on the diagram, there is clearly no relationship, or a very weak one, between the two variables.

Having plotted the scores on a scattergram, you may want to approximate these points by some sort of a best-fitting curve. One way of doing this is to draw a curve (in this case a straight line) by inspection. There are other more precise methods of doing this, however. One of these is the method of least squares which will be discussed in the present section. Our goal is now somewhat different from that of regression analysis in which we traced the path of the mean of the Y's. Here we want to approximate a number of dots by a best-fitting curve of some type.

In order to use least-squares theory, we must postulate the form of the curve to be used in fitting the data. In the case of regression analysis the form of the curve would actually be determined by the path of the means, assuming that data for the entire population were available. Again we shall take the simplest possible curve, the straight line, as our least-squares curve. This means that we shall fit the data with a best-fitting straight line according to the least-squares criterion, getting an equation of the form

$$Y = a + bX \qquad (17.2)$$

It will then turn out that the a and b obtained by this method are the most efficient unbiased estimates of the population parameters α and β if the regression equation actually is a straight line and if we assume (1) random sampling, (2) that $E(\epsilon_i) = 0$, and (3) that X_i and ϵ_i are statistically independent.

The least-squares criterion involves our finding the unique straight line which has the property that the sum of the squares of the deviations of the actual Y values from this line is a minimum. Thus if we draw vertical lines from each of the points to the least-squares line, and if we square these distances and add, the resulting sum will be less than a comparable sum of squares from any other possible straight line. (See

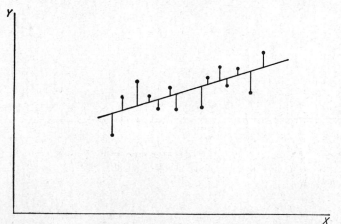

Figure 17.5 Least-squares equation minimizing sum of squares of vertical distances and estimating the regression of Y on X.

Fig. 17.5.) Notice that it is the vertical distances rather than the perpendicular or horizontal distances that are being considered here. It would be possible to minimize the sum of the squares of perpendicular distances (referred to as orthogonal least squares), but the resulting equations are not nearly as convenient. If the horizontal distances were used, the resulting least-squares line could be used to estimate the regression of X on Y. You should convince yourself that minimizing the sum of squares of vertical distances does not necessarily minimize the sum of squares of horizontal distances. Thus we can obtain several distinct least-squares lines. Only if all points lie exactly on a single line will these lines coincide. It also turns out that by minimizing the sum of squares of vertical distances, we are in effect finding the straight line having the property that the sum of the positive and negative vertical distances will be zero *and* the standard deviation of the points from the line will be a minimum. The notion of the standard deviation of the Y's from the line will be discussed below in more detail.

In order to obtain the least-squares line, then, we need to compute the a and b that determine the line with the desired property. This type of problem can easily be solved by means of calculus and leads to the following computing formulas for a and b.[3]

$$a = \frac{\sum\limits_{i=1}^{N} Y_i - b \sum\limits_{i=1}^{N} X_i}{N} = \bar{Y} - b\bar{X} \qquad (17.3)$$

and

$$b = \frac{\sum\limits_{i=1}^{N} (X_i - \bar{X})(Y_i - \bar{Y})}{\sum\limits_{i=1}^{N} (X_i - \bar{X})^2} = \frac{\sum\limits_{i=1}^{N} x_i y_i}{\sum\limits_{i=1}^{N} x_i^2} \qquad (17.4)$$

where $x_i = X_i - \bar{X}$ and $y_i = Y_i - \bar{Y}$. Notice that in these equations a and b are the unknowns, the other quantities being determined directly from the data. Once b has been obtained, a can readily be computed from the first of the two formulas. We can thus focus our attention on the computation of b.

The numerator of b involves the expression $\sum\limits_{i=1}^{N} (X_i - \bar{X})(Y_i - \bar{Y})$ which is referred to as the *covariation* of X and Y. This quantity is directly analogous to the sums of squares for either X or Y except that instead of squaring $(X - \bar{X})$ or $(Y - \bar{Y})$, we take the product of these two terms. We thus get a measure of how X and Y vary together, hence the name *covariation*. If we divide this expression by N, we obtain, by analogy, what is called the *covariance*. It will be seen immediately that b can be set equal to the ratio of the covariance to the variance in X.

---

[3] For those students familiar with elementary calculus we may outline the nature of the derivation. We begin with the equation $Y_i = a + bX_i + e_i$, where e_i is a residual term that can be used to estimate the residual ϵ_i of the regression equation. We wish to minimize the sum of the squares of these residuals, i.e., the quantity $\Sigma e_i^2 = \Sigma(Y_i - a - bX_i)^2$ with respect to the two quantities a and b, which are here treated as unknowns. We take partial derivatives with respect to a and b, set these equal to zero, and solve the two resulting equations (referred to as *normal equations*) for a and b. This same procedure generalizes to the multivariate case.

Taking a more detailed look at the covariation of X and Y, we see that, unlike a sum of squares, the covariation can take on both positive and negative values. If X and Y are positively related, then large values of X will ordinarily be associated with large values of Y. Thus if $X > \bar{X}$, it will usually be true that $Y > \bar{Y}$. Also, for a positive relationship, if $X < \bar{X}$, we shall ordinarily have $Y < \bar{Y}$. Therefore the product of $(X - \bar{X})$ and $(Y - \bar{Y})$ will usually be positive, and the sum of these products will also be positive. Similarly, if X and Y are negatively related, we would expect that if $X > \bar{X}$, then Y will be less than $\bar{Y}$, and the resulting sum of the products will be negative. If there is no relationship, then about half of the products will be positive and half negative since X and Y will vary independently. In this case b will be zero or close to zero. Therefore the higher the degree of the relationship, regardless of direction, the larger the numerical value of the covariation. As we shall see shortly, the covariation also appears in the numerator of the correlation coefficient, our measure of degree of association. In the case of b, we take the covariation and divide this by the sum of squares in X in order to obtain our estimate of the slope of the regression equation.

It is more convenient to use a computing formula for the covariation which is directly analogous to the computing formula for the sum of squares and which can be derived in a similar manner. We can write the computing formula for b as follows:

$$b = \frac{N\Sigma XY - (\Sigma X)(\Sigma Y)}{N\Sigma X^2 - (\Sigma X)^2} \qquad (17.5)$$

In Eq. (17.5) both numerator and denominator have been multiplied by N in order to reduce rounding errors due to division and in order to facilitate machine computation.[4]

PROBLEM Suppose we have the data given in Table 17.1, with X representing the percentage of Negroes in large Midwestern cities and Y indicating the difference between white and Negro median incomes as an indicator of economic discrimination.[5]

From the raw data we can compute five sums which, together with N, are all that we need in order to handle regression and correlation problems. All but one of these sums will be used in computing a and b. Computa-

[4] In this and succeeding formulas the subscripts have been dropped since we are always summing over the total number of cases N.
[5] Although the word "Negro" will be offensive to some readers, it is necessary to retain this terminology when referring to actual census data, as contrasted with hypothetical data or data obtained from other sources.

Table 17.1 Data for correlation problem

Per cent Negro X	Income difference Y	Per cent Negro X	Income difference Y
2.13	$809	4.62	$859
2.52	763	5.19	228
11.86	612	6.43	897
2.55	492	6.70	867
2.87	679	1.53	513
4.23	635	1.87	335
		10.38	868

tions can be summarized as follows:

$$N = 13 \qquad\qquad \Sigma Y = 8{,}557$$

$$\Sigma X = 62.88 \qquad\qquad \Sigma Y^2 = 6{,}192{,}505$$

$$\Sigma X^2 = 432.2768 \qquad \Sigma XY = 43{,}943.32$$

The only new quantity here is ΣXY. Placing these values in the formulas for a and b we now get

$$b = \frac{N\Sigma XY - (\Sigma X)(\Sigma Y)}{N\Sigma X^2 - (\Sigma X)^2}$$

$$= \frac{13(43{,}943.32) - (62.88)(8{,}557)}{13(432.2768) - (62.88)^2} = \frac{33{,}199.0}{1{,}665.7} = 19.931$$

and
$$a = \frac{\Sigma Y - b\Sigma X}{N}$$

$$= \frac{8{,}557 - (19.931)(62.88)}{13} = 561.83$$

The resulting linear equation is therefore

$$Y_p = a + bX = 561.83 + 19.931X$$

where we have used Y_p to indicate that the Y value has been predicted from a least-squares equation. As previously mentioned, the a and b obtained by this method are the most efficient unbiased estimates of α and β, the actual regression coefficients, provided that the disturbance term ϵ_i in the equation $Y_i = \alpha + \beta X_i + \epsilon_i$ has an expected value of zero and is unrelated to X, and provided that we actually have a random sample from the population under consideration. The least-squares line therefore will be the best estimate of the true regression if the regression equation actually is linear.

The least-squares equation also has the property of passing through the point $(\bar{X}, \bar{Y})$ representing the means of both X and Y. This can be seen from Eq. (17.3). Since

$$a = \bar{Y} - b\bar{X}$$

we have

$$\bar{Y} = a + b\bar{X}$$

indicating that these values of X and Y satisfy the equation. Therefore the point $(\bar{X}, \bar{Y})$ lies exactly on the line.

In the above problem if we know the X value (per cent Negro) for any given Midwestern city, our best guess as to the Y value would be that value of Y on the least-squares equation corresponding to the given X. Since income-difference scores indicate differences (in dollars) between the median incomes of whites and blacks, we see that an increase of 1 per cent Negro corresponds to a difference of \$19.93 in the median incomes of whites and blacks. A scattergram and the least-squares equation have been drawn in Fig. 17.6. To illustrate the use of such a

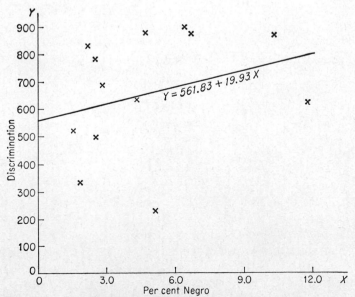

Figure 17.6 Scattergram and least-squares line for data of Table 17.1.

6291980

prediction equation, if we knew that there were 8 per cent Negro in a given city, the estimated median income differential would be

$$Y_p = a + b(8) = 561.83 + (19.931)(8) = \$721.28$$

We can see from the figure that approximately the same result would have been obtained graphically. Incidentally, by setting $X = 8$ and solving for Y, we have located a second point on the line which can then be used for the purpose of drawing in the line on the scatter diagram.

17.2 Correlation

Henceforth let us suppose that X is stochastic and therefore not subject to the control of the investigator. Not only do we want to know the *form* or nature of the relationship between X and Y so that one variable can be predicted from the other, but also it is necessary to know the *degree* or strength of the relationship. Obviously, if the relationship is very weak, there is no point in trying to predict Y from X. Sociologists are often primarily interested in discovering *which* of a very large number of variables are most closely related to a given dependent variable. In exploratory studies of this sort, regression analysis is of secondary importance. As a science matures and as important variables become identified, attention can then be focused on methods of exact prediction. Some statisticians are of the opinion that entirely too much attention has been given to correlation and too little to regression analysis. Whether or not this is true depends, of course, on the state of knowledge in the science concerned.

The correlation coefficient r to be discussed in this section was introduced by Karl Pearson and is often referred to as product-moment correlation in order to distinguish it from other measures of association. This coefficient measures the amount of spread about the *linear* least-squares equation. There is a comparable population coefficient rho (ρ) that measures the goodness of fit to the true regression equation. We obtain an estimate r of this parameter by measuring deviations from the line computed by least squares.

Since the regression equation represents the path of the means of Y's for given X's, it would also be possible to measure spread about this line by taking a standard deviation from the line.[6] Researchers in most

[6] The exact nature of such a measure will be discussed below. For the present we can simply point out that it represents an extension of the notion of a standard deviation, where the mean of the Y's is no longer taken as fixed but is considered to be a function of X.

applied fields have become accustomed to the correlation coefficient, however, and the correlation coefficient is probably here to stay. It has the advantage of being easily interpreted, and its range is from -1.0 to 1.0, a fact that is appealing to most practitioners. As we shall see, the relationship between the correlation coefficient and the standard deviation about the least-squares line is a very simple one, and this fact can be used to provide an interpretation for r.

It has been mentioned that r has an upper limit of 1.0. If all points are exactly on the straight line, r will be either 1.0 or -1.0 depending on whether the relationship is positive or negative. If the dots are randomly scattered, r will be zero. The better the fit, the larger the magnitude of r. This is indicated in Fig. 17.7.

Notice that r is a measure of *linear* relationship, being a measure of the goodness of fit of the least-squares straight line. You should not jump to the conclusion that if $r = 0$ (or if $\rho = 0$), there is no relationship whatsoever. If there is no relationship, it follows that r will be approxi-

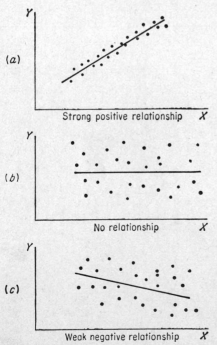

Figure 17.7 Scattergrams showing different strengths and directions of relationships between X and Y.

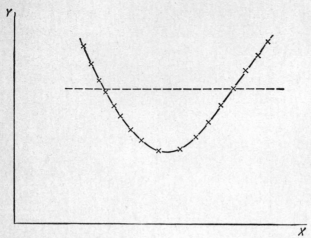

Figure 17.8 Scattergram for perfect nonlinear relationship
for which $r = 0$.

mately zero and there will be a random scatter of points. There may,
however, be a perfect curvilinear relationship and yet r can be zero,
indicating that there is no *straight line* that can fit the data. In Fig. 17.8
this is actually the case. Therefore, if a researcher finds a correlation
of zero, he should be careful not to infer that the two variables are unrelated. Usually, inspection of the scattergram will indicate whether there
is in fact no relationship or whether the relationship is sufficiently nonlinear to produce a zero correlation. In most sociological problems relationships can be approximated reasonably well with straight lines. This
does not mean that one should not be alert for possible exceptions,
however.

We have not as yet defined the correlation coefficient, but we can easily
do so in terms of the formula

$$r = \frac{\Sigma(X - \bar{X})(Y - \bar{Y})}{\sqrt{[\Sigma(X - \bar{X})^2][\Sigma(Y - \bar{Y})^2]}} = \frac{\Sigma xy}{\sqrt{(\Sigma x^2)(\Sigma y^2)}} \qquad (17.6)$$

In words, the correlation coefficient is the ratio of the covariation to the
square root of the product of the variation in X and the variation in Y.
Dividing numerator and denominator by N, writing this quantity as N^2
within the radical, we see that r can also be defined as the ratio of the
covariance to the product of the standard deviations of X and Y. The
covariance is a measure of the joint variation in X and Y, but its magni-

tude depends on the total amount of variability in both variables. Since the numerical value of the covariance can be considerably greater than unity, it is inconvenient to use it directly as a measure of association. Instead, we standardize by dividing by the product of the two standard deviations, thereby obtaining a measure that varies between -1.0 and 1.0.

We have already seen that the covariance will be zero whenever X and Y are unrelated. It can also easily be shown that the upper limit of r is unity. Let us take the case where b is positive and where all points lie exactly on the line. Then for every Y we can write $Y = a + bX$. Since $(\bar{X}, \bar{Y})$ also lies on the line, we have $\bar{Y} = a + b\bar{X}$. Therefore for all points on the line

$$Y - \bar{Y} = (a + bX) - (a + b\bar{X}) = b(X - \bar{X})$$

Hence
$$\Sigma(X - \bar{X})(Y - \bar{Y}) = b\Sigma(X - \bar{X})^2$$

and
$$\Sigma(Y - \bar{Y})^2 = b^2\Sigma(X - \bar{X})^2$$

Inspection of the numerator and denominator of r now indicates that under these conditions $r = 1.0$. Similarly, it can be shown that if all points lie exactly on a line with negative slope, the resulting r will be -1.0.

The relationship between the correlation coefficient and the slopes of the two least-squares equations should also be noted. If we let b_{yx} be the slope of the least-squares equation estimating the regression of Y on X, and if we let b_{xy} indicate the slope of the estimate of the regression of X on Y, we have by symmetry that

$$b_{xy} = \frac{\Sigma(X - \bar{X})(Y - \bar{Y})}{\Sigma(Y - \bar{Y})^2}$$

where
$$X = a_{xy} + b_{xy}Y$$

Thus, r has the same numerator as both b's. If the b's are zero, it follows that r must also be zero and vice versa.

For given sums of squares in X and Y, the value of b_{yx} (or b_{xy}) will be proportional to r. This might seem to lead to the conclusion that the strength of relationship is proportional to the slope of the least-squares line. This will be true only if the denominator remains fixed, however. Thus b is a function not only of the strength of the relationship but also of the standard deviations.[7] If there is considerable variability in X relative to Y, the value of b will be relatively small, indicating that it takes a large change in X to produce a moderate change in Y. As will be discussed below, the numerical values of the b's therefore depend on the size of the units of measurement.

The value of r has been standardized so that it is independent of the relative sizes of the standard deviations in X and Y. It would indeed be unfortunate if this were not the case, since we would hardly want a measure that varied according to whether we selected dollars or pennies as our monetary unit. It will be noted from the formulas for r and the b's that r^2 can be expressed in terms of the b's. Thus

$$r^2 = b_{yx}b_{xy} = \frac{[\Sigma xy]^2}{\Sigma x^2 \Sigma y^2} \tag{17.7}$$

You should verify that when r is 1.0 (or -1.0), $b_{yx} = 1/b_{xy}$ and that this means that the two least-squares equations coincide. Generally, as r approaches zero, the angle between the two lines becomes larger and larger until when $r = 0$, the lines have become perpendicular.

Finally, we can introduce a computing formula for r that involves the five sums previously obtained in connection with the computations of a and b. The formula is

$$r = \frac{N\Sigma XY - (\Sigma X)(\Sigma Y)}{\sqrt{[N\Sigma X^2 - (\Sigma X)^2][N\Sigma Y^2 - (\Sigma Y)^2]}} \tag{17.8}$$

The numerator has, of course, already been computed, and so has part of the denominator. Thus the correlation between per cent Negro and the index of discrimination is

$$r = \frac{13(43{,}943.32) - (62.88)(8{,}557)}{\sqrt{[13(432.2768) - (62.88)^2][13(6{,}192{,}505) - (8{,}557)^2]}}$$
$$= \frac{33{,}199}{110{,}120} = .301$$

It should be noted that one can add or subtract values from either X or Y without affecting the value of the correlation coefficient. Likewise, r will be unaffected by a change of scale in either variable. This says in effect that the correlation between income and education is the same

[7] Except in cases where confusion might arise, we shall continue to make use of b without the subscripts to represent b_{yx}.

regardless of whether income is measured in dollars or pennies. But although the correlation coefficient is invariant under transformations of this sort, the least-squares equation is not. Adding or subtracting values will affect the numerical value of a. A change of scale will affect the slope of the line. For example, if every X is divided by 10 while Y is kept fixed, the resulting b will be multiplied by 10. You should verify that these properties hold by examining the formulas for r, a, and b. These facts may be used in order to simplify computations. For example, if X involves either a very large number or a very small decimal, a change of scale may reduce the risk of computing errors. Or if the X variable consists of values such as 1,207, 1,409, 1,949, and 1,568, it would probably be advisable to subtract 1,000 from each score. Certain computing routines require that all values be positive. In computing r, therefore, it may be necessary to add to each value a number that is slightly larger than the largest negative score.

Another fact about the correlation coefficient should be noted at this time. Since this measure involves both variances and covariances, it is highly affected by a few extreme values of either variable. Furthermore, the magnitude of r depends on the degree of general variability in the independent variable. Figure 17.9 illustrates these points. In Fig. 17.9a, the effect of one or two extreme values is to produce a moderately high correlation where none exists among the remaining cases. In Fig. 17.9b, we have a moderately high linear relationship except for the fact that extreme cases are out of line with the rest. In this latter instance we probably have an example of a nonlinear relationship. A scattergram will always be helpful in indicating the nature of the situation in any given problem. Let us now discuss what can be done if either of these situations should occur.

Figure 17.9a illustrates the point made above that the magnitude of the correlation coefficient depends on the range of variability in both variables. Had there been a larger number of extreme cases, the resulting distribution might have been as in Fig. 17.10. In this instance the overall correlation may be high, but within any limited range of X's the correlation may be close to zero. In effect, this indicates that there is insufficient variability in X within this limited range to counteract the effects of numerous uncontrolled variables. In reality, X is almost being held constant. Therefore, if a scattergram turns out to be similar to the one in Fig. 17.9a, one should always attempt to extend the range of variability in X by finding more extreme cases.

If extending the range of variability is not feasible empirically or if the researcher's interest is focused primarily on less extreme cases, it may be more sensible to exclude the extreme cases from the analysis

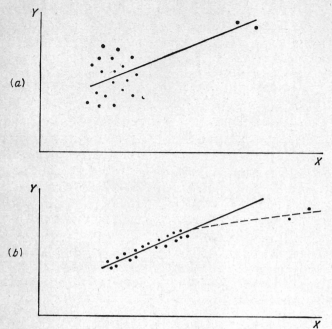

Figure 17.9 Scattergrams showing possible effects of extreme *X* values.

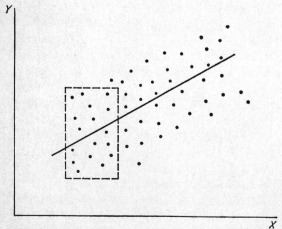

Figure 17.10 Scattergram showing no relationship within a limited range of variation in *X* but a positive relationship over the total range.

altogether. For example, suppose X is size of city and New York City appears in the sample. Unless there are a large number of cities of comparable size, and there are not, it may become necessary to confine one's attention to cities of less than 500,000. In some instances it would seem advisable to compute r both with and without the extreme cases. Obviously the decision made will depend upon the nature of the problem and the research interests of the social scientist. You should be cautioned that one or two extreme scores can have a very pronounced effect on the magnitude of r, and you should always take this into consideration in some manner. The range of variability should therefore be reported with correlation coefficients. This is another illustration of the important point that a single summarizing measure, no matter how superior it may be to other measures, can often be misleading.

If the data turn out to be as in Fig. 17.9b, we would obviously suspect nonlinearity. Again, if possible, additional extreme cases should be obtained. If there are only one or two extremes, it may be advisable to exclude these from the analysis. Situations of this sort illustrate the fact that within a limited range of variation a relationship may be approximately linear, but when extended, the linear model may be inappropriate. You should therefore be careful not to generalize beyond the limits of the data. A cautious statement such as, "Within the limits of _____ and _____ the relationship appears to be approximately linear" would be most appropriate.

A comparison of correlations and slopes The foregoing remarks about the sensitivity of correlation coefficients to differences in the amount of variation in X, relative to the scatter produced by extraneous factors, points to one of the fundamental problems with *any* measure of degree of association. Our primary focus should be on the nature of the law relating X and Y, on whether or not the relationship is linear, and if linear, on the magnitude of the slope. In comparing the results for two studies or several subsamples, we must recognize that it is entirely possible to obtain rather substantial differences among the correlation coefficients even though the same laws (as measured by the slopes) are operative. That is, the r's may differ even though the slopes do not, and this may be solely due to differences in the amount of variation in the independent variable X, or in differences in the extent to which extraneous factors producing random variation in Y have been brought under control. As we shall see in connection with analysis of covariance, in testing for interaction we are in effect testing for a difference among slopes, rather than correlations. In the next chapter we shall briefly consider tests for differences among correlations, but the reader should be alerted to the fact that such differences, if found, can be easily misinterpreted.

It may be useful to conceive of a correlation coefficient r_{xy} as a function of two kinds of variables, the slope b_{yx} and a factor s_x/s_y involving the ratio of the two standard deviations that apply to the particular sample or subsample of immediate concern. Thus

$$r_{xy} = b_{yx}(s_x/s_y)$$

Of course the numerical value of b_{yx} is determined not only by the law connecting X and Y but also by the investigator's choice of measurement units. The factor s_x/s_y is also a function of these units, which are of course known in advance of the population or sample data. But the ratio s_x/s_y will also be unique to each sample (and σ_x/σ_y to each population) and is used to produce the standardized measure r_{xy}. A correlation coefficient has the advantage of being standardized so as to be independent of one's choice of units of measurement, but unfortunately it must be standardized in terms of something, which turns out to be a quantity that is not invariant across samples or populations. This fact should be clearly understood, and the unstandardized slopes should *always* be reported so that replications will not be misleading in this respect.

Putting the matter in a slightly different way, we may recognize a hierarchy of scientific goals in statistical inference and estimation. We first test for significance to decide whether we have found a relationship that cannot readily be explained away by chance mechanisms. In this connection we noted that the probability or significance level is a function of the degree of relationship and sample size. If we have a very large sample, we may obtain a small probability level even with a very weak relationship that may have no practical importance. But having found at least a moderately strong relationship, we again have another more important task, namely that of estimating the nature of this relationship, as measured by a regression coefficient in the linear case. When correlations are moderately strong, instead of comparing these r's directly, we estimate the slopes and compare these slopes in our tests for interaction. We can diagram the process as follows:

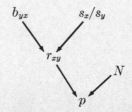

where the direction of the arrows represents the "causal flow" (e.g., probabilities influenced by magnitudes of relationships and sample sizes), which is often in the opposite direction from that of one's procedural steps in a statistical analysis. The diagram indicates that p is a function of two variables, one of which (sample size) is of no inherent interest, and that the correlation r_{xy} is also a function of two factors, one of which (s_x/s_y) is of no interest. Our objective is to move the analysis upward in the diagram to the estimation of the regression coefficients, rather than becoming fixated either with probability levels or with statements about correlation coefficients.

It turns out that whenever we are dealing with ordinal measures of association, such as those to be discussed in the next chapter, this distinction between slopes and measures of association breaks down. However, in the case of dichotomies it can be shown that if one follows the rule of placing the independent variable across the top of a table, computing proportions (or percentages) so as to add to 1.00 (or 100) downwards, and then comparing across, the resulting difference of proportions can be considered a special case of the slope b_{yx}, whereas ϕ becomes a special case of r_{xy}. If proportions are computed in the other direction, the difference of proportions becomes a special case of b_{xy}, so that we have an additional rationale for following the rule of thumb previously suggested. These results may be obtained by the simple device of assigning scores of 0 and 1 to both X and Y and then using the basic formulas for the computation of r_{xy} and b_{yx}.

*Computations from grouped data   If the number of cases is large and if computer facilities or desk calculators are unavailable, the computation of correlation coefficients may become extremely tedious. In such instances it may be more convenient to make use of grouped data in spite of the fact that certain slight inaccuracies will be introduced. In principle, these computations for grouped data are straightforward extensions of the procedures used in obtaining the mean and standard deviation. We now have two variables which must be cross classified as in Table 17.2. We shall guess a mean for each variable, taking step deviations from each guessed mean, and make use of correction factors in each case. In addition, we shall need a cross-product term equivalent to Σxy. Since deviations for both X and Y will be taken from the respective *guessed* means, we shall need to make use of a correction factor to be subtracted from the estimated cross-product term. We can then modify the computing formulas for r and b so as to take into consideration the fact that we have used guessed means rather than correct means.

You will recall that one of the computing formulas for s using grouped data was (dropping the subscripts)

$$s = \frac{i}{N} \sqrt{N\Sigma fd'^2 - (\Sigma fd')^2}$$

Since we now have two variables X and Y, we shall make use of subscripts in order to distinguish frequencies and step deviations for X (that is, f_x and d'_x) from those for Y (that is, f_y and d'_y). In computing the cross-product term we also need to obtain the frequencies f_{xy} in each subcell. These latter frequencies will ordinarily be smaller than either f_x or f_y. Thus, although there are 24 cases in the category 40.0 to 49.9 for the X variable and 30 cases in the 15.0 to 19.9 category of Y, there are only 6 cases in the subcell corresponding to both of these categories. You should convince yourself that the computing formula for r [Eq. (17.8)] can be modified as follows

$$r = \frac{N\Sigma f_{xy}d'_x d'_y - (\Sigma f_x d'_x)(\Sigma f_y d'_y)}{\sqrt{[N\Sigma f_x d'^2_x - (\Sigma f_x d'_x)^2][N\Sigma f_y d'^2_y - (\Sigma f_y d'_y)^2]}} \qquad (17.9)$$

Similarly, the formula for b becomes

$$b = \frac{N\Sigma f_{xy}d'_x d'_y - (\Sigma f_x d'_x)(\Sigma f_y d'_y)}{N\Sigma f_x d'^2_x - (\Sigma f_x d'_x)^2} \frac{i_y}{i_x} \qquad (17.10)$$

where i_y and i_x represent the interval widths for Y and X respectively. The value of a can now be computed from the equation

$$a = \frac{\Sigma Y - b\Sigma X}{N} = \bar{Y} - b\bar{X}$$

where $\bar{X}$ and $\bar{Y}$ can be obtained using the usual formula for grouped data.

Let us now compute the values of these coefficients for data on 150 Southern counties given in Table 17.2. We shall take as the dependent variable Y the percentage of females in the labor force, the independent variable being the percentage of the population classed as rural-farm. It will be convenient to make use of a computing form such as that given in Table 17.3. In this table the class limits and midpoints are given across the top (for Y) and down the left-hand margin (for X). Focus your attention on the boxed-in area of the table. Notice that there are three numbers in each subcell. In each subcell the top number represents the number of cases in the subcell as given in Table 17.2. The remaining numbers in the subcell are used in computing the cross-product term.

Table 17.2   Data cross classified for obtaining correlations from grouped data

Per cent rural-farm, X	Percentage of females in the labor force, Y							Totals
	10.0–14.9	15.0–19.9	20.0–24.9	25.0–29.9	30.0–34.9	35.0–39.9	40.0–44.9	
0.0– 9.9	0	0	0	1	8	4	0	13
10.0–19.9	1	2	0	2	4	1	3	13
20.0–29.9	2	5	1	2	3	3	0	16
30.0–39.9	2	0	5	5	7	3	0	22
40.0–49.9	4	6	6	7	1	0	0	24
50.0–59.9	3	10	9	6	2	0	0	30
60.0–69.9	2	4	3	7	4	0	0	20
70.0–79.9	2	3	4	1	0	0	0	10
80.0–89.9	1	0	1	0	0	0	0	2
Totals	17	30	29	31	29	11	3	150

SOURCE: U.S. Census, 1950.

The middle figure in each subcell represents the product of the step deviations $d_x' d_y'$. Thus in the bottom left-hand subcell (corresponding to the categories 80.0 to 89.9 and 10.0 to 14.9) the entry -12 is the product of 4 and -3. In other words, the category 80.0 to 89.9 is four step deviations *above* the hypothesized mean of X, and the category 10.0 to 14.9 is three step deviations *below* the hypothesized mean of Y. Finally, the bottom number in each subcell represents the product of the first two numbers and therefore can be represented symbolically as $f_{xy} d_x' d_y'$. The sum of these bottom figures over the entire number of subcells therefore gives us the cross-product term uncorrected for errors introduced by using guessed means. This sum, which will be used in the first term of the numerator of r, is numerically equal to -200 and has been placed in the lower right-hand corner of the table.

The remaining quantities needed in the computation of r and b can be obtained in the familiar manner. The last four columns in the table are used to obtain f_x, d_x', $f_x d_x'$, and $f_x (d_x')^2$, the sums of the latter two quantities being directly used in the formula for r. Notice that in computing the numerical values for these last four columns, we are completely ignoring the Y values. Thus, if we ignore the boxed-in area, we have exactly the same kind of table as was used in computing the mean and standard deviation from grouped data. Similarly, the bottom four rows can be used to obtain comparable sums for the Y variable. All of the quantities needed in the formulas for r and b can now be placed in the bottom right-hand cell of the larger table.

Table 17.3   Computations for correlations from grouped data*

Class limits	Y	10.0–14.9	15.0–19.9	20.0–24.9	25.0–29.9	30.0–34.9	35.0–39.9	40.0–44.9	f_x	d'_x	$f_x d'_x$	$f_x(d'_x)^2$
X	Mid-points	12.45	17.45	22.45	27.45	32.45	37.45	42.45				
0.0–9.9	4.95				1 / 0 / 0	8 / −4 / −32	4 / −8 / −32		13	−4	−52	208
10.0–19.9	14.95	1 / +9 / 9	2 / +6 / 12		2 / 0 / 0	4 / −3 / −12	1 / −6 / −6	3 / −9 / −27	13	−3	−39	117
20.0–29.9	24.95	2 / +6 / 12	5 / +4 / 20	1 / +2 / 2	2 / 0 / 0	3 / −2 / −6	3 / −4 / −12		16	−2	−32	64
30.0–39.9	34.95	2 / +3 / 6		5 / +1 / 5	5 / 0 / 0	7 / −1 / −7	3 / −2 / −6		22	−1	−22	22
40.0–49.9	44.95	4 / 0 / 0	6 / 0 / 0	6 / 0 / 0	7 / 0 / 0	1 / 0 / 0			24	0	0	0
50.0–59.9	54.95	3 / −3 / −9	10 / −2 / −20	9 / −1 / −9	6 / 0 / 0	2 / +1 / 2			30	1	30	30
60.0–69.9	64.95	2 / −6 / −12	4 / −4 / −16	3 / −2 / −6	7 / 0 / 0	4 / +2 / 8			20	2	40	80
70.0–79.9	74.95	2 / −9 / −18	3 / −6 / −18	4 / −3 / −12	1 / 0 / 0				10	3	30	90
80.0–89.9	84.95	1 / −12 / −12			1 / −4 / −4				2	4	8	32
f_y		17	30	29	31	29	11	3	N = 150		−37	643
d'_y		−3	−2	−1	0	1	2	3				
$f_y d'_y$		−51	−60	−29	0	29	22	9	−80		$\Sigma f_{xy} d'_x d'_y$ = −200	
$f_y(d'_y)^2$		153	120	29	0	29	44	27	402			

* This computing form has been taken, with slight adaptations, from [1], table 19.4, p. 476, with the kind permission of the publisher.

We now obtain the values of r and b as follows:

$$r = \frac{150(-200) - (-37)(-80)}{\sqrt{[150(643) - (-37)^2][150(402) - (-80)^2]}} = \frac{-32,960}{71,590} = -.460$$

$$b = \frac{150(-200) - (-37)(-80)}{150(643) - (-37)^2} \frac{5.0}{10.0} = \frac{-32,960}{95,081} \frac{1}{2} = -.1733$$

Since the values of $\bar{X}$ and $\bar{Y}$ are 42.48 and 24.78 respectively, we get

$$a = \bar{Y} - b\bar{X} = 24.78 - (-.1733)(42.48) = 32.14$$

and the least-squares equation can be written as

$$Y_p = 32.14 - .1733X$$

Interpretations for the correlation coefficient In order to obtain an interpretation for r which will be meaningful when r is neither 0 nor 1.0, let us return to the notion of variability about the regression equation. We have defined the variance about the mean of Y as

$$\sigma_y{}^2 = \frac{\Sigma(Y - \mu_y)^2}{M}$$

where M represents the size of the population (as contrasted with the sample size N) and where we use the subscripts to emphasize the fact that there are now two variables that must be distinguished. Thus the ordinary concept of variance involves deviations from a *fixed* measure of central tendency, the overall mean. But we can also obtain the mean of Y's for a fixed X, and we are assuming that these values vary with X so as to produce a linear regression. We thus can generalize the concept of a mean, obtaining a kind of conditional mean of Y for a given X which we may symbolize as $\mu_{y|x}$ or as $E(Y|X)$.

If we generalize the concept of variance in a similar manner, we can obtain as a measure of spread about the regression equation

$$\sigma_{y|x}^2 = \frac{\Sigma(Y - \mu_{y|x})^2}{M} \tag{17.11}$$

where the symbol $\sigma_{y|x}^2$ is used to emphasize the fact that the magnitude of variability about the regression equation, as well as the mean of Y, depends upon the value of X. In other words, for each X there is both a mean of the Y's and a variance about this mean. The amount of spread about the line need not always be the same for each X although we are going to *assume* the property of homoscedasticity or equal variances.

We now have two measures of variability for Y. The first measures the spread about that value of Y, the grand mean μ_y, which would be the best single predicted value of Y if X were not known. In other words, if one were asked to predict Y with no knowledge of X, his best guess would be μ_y (or $\bar{Y}$ if only sample data were available). Knowing X, however, he would predict the corresponding value of Y that lay on the regression equation. Unless there were no relationship between X and Y, knowledge of X should help him predict the value of Y. If the relationship were perfect, he could predict Y exactly since all points would be exactly on the line. Ordinarily, he will not be able to do this well, but since we are assuming a normal distribution of Y's and a fixed standard deviation $\sigma_{y|x}$, we can make probability statements about the risks and magnitude of his error. More important for our purposes, we can compare the two standard deviations (or variances) and obtain a measure of how much the prediction has been improved by knowing X. In so doing, we can make use of processes familiar from analysis of variance.

In analysis of variance we took the total variation or sum of squares and divided this quantity into explained and unexplained portions. We shall now use exactly the same procedure, obtaining almost as a by-product the values of $\sigma_{y|x}^2$ and r^2. We can then give a meaningful interpretation for the correlation coefficient. First, we express the deviations of each Y from $\bar{Y}$ as the sum of two quantities $(Y - Y_p) + (Y_p - \bar{Y})$. (See Fig. 17.11.) The first of these quantities represents the deviation of the Y value from the least-squares line and indicates the amount of error made when Y_p is used to predict Y. The second expression indicates the deviation of the least-squares line (at a given X) from $\bar{Y}$. For most cases, this quantity will represent the amount by which the error is reduced when Y_p is known. If we now square both sides of the equation and then sum over all cases, we obtain

$$\Sigma(Y - \bar{Y})^2 = \Sigma(Y - Y_p)^2 + 2\Sigma(Y - Y_p)(Y_p - \bar{Y}) + \Sigma(Y_p - \bar{Y})^2$$

Fortunately, the middle term again drops out, and we are left with

$$\Sigma(Y - \bar{Y})^2 = \Sigma(Y - Y_p)^2 + \Sigma(Y_p - \bar{Y})^2 \qquad (17.12)$$

Total SS = unexplained SS + explained SS

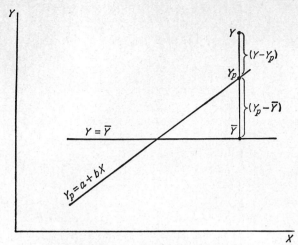

Figure 17.11 Geometric representation showing deviations from mean $\bar{Y}$ as a sum of deviations from least-squares line and deviations of least-squares line from $\bar{Y}$.

The first quantity on the right-hand side of the equation represents the sum of the squares of the deviations of the actual Y values from the least-squares line. This quantity is unexplained since it indicates the amount of error in prediction. The remaining quantity indicates what we have gained in using Y_p in preference to $\bar{Y}$ and can be referred to as the explained sum of squares. By *explained*, of course, we do not imply a causal explanation but merely an association between the two variables. Let us now look more closely at each of these quantities.

If we take the unexplained sum of squares and divide by the total number of cases, we obtain the sample variance $s_{y|x}^2$ about the least-squares line. Thus

$$s_{y|x}^2 = \frac{\Sigma(Y - Y_p)^2}{N} \tag{17.13}$$

If we wish to obtain an unbiased estimate of the population variance $\sigma_{y|x}^2$ about the true regression, we must divide not by N but by the appropriate degrees of freedom. In this case we have lost 2 degrees of freedom in calculating a and b as estimates of α and β. Therefore, if we wish to estimate $\sigma_{y|x}^2$, we would use

$$\hat{\sigma}_{y|x}^2 = \frac{\Sigma(Y - Y_p)^2}{N - 2} \tag{17.14}$$

Thus the unexplained sum of squares can readily be converted into an estimate of the variance about the regression equation. You should convince yourself that what we have done is directly parallel to our earlier treatment of analysis of variance. The variability about the least-squares equation has replaced the notion of variability *within* categories of X.

Turning now to the explained sum of squares $\Sigma(Y_p - \bar{Y})^2$, we can easily show that this quantity is equivalent to $r^2[\Sigma(Y - \bar{Y})^2]$ or $r^2\Sigma y^2$. Since $Y_p = a + bX$ and $\bar{Y} = a + b\bar{X}$, we have

$$(Y_p - \bar{Y}) = b(X - \bar{X})$$

Therefore

$$
\begin{aligned}
\Sigma(Y_p - \bar{Y})^2 &= b^2\Sigma(X - \bar{X})^2 = b^2\Sigma x^2 \\
&= \frac{(\Sigma xy)^2}{(\Sigma x^2)^2}(\Sigma x^2) = \frac{(\Sigma xy)^2}{\Sigma x^2} \\
&= \frac{(\Sigma xy)^2}{\Sigma x^2 \Sigma y^2}(\Sigma y^2) = r^2\Sigma y^2 \\
&= r^2\Sigma(Y - \bar{Y})^2
\end{aligned}
$$

We have thus shown that

$$r^2 = \frac{\Sigma(Y_p - \bar{Y})^2}{\Sigma(Y - \bar{Y})^2} = \frac{\text{explained SS}}{\text{total SS}}$$

By a similar argument we could show that r^2 represents the ratio of explained variation in X to total variation in X. The square of the correlation coefficient can therefore be interpreted as the proportion of the total variation in the one variable explained by the other. The quantity $\sqrt{1 - r^2}$, sometimes referred to as the *coefficient of alienation*, represents the square root of the proportion of the total sum of squares that is unexplained by the independent variable.

It should be noted that there is no direct and simple interpretation for r itself. In fact it is possible to be misled by values of r since these values will be numerically larger than those of r^2 (unless r is 0 or ± 1.0). Thus it might appear that an r of .5 is half as good as a perfect correlation whereas we see that in this case we are explaining only 25 per cent of the variation. A correlation of .7 indicates that slightly less than half of the variation is being explained. We see also that correlations of .3 or less mean that only a very small fraction of the variation is being explained. Table 17.4 indicates the relationships among the various quantities.

Since $1 - r^2$ represents the proportion of unexplained variation, we have

$$(1 - r^2)[\Sigma(Y - \bar{Y})^2] = \Sigma(Y - Y_p)^2$$

Therefore $$(1 - r^2)\frac{\Sigma(Y - \bar{Y})^2}{N} = \frac{\Sigma(Y - Y_p)^2}{N}$$

or $$(1 - r^2)s_y{}^2 = s_{y|x}^2$$

Hence $$s_{y|x} = \sqrt{1 - r^2}\, s_y$$

This result gives us an indication of how much we can reduce the standard deviation by knowing X. (See the last column of Table 17.4.) When r

Table 17.4   Numerical relationships between r, r^2, $1 - r^2$, and $\sqrt{1 - r^2}$

r	r^2	$1 - r^2$	$\sqrt{1 - r^2}$
.90	.81	.19	.44
.80	.64	.36	.60
.70	.49	.51	.71
.60	.36	.64	.80
.50	.25	.75	.87
.40	.16	.84	.92
.30	.09	.91	.95
.20	.04	.96	.98
.10	.01	.99	.995

is zero, the two standard deviations are equal. This fact is obvious, of course, when we realize that the least-squares line would in this case be a horizontal line having the equation $Y = \bar{Y}$. When r^2 is unity, $s_{y|x}$ will of course be zero since all points will be exactly on the line. From Table 17.4 we see that the magnitude of r has to be large before we get a substantial reduction in standard deviations. For an r of .80 the standard deviation about the least-squares line is .60 of the ordinary standard deviation, but for an r of .40 we see that we have not gained a great deal in estimating Y from X.

Glossary
Bivariate normal distribution
Coefficient of alienation
Coefficient of correlation
Covariance

Intercept
Least-squares equation
Regression of Y on X
Slope

Exercises
1. The following data for 29 non-Southern cities of 100,000 or more are taken from
R. C. Angell's study of moral integration of American cities. The moral-integration
index was derived by combining crime-rate indices with those for welfare effort.
Heterogeneity was measured in terms of the relative numbers of nonwhites and
foreign-born whites in the population. A mobility index measuring the relative
numbers of persons moving in and out of the city was also computed as a second
independent variable.

City	Integration index	Heterogeneity index	Mobility index
Rochester	19.0	20.6	15.0
Syracuse	17.0	15.6	20.2
Worcester	16.4	22.1	13.6
Erie	16.2	14.0	14.8
Milwaukee	15.8	17.4	17.6
Bridgeport	15.3	27.9	17.5
Buffalo	15.2	22.3	14.7
Dayton	14.3	23.7	23.8
Reading	14.2	10.6	19.4
Des Moines	14.1	12.7	31.9
Cleveland	14.0	39.7	18.6
Denver	13.9	13.0	34.5
Peoria	13.8	10.7	35.1
Wichita	13.6	11.9	42.7
Trenton	13.0	32.5	15.8
Grand Rapids	12.8	15.7	24.2
Toledo	12.7	19.2	21.6
San Diego	12.5	15.9	49.8
Baltimore	12.0	45.8	12.1
South Bend	11.8	17.9	27.4
Akron	11.3	20.4	22.1
Detroit	11.1	38.3	19.5
Tacoma	10.9	17.8	31.2
Flint	9.8	19.3	32.2
Spokane	9.6	12.3	38.9
Seattle	9.0	23.9	34.2
Indianapolis	8.8	29.2	23.1
Columbus	8.0	27.4	25.0
Portland (Ore.)	7.2	16.4	35.8

SOURCE: R. C. Angell, "The Moral Integration of American Cities," *American Journal of Sociology*, vol. 57, part 2, p. 17, July 1951, with the kind permission of the author and publisher.   (Copyright 1951 by the University of Chicago.)

a. Draw a scattergram, relating moral integration to heterogeneity.

b. Compute r, a, and b for these same variables and draw in the least-squares line on your scattergram, taking moral integration as Y. (*Ans.* $r = -.156$; $a = 13.9$; $b = -.049$)

c. How large is the standard deviation about the least-squares line as compared with the standard deviation about Y?

2. You will need to obtain the correlations between moral integration and mobility and between heterogeneity and mobility in order to handle exercises in Chap. 19. Compute both of these r's. (*Ans.* $r = -.456$; $r = -.513$)

3. Group the moral-integration and heterogeneity indices into intervals and compute r, a, and b using formulas for grouped data. Compare results with those for ungrouped data.

References

1. Blalock, H. M.: *Causal Inferences in Nonexperimental Research*, University of North Carolina Press, Chapel Hill, 1964, chaps. 2 and 3.

2. Christ, Carl: *Econometric Models and Methods*, John Wiley & Sons, Inc., New York, 1966, Part III.

3. Croxton, F. E., and D. J. Cowden: *Applied General Statistics*, 3d ed., Prentice-Hall, Inc., Englewood Cliffs, N.J., 1967, chaps. 19 and 20.

4. Hagood, M. J., and D. O. Price: *Statistics for Sociologists*, Henry Holt and Company, Inc., New York, 1952, chap. 23.

5. Hays, W. L.: *Statistics*, Holt, Rinehart and Winston, Inc., New York, 1963, chap. 15.

6. Johnston, J.: *Econometric Methods*, McGraw-Hill Book Company, New York, 1963, Part II.

7. McCollough, C., and L. Van Atta: *Introduction to Descriptive Statistics and Correlation*, McGraw-Hill Book Company, New York, 1965, chaps. 5–8.

8. Mueller, J. H., K. Schuessler, and H. L. Costner: *Statistical Reasoning in Sociology*, 2d ed., Houghton Mifflin Company, Boston, 1970, chap. 11.

9. Wallis, W. A., and H. V. Roberts: *Statistics: A New Approach*, The Free Press of Glencoe, Ill., Chicago, 1956, chap. 17.

10. Weinberg, G. H., and J. A. Schumaker: *Statistics: An Intuitive Approach*, Wadsworth Publishing Company, Inc., Belmont, Calif., 1962, chaps. 16–18.

Correlation and Regression (Continued)

In the present chapter we shall continue the discussion of correlation and regression. First, various tests of significance will be considered. We shall then turn to nonlinear relationships, a subject which will also be discussed briefly in Chap. 19. Next we consider the effects of measurement errors on slopes and correlations. Finally, the topic of rank-order correlation will be treated.

18.1 Significance Tests and Confidence Intervals

Significance test for r and b   Since r and the least-squares coefficients a and b are descriptive only of the sample data, our interest usually centers on the comparable population parameters ρ, α, and β. In particular, we may wish to test the null hypothesis that there is no (linear) relationship in the population, or we may want to obtain confidence intervals for ρ or the regression coefficients. We shall first consider a test of the null hypothesis that there is no linear relationship in the population. As we shall see, if we can assume a normal distribution of Y about X and homoscedasticity, we can use analysis of variance to test the hypothesis that $\rho = \beta = 0$.

Let us make use of the fact that since r and b (and therefore ρ and β) have the same numerators, a test of the hypothesis $\rho = 0$ is also a test of the hypothesis $\beta = 0$ and vice versa. In other words, if there is no linear association in the population, the slope of the regression equation will be zero, and the line will therefore be horizontal. Remembering that a regression equation represents the path of the means of the Y's for fixed values of X, we see immediately that whenever $\beta = 0$, the means of the Y's must be the same for every value of X (see Fig. 18.1). This assumes, of course, that the regression equation is actually linear in form. In

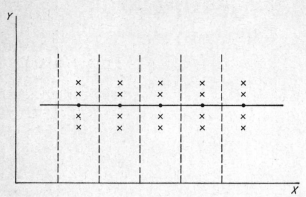

Figure 18.1 Geometric representation of the fact that the
hypothesis that $\beta = 0$ is equivalent to the hypothesis
$\mu_1 = \mu_2 = \cdots = u_k$.

particular, if we were to divide the X axis into a number of categories, we
would find the population category means to be exactly equal. Thus we
can translate the hypothesis that $\beta = \rho = 0$ into the statement that the
means for Y will be equal for each of the X categories. If we imagine an
infinite population, as will be necessary in order to meet the assumption of
normality, we can imagine the X axis being divided into an indefinitely
large number of categories each having identical means in Y. Thus our
null hypothesis becomes $\mu_{y1} = \mu_{y2} = \mu_{y3} = \cdots$, where we have used
the double subscript to emphasize that it is the means of the Y's with
which we are concerned and that we have an indefinitely large number of
X categories.

The above line of reasoning obviously suggests an extension of the
analysis-of-variance test covering an indefinitely large number of categories of the nominal-scale variable (now X). Let us recall the assumptions required in analysis of variance. In addition to the null hypothesis
and the assumption that cases have been selected randomly and independently from each of the categories, we must also assume normal
populations and equal variances within each category. Provided we can
also assume random sampling, we see that all of these assumptions can
be met if we assume the joint distribution of X and Y to be bivariate
normal. You will recall that this latter assumption simultaneously assures
us of a linear regression equation, normality of the Y's for each fixed
value of X, and equal variances for all values of X. In effect, therefore,
the assumptions of random sampling and bivariate normality enable us
to make use of analysis of variance to test the hypothesis that $\rho = \beta = 0$,

though the normality of the X's is not required as long as the ϵ_i have an approximately normal distribution.

Previously we found it necessary to obtain the total and between sums of squares and then to subtract in order to get the within sum of squares. In testing the hypothesis that $\rho = 0$, the process is substantially simplified, however. We have already seen that the *proportion* of the total sum of squares of Y explained by X is given by r^2. Likewise, the proportion left unexplained by X will be $1 - r^2$. Since the total sum of squares can be symbolized by Σy^2, the explained and unexplained sums of squares therefore become $r^2 \Sigma y^2$ and $(1 - r^2)\Sigma y^2$ respectively.

The degrees of freedom associated with the total sum of squares is of course $N - 1$. In computing the unexplained sum of squares we take the sum of the squared deviations about the least-squares line rather than about the grand mean of the Y's. But in order to obtain the least-squares line we have had to make use of two coefficients a and b. We have therefore lost 2 degrees of freedom, one more than was lost in taking deviations about the single value $\bar{Y}$. We thus can associate $N - 2$ degrees of freedom with the unexplained sum of squares, and by subtraction we see that there is 1 degree of freedom to be associated with the explained sum of squares.

The results can now be summarized as in Table 18.1. The advantage

Table 18.1 Analysis of variance test of the hypothesis that $\rho = 0$

	Sums of squares	Degrees of freedom	Estimates of variance	F
Total	Σy^2	$N - 1$		
Explained	$r^2 \Sigma y^2$	1	$\dfrac{r^2 \Sigma y^2}{1}$	$\dfrac{r^2(N-2)}{(1-r^2)}$
Unexplained	$(1 - r^2)\Sigma y^2$	$N - 2$	$\dfrac{(1 - r^2)\Sigma y^2}{N - 2}$	

of inserting symbols rather than numbers into the table is that we see immediately that the quantity Σy^2 disappears when we take the ratio of the explained to unexplained estimates. In other words, the total sum of squares cancels out, and we can write a formula for F in terms of the *proportions* of explained and unexplained sums of squares. The formula for F then involves only the quantities r^2 and $1 - r^2$ along with the degrees of freedom of $N - 2$ and 1. Therefore we may use the formula

$$F_{1,N-2} = \frac{r^2}{1 - r^2} (N - 2) \tag{18.1}$$

without actually bothering to construct an analysis-of-variance table as was necessary in the previous chapter. Since the tables for F only permit tests at the .05, .01, and .001 levels, it may be preferable to take the positive square root of (18.1) and utilize the t distribution with $N - 2$ degrees of freedom.

We can illustrate the use of this analysis-of-variance test for the significance of r with the data of Table 17.1. We obtained a correlation of $r = .301$ between per cent Negro and our index of discrimination. In testing for the significance of r we are asking the very important question, "How likely is it that we would obtain an r of .301 or larger (in absolute value) if there were actually no linear association in the population?" In order to make the F test, we simply compute r^2 and $1 - r^2$ and make use of Eq. (18.1). Thus, since r was based on 13 cases,

$$F_{1,11} = \frac{(.301)^2}{[1 - (.301)^2]} 11 = \frac{.0906}{.9094} 11 = 1.10$$

Referring to the F table, we see that for 1 and 11 degrees of freedom we would need an F of 4.84 or larger in order to reject at the .05 level, assuming direction had not been predicted in advance. We therefore decide not to reject the null hypothesis that $\rho = 0$. Apparently, we could have gotten an r of .301 or larger fairly frequently by chance even if there were absolutely no association in the population.

Again, it is necessary to emphasize the distinction between a test of significance and a measure of the degree of relationship. Had we obtained an r of .301 with a sample size of 50, we would have gotten

$$F_{1,48} = \frac{.0906}{.9094} 48 = 4.78$$

a value that is significant at the .05 level. In both cases we have explained approximately 9 per cent of the total sample variation, but in the latter case we have more faith that there is a relationship, however slight, in the population.

Confidence intervals   Whenever a bivariate normal population can be assumed or approximated, it is possible to construct confidence intervals for ρ, β, and the regression line. The standard error of r is given by the formula

$$\sigma_r = \frac{1 - \rho^2}{\sqrt{N - 1}}$$

Unfortunately, the sampling distribution of r will not in general be symmetrical except in the special case where $\rho = 0$. In fact, the sampling distribution becomes more and more skewed as the absolute value of ρ approaches unity. In addition, we note that in order to make use of the above formula for the standard error of r, it would be necessary to know or estimate the value of ρ. Both of these complications make it difficult to obtain confidence intervals for ρ in the straightforward manner.

In computing a confidence interval about r, we shall first convert r to a new statistic z which has a sampling distribution that is approximately normal. We then place a confidence interval about z in the usual way. Finally, after noting the upper and lower confidence limits for z, we reconvert these particular z values back into r's, thereby obtaining the confidence limits for r.

We transform r into z by means of the formula

$$z = 1.151 \log \frac{1 + r}{1 - r}$$

where z can take on values from zero to infinity. It should be called to your attention that the z value computed from the above formula has absolutely no connection with the Z values we have been using with the standard normal curve. Values of z can be directly obtained from Table K, Appendix 2, rather than making use of logarithms. The first two digits of r are located by going down the left-hand margin, the third being given across the top. The corresponding z values are given in the body of the table. For example, a z of 0.3228 corresponds to an r of .312; a z of 1.3892 corresponds to an r of .883. In using Table K we ignore the sign of r, affixing to z the proper sign after its numerical value has been located. Notice that the numerical values of z are only slightly larger than r whenever $|r| \leq .40$ but that as r increases z begins to take on values greater than unity.

We may now make use of the z transformation in a confidence-interval problem. The sampling distribution of z is approximately normal even for relatively small N's and moderate departures from bivariate normality. Its standard error is given by

$$\sigma_z = \frac{1}{\sqrt{N - 3}} \tag{18.2}$$

Not only does this make it possible to use the normal table, but we have eliminated the necessity of estimating ρ since the standard error of z depends only on N. Taking as our numerical example the correlation of .301 between per cent Negro and discrimination, we find that the corresponding z value is 0.3106. Since there were only 13 cases,

$$\sigma_z = \frac{1}{\sqrt{13 - 3}} = \frac{1}{\sqrt{10}} = 0.3162$$

Suppose we wish to obtain a 95 per cent confidence interval for ρ. We first compute such an interval in terms of z values. Thus we would take

$$z \pm 1.96\sigma_z = 0.3106 \pm 1.96(0.3162)$$

$$= 0.3106 \pm 0.6198$$

Therefore the confidence interval about z runs from $-.3092$ to $+.9304$. Notice that in obtaining the lower limit we had to subtract a number that was numerically larger than 0.3106. This yields a negative result which, in turn, means that the value of r corresponding to this lower limit must also be taken as negative. Looking up the values of r corresponding to the two confidence limits for z, we get values of $-.300$ and .731 for the lower and upper limits respectively.

Notice that the interval is not quite symmetrical about the obtained r of .301. In this case the upper limit is somewhat closer to r than is the lower limit. Had we found an r of .80, the resulting interval would have been even more highly skewed in the same direction. This can be seen to make sense intuitively when we realize that whenever we begin to approach the upper limit of unity, we also place a restriction on the upper limit of the confidence interval. Thus it would be impossible to obtain a confidence interval of .86 $\pm$ .16. If r happens to be negative, the direction of skewness will of course be opposite to that obtained above. Only when $r = 0$ will the interval be exactly symmetrical about r.

We can interpret this confidence interval in the usual manner. Our procedure is such that in the long run we can expect to obtain intervals that will include the (fixed) value of ρ about 95 per cent of the time. We may also use such confidence intervals as implicit tests of hypotheses. In the above problem we have already noted that the lower limit of the interval is negative. Since zero is included in the interval, we know immediately that we would not reject the null hypothesis that $\rho = 0$. If we ever wanted to test any other specific hypothesized value of ρ, we could also do so. For example, had we hypothesized that $\rho = .80$, we would have rejected at the .05 level since this value is beyond the upper limit of .731.

It would also be desirable to compute confidence intervals about other measures of degree of relationship. Unfortunately, too little is known about the sampling distributions of some measures of association for contingency problems to permit the construction of confidence intervals for these measures. Haggard [11] suggests a method for computing confidence intervals about the intraclass correlation r_i, and Goodman and Kruskal [10] discuss the sampling distributions of various nominal and ordinal measures.

One may also want to put a confidence interval about b, or he may need to find a band within which the true regression equation can be expected to lie. In both cases we can make use of the t distribution in a fairly straightforward manner. The estimate of the standard error of b is given by

$$\hat{\sigma}_b = \frac{\hat{\sigma}_{y|x}}{\sqrt{\sum_{i=1}^{N} (X_i - \bar{X})^2}} \tag{18.3}$$

where you will recall that

$$\hat{\sigma}_{y|x} = \sqrt{\sum_{i=1}^{N} \frac{(Y_i - Y_p)^2}{N - 2}}$$

For computational purposes it can be shown algebraically that

$$\hat{\sigma}_{y|x} = \sqrt{\frac{\sum_{i=1}^{N} (Y_i - \bar{Y})^2 - b \sum_{i=1}^{N} (X_i - \bar{X})(Y_i - \bar{Y})}{N - 2}} \tag{18.4}$$

We can now make use of numerical computations already obtained for the discrimination data of Table 17.1, getting

$$\hat{\sigma}_{y|x} = \sqrt{\frac{560{,}024 - 19.931(2{,}553.77)}{11}} = \sqrt{46{,}284} = 215.1$$

and
$$\hat{\sigma}_b = \frac{215.1}{\sqrt{128.131}} = \frac{215.1}{11.32} = 19.00$$

If we wish to compute the 99 per cent confidence interval, we go directly to the t table using $N - 2$ or 11 degrees of freedom. We thus get

$$b \pm (3.106)(19.00) = 19.931 \pm 59.014$$

*In estimating the regression equation we have seen that our best single ("point") estimate is the least-squares line. Since the quantity we are now estimating is no longer a single value but rather an entire line, our *interval estimate* will no longer be an interval but instead will consist of a band on either side of the least-squares line. At first thought, one might expect such a band to consist of two lines running parallel to the least-squares line. But such a band would imply that we knew the correct slope and that the only source of error came in estimating α. We must remember that there are now two quantities being estimated (α and β) and therefore two sources of error. You should convince yourself that since the slope may also have been estimated incorrectly, the farther one goes in either direction from the point $(\bar{X},\bar{Y})$, the greater the possible inaccuracy. The confidence band takes the general form shown in Fig. 18.2.

*In order to draw such a confidence band, it will be necessary to compute the standard error of Y_p for various values of X. The estimated standard error is given by the formula

$$\hat{\sigma}_{y_p} = \hat{\sigma}_{y|x} \sqrt{\frac{1}{N} + \frac{(X - \bar{X})^2}{\sum\limits_{i=1}^{N}(X_i - \bar{X})^2}} \qquad (18.5)$$

where the particular value of X to be used in $(X - \bar{X})^2$ may be set at any desired location on the X axis. Notice, incidentally, that the farther X is from $\bar{X}$, the larger the numerical value of the standard error. Suppose we wish to obtain the estimated standard error when $X = 10.0$.

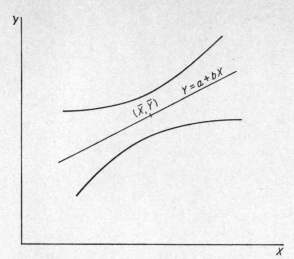

Figure 18.2 Confidence band about least-squares line.

Since $\bar{X} = 4.837$ we get

$$\hat{\sigma}_{y_p} = 215.1 \sqrt{\frac{1}{13} + \frac{(10.0 - 4.837)^2}{128.131}} = 215.1 \sqrt{.28496} = 114.86$$

*Again using the t table and a 99 per cent interval about Y_p *computed for this fixed value of X*, we would get

$$Y_p \pm (3.106)(114.86) = Y_p \pm 356.8$$

Once we have obtained similar intervals about Y_p for other particular values of X, we can plot the entire band. Needless to say, the procedure involved would become quite tedious without the aid of high-speed computers if the entire band were desired.

Testing the difference between two correlations As noted previously, it usually makes more sense theoretically to compare two or more slopes than to compare correlations, and such a comparison of slopes will occupy our attention in Chap. 20 on analysis of covariance. However, it is sometimes the case that one has obtained several correlations and wishes to establish that one is significantly higher than another. As long as he is content to describe relationships within his particular sample, he may simply compare the relative sizes of the two r's and note the magnitude of the difference. If he wishes to generalize to some larger population, however, the question will arise as to whether or not the obtained difference is likely to be due to chance. Suppose, for example, that he has obtained one r of .50 and another of .30. He may wish to test the null hypothesis that the two population correlations are identical, that is, $\rho_1 = \rho_2$.

Two different situations in which such tests might be made come to mind. First, one may have *two independent samples* and may wish to compare the degrees of relationships between X and Y within each of these samples. For example, the relationship between per cent Negro and discrimination may not be the same in the South as it is in the North. One might set up the research hypothesis that ρ_{xy} is higher in the South than in the North, testing the null hypothesis that the two correlations are equal. A second type of situation, likely to be confused with the first, may occur when one has a *single sample*. There may be a single dependent variable (say, discrimination) and two independent variables (say, per cent Negro and percentage of the labor force engaged in manufacturing). One may wish to establish that one of these independent variables is more highly related to the dependent variable than is the other. Referring to the second independent variable as Z, he would then be interested in testing the null hypothesis that $\rho_{xy} = \rho_{zy}$. Let us first see how the former type of situation can be handled, then turn to the single-sample test.

If the two correlations are based on independent samples, we can transform each of the r's into z's and then make use of a formula for the standard error of the difference between two z's, which is analogous to that for the standard error of a difference between means, and which is as follows

$$\sigma_{z_1 - z_2} = \sqrt{\frac{1}{N_1 - 3} + \frac{1}{N_2 - 3}} \tag{18.6}$$

We can then either put a confidence interval about $(z_1 - z_2)$ or look up the value of

$$Z = \frac{(z_1 - z_2) - 0}{\sigma_{z_1 - z_2}}$$

in the normal table. Zero appears in the above formula because of the fact that our null hypothesis takes the form $\rho_1 = \rho_2$.

Suppose that for 17 Southern cities the correlation between per cent Negro and discrimination turns out to be .567 as compared with the result of .301 for Northern cities.

Thus $r_1 = .301$ $r_2 = .567$

$z_1 = 0.3106$ $z_2 = 0.6431$

and $\sigma_{z_1 - z_2} = \sqrt{\tfrac{1}{10} + \tfrac{1}{14}} = \sqrt{.1000 + .0714} = .414$

Therefore $Z = \dfrac{.3106 - .6431}{.414} = \dfrac{-.3325}{.414} = -.803$

and we see that this difference in r's is not significant at the .05 level. Thus even though the correlation is higher for Southern cities, such a difference could have occurred fairly frequently by chance.

In the second type of situation mentioned above, we do not have two independent samples and cannot use the same formula for the standard error of $z_1 - z_2$. A method of handling this latter type of problem is available, provided we are interested in generalizing only to a subpopulation of all possible samples for which X and Z (the two independent variables) have the same sets of values as those in the particular sample we have obtained. For most practical purposes, this restriction can be safely ignored unless there is reason to suspect that the range of variation is much greater in the population than in the sample studied—in which case one should be very much on guard against generalizing anyway.

If we are testing the null hypothesis that $\rho_{xy} = \rho_{zy}$, we form t as follows

$$t = (r_{xy} - r_{zy}) \sqrt{\frac{(N-3)(1 + r_{xz})}{2(1 - r_{xy}^2 - r_{xz}^2 - r_{zy}^2 + 2r_{xy}r_{xz}r_{zy})}} \qquad (18.7)$$

We can then look up the value of t in the table, using $N - 3$ degrees of freedom. In our numerical example, suppose the correlation for Northern cities between X and Z turns out to be .172 and that between Y and Z is .749. We would then get

$$t = (.301 - .749) \sqrt{\frac{10(1 + .172)}{2[1 - .301^2 - .172^2 - .749^2 + 2(.301)(.172)(.749)]}}$$
$$= -1.72$$

Since we have 10 degrees of freedom, we see that we cannot reject the null hypothesis of no difference between the population correlations of each of the independent variables with discrimination.

18.2   Nonlinear Correlation and Regression

Up to this point we have assumed that the regression equation was linear in form. In many practical sociological problems the linear model, although perhaps not exact, yields a close enough approximation to the true form of the equation that we need not concern ourselves with alternative more complicated models. This will especially be true for exploratory studies in which the degree of fit is not too exact. There are instances, however, when inspection of the scattergram may clearly indicate a nonlinear relationship or when one's theory has predicted such a relationship. Whenever such a nonlinear relationship does exist, the product-moment coefficient will obviously underestimate the true degree of relationship since this coefficient measures only the goodness of fit of the best single straight line. You have already seen that in the case of a U-shaped curve it is possible to have a strong relationship with an r of approximately zero, and you have been cautioned that it is therefore incorrect to conclude that two variables are independent merely because r is zero. If the scattergram indicates a more or less random distribution of dots, we may then conclude that no relationship exists, but we must also be on the lookout for nonlinear relationships. This is of course all the more reason why one should form the habit of always drawing scattergrams before proceeding with his analysis.

The general topic of nonlinear correlation and regression is too complex to be covered adequately in this text. The reason for the complexity of nonlinear analysis is that once we get beyond the equation of the straight line, there are numerous types of equations representing the different possible forms that nonlinear relationships can take. Only the simplest of these equations can be treated here. Fortunately, these relatively simple equations will usually be adequate for the kinds of nonlinear relationships arising from sociological research. One general type of nonlinear function can be represented in terms of polynomials of the nth degree which have equations of the form

$$Y = a + bX + cX^2 + dX^3 + \cdots + kX^n$$

Our discussion of nonlinear relationships of this general type will be postponed until the next chapter when we take up multiple-regression problems. Once these latter regression problems are understood, we shall have a relatively simple method for handling those types of nonlinear relationships that can adequately be described by means of polynomials. Certain other relatively simple types of nonlinear relationships can often be handled by a transformation of variables that permits the use of the familiar linear model. The process can be illustrated in the case of logarithmic functions represented by equations of the type

$$Y = a + b \log X$$

and which have the general form illustrated in Fig. 18.3. In an equation of this type, Y is actually a linear function not of X itself but of $\log X$.

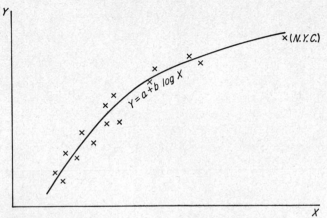

Figure 18.3 Logarithmic least-squares equation of the form $Y = a + b \log X$.

This suggests that if we transform each of the X scores into a new variable $Z = \log X$, we can write Y as a linear function of Z. Thus

$$Y = a + b \log X = a + bZ$$

We can now compute the correlation between Y and Z (that is, Y and $\log X$) in the usual manner. If we plot the distribution of scores on the Y and Z axes, the result should be approximately linear in form. If we wish, we can compare the degree of relationship between Y and Z with that between Y and X. If r_{yz} is substantially larger than r_{xy}, the logarithmic model gives a better fit than the linear model between X and Y.

Logarithmic models of the above sort often arise in instances where the independent variable X takes on a wide range of values but where once a certain value has been reached, further increases produce less and less effect on the dependent variable. City size is a variable that often has this kind of effect. Cities over 500,000 may all have very much the same scores on Y. But if New York is included in the sample, the value of X for this city will be so much greater than those of the remainder that the net effect will be to bend the relationship in much the same manner as shown in Fig. 18.3. In such a case, it may be preferable to relate Y to log X since taking the logarithm of city size will have the effect of bunching together the extremely large scores and lessening the "bending effect" of these large cities.

In a number of instances the researcher may have no real interest in finding the exact form of the prediction equation that best fits his data. He may simply wish to show that the relationship is nonlinear in form or to obtain a measure for the degree of relationship regardless of its form. When a simple transformation such as the log transformation can be made, it will undoubtedly be to one's advantage to make use of such a procedure. Even so, he may still wish to test whether or not the measure he has obtained is a good approximation of the result he would have gotten had the best possible fit been found. In order to handle problems of this sort we can use the basic principles of analysis of variance and one of the measures of degree of association developed in the chapter on analysis of variance.

You will recall that in obtaining the within sum of squares for one-way analysis of variance, we took the sum of the squared deviations from each of the category means. Now let us suppose that X were subdivided into a number of categories and that the sum of squares in Y were analyzed in the usual manner. We know for any given category of X that the sum of squares about the category mean will produce a smaller numerical result than the sum of squares about any other number. In particular, it follows that the within sum of squares will be smaller than the sum of squared deviations about those points on the least-squares line falling at the midpoints of the intervals (see Fig. 18.4).

If the regression equation happens to be linear in form, we can expect that the $\bar{Y}_{.j}$ will all fall approximately on the least-squares line so that it will make little difference whether deviations are taken about the category means or the least-squares line. On the other hand, if the relationship is actually nonlinear, then for at least some of the categories, the sum of squares about the category mean will be quite a bit smaller than that about the least-squares line. In other words, the within or unexplained sum of squares will be minimized by using the category means, and therefore the between or explained sum of squares will be maximized. Thus the proportion of variation explained by the categories, as measured by the correlation ratio E^2, will be larger than the proportion explained by

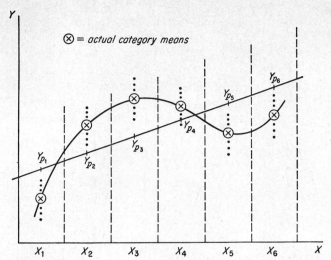

Figure 18.4 Comparison of deviations about least-squares line with deviations about category means.

the least-squares line unless the true relationship is actually linear in form.

We can take advantage of this fact in making a test for nonlinearity. If we form the quantity $E^2 - r^2$, we get the proportion of variation explained assuming any form of relationship which is not explained by a linear relationship. It is clear that in obtaining E^2 we are permitting the relationship to take any possible form, since we have simply taken deviations about category means, regardless of where these category means happen to lie. We are essentially asking ourselves how much we can improve our ability to predict values of Y if we do not restrict ourselves to the linear model. If the improvement is greater than we would expect by chance under the assumption that the regression equation is actually linear, then we may conclude that the relationship is nonlinear.

The analysis-of-variance test we shall use to test for nonlinearity takes a form that will soon become familiar. We first find the amount of variation that can be explained using the linear model. Algebraically, this quantity can be represented as $r^2 \Sigma y^2$. Of the variation left unexplained by the linear model, $(1 - r^2) \Sigma y^2$, we then find out how much can be accounted for by the general model. Since $E^2 \Sigma y^2$ gives us the sum of squares that can be explained by Y when there is no restriction placed on the form of the relationship, the quantity $(E^2 - r^2) \Sigma y^2$ represents the explained increment due to nonlinearity. Assuming no grouping errors, this quantity should always be positive.[1] Since the quantity $(1 - E^2)\Sigma y^2$ gives us the sum of squares that is unexplained by even the best-fitting model, we can make an F test as indicated in Table 18.2. As usual, the denominator of F is the error term, and since we are testing for departures from linearity, we take as the numerator the estimate of variance based on $(E^2 - r^2)\Sigma y^2$ or the amount explained by the best general model which has not already been explained by the linear model. The degrees of freedom associated with the numerator can be obtained by subtraction.

Again we note that the total sum of squares cancels, leaving us with the following formula for F:

$$F_{k-2,N-k} = \frac{(E^2 - r^2)(N - k)}{(1 - E^2)(k - 2)} \tag{18.8}$$

where k represents the number of categories into which X has been divided.

Table 18.2   Analysis-of-variance test for nonlinearity

	Sums of squares	Degrees of freedom	Estimates of variance	F
Total	Σy^2	$N - 1$		
Explained by linear model	$r^2\Sigma y^2$	1		
Additional explained by nonlinear model	$(E^2 - r^2)\Sigma y^2$	$k - 2$	$\dfrac{(E^2 - r^2)\Sigma y^2}{k - 2}$	$\dfrac{(E^2 - r^2)(N - k)}{(1 - E^2)(k - 2)}$
Unexplained	$(1 - E^2)\Sigma y^2$	$N - k$	$\dfrac{(1 - E^2)\Sigma y^2}{N - k}$	

Let us illustrate the test for nonlinearity with the data that were grouped in Table 17.2. As can easily be verified, the total and between sums of squares in Y are as follows:

$$\text{Total SS} = 101,115.38 - 92,132.04 = 8,983.34$$

$$\text{Between SS} = 94,792.59 - 92,132.04 = 2,660.55$$

[1] Whenever N is small and therefore one can only use a small number of categories, the assumption that X scores are clustered at the midpoints of each interval becomes unrealistic. This may lead to grouping errors producing a value of E^2 that is less than r^2.

where we have treated all Y scores as though they were at the midpoints of their respective intervals and where we have made use of procedures for grouped data (see Sec. 6.4). Therefore,

$$E^2 = \frac{\text{between SS}}{\text{total SS}} = \frac{2,660.55}{8,983.34} = .2962$$

Since we previously found an r of $-.460$ assuming a linear relationship, we get

$$F_{7,141} = \frac{.2962 - (-.460)^2}{1 - .2962} \frac{150 - 9}{9 - 2} = \frac{.0846}{.7038} \frac{141}{7} = \frac{11.929}{4.927} = 2.42$$

and we see that at the .05 level we may reject the null hypothesis of a linear relationship between the percentage of persons classified as rural-farm and the percentage of females in the labor force.

If a relationship turns out to be nonlinear in form, it is quite possible that r will not be statistically significant whereas E will be. The significance of E can of course be tested by a straightforward analysis of variance by taking the ratio of the explained and unexplained estimates of variance. There are thus three distinct tests that can be made: (1) the significance of r, (2) the significance of departures from linearity ($E^2 - r^2$), and (3) the significance of E.

If a nonlinear relationship is found and an estimate of the degree of relationship in the population desired, it is preferable to use the unbiased correlation ratio ϵ discussed in Chap. 16 and given by the formula

$$\epsilon^2 = 1 - \frac{V_w}{V_t}$$

since the numerical value of E is a function of the number of categories used and will generally slightly overestimate the relationship in the population. If E has already been computed, the value of ϵ can be computed from the formula

$$\epsilon^2 = \frac{E^2(N - 1) - (k - 1)}{N - k} \tag{18.9}$$

18.3 The Effects of Measurement Errors

If there is either random or systematic measurement error in X or Y we can expect our results to be altered. Of course this also applies to *all* of the tests and measures we have discussed, including the nonparametric procedures. In fact, one of the most common kinds of measurement errors in sociology, political science, and most of the remaining social sciences would seem to result from our use of rather crude categories such as the simple dichotomies of *high* and *low* or *present* and *absent*. The implications of measurement errors of various kinds are poorly understood, but the most systematic work on the subject has been done on interval scales and problems involving correlation and regression analysis. Unfortunately, the topic is too technical to be covered in the present text, but it will at least be possible to introduce a few words of caution.

If there is systematic or nonrandom measurement error then almost any kind of distortion is possible, and it becomes necessary to specify the sources of nonrandom error and how they operate. For example, if one is comparing the means of three samples and if the measurement error is such as to bring the means of the second and third samples closer to that of the first, then one may not obtain statistical significance when, with more accurate measurement, the null hypothesis could easily be rejected. But if the measurement errors are strictly random, it is possible to be more definite about the effects of measurement errors. In general, measures of association will be attenuated by random measurement errors in either variable. For example, in analysis of variance situations, random measurement errors in the interval scale will increase the variation *within* categories but will not systematically affect variation between categories, and this will lower both the F value obtained and also the intraclass correlation.

In the case of two interval scales, random measurement error in either variable will reduce the magnitude of the correlation coefficient. Corrections for attenuation are sometimes discussed in elementary statistics texts, but these rely on special assumptions that are often inappropriate in sociological research. (See [3].) In general, whenever there are two or more measures of each variable, it is possible to obtain corrected estimates under varying sets of assumptions. (See [2], [6], and [14].)

If there is random measurement error in Y but not in X, we can conceptualize this as contributing only to the error term in the equation $Y_i = \alpha + \beta X_i + \epsilon_i$, and it can be shown that there will be no systematic effect on the slope estimate b_{yx} except that the standard error of this estimate will be increased owing to the larger error variance. But if there is also random measurement error in X—and of course this is very likely in realistic research—the slope estimate b_{yx} will also be attenuated. In the case of large samples an approximate formula for the expected value of the slope $b_{yx'}$ is

$$E(b_{yx'}) = \beta \frac{\sigma_x^2}{\sigma_{x'}^2} = \beta \frac{\sigma_x^2}{\sigma_x^2 + \sigma_u^2}$$

where X' represents the measured value of X, as represented by the equation $X' = X + u$, where u is assumed to be a strictly random component with expected value equal to zero and with there being no correlation between u and X. The reason for the attenuation is that the variance of the measured value X' will be greater than the true variance of X according to the formula

$$\sigma_{x'}^2 = \sigma_x^2 + \sigma_u^2$$

Thus we see that the attenuation in a slope estimate is a function of the measurement-error variance *relative* to the variance in X.

This fact has a number of important implications practically. It means that whenever there is random measurement error in an independent variable, we cannot count on equal estimated slopes even when the true slopes are in fact equal. If several populations (or samples) differ with respect to the amount of variation in X, then even with the same measurement-error variances, the slope attenuations will also differ. It is well to keep this in mind whenever one is making comparisons of results of several different studies. This difficulty also applies to all measures of association and cannot be considered a unique defect of regression analysis.

18.4 Ordinal Scales: Rank-order Correlation

We have now taken up measures of association which can be used to relate two nominal scales (ϕ^2, τ_b, etc.), a nominal and an interval scale (intraclass correlation), and two interval scales (r). The three measures to be discussed in this section, Spearman's r_s, Kendall's tau and gamma, can be used to correlate two ordinal scales. As long as both variables can be ranked, these latter measures can be used to give correlations which are somewhat analogous to product-moment correlations.

The ordinal measures discussed in this section are appropriate whenever the relationship between X and Y is what is termed either *monotonic increasing* or *monotonic decreasing*. The notion of linearity is of course inappropriate in the case of ordinal scales, as is the notion of a distance between X (or Y) values. We can, however, speak of relationships that are always increasing (or decreasing). A monotonic increasing function is one that either always increases or remains constant as X increases. In other words, as X increases, Y does not decrease. A linear function is a special case of a monotonic increasing (or decreasing) function, but so is a logarithmic function such as $Y = a + b \log X$. We recognize two kinds of nonlinear relationships, namely those that are monotonic and those that are not. The latter kind of nonlinear relationship will of course have one or more bends or reversals of direction, as exemplified by a parabola or third-degree equation.

We frequently encounter theoretical propositions of the form "the greater the X, the greater the Y (or the less the Y)." These statements imply that the relationship between X and Y is monotonic, but they do not specify the form. Ordinal measures are appropriate in connection with propositions of this nature. Of course it would be preferable to refine our theories so as to specify either linearity or some particular kind of nonlinearity (e.g., logarithmic), but if measurement has not been stronger than ordinal level, it will be impossible to distinguish empirically among linear and nonlinear alternatives. (See [22].)

Spearman's r_s   The principle behind Spearman's measure is very simple. We compare the rankings on the two sets of scores by taking the differences of ranks, squaring these differences and then adding, and finally manipulating the measure so that its value will be $+1.0$ whenever the rankings are in perfect agreement, -1.0 if they are in perfect disagreement, and zero if there is no relationship whatsoever. If we symbolize the difference between any pair of ranks as D_i, we then find the value of $\sum_{i=1}^{N} D_i^2$ and compute r_s by means of the formula

$$r_s = 1 - \frac{6 \sum_{i=1}^{N} D_i^2}{N(N^2 - 1)} \qquad (18.10)$$

This formula for r_s is derived by taking the formula for a product-moment correlation and applying it to ranks, rather than raw scores, and thus Spearman's measure can be interpreted as the product-moment correlation between the ranks on X and the ranks on Y.

Let us illustrate with some data collected by the author. Members of a work-camp group were ranked from high to low with respect to both popularity, as measured by friendship choices, and participation in group discussions. For both variables a rank of one indicates a high score. Tied ranks are computed by giving each tied score the arithmetic mean of the scores that would have been received had there been no ties. The values of D_i are then computed as indicated in Table 18.3. If the number of ties is relatively small, as is presently true, we need make no modifications in the formula for r_s. If there is a substantial number of ties, a

Table 18.3 Computation of Spearman's coefficient of rank-order correlation

Person	Popularity rank	Participation rank	D_i	D_i^2
Ann	1	5.5	4.5	20.25
Bill	2.5	5.5	3.0	9.00
Jim	2.5	1	−1.5	2.25
Hans	4	2	−2.0	4.00
Marcia	5	3	−2.0	4.00
Joan	6	9.5	3.5	12.25
Ruth	7	5.5	−1.5	2.25
Doris	8	13.5	5.5	30.25
Barbara	9	9.5	0.5	0.25
Cynthia	10	16	6.0	36.00
Ellie	11.5	5.5	−6.0	36.00
Flo	11.5	11.5	0.0	0.00
Nancy	13.5	8	−5.5	30.25
Mart	13.5	15	1.5	2.25
Stan	15	11.5	−3.5	12.25
Sarah	16	13.5	−2.5	6.25
Total			0.0	207.50

correction factor can be computed (see [19], pp. 206 to 210). We get

$$r_s = 1 - \frac{6(207.50)}{16(255)} = 1 - .305 = .695$$

Notice that if the rankings agree perfectly, $\sum_{i=1}^{N} D_i^2$ will be zero and the value of r_s will be unity. Although direct inspection of the formula does not immediately give us the values of r_s for independence and perfect negative association, it turns out that for perfect negative association, the value of the second term will be −2.0 and therefore r_s will be −1.0. For no association, the second factor will be exactly unity.

If $N \geq 10$ the sampling distribution of r_s is approximately normal with standard deviation $1/\sqrt{N - 1}$. In the above example the standard error will therefore be $1/\sqrt{15}$. As a test of the null hypothesis of no relationship in the population, we can compute Z as follows

$$Z = \frac{r_s - 0}{1/\sqrt{N - 1}} = .695 \sqrt{15} = 2.69$$

Making use of the normal table we see that the relationship is significant at the .01 level.

Kendall's Tau   In computing Spearman's r_s we made use of the squares of the differences in ranks. Kendall's tau, which also varies between -1.0 and 1.0, is based upon a somewhat different operation. We first compute a statistic S by looking at all possible pairs of cases and noting whether or not the ranks are in the same order. For example, suppose we have the following sets of ranks:

	a	b	c	d
A	1	2	3	4
B	2	3	1	4

Since the scores of set A have been given in ascending order, we may compute S by examining each of the B rankings in turn. Focusing on the first value in the B row (individual a), we see that the B score is in the proper order for the pairs (a,b) and (a,d). In other words, individual a has received a lower rank than b and d on both variables A and B. On the other hand, the B score is out of order (with respect to the A score) for pair (a,c) since a has a lower rank than c for A but the opposite is true for B.

Let us make use of $+1$ every time a given pair is ordered the same way (referred to as a "concordant" pair) for both A and B and -1 whenever they are ordered oppositely (referred to as a "discordant" pair). The value of S is obtained by summing these $+1$'s and -1's for all possible pairs. Therefore S is equal to the number of concordant pairs C minus the number of discordant pairs D. The contribution of pairs (a,b), (a,c), and (a,d) is therefore $+1 - 1 + 1 = (2 - 1) = 1$. In order to pick up the remaining pairs we move across the table. We see that the contribution of pairs (b,c) and (b,d) is $-1 + 1$ or 0. Finally, the contribution of pair (c,d) is $+1$. Notice that in effect we can obtain the total value of S by first arranging A in the proper order and then examining successive rankings in the B row, each time counting the number of ranks falling to the right which are in the proper order and subtracting out those which are in the opposite order. Thus in this simple example we get

$$S = C - D = (2 - 1) + (1 - 1) + (1 - 0) = 2$$

If we now divide S by the maximum possible value that it could have, that is, $(N - 1) + (N - 2) + \cdots + 2 + 1 = N(N - 1)/2$, we obtain a coefficient which can vary from -1 to $+1$. Thus we define the coefficient tau$_a$ (following Kendall [16]), appropriate when there are no ties, as follows:[2]

$$\tau_a = \frac{S}{\frac{1}{2}N(N-1)} = \frac{C-D}{\frac{1}{2}N(N-1)} \tag{18.11}$$

Clearly, if there is perfect disagreement between the two ranking systems (i.e., if B were ranked 4, 3, 2, 1), the value of S would be $-\frac{1}{2}N(N-1)$ and τ would be -1.0. Also, if the two variables are completely unrelated, the positive and negative contributions to S will exactly cancel, and τ will be zero.

In order to illustrate the case of tied rankings let us again make use of the work-camp example. Let us arrange the individuals in horizontal arrays and substitute letters for names. Our rankings thus become

	a	b	c	d	e	f	g	h	i	j	k	l	m	n	o	p
A	1	2.5	2.5	4	5	6	7	8	9	10	11.5	11.5	13.5	13.5	15	16
B	5.5	5.5	1	2	3	9.5	5.5	13.5	9.5	16	5.5	11.5	8	15	11.5	13.5

We must now follow the rule that whenever any pair involves a tie in either the A or B score, its contribution to S will be zero. Looking first at all pairs that can be made with a, we see that pairs (a,b), (a,g), and (a,k) will contribute nothing to S since the B scores for all these individuals are tied at 5.5. Therefore, the contribution from all the remaining a pairs will be

(a,c) (a,d) (a,e) (a,f) (a,h) (a,i) (a,j) (a,l) (a,m) (a,n) (a,o) (a,p)

-1 -1 -1 $+1$ $+1$ $+1$ $+1$ $+1$ $+1$ $+1$ $+1$ $+1$

$$= 9 - 3 = 6$$

We next compare the b ranks with each of the ranks to the right of b. Notice, however, that b and c are tied with respect to A. Since b and c could therefore just as well have been given in the reverse order, we must eliminate the pair (b,c). Likewise, the pairs (b,g) and (b,k) are tied on B and will also make no contribution to S. Thus for b pairs we get a sum of

---

[2] This coefficient derived from sample data is sometimes denoted by t, whereas tau is reserved for the population counterpart. However, we shall follow the more conventional usage. Kendall's tau should not be confused with Goodman and Kruskal's tau$_a$ and tau$_b$, which are appropriate for nominal data.

9 − 2 or 7. Continuing across we get

$$S = C - D = (9 - 3) + (9 - 2) + (13 - 0) + (12 - 0) + (11 - 0)$$
$$+ (6 - 3) + (8 - 0) + (2 - 5) + (5 - 2) + (0 - 6) + (4 - 0)$$
$$+ (2 - 1) + (2 - 0) + (0 - 2) + (1 - 0) = 60$$

We must now make an adjustment in the denominator of tau in order to correct for ties. Such an adjustment has an effect of increasing the numerical value of tau, although the increase will be slight unless the number of ties is quite large. The formula for tau (which Kendall denoted as τ_b) can be generalized as follows

$$\tau_b = \frac{S}{\sqrt{\frac{1}{2}N(N - 1) - T}\ \sqrt{\frac{1}{2}N(N - 1) - U}} \tag{18.12}$$

where $T = \frac{1}{2}\Sigma t_i(t_i - 1)$, t_i being the number of ties in each set of ties in A, and $U = \frac{1}{2}\Sigma u_i(u_i - 1)$, u_i being the number of ties in each set of ties in B. In the above example, we have three ties of two each in variable A (popularity). Thus

$$T = \frac{1}{2}[2(1) + 2(1) + 2(1)] = 3$$

Similarly, there are three ties of two each and one score with four ties in variable B (participation). Therefore

$$U = \frac{1}{2}[2(1) + 2(1) + 2(1) + 4(3)] = 9$$

Hence

$$r_b = \frac{60}{\sqrt{[8(15) - 3][8(15) - 9]}} = \frac{60}{\sqrt{(117)(111)}} = \frac{60}{114.0} = .526$$

Tests of significance for tau   Kendall [16] has shown that for sample sizes of 10 or more, the sampling distribution of S under the null hypothesis will be approximately normal with mean of zero and variance given by

$$\sigma_s{}^2 = \frac{1}{18}N(N - 1)(2N + 5) \tag{18.13}$$

Strictly speaking, the above formula holds only when there are no ties but can safely be used when the number of ties is relatively small. If there is a very large number of ties a rather lengthy correction factor

To test for the significance of tau for the work-camp data, we first compute σ_s^2 as follows

$$\sigma_s^2 = \tfrac{1}{18}(16)(15)(37) = 493.3$$

Taking the square root we get

$$\sigma_s = 22.21$$

which can be used in the denominator of Z in testing the null hypothesis that A and B are unrelated. Thus

$$Z = \frac{S - 0}{\sigma_s} = \frac{60.0}{22.21} = 2.70$$

and we see that a value of tau of .526 is significant at the .01 level.

Ordinal measures for grouped data: tau$_c$, gamma, d_{yx} and d_{xy} One advantage of tau over r_s is that the former measure can readily be used when there are very large numbers of ties. Although the computing routine described above would become rather tedious in such instances, we may greatly simplify the procedure whenever both variables have been grouped into several rather crude categories. For example, persons may have been placed into one of five social classes, with all those in the same class being considered as tied with respect to status. If the second variable has been categorized in a similar manner, we can use a modified formula for tau and thereby make use of the information that the data have actually been ranked rather than simply categorized.

We can compute $S = C - D$ by a procedure to be described below. Using the formulas given above, we would find that the upper limit of tau$_b$ would be unity only when the number of rows and columns were equal. In order to correct for the possibility that $r \neq c$ we form the ratio

$$\tau_c = \frac{S}{\tfrac{1}{2}N^2[(m-1)/m]}$$

(18.14)

where $$m = \text{Min}\ (r,c)$$

We again follow Kendall in using the symbol τ_c in order to distinguish Eq. (18.14) from the previous two formulas. Let us now see how τ_c is computed.

Table 18.4   Data cross classified for computing Kendall's tau from grouped data

Strength of desire to join organizations (A)	Concern with proper behavior (B)				Total
	High	Moderately high	Moderately low	Low	
High	18	19	12	8	57
Moderately high	16	16	12	10	54
Moderately low	11	14	18	16	59
Low	5	5	15	22	47
Total	50	54	57	56	217

The data given in Table 18.4 represent rankings given to 217 students of introductory sociology at the University of Michigan. Variable B involves the student's general concern that he engage in the "proper" or "correct" forms of behavior in conventional settings. Variable A involves his desire to join organizations merely to improve his social status. Since the measurement of both variables was somewhat crude, it was decided to divide each variable into four categories: high, moderately high, moderately low, and low. Thus, although each variable involves an ordinal scale with large numbers of ties, the results can be summarized in the form of a contingency table.

In computing S it will be convenient to obtain C and D separately since these quantities will also be used for the other measures discussed in this section. We first note that scores on A have again been ranked from high to low except that we now have 57 individuals "tied" for high, 54 for moderately high, 59 for moderately low, and 47 for low. Looking first at those with high scores on A, we see that 18 are also high on B, 19 moderately high, and so forth. In obtaining the contributions to C and D (and thus S), we note that since all of the individuals in the high category of A are tied, none of these pairs will contribute to C or D. Likewise, none of the pairs in the same column will contribute to C or D because of the fact that they are all tied with respect to B. If we look at any given cell, all the scores which are below it and to the right will contribute to the number of concordant pairs C, whereas those below it and to the left will contribute to D. For example, each of the 18 individuals in the top left-hand cell will produce concordant pairs with any of the

$$16 + 14 + 5 + 12 + 18 + 15 + 10 + 16 + 22$$

scores which are beneath and to the right of this cell. In total, then, the contribution from this cell to C will be

$$18(16 + 14 + 5 + 12 + 18 + 15 + 10 + 16 + 22) = 18(128)$$

Next, we focus on the 16 cases immediately below the top left-hand corner. Each of these individuals also has high B scores. In order to count the number of pairs with contributions to C we again add the quantities appearing below and to the right. Multiplying by the number of cases in this cell we get

$$16(14 + 5 + 18 + 15 + 16 + 22) = 16(90)$$

When we move over into the second and subsequent columns, we begin to find contributions to both C and D since columns to the left will have higher B scores. Thus for the first cell in the second column we get as the contribution to C

$$19(12 + 18 + 15 + 10 + 16 + 22) = 19(93)$$

and as the contribution to D the quantity $19(16 + 11 + 5) = 19(32)$. By counting down and across the table in a similar manner we can obtain S rather simply, as follows:

$$\begin{aligned} C = {} & 18(128) + 16(90) + 11(42) + 19(93) + 16(71) + 14(37) + 12(48) \\ & + 12(38) + 18(22) \\ = {} & 9055 \end{aligned}$$

$$\begin{aligned} D = {} & 19(32) + 16(16) + 14(5) + 12(67) + 12(35) + 18(10) + 8(112) \\ & + 10(68) + 16(25) \\ = {} & 4314 \end{aligned}$$

Therefore $\qquad S = 9055 - 4314 = 4741$

Thus $\qquad \tau_c = \dfrac{4741}{\frac{1}{2}(217)^2[(4 - 1)/4]} = .268$

Notice that the denominator of τ_c depends only on the number of rows and columns and not on the marginal distributions, which of course determine the number of ties. This makes τ_c difficult to interpret and in this respect less satisfactory than τ_b.[3] There are also several other mea-

---

[3] It can be shown that in the $k \times k$ case in which *all* marginal totals are exactly N/k, τ_b and τ_c will be equal. Otherwise, in the $k \times k$ case τ_c will generally be smaller than τ_b in numerical value, though it may be greater than τ_b in the $r \times c$ case.

sures that differ with respect to the handling of ties in the denominator. The best known such measure is gamma (γ) which excludes ties altogether from the denominator, and which may also be applied to ungrouped data. The formula for gamma is

$$\gamma = \frac{C - D}{C + D}$$

In the example under consideration we thus obtain

$$\gamma = \frac{9055 - 4314}{9055 + 4314} = .354$$

In Chap. 15 it was pointed out that Yule's $Q = (ad - bc)/(ad + bc)$ is a special case of gamma. Therefore we might expect that gamma will behave essentially the same way in instances where marginal distributions are very unequal, and the same cautions that applied to Q should be noted. Since gamma, τ_a, and τ_b all have the same numerators, and since the denominator of gamma excludes all ties, it can easily be seen that $|\gamma| \geq |\tau_b| \geq |\tau_a|$. In general, to the degree that the marginal totals for A and B are very different, gamma can exceed τ_b by a substantial amount. For example in the case of the following hypothetical table

		B		
A	High	Medium	Low	Total
High	100	80	0	180
Medium	0	20	80	100
Low	0	0	20	20
Total	100	100	100	300

we note that there are no discordant pairs, so that $\gamma = 1.0$. However, $\tau_b = .77$ and $\tau_c = .68$. Whether or not one wishes to refer to the above association as "perfect" will depend upon his assumptions concerning why the marginal distributions are not identical.

In addition to the tau's and gamma, we have two asymmetric measures d_{yx} and d_{xy} introduced by Somers [20] and defined as follows:

$$d_{yx} = \frac{C - D}{C + D + T_y}$$

and

$$d_{xy} = \frac{C - D}{C + D + T_x}$$

where T_x is the number of pairs that are tied on X but not on Y, and T_y is the number of pairs tied on Y but not X. If we let T_{xy} refer to the number of pairs tied on both X and Y, then referring back to Eq. (18.12) for τ_b we see that $T = T_x + T_{xy}$, and $U = T_y + T_{xy}$ and therefore since the total number of pairs $\frac{1}{2}N(N-1) = C + D + T_x + T_y + T_{xy}$, we have $C + D + T_y = \frac{1}{2}N(N-1) - (T_x + T_{xy}) = \frac{1}{2}N(N-1) - T$. Similarly, the denominator of d_{xy} is $C + D + T_x = \frac{1}{2}N(N-1) - U$. Thus the product $d_{yx}d_{xy} = \tau_b{}^2$. In this sense the asymmetric measures can be thought of as *slope analogs*. However, since their asymmetry is a function of numbers of ties, which usually depend on classification procedures, the analogy with the slopes b_{yx} and b_{xy} is at best a very tenuous one.

Costner [5] has pointed out that gamma can be given a proportional-reduction-in-error interpretation that is similar to the interpretation given to Goodman and Kruskal's τ_b or λ_b. Suppose we wish to predict the order of a pair of cases with respect to B. If we rule out ties, our probability of making an error knowing nothing else would be .5. But if we know the order with respect to A, it turns out that the absolute value of gamma is equal to the number of expected errors knowing A, less the number expected not knowing A, divided by the number expected not knowing A.

Thus we have available a number of ordinal measures that differ only with respect to the treatment of ties in the denominator. Unfortunately, however, we ordinarily have no clearcut decision rule for choosing among such measures since the reasons for the ties often remain obscure. Wilson [23] has shown that the proportional-reduction-in-error property of gamma breaks down if one admits that errors can be made when one predicts an ordering with respect to B if, in fact, the pair is tied with respect to B.[4] It seems as though this problem of handling ties admits of no simple solution. Perhaps the best rule of thumb is to use as many categories as possible of each variable, thereby reducing the number of ties and reducing the differences among the various measures.

Kruskal [17] has shown that Spearman's measure r_s can be interpreted in terms of triplets of observations, instead of pairs, by asking what the

---

[4] Wilson [23] notes that such ties are not excluded from the analysis in regression models. Thus if two cases are extremely close together with respect to their X scores, we would also predict their Y scores to be extremely close. In this sense, if a pair is tied with respect to X, we should predict them to be tied with respect to Y, and we would be making an error if they were *not* tied with respect to Y. But how serious is this "error" of predicting ties incorrectly as compared with the error of making the wrong prediction in the untied case? It can be seen that this whole question of the exclusion of ties, a procedure which tends to favor gamma over the other measures, is not a simple matter. Therefore the greater the number of ties due to crudity of measurement, the more ambiguous one's choices among the measures and the greater the sensitivity of the results to this choice.

probability is that at least one of the three is concordant with *both* the other two. Such an interpretation has much less intuitive appeal than interpretations for pairs, and in addition more is known about the sampling errors of tau and gamma. For these reasons, Kruskal prefers tau to r_s. However, if the underlying distribution of the two variables is really bivariate normal, the absolute value of r_s will be greater than that of tau, and its behavior seems to be much more similar to that of the product-moment correlation. Preliminary unpublished work shows that the behavior of *partial* r_s's (when corrected for ties) is very similar to that of partial correlations when the true relationships are multivariate normal (see next chapter for definition), so that it is at present an open question as to which measure is preferable. In view of this, an investigator should compute several different measures to see whether or not they behave similarly for the data under consideration.

Finally, we should make note of an argument by Wilson [22] that *no* ordinal measure that involves the notion of pairs (or triplets) can have fully desirable properties. Wilson's basic point is that theoretical reasoning ordinarily specifies laws that are appropriate to *single* cases, as for example where we specify that a change in X of one unit should produce a change in Y of b_{yx} units. In terms of these theories it does not make sense to think in terms of ordered pairs that, of necessity, force one to make comparisons across cases. For example, if one's theory specifies that changes in per cent Negro will produce changes in discrimination rates, one is presumably referring to a "law" that operates *within* a single community (or other unit of observations). It does not directly concern comparisons across pairs of observations. Of course as long as one defines his task as a simple generalization to fixed populations, this kind of conceptual difficulty will not arise. The reader is referred to Wilson for a more complete discussion. Clearly, the use of ordinal measures involves a number of difficulties that have not as yet been adequately resolved.

Exercises

1. In Exercises 1 and 2, Chap. 17, you computed three correlation coefficients.
   a. For each of these coefficients, make use of analysis of variance to test the null hypothesis that $\rho = 0$. (*Ans.* $F = .67$; $F = 7.09$; $F = 9.6$)
   b. Place 99.9 per cent confidence intervals about all three r's.
   c. Test the relationship between moral integration and heterogeneity for nonlinearity.
   d. Convert these same data to ranks and obtain both Kendall's tau and Spearman's r_s for all three correlations.
   e. Test each of these rank-order coefficients for significance.

2. In Exercise 3, Chap. 17, you grouped both the moral integration and heterogeneity indices. Compute Kendall's tau_c and gamma for these grouped data and compare the result with that obtained above in Exercise 1*d*.

References

1. Anderson, T. R., and M. Zelditch: *A Basic Course in Statistics*, 2d ed., Holt, Rinehart and Winston, Inc., New York, 1968, chaps. 7 and 8.
2. Blalock, H. M.: "Estimating Measurement Error Using Multiple Indicators and Several Points in Time," *American Sociological Review*, vol. 35, pp. 101–111, 1970.
3. Bohrnstedt, G. W.: "Observations on the Measurement of Change," in Edgar Borgatta (ed.), *Sociological Methodology 1969*, Jossey-Bass Inc., Publishers, San Francisco, 1969, chap. 4.
4. Christ, Carl: *Econometric Models and Methods*, John Wiley & Sons, Inc., New York, 1966, Part III.
5. Costner, H. L.: "Criteria for Measures of Association," *American Sociological Review*, vol. 30, pp. 341–353, 1965.
6. Costner, H. L.: "Theory, Deduction and Rules of Correspondence," *American Journal of Sociology*, vol. 75, pp. 245–263, 1969.
7. Croxton, F. E., and D. J. Cowden: *Applied General Statistics*, 3d ed., Prentice-Hall, Inc., Englewood Cliffs, N.J., 1967, chap. 20.
8. Goodman, L. A., and W. H. Kruskal: "Measures of Association for Cross Classifications," *Journal of the American Statistical Association*, vol. 49, pp. 732–764, 1954.
9. Goodman, L. A., and W. H. Kruskal: "Measures of Association for Cross Classifications. II: Further Discussion and References," *Journal of the American Statistical Association*, vol. 54, pp. 123–163, 1959.
10. Goodman, L. A., and W. H. Kruskal: "Measures of Association for Cross Classifications. III: Approximate Sampling Theory," *Journal of the American Statistical Association*, vol. 58, pp. 310–364, 1963.
11. Haggard, E. A.: *Intraclass Correlation and the Analysis of Variance*, The Dryden Press, Inc., New York, 1958, pp. 22–26.
12. Hagood, M. J., and D. O. Price: *Statistics for Sociologists*, Henry Holt and Company, Inc., New York, 1952, chap. 23.
13. Hays, W. L.: *Statistics*, Holt, Rinehart and Winston, Inc., New York, 1963, chap. 16.
14. Heise, D. R.: "Separating Reliability and Stability in Test-Retest Correlation," *American Sociological Review*, vol. 34, pp. 93–101, 1969.
15. Johnston, J.: *Econometric Methods*, McGraw-Hill Book Company, New York, 1963, Part II.
16. Kendall, M. G.: *Rank Correlation Methods*, Hafner Publishing Company, Inc., New York, 1955, chaps. 1, 3, and 4.
17. Kruskal, W. H.: "Ordinal Measures of Association," *Journal of the American Statistical Association*, vol. 53, pp. 814–861, 1958.
18. Mueller, J. H., K. Schuessler, and H. L. Costner: *Statistical Reasoning in Sociology*, 2d ed., Houghton Mifflin Company, Boston, 1970, chap. 10.
19. Siegel, Sidney: *Nonparametric Statistics for the Behavioral Sciences*, McGraw-Hill Book Company, New York, 1956, chap. 9.
20. Somers, R. H.: "A New Asymmetric Measure of Association for Ordinal Variables," *American Sociological Review*, vol. 27, pp. 799–811, 1962.
21. Wallis, W. A., and H. V. Roberts: *Statistics: A New Approach*, The Free Press of Glencoe, Ill., Chicago, 1956, chap. 17.
22. Wilson, T. P.: "A Critique of Ordinal Variables," *Social Forces*, vol. 49, pp. 432–444, 1971.
23. Wilson, T. P.: "A Proportional-Reduction-in-Error Interpretation for Kendall's tau-b," *Social Forces*, vol. 47, pp. 340–342, 1969.

Multiple and Partial Correlation

19

In the last two chapters we have been concerned with the relationship between two interval scales, between a dependent variable and a single independent variable. Correlation and regression analysis can readily be extended to include any number of interval scales, one of which can be taken as dependent and the remainder independent. The problem can be conceptualized as a prediction problem in which we attempt to predict a dependent variable Y from the variables $X_1, X_2, \ldots, X_k$. We shall again have to make use of a very simple model which will be directly analogous to linear regression except for the fact that there will be more than two dimensions.

The notion of correlation will be generalized in two ways. We shall use the term *partial correlation* to refer to the correlation between any two variables when the effects of the other variables have been controlled. *Multiple correlation*, on the other hand, will be used to indicate how much of the total variation in the dependent variable can be explained by all of the independent variables acting together. You will find that, for the most part, the materials discussed in the present chapter involve straightforward extensions of arguments previously presented. Once we have extended the notions of correlation and regression, we shall be in a position in the next chapter to take up analysis of covariance which involves a combination of regression techniques with analysis of variance.

19.1 Multiple Regression and Least Squares

In multiple regression we attempt to predict a single dependent variable from any number of independent variables. If there are a large number of interval-scale variables that are to be interrelated, it will of course be possible to predict any particular variable from any combination of the others. It will usually be clear from context which variables are to be taken as independent and which dependent.[1] For example, one may want to predict performance in college from a series of aptitude scores and performance in high school. Or it may be possible to predict the rate of growth of cities knowing such factors as present size, percentages of the labor force in various types of occupations, or the size and distance of the nearest large city.

In multiple regression analysis we define the regression equation as the path of the mean of the dependent variable Y for all combinations of X_1, X_2, . . . , X_k. In other words, for every combination of fixed X's there will be a distribution of Y's. Each distribution will have a mean $\mu_{Y|X_1, X_2, \ldots, X_k}$ and a standard deviation $\sigma_{Y|X_1, X_2, \ldots, X_k}$, and we shall again assume that these distributions are all normal and that the standard deviations are equal (homoscedasticity). The path of the means will no longer be a curve in two-dimensional space. Instead it will be a kind of hypersurface in $(k + 1)$-dimensional space. Obviously, we shall no longer be able to represent such a path except in the case where there are only two independent variables X_1 and X_2.

In the previous chapter we assumed a linear regression equation of the form $Y = \alpha + \beta X$. We shall again have to assume a simple form for the regression equation. Let us assume that the path of the means of Y takes the form

$$Y = \alpha + \beta_1 X_1 + \beta_2 X_2 + \cdots + \beta_k X_k \qquad (19.1)$$

where α, β_1, β_2, . . . , and β_k are constants. This is the simplest possible type of multiple regression equation and is directly analogous to linear regression in the two-variable case. In fact, if all β's except one are zero, the problem reduces to the two-dimensional case.

If we can assume a "multivariate normal" population in which each variable is distributed normally about all of the others, then we can meet all three of our required assumptions. In other words, a multivariate normal distribution assures us that regression equations will be of the above form, that the distributions of Y's for fixed X's will all be normal, and that the variances are also equal. This is an obvious generalization of the properties of the bivariate normal distribution. Needless to say, the multivariate normal distribution cannot be pictured geometrically (though it has a definite algebraic equation) since we have already used three dimensions in depicting the bivariate case.

---

[1] Whenever there is thought to be reciprocal causation or feedback from the "dependent" variable to any of the others, simultaneous equations should be used instead of least squares. See [4] and [12].

In order to give you a better intuitive grasp of the nature of the extensions involved, it will be helpful to examine the case where there are only two independent variables (see Fig. 19.1). The regression equation $Y = \alpha + \beta_1 X_1 + \beta_2 X_2$ can in this instance be represented by a *plane* in three-dimensional space. If we let $X_1 = X_2 = 0$, we get $Y = \alpha$, indicating that the regression plane crosses the Y axis at a height α. In order to get an interpretation for the β's, we take the intersections of the regression plane with planes perpendicular to the X_1 and X_2 axes. For example, if we take a plane perpendicular to the X_2 axis, we are in effect holding X_2 constant since all points on this plane will have the same X_2 value. This

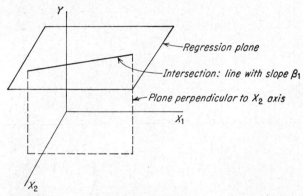

Figure 19.1 Geometrical interpretation of multiple regression of Y on X_1 and X_2.

plane intersects the regression plane in a straight line, and the slope of this line will be β_1. In other words, if we hold X_2 at a fixed value, β_1 represents the slope of the regression *line* of Y on X_1. Similarly, holding X_1 constant gives us a plane which intersects the regression plane in a line with slope β_2.

It should be noted that the β's used in multiple regression will not ordinarily be the same as those obtained in the two-variable case. Referring to the two-variable case as *total* regression, we see that the β used in total regression is obtained by *ignoring* other independent variables, not by holding them constant. The β's obtained in multiple regression equations are referred to as *partial* coefficients, since they involve slopes that would be obtained by controlling for each of the remaining independent variables considered in the regression equation.

The concept of least squares can be extended in a similar manner.

Since it is practically always necessary to estimate a regression equation by fitting an equation to empirical data, we shall again require that the estimating equation have a particular form and use the least-squares criterion to obtain the "best" fit. We shall use a least-squares equation of the form

$$Y_p = a + b_1X_1 + b_2X_2 + \cdots + b_kX_k \qquad (19.2)$$

and it will again turn out that, provided the true regression equation is actually of this same form, the least-squares equation represents the best estimate of the regression equation. In other words if we use a to estimate α and b_i to estimate β_i, these estimates will be unbiased and also of maximum efficiency. Our attention can then be focused on least-squares analysis as a practical method for estimating a theoretical equation which applies to the population. If there are only two independent variables, we shall be fitting a series of points in three-dimensional space with a best-fitting plane. In a $(k + 1)$-dimensional space we shall be fitting points with a k-dimensional hyperplane, if such a figure can be imagined.

Taking the three-dimensional case, we shall minimize the quantity $\Sigma(Y - Y_p)^2$ which represents the sum of the squared deviations from the least-squares plane in the vertical Y dimension (see Fig. 19.2). The result will be a unique best-fitting plane determined by specific values of

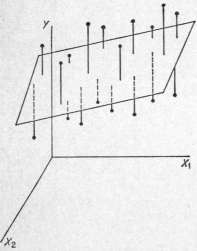

Figure 19.2 Least-squares plane minimizing sum of squared deviations in vertical Y dimension.

a, b_1, and b_2. As we shall see, a multiple correlation coefficient can then be used to measure the goodness of fit of the points to the least-squares plane. It would, of course, also be possible to measure goodness of fit by means of a standard deviation about the plane, and we could compare this standard deviation with the standard deviation about the fixed $\bar{Y}$ (now represented as a *plane* perpendicular to the Y axis). Algebraically, the more general case is a straightforward extension of the three-variable one. The quantity $\Sigma(Y - Y_p)^2$ is minimized, and there will be $(k + 1)$ coefficients to compute, that is, a, b_1, b_2, . . . , b_k. The actual computation of these coefficients will be discussed later after we have taken up partial correlation.

19.2 Partial Correlation

We can make use of this multiple regression model to obtain measures of the degree of relationship between a dependent variable Y and any of the independent variables, controlling for one or more other independent variables. The term *partial correlation* is used to refer to this type of controlling procedure which, as we shall see, is basically very similar to that involved in two-way analysis of variance. In partial correlation we control by adjusting values of the dependent and independent variables in order to take into consideration the scores of the control variables. In order to understand the nature of partial correlation and the adjusting procedure, we shall for the present confine our attention to the simplest problems involving only three variables and shall assume linear regression models between all three combinations of variables taken two at a time.

Let us assume we wish to measure the degree of relationship between a dependent variable Y and an independent variable X_1, controlling for a second independent variable X_2. To make use of a concrete example, we may be interested in predicting the rate of economic discrimination against blacks, as measured by income differentials, and degree of urbanization, as indicated by the percentage of a county which is defined as urban. It is certainly expected that the percentage of Negroes in the county will also affect the rate of discrimination, and it is therefore decided to use per cent Negro as a control variable.

Suppose that the least-squares lines between discrimination Y and per cent Negro X_2 and between per cent urban X_1 and per cent Negro are as indicated in Fig. 19.3. The relationship between discrimination and per cent Negro is positive, indicating that high rates of discrimination are associated with high minority percentages. On the other hand, the relationship between the urbanization index and per cent Negro is nega-

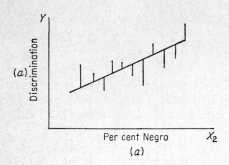

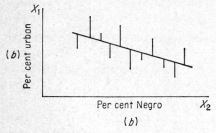

Figure 19.3   Least-squares lines indicating residuals between (a) Y and X_2 and
(b) X_1 and X_2.

tive. On the basis of this information alone, we would predict a negative
relationship between discrimination rates and urbanization. In other
words, urban areas might have low rates simply because of the fact that,
on the average, they have relatively fewer blacks. Suppose, however,
that we could in some manner "force" all counties to have the same
minority percentages. We could then cancel out the disturbing effect
of this third variable. In reality, of course, we cannot actually make all
minority percentages equal, but we can at least make adjustments for the
fact that they differ. Since we know (or can estimate) the relationship
between the control variable and each of the other two variables, we
can predict how both variables would behave with changes in the control
variable.[2] In fact, the least-squares equations represented in Fig. 19.3
are our prediction equations and can be used in the adjusting process.

In relating discrimination Y to per cent Negro X_2, we can think of
variation in discrimination rates as being the result of two components:

[2] Again it should be emphasized that the rationale for interpreting results of such a
paper-and-pencil adjustment involves an implicit causal assumption that the control
variable may affect both other variables.

the first component being per cent Negro and the remainder being due to other factors, one of which may be urbanization. As we have already seen, this second component can be represented as *deviations* from the least-squares equation involving Y and X_2. In terms of X_2, these deviations or *residuals* represent error. Even if X_2 were held constant, they would still remain. It is these residuals, therefore, in which we are really interested since they represent the amount of variation in discrimination left after per cent Negro has explained all of the variation it could.

Similarly, we shall be interested in the residuals or deviations from the equation used to predict per cent urban from per cent Negro. In other words, we let per cent Negro explain all of the variation it can in both of the other variables. If we now correlate the residuals, we obtain a measure of the relationship between Y and X_1 which is independent of the effects of X_2. *The partial correlation between Y and X_1, controlling for X_2, can be defined as the correlation between the residuals of the regressions of Y on X_2 and X_1 on X_2.* In a sense, then, a partial correlation represents the correlation between "errors" with respect to the control variable.

The reason why it makes sense to control for X_2 by correlating residuals may still be obscure. Perhaps the explanation will be made intuitively more appealing if we look more closely at a hypothetical relationship between these residuals. Suppose, for example, that for county A we find a large negative residual when we correlate Y with X_2. This means that county A has considerably less discrimination than would be expected knowing only its minority percentage. The point representing this particular county would of course be somewhere below the least-squares line. Suppose, also, that the residual for this same county was positive when we correlated X_1 with X_2. In this case we know that the county is more highly urbanized than would be expected knowing only its minority percentage. We therefore have a relatively urbanized county with low discrimination rates, and furthermore we know that these values are high and low respectively as compared with other counties having the same minority percentage. We thus cannot attribute the negative relationship between the residuals to the fact that the per cent Negro figure happens to be either large or small. Similarly, county B may have large positive residuals for Y but negative ones for X_1. This county, therefore, would be one with higher discrimination rates than expected but would also be less urbanized than other counties having the same minority percentage. Clearly, if most counties are similar to either county A or B, we shall obtain a negative correlation between the residuals, indicating a negative relationship between discrimination and urbanization, adjusting for per cent Negro.

Partial correlation yields a single measure summarizing the degree of relationship between two variables, controlling for a third. As we shall see when we discuss computational procedures, the argument can readily be extended to additional control variables. We can visualize several multiple regression equations, one involving Y and all the control variables, and the other relating X_1 to these same control variables. Residuals can be obtained from each of these multiple regression equations and then correlated. We shall thus be adjusting for all of the control variables simultaneously. The important point, here, is that we obtain only one partial correlation, whereas in controlling with contingency tables (allowing for interaction) we obtained a separate measure for each of the categories of the control variables.

In Chap. 15 we saw that the degree of relationship between two variables may vary from one category of the control variable to the next. Thus, had per cent Negro been categorized, it is entirely possible that we might have found interaction, perhaps even involving a high negative correlation between discrimination and urbanization for counties with very low minority percentages but a positive correlation at the opposite end of the per cent Negro continuum. The fact that we have obtained a single summarizing measure in the partial correlation may therefore obscure certain information about interaction.

It turns out that the partial correlation coefficient can also be interpreted as a *weighted average* of the correlation coefficients that would have been obtained had the control variable been divided into very small intervals and separate correlations computed within each of these categories. The exact nature of this weighting procedure is unimportant since in practice we never make use of it. Consequently, it does not make sense to think of partial correlations as relating two variables "holding constant" a third variable, since the strength of their relationship may vary according to the particular value at which the control variable is held constant.

In the case of the multivariate normal distribution we know that all regression equations will have the special form described by Eq. (19.1). The multivariate normal distribution has another remarkable property as well. The strength of relationship between two variables will be the same regardless of the values of the control variables. In other words, if a large number of categories of a control variable were selected and correlations obtained within each of these categories, all correlations would have the same value. Therefore the partial correlation would have the same value as each of these within-category correlations. In this special case it would thus make some sense to think in terms of holding the third variable constant. Since the multivariate normal distribution can at best be only approximated with real data, however, it is safer to think of a partial correlation as a weighted average or as involving an adjustment for the control variable.

Computation of partial correlation coefficients The computation of partial correlations is extremely simple unless it is desired to control for three or more variables simultaneously. Before presenting the formula for partial correlation we must introduce a change of notation. Unfortunately, what is a convenient notation for one purpose may not be so for another, nor is conventional usage completely consistent. We have been letting the dependent variable be represented by Y and the independent variables by $X_1, X_2, \ldots, X_k$. In recognition of the fact that sometimes the choice of dependent variable depends on the subset of variables under immediate consideration, and that we may therefore want to compute partial correlations between various combinations of variables, it will be convenient simply to renumber the variables from 1 to $k + 1$ and to represent the correlation between variables 1 and 2 controlling for 3 by $r_{12 \cdot 3}$. Similarly, the correlation between variables 2 and 3 controlling for 1 would be $r_{23 \cdot 1}$.

This notation can readily be extended to any number of control variables by adding further numbers to the right of the center dot. Thus the relation between variables 5 and 7, controlling for variables 1, 2, 3, 4, and 6, would be given by $r_{57 \cdot 12346}$. The order of the two variables to the left of the dot is irrelevant as is that to the right. To distinguish between partials with differing numbers of control variables, we refer to the number of controls as the *order* of the correlation. Thus, a first-order partial will have one control, a second-order two controls, and so on. In keeping with this terminology, a correlation with no controls is often referred to as a zero-order correlation. As indicated above, the term *total correlation* is also used to indicate a correlation between two variables with no controls.

We can now give the formula for the general first-order partial $r_{ij \cdot k}$:

$$r_{ij \cdot k} = \frac{r_{ij} - (r_{ik})(r_{jk})}{\sqrt{1 - r_{ik}^2} \sqrt{1 - r_{jk}^2}} \tag{19.3}$$

Notice that the first correlation in the numerator is the total correlation between the two variables to be related (i and j). The control variable appears in the second expression of the numerator, where it is related to each of the other variables, and in both terms of the denominator. Any particular partial correlation can be obtained from this general formula by substituting the proper numbers for i, j, and k. Thus

$$r_{13 \cdot 2} = \frac{r_{13} - (r_{12})(r_{23})}{\sqrt{1 - r_{12}{}^2}\ \sqrt{1 - r_{23}{}^2}}$$

In a study of 150 Southern counties [3], the correlation between income discrimination and per cent Negro was .536; that between income discrimination and per cent urban was .139; and the correlation between per cent Negro and per cent urban was $-.248$. Calling the discrimination index variable 1, per cent Negro variable 2, and per cent urban variable 3, we can obtain the partial between discrimination and per cent urban, controlling for per cent Negro. We have

$$r_{13 \cdot 2} = \frac{.139 - (.536)(-.248)}{\sqrt{1 - (.536)^2}\ \sqrt{1 - (-.248)^2}} = \frac{.2719}{.8178} = .332$$

The result can be interpreted as the correlation between discrimination and per cent urban after per cent Negro has been allowed to explain all it can of both variables.

Although it will not be immediately apparent that the above formula can be derived from the definition of a partial correlation in terms of a correlation of residuals, the computing formula at least makes sense. In the numerator we are essentially subtracting a correction factor from the total correlation. The denominator consists of two correction factors, neither of which can be greater than unity, which take into consideration the fact that the control variable explains a certain proportion of the variation in the other variables. If we square the partial correlation coefficient, the resulting number will represent the proportion of variation in variable 1 (discrimination) which is left unexplained by 2 (per cent Negro) but which can be explained by the adjusted values of X_3 (per cent urban).

Let us examine Eq. (19.3) more carefully to note how the partial correlation behaves in relation to the three total correlations. For the sake of simplicity, let us first assume that r_{ij} is positive. If r_{ik} and r_{jk} both have the same signs (either positive or negative), their product will be positive, and either the numerator will be a smaller positive number than r_{ij} or it may even be zero or negative. On the other hand, the denominator will always be less than unity unless $r_{ik} = r_{jk} = 0$. The resulting fraction may therefore be almost any number between -1.0 and $+1.0$, depending on the magnitudes of the three total correlations. We shall see later just what we can and cannot say about the behavior of the partial under these circumstances.

Now suppose the correlations with the control variable are of opposite signs. We then obtain a negative product to be subtracted from a positive number, and the result will be a larger positive number. This means that if we start with two variables which are positively related, and if we can find a control variable which is negatively related to one of them but positively related to the other, the resulting partial will be larger than the zero-order correlation. If the correlation of the control variable with either of the other variables happens to be zero, the correction factor in the numerator will be zero. But if the control variable is correlated either positively or negatively with the remaining variable, the denominator will be less than unity, and the partial correlation will again be larger than the total correlation.

If we had started with a negative total correlation, a control variable related to each of the other two in the same direction (either positive or negative) would produce a larger negative correlation. If it were related oppositely to the other two, however, the result would be analogous to the one first described (where the total correlation was positive and the correction factor also positive). Why? Again, if the control variable were not related to one of the other variables, the result would be a partial correlation with a larger absolute value than the total correlation. If the control variable were unrelated to *both* other variables, the partial correlation would of course exactly equal the total correlation. After we have discussed the relationship between partial correlation and causal interpretations, we shall be able to give an intuitive justification for the behavior of partial correlations under these various conditions.

The formulas for second- and higher-order partials are directly analogous to that for the first-order partial. We simply add successive control variables, each time starting with the partial of order one less than that desired. Thus the formulas for $r_{ij \cdot kl}$ and $r_{ij \cdot klm}$ will be

$$r_{ij \cdot kl} = \frac{r_{ij \cdot k} - (r_{il \cdot k})(r_{jl \cdot k})}{\sqrt{1 - r_{il \cdot k}^2}\sqrt{1 - r_{jl \cdot k}^2}} \tag{19.4}$$

and

$$r_{ij \cdot klm} = \frac{r_{ij \cdot kl} - (r_{im \cdot kl})(r_{jm \cdot kl})}{\sqrt{1 - r_{im \cdot kl}^2}\sqrt{1 - r_{jm \cdot kl}^2}} \tag{19.5}$$

Notice in Eq. (19.4) that we assume we have already controlled for variable X_k. Therefore, k appears to the right of each dot in all three first-order partials. Similarly, in Eq. (19.5) we have previously controlled for both X_k and X_l, and therefore both of these quantities appear in each of the second-order partials.

Fourth- and fifth-order partials could be obtained in a similar manner, and it will be instructive to try to write out formulas for these higher-order partials. The form for computing higher-order partials is thus identical to that used for the first-order case. But the work involved rapidly becomes tedious. For example, in order to obtain a third-order partial by this method, one must first have obtained three second-order partials, each of which in turn must be obtained by computing first-order partials from the zero-order correlations. If you were to try to express the formula for third-order partials directly in terms of the zero-order correlations, you would begin to see how much work would actually be involved.

Fortunately, it is seldom necessary in sociological research to go much beyond second- or third-order partials. Usually, the addition of controls beyond a second or third control produces very few new insights. If it is necessary to make use of higher-order partials, or of multiple regression equations involving four or more variables, there are computing routines that can do the job much more simply. You may wish to refer to either the abbreviated Doolittle method or Dwyer's square-root method in order to handle such problems (see [9] and [11]). Of these two methods, the latter is perhaps more satisfactory in that it enables one to obtain directly the successive partials $r_{12 \cdot 3}$, $r_{12 \cdot 34}$, $r_{12 \cdot 345}$, etc.

Partial rank-order correlation   The theory with respect to partial rank-order correlations is less well developed. Kendall's tau can be extended in the case of first-order partials although the interpretation for the partial tau is not as intuitively appealing as in the case of product-moment correlation. If there are no ties, it turns out that the formula for partial tau's is identical to that we have been using. (See [13] and [19].) Thus

$$\tau_{ij \cdot k} = \frac{\tau_{ij} - (\tau_{ik})(\tau_{jk})}{\sqrt{1 - \tau_{ik}^2} \sqrt{1 - \tau_{jk}^2}} \tag{19.6}$$

An alternative procedure that can be used in the case of large numbers of ties has been suggested by Davis [7] for the case of gamma, but the principle can be extended to any of the tau measures or to d_{yx} and d_{xy}. If we are controlling for W, we simply categorize W, compute gammas (or other measures) within categories of W, and obtain a weighted average of these gammas. But instead of weighting by the number of cases in each category, we weight by the number of *pairs* involved. Thus in the case of a partial gamma we are considering only the pairs that are not tied on either X or Y but that *are* tied with respect to the category of the control variable. Davis shows that such a weighted average can be given a simple proportional-reduction-in-error interpretation. Quade [16] offers a similar weighted-average procedure in the case of tau$_a$ and also provides a test of significance for this partial.

In exploratory research it probably makes sense to utilize multiple control variables, either by extending formula (19.6) or by breaking the control variables into multiple subcategories. However, the theoretical underpinnings of such procedures are not very strong, particularly in the presence of numerous ties. (See [20].) Somers [19] has noted that in the case of marked nonmonotonic relationships the procedure suggested by Davis can be misleading. As a general practice, in view of our ignorance of the properties and behavior of partial ordinal measures, it would seem wise to utilize them with caution and to supplement them with product-moment measures even where legitimate interval scales are not fully justified. Ideally, of course, one should attempt to improve his measurement procedures so as to justify the use of more powerful parametric tests and measures.

As has been implied in our previous considerations of ordinal data, one of the fundamental reasons why it is difficult to reach definitive conclusions concerning the appropriateness of alternative measures is that these answers seem to depend on one's conception of the "underlying reality" behind the data. We have already noted this in connection with the handling of ties and, by implication, with the categorization process. One very promising mode of attack on this difficult problem involves the construction of a "reality" with known properties through the use of computer-generated or simulated data. For example one may construct variables with normal, rectangular, or skewed frequency distributions. One may use linear or nonlinear models, vary the relative magnitudes of the error variances, and construct multivariate data sets with known causal structures (for example, X and Y spuriously related due to Z, or due to several Z_i). Data may then be collapsed in different ways, different kinds of partialling procedures used, and the several ordinal measures compared in terms of their conformity to desired behavior. For example, does the partial between X and Y reduce to almost zero when one controls for Z, where the data have been constructed to conform to the model $X \leftarrow Z \rightarrow Y$?

Reynolds [17] has obtained some rather encouraging results using a variety of models, types of frequency distributions, nonlinear models, and introducing a number of additional complications. In general he has found that if one uses at least five levels of each variable (though preferably as many as ten), one can obtain very good approximations to the behavior of product-moment partials using several different partialling procedures and any of the measures τ_b, τ_c, d_{yx}, or r_s corrected for ties. If there are a sizable number of ties, the numerical values of total associations using τ_a (which does not correct for ties) tend to be so low that it is difficult to distinguish their values from those of partials. For example, if the total τ_a is only .20, then sampling error may be sufficient to make it difficult to determine whether or not there has been a large enough reduction in the partial to support the hypothesis that the relationship is spurious.

Reynolds also found that gamma does not behave as well under partialling as the other measures, probably because of its extreme sensitivity to unequal marginals. In instances where the correct model involves a spurious relationship between X and Y due to W, controls for W did not reduce the partial gamma to zero. Reynolds' data also seem to favor the use of weighted-average partialling procedures as compared with the use of a partialling formula such as Eq. (19.6), though Somers' caution regarding nonmonotonic control variables should be kept in mind. Finally, and importantly, Reynolds found that partialling (using weighted averages) with τ_b, τ_c, and d_{yx} gave excellent results in the case of monotonic but nonlinear relationships, whereas product-moment or parametric procedures did not. In the latter case, if one knew the actual true scores it would be preferable to work with explicit nonlinear models and parametric procedures. But in the absence of such knowledge, the use of parametric procedures with arbitrarily assigned numerical scores (preserving the ordering) gave misleading results.

Finally, it should be noted that the problem of developing measures of multiple correlation using ordinal techniques has not been systematically studied. Morris [15] has even found that both gamma and d_{yx} have the undesirable property that, if one forms measures of multiple correlation by entirely reasonable procedures, the addition of further explanatory values may actually result in *decreased* values of these two measures. He suggests an alternative measure γ_k which is a multivariate generalization of Somers' d_{xy} (not d_{yx}) as a more appropriate asymmetric measure of multiple-ordinal association.

19.3   Partial Correlation and Causal Interpretations

It has been pointed out that correlation analysis cannot be directly used to establish causality because of the fact that correlations merely measure covariation or the degree to which several variables vary together. One of the basic aims of any science, however, is to establish causal relations. Regardless of one's philosophical reservations concerning the notions of cause and effect, it is extremely difficult to think theoretically in any other terms. In Chap. 2 it was pointed out that there is a very real gap between the theoretical language used for thinking and the operational language used for testing hypotheses. The thorny problem of causality is just another indication of the existence of this gap. We often think in terms of causal relationships involving *necessary* temporal sequences. Thus, if A is a cause of B, then A must necessarily be followed by B; and if A is absent, B must also be absent. This conceptualization of causality is of course greatly oversimplified. For one thing, other variables have not been considered, and it makes sense to think of cause and effect only when certain assumptions can be made about these other factors. Also, both A and B may vary in degree rather than merely being present or absent.

Empirically, we can of course never prove that the connection between two variables is necessary. We can ascertain, however, the degree to which they vary together, and it is also sometimes possible to note the time sequence involved. From these two pieces of information we may make causal inferences if we wish. If our theory is able to show a logical connection between the two variables or to predict that A will be followed by B, we need not be too unhappy about making the intellectual leap to a causal interpretation. On the other hand, if we can find no theoretical reason for directly connecting the two events, we are ordinarily more hesitant. We are more likely to think of the relationship as *spurious*, for example. Unfortunately, there is nothing in correlational analysis that can help us make this decision *unless* we are willing to make certain assumptions about these particular variables and about others that might also be operating. Let us see what these assumptions will have to be.

Suppose we are investigating the relationship between the per capita consumption of ice cream and rates of juvenile delinquency. We are likely to find a negative relationship. One possible causal interpretation would be that ice cream is so good for children that it prevents delinquency. Another would be that high rates of delinquency cause children to lose their appetites for sweets. We would of course immediately reject both these interpretations as absurd—although many similarly absurd ones have at one time or another been taken seriously. It would probably be argued that the relationship found was *spurious* in that a third variable, let us say income, was causing both variables to vary in such a manner that a negative correlation was obtained.

One test for spuriousness, and a valid one if used properly, is to control for income level. If when controlling for income the partial correlation between ice-cream consumption and delinquency is reduced to zero, or approximately zero, we may conclude that there is no causal relationship between the two variables. Or can we? Let us take another very similar example. Suppose we find a negative relationship between income level and delinquency and decide to control for the percentage of broken homes in the area. Again we may find that the partial is reduced to zero. Is this relationship therefore also spurious? This time we are not so sure, although there may be absolutely nothing about the magnitude of the correlations or the behavior of the partials which is in any way different from the first case. In order to come to grips with the basic problem we are facing here, let us go back and look at it somewhat more systematically.

Confining our attention for the moment to the three-variable case, we note that there are six possible causal connections between the three variables. Calling the variables X, Y, and Z and indicating the direction of causality by means of an arrowhead, we can diagram these possible connections as in Fig. 19.4. In any given problem, of course, some of

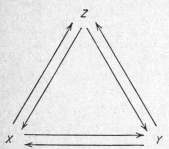

Figure 19.4 The six possible causal arrows among X, Y, and Z.

these arrows will have to be erased. Let us do away with the possibility of two-way causation by arguing that if discrete events are selected, the time sequence must be in one direction or the other but not both simultaneously.[3] For example, rather than claim that unemployment causes a recession and vice versa, let us say that Jones's unemployment causes him to spend less money, which in turn puts Smith out of work, etc. We are then left with only certain possible causal relationships which have been indicated in Fig. 19.5. In order to reduce the number of figures shown in Fig. 19.5, it was arbitrarily decided to select Y as the dependent variable or the one which must occur last in any time sequence. Therefore, no arrows have been drawn from Y to either X or Z. Of all these

---

[3] Of course most empirical situations are much more complex than this simple illustration suggests and more advanced techniques may be required, procedures that apply when the assumptions appropriate for ordinary least squares cannot be met. See [2], [4], and [12] for more extensive discussions of this problem.

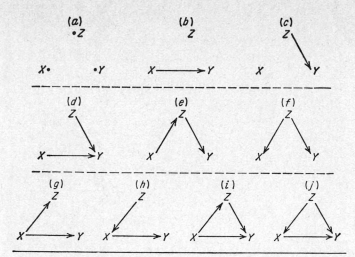

Figure 19.5 Possible causal relationships among X, Y, and Z taking Y as the dependent variable and ruling out two-way causation.

possible relationships, the first three are uninteresting and call for no further comment. Also, in order to simplify matters, let us confine our attention to those figures in which only two causal arrows have been drawn (*d*, *e*, *f*, *g*, and *h*).

Can we distinguish among these various models by looking at the relative magnitudes of the correlation coefficients? The answer is in the affirmative if we are willing to make two kinds of assumptions. Simon [18] has shown mathematically what these assumptions must be. *First*, we must be able to eliminate certain of the models by postulating that at least some of the possible causal relationships do *not* hold. This we have already done to some extent in that all double arrows were ruled out, and also Y was taken to be the dependent variable, i.e., was assumed to cause neither X nor Z. Still further assumptions will have to be made, but we shall postpone their consideration until later.

The *second* type of general assumption we have to make concerns other variables that might be operating. We shall assume, following Simon, that all other variables influencing X are uncorrelated with all other variables affecting Y and Z, and so forth. In other words, we can admit the existence of other uncontrolled variables, but we have to assume that they have essentially a random effect on X, Y, and Z. Notice that this really involves a weaker set of assumptions than is usually implied in the model of the ideal experiment in which it is assumed that all "relevant" variables have been controlled. We recognize the disturbing influence of other variables in that we do not expect correlations to be perfect. On the other hand, we must assume that they operate in such a manner as not to disturb the *pattern* of relationships among X, Y, and Z. This condition can be approximated empirically if there are a large number of other variables operating, no one of which has a very great effect on more than one of the variables under consideration.

If an outside variable with a disturbing effect does exist, it should be brought into the model as a fourth variable. Simon argues that this is essentially what we always must do and that our failure to be satisfied with a causal explanation in the two-variable case is the reason why we introduce the notion of a spurious relationship. For example, if we were satisfied that there were no such variable operating to disturb the relationship between ice-cream consumption and delinquency, and if we could rule out the possibility that delinquency caused ice-cream sales to go down, then we would be satisfied with the explanation that the causal arrow went in the opposite direction. We introduce the income factor precisely because we expect this latter variable to affect the relationship between the first two. Similarly, we could add a fourth or fifth variable to the system, but at some point we must be willing to stop. At this point, if we are to make any causal inferences at all, we must assume the system to be *closed* in the sense we have been describing.

Notice that we are in the familiar position of having to make certain assumptions that cannot be verified empirically by the statistical analysis. It will therefore not be possible to establish the correctness of any particular causal model, but we can proceed by elimination. For example, one of the models indicated in Fig. 19.5 may appear to work. But the correct model might actually involve four or more variables and the picture might look quite different indeed. Having made these assumptions, however, we can make use of the mathematical analysis formalized by Simon to lead to certain predicted relationships that ought to hold among the correlations if the particular model is actually correct. As we shall see, exactly the same empirical relationships among correlations are predicted by several of the above models, making it necessary to choose between them on other grounds. It is here that we must make use of the first kind of assumption discussed above, namely, that certain causal relationships do *not* hold. Let us first examine the mathematical predictions concerning the interrelations among correlation coefficients, however.

If we look at Fig. 19.5g, we see that the relationships between X and Y and between X and Z are direct, whereas that between Y and Z is only indirect. The same is also true in Fig. h. In both these cases, common sense would suggest that if all other variables were operating in an essentially random manner, we would expect to find that the correlation between Y and Z is smaller in magnitude than either of the other two correlations. Similarly, in Fig. 19.5e and f we would expect the correlation between X and Y to be the smallest of the three, ignoring signs. As the mathematics works out, we can be even more definite than this. It is possible to show for both (g) and (h) where the relationship between Y and Z is indirect that

$$\rho_{YZ} = \rho_{XY}\rho_{XZ}$$

We have used ρ's to indicate the fact that these exact relationships can be expected only in the population and that the values of sample r's will ordinarily deviate from this predicted relationship because of sampling fluctuations. Similarly, it can be shown that for cases (e) and (f) we shall have

$$\rho_{XY} = \rho_{XZ}\rho_{YZ}$$

Since the absolute values of correlation coefficients cannot be greater than unity, it is clear that in the first case the numerical value of ρ_{YZ} must be less than that of either of the other coefficients unless one of these values happens to be unity. In this special case, of course, one of the variables can be predicted exactly from one of the others, and we have essentially a problem in only two variables.

Looking more closely at the first of these equations which applies to Fig. 19.5g and h, we see immediately that were this equation to hold, the partial correlation (in the population) between Y and Z controlling for X would vanish since the numerator in the formula for the partial would then be zero. Thus if either (g) or (h) should hold, the value of $r_{YZ \cdot X}$ should be zero or very close to zero if we allow for sampling errors. Similarly, for both (e) and (f) the partial between X and Y controlling for Z can be expected to be approximately zero. What are the implications of these facts? Confining our attention to a comparison between (e) and (f), since the relationship between (g) and (h) is directly comparable if X and Z are interchanged, we see that in (f) we would interpret the relationship between X and Y as spurious since Z is operating to cause variation in both X and Y. This situation is exactly that described in the example of ice-cream consumption X and delinquency rates Y. Because we suspect the relationship between these two variables to be due to a third variable, income level Z, we control for Z to see if the correlation between X and Y is reduced to approximately zero. If (f) in fact is the correct model, we have just seen mathematically that this will happen.

We have also seen, however, that the partial would have been zero if

Fig. 19.5*e* were the correct model. In (*e*) we have *Z* operating as an intervening variable in the sense that *X* causes *Z* which in turn causes *Y*. But does it make any sense to control for *Z* under these circumstances? In one sense it does not. For if *X* is actually a cause of *Z*, how can we imagine ourselves holding *Z* constant while still varying *X*? Certainly it does not make sense to think in terms of obtaining residuals by taking out that portion of the variation in *X* that is "due to" *Z* when *Z* is an effect of *X*. Nevertheless, it may make sense to control for *Z* if we are trying to show that there is no causal connection between *X* and *Y* except through the intervening variable *Z*. The manipulation of statistical formulas is no substitute for knowing what one is doing. In this instance, knowing what one is doing consists of being in a position to choose between models (*e*) and (*f*) by going beyond the statistical information available and making an assumption about the direction of the arrow between *X* and *Z*.

So far we have ignored situation (*d*) in Fig. 19.5 in which arrows go from both *X* and *Z* to *Y* but in which there is no direct relationship between *X* and *Z*. What happens in this case if we control for *Z*? We note first that it makes sense to control for *Z* here because *Z* is conceived as a completely independent variable that also affects *Y*. From the point of view of the relationship between *X* and *Y*, it is operating as a disturbing influence. It is an "outside" variable that is producing essentially random variations in *Y* with respect to variations in *X*. We would therefore expect that controlling for *Z* would increase the magnitude of the relationship between *X* and *Y*. It can be shown mathematically that if we make the required assumptions about other variables, the correlation in the population between *X* and *Z* will be zero. Incidentally, this fact will enable us to distinguish (*d*) empirically from each of the other situations we have been discussing. This is the situation, then, in which the control variable is unrelated to one of the other variables, and we have already seen that in this case the partial will be larger in absolute value than the total correlation, a fact which is consistent with common sense. It is also the situation we faced in two-way analysis of variance, where the condition of equal subcells implied complete independence between the row and column variables, and where we also saw that a control for one of the variables reduced the unexplained sum of squares without reducing the variation explained by the other independent variable.

There is another type of control situation which has not been discussed but which can be handled briefly since there are few, if any, instances in which one would be tempted to use a control. Suppose in either of situations (*e*) or (*h*) that one were to relate the intervening and dependent variables, controlling for the independent variable. In (*h*), for example, what would happen if we were to obtain the partial correlation between X and Y controlling for Z? It can be shown algebraically that the resulting partial would be smaller in magnitude than the total correlation. This is consistent with the intuitive notion that in holding constant the independent variable one necessarily reduces the variation in the intervening variable, thereby weakening the relationship with the dependent variable. Again, it would make little sense to carry out such an operation. Our interest would ordinarily center on the question of whether or not a direct link exists between X and Y and not on the problem of antecedent causes of X. It can be shown, however, that if we had inadvertently controlled for Z in (h), we would *not* have systematically affected the *slope* estimate b_{yx} except for the fact that we would have increased its sampling error.

Extensions to four or more variables are straightforward, provided we restrict ourselves to one-way causation. It can be shown that whenever there is no direct link between two variables, there will be some higher-order partial between these variables that should be approximately zero, except for sampling errors. In general, we should control for all antecedent and intervening variables in order to make the appropriate partial correlation disappear, but we should be careful to avoid controlling for variables that are assumed to be dependent on both variables under consideration. For example, in the model

$$
\begin{array}{ccc}
X_1 & \!\!\!\to\!\!\! & X_2 \\
\downarrow & & \downarrow \\
X_3 & \!\!\!\to\!\!\! & X_4
\end{array}
$$

it will be necessary to control for *both* X_2 and X_3 in order to reduce the partial $r_{14 \cdot 23}$ to zero. Likewise the model predicts that $r_{23 \cdot 1} = 0$ (except for sampling error) but we would *not* expect that $r_{23 \cdot 14} = 0$, nor would it make any sense to control for X_4 in this instance. (For further discussion see [2].)

Several notes of caution are again necessary. As in the three-variable case, there will always be alternative models that predict exactly the same empirical intercorrelations, and one will have to rely on knowledge of temporal sequences or a priori assumptions in order to choose among these alternatives. Also, the existence of random and nonrandom measurement errors will invalidate the predictions of any given model. As we noted in the previous chapter, random measurement error in an independent variable will attenuate correlations between it and other variables. In the case of *multiple* regression, whenever independent variables are highly intercorrelated, random measurement errors in some will tend to *increase* the apparent effects of those variables with which they are most highly intercorrelated. Thus measurement errors in the presence of high intercorrelations among independent variables are especially likely to lead to faulty inferences.

From the above remarks it should be obvious that if one adds variables to a regression equation, the partial correlations can be expected to change according to the nature of the intercorrelations among the independent variables. Exactly the same applies to the partial slopes and standardized slopes that are discussed in the following section. We are assuming that the error or residual term for the regression equation is uncorrelated with each of the independent variables in that equation. In causal terms, this means that factors that are major causes of the dependent variable are assumed not to be systematically related to the independent variables. If we are able to locate variables contributing to this error term, and if we bring them into the equation explicitly, then such variables should *not* be related to the original independent variables, apart from sampling error, and the partial slopes should not be systematically affected. The partial correlations, on the other hand, will be increased in numerical value owing to the fact that some of the unexplained variance will have been removed. However, if the additional variables brought into the equation are systematically related to the original independent variables, then we must anticipate that all coefficients will be affected.

19.4   Multiple Least Squares and the Beta Coefficients

We have been using partial correlations to indicate the degree of relationship between a dependent variable and an independent variable, controlling for one or more other independent variables. If we have a large number of independent variables, we can obtain a measure of their individual direct effects by relating the dependent variable to each independent variable in turn, always controlling for the remaining variables. Earlier, in our discussion of multiple regression and least squares, we also noted that the b's and β's which appear in our equations linking Y to the independent variables can in a sense be interpreted as partials. It will be recalled that they represent the slopes of the regression or least-squares equations in the dimension of the appropriate independent variable, i.e., with all other independent variables held constant. Each coefficient therefore represents the amount of change in Y that can be associated with a given change in one of the X's with the remaining independent variables held fixed. Noting this similarity to the partial correlation coefficients, it would not be surprising if the formulas used in computing these partial b's turned out to be very similar to those used in obtaining partial r's and if, furthermore, these slopes could be used to give a measure of the direct effects of each of the independent variables in determining variation in Y.

We must again modify our notation in order to distinguish among the large number of possible combinations of slopes. Referring to our variables simply as 1, 2, 3, and so forth, we use the symbol $b_{12 \cdot 3}$ if we are predicting variable 1 from variables 2 and 3 and if we are referring to the coefficient of the second variable. The coefficient $b_{12 \cdot 3}$ must be distinguished from $b_{21 \cdot 3}$ which would be used if the second variable were taken as the dependent variable. In both cases, the fact that the number three has been placed to the right of the dot indicates that the third variable has been controlled. Similarly, $b_{12 \cdot 34}$ is used to indicate the coefficient of the second variable in a prediction equation in which the first variable is taken as the dependent variable and which involves two control variables. In this case, the least-squares equation would be given as

$$X_1 = a_{1 \cdot 234} + b_{12 \cdot 34}X_2 + b_{13 \cdot 24}X_3 + b_{14 \cdot 23}X_4$$

where the subscript for a indicates that we are predicting to variable 1 from variables 2, 3, and 4. The reason we have found it convenient to depart from the practice of denoting the dependent variable by Y is to make use of a simpler set of subscripts for keeping track of the various b's.

The computing formulas for $a_{i \cdot jk}$ and $b_{ij \cdot k}$ are as follows

$$a_{i \cdot jk} = \bar{X}_i - b_{ij \cdot k} \bar{X}_j - b_{ik \cdot j} \bar{X}_k \qquad (19.7)$$

and
$$b_{ij \cdot k} = \frac{b_{ij} - (b_{ik})(b_{kj})}{1 - b_{jk}b_{kj}} \qquad (19.8)$$

Notice that although the denominator in (19.8) differs in form from that in the formula for $r_{ij \cdot k}$, the numerator is essentially similar in nature. Recalling that

$$r_{jk}{}^2 = b_{jk}b_{kj}$$

we see that even the denominators are not too dissimilar in form. In using this formula to obtain the partial b's, however, you must be careful to distinguish between b_{jk} and b_{kj} since the subscripts can no longer be interchanged.

The extension to higher-order partials is straightforward (see [5]).

The equations for $a_{i \cdot jkl}$ and $b_{ij \cdot kl}$ can be written as

$$a_{i \cdot jkl} = \bar{X}_i - b_{ij \cdot kl}\,\bar{X}_j - b_{ik \cdot jl}\,\bar{X}_k - b_{il \cdot jk}\,\bar{X}_l \qquad (19.9)$$

and
$$b_{ij \cdot kl} = \frac{b_{ij \cdot k} - (b_{il \cdot k})(b_{lj \cdot k})}{1 - b_{jl \cdot k}(b_{lj \cdot k})} \qquad (19.10)$$

Similarly true in computing higher-order partial correlations, as the number of variables becomes large, the use of these formulas may involve considerably more work than that required by the abbreviated Doolittle or Dwyer square-root method. Usually, of course, it will be most convenient to rely on standard computer programs to obtain these coefficients.

A partial slope can be interpreted as the hypothetical change that would occur in the dependent variable *if* one of the independent variables were to change by one unit and *if* the other independent variables were to remain constant. This can therefore be interpreted as a measure of the direct effect of the independent variable on the dependent variable; if the partial slope is zero, this would imply no direct effect. But since we have not specified the causal connections among the independent variables themselves, having merely allowed for their intercorrelations, we cannot say anything about the total impact of each variable. For example, if the first independent variable is a cause of the second, then a change in this first variable would also produce a change in the second, so that there could be a direct as well as an indirect effect. Thus we cannot assess the relative importance of each variable unless we know more about the causal structure of the total system. This would require working with an entire *set* of equations, one equation for each variable that is taken as dependent on any of the others. Unfortunately, ordinary least squares are generally not appropriate for such systems of equations. (See [4] and [12].)

As long as one is not interested in generalizing beyond the limits of a single population, it is sometimes desirable to obtain an asymmetric measure of the direct effects of each independent variable that does not depend on the units of measurement used. In effect, we thereby obtain a measure of the actual direct effect in the particular case of the population under study, given that some independent variables will vary more than others. One variable may be measured in terms of dollars, another in years. It would therefore be meaningless to compare a unit change in one variable with a unit change in another.

If each variable is standardized by dividing by its standard deviation, just as was done in obtaining the standard normal curve, we can obtain adjusted slopes which are comparable from one variable to the next. We thus measure changes in the dependent variable in terms of standard-deviation units for each of the other variables, a fact which assures us of the same variability in each of these variables. These adjusted partial slopes are thus standardized b's which are often called *beta weights* and in most simple causal models involving one-way causation are equivalent to what have been called *path coefficients*. (See [14].)

Unfortunately we are again involved with notational inconsistencies. These beta weights are not the same as the β's in the regression equation which refer to characteristics of the *population* and which have not been adjusted for differences in variability. The beta weights are obtained from *sample* data and are simple functions of the partial b's. The general formulas for $\beta_{ij \cdot k}$ and $\beta_{ij \cdot kl}$ are

$$\beta_{ij \cdot k} = b_{ij \cdot k} \frac{s_j}{s_i} \tag{19.11}$$

and
$$\beta_{ij \cdot kl} = b_{ij \cdot kl} \frac{s_j}{s_i} \tag{19.12}$$

Thus the beta weight can be obtained by multiplying the comparable b by the ratio of the standard deviation of the independent variable (not controlled) to that of the dependent variable.

The comparability of beta weights and partial correlation coefficients can be seen from the formula

$$\beta_{ij \cdot k} = \frac{r_{ij} - r_{ik} r_{jk}}{1 - r_{jk}^2} \tag{19.13}$$

The two measures differ only in their denominators. In fact, we see immediately that

$$r_{ij \cdot k}^2 = (\beta_{ij \cdot k})(\beta_{ji \cdot k})$$

since $\beta_{ji \cdot k}$ differs from $\beta_{ij \cdot k}$ only in that the denominator of r_{jk}^2 will be replaced by r_{ik}^2. Since the beta weights and partial correlations represent somewhat different types of measures of association, they will not give exactly the same result although usually they will rank variables in the same order of importance. The partial correlation is a measure of the *amount of variation explained* by one independent variable after the others have explained all they could. The beta weights, on the other hand, indicate *how much change* in the dependent variable is produced by a standardized change in one of the independent variables when the others are controlled.

19.5   Multiple Correlation

Since our primary interest may be in the explanatory power of a number of independent variables taken together, rather than in the relationship between the dependent variable and each of the independent variables taken separately, we may prefer to make use of the *multiple correlation coefficient* which is a measure of the goodness of fit of the least-squares surface to the data. Just as the square of the zero-order correlation coefficient indicated the percentage of the variation explained by the best-fitting straight line, the square of the multiple correlation coefficient can be used to give the percentage of variation explained by the best-fitting equation of the form

$$Y_p = a + b_1 X_1 + b_2 X_2 + \cdots + b_k X_k$$

Another way of conceiving of multiple correlation is that it represents the *zero-order correlation* between the *actual values* obtained for the dependent variable and those *values predicted from the least-squares equation*. If all of the points are exactly on the least-squares surface, the actual and predicted values will coincide, and the multiple correlation will be unity. The greater the scatter about the least-squares equation, the lower the correlation between actual and predicted values.

The formula for the multiple correlation can easily be developed using the fact that the square of the multiple will be equal to the percentage of the variation explained by all of the independent variables. Again, it should be emphasized that a linear-type model is being assumed. In writing a formula for the multiple correlation, we first let one of the independent variables do all of the explaining it can. We then permit the second independent variable to go to work on that portion of the variation left unexplained by the first. In order to avoid duplication, however, we must control for this first independent variable. We then permit the third to explain all it can of the remainder, now controlling for both of the first two independent variables. The process can be extended indefinitely. For the present, we confine ourselves to the three-variable case involving only two independent variables. If we take the first variable as the dependent variable and denote the multiple correlation coefficient by $R_{1\cdot 23}$, we may write

$$R_{1\cdot 23}^2 \quad = \quad r_{12}^2 \quad + \quad r_{13\cdot 2}^2 \quad (1 - r_{12}^2)$$

$$\begin{pmatrix} \text{Proportion} \\ \text{explained} \\ \text{by 2 and 3} \end{pmatrix} = \begin{pmatrix} \text{proportion} \\ \text{explained} \\ \text{by 2} \end{pmatrix} + \begin{pmatrix} \text{additional} \\ \text{proportion} \\ \text{explained} \\ \text{by 3} \end{pmatrix} \begin{pmatrix} \text{proportion} \\ \text{unexplained} \\ \text{by 2} \end{pmatrix} \quad (19.14)$$

Notice that multiple correlations have only one figure to the left of the dot, this figure indicating the dependent variable. Numbers to the right indicate those independent variables which are being used to explain the variation in the dependent variable. The general formula (for three variables) can thus be written

$$R^2_{i \cdot jk} = r^2_{ij} + r^2_{ik \cdot j}(1 - r^2_{ij})$$

$$= r^2_{ik} + r^2_{ij \cdot k}(1 - r^2_{ik}) \tag{19.15}$$

It is irrelevant, of course, which of the two independent variables is used first in the formula, as long as this variable is controlled in subsequent terms in the equation.

We are dealing with the squares of both the total and partial correlations since we are obtaining the percentages of variation explained. Therefore we do not have to worry about the sign of these correlations. In fact, the direction of the multiple has no meaning since it involves correlations with a number of variables, some of these correlations being positive and others possibly negative. By convention we always take the positive square root of R^2 in denoting the multiple correlation coefficient.

If we solve Eq. (19.14) for the partial $r^2_{13 \cdot 2}$ we obtain

$$r^2_{13 \cdot 2} = \frac{R^2_{1 \cdot 23} - r^2_{12}}{1 - r^2_{12}} \tag{19.16}$$

This enables us to see the relationship between multiple and partial correlation coefficients in a somewhat different perspective. In the numerator we have subtracted the proportion of the variation in 1 explained by 2 alone from the proportion explained by both 2 and 3 acting together $(R^2_{1 \cdot 23})$. The result is the increment explained by 3 after 2 has been allowed to operate. When this increment is divided by the proportion of variation left unexplained by 2, we obtain the partial between 1 and 3 controlling for 2. This is consistent with our earlier interpretation of the partial correlation coefficient.

Several alternative but equivalent formulas for $R^2_{1 \cdot 23}$ can be derived from Eq. (19.14). Subtracting both sides of this equation from unity we get

$$1 - R^2_{1 \cdot 23} = 1 - r^2_{12} - r^2_{13 \cdot 2}(1 - r^2_{12})$$

$$= (1 - r^2_{12})(1 - r^2_{13 \cdot 2}) \tag{19.17}$$

This equation indicates that we can write the proportion of variation left unexplained by 2 and 3 together as the product of the proportion unexplained by 2 and that unexplained by 3, controlling for 2.

The formula for the multiple can also be written completely in terms of zero-order correlations. Making use of Eq. (19.3) for $r_{13 \cdot 2}$ in terms of zero-order coefficients and simplifying the resulting algebraic expression, we get

$$R^2_{1 \cdot 23} = \frac{r^2_{12} + r^2_{13} - 2r_{12}r_{13}r_{23}}{1 - r^2_{23}}$$

or, in general

$$R^2_{i \cdot jk} = \frac{r^2_{ij} + r^2_{ik} - 2r_{ij}r_{ik}r_{jk}}{1 - r^2_{jk}} \tag{19.18}$$

In particular, if the correlation between the two independent variables j and k happens to be zero, we get

$$R^2_{i \cdot jk} = r^2_{ij} + r^2_{ik}$$

Several relationships between the multiple R and the various total correlations can now be noted. Obviously, R cannot be less in magnitude than any of the total correlations since it is impossible to explain *less* variation by adding further variables. Ordinarily, of course, the multiple R will be larger than any of the total r's. Its maximum value relative to the total coefficients usually occurs when the intercorrelations between independent variables are all zero. As we have just seen, the square of the multiple correlation will in this case be equal to the sum of the squares of the other correlations. On the other hand, if the intercorrelations among independent variables are quite high in magnitude, the multiple R will ordinarily not be much larger than the largest total correlation with the dependent variable. In other words, if we wish to explain as much variation in the dependent variable as possible, we should look for independent variables which are relatively unrelated to each other but which have at least moderately high correlations with the dependent variable. Put another way, if we have two highly interrelated independent variables, the second will be explaining essentially the same variation as the first since there will be considerable overlap. If they are uncorrelated, they will each explain a different portion of the total variation.

There is another important reason for preferring independent variables that are not highly intercorrelated. Not only will there be less overlap in explained variance and therefore less ambiguity in our causal interpretations of their supposed effects, but to the degree that independent variables are highly intercorrelated both partial correlations and slope estimates will be increasingly sensitive to sampling and measurement errors. This difficulty is referred to as *multicollinearity* in the econometrics literature. (See [4] and [12].) The problem is especially serious when blocks of highly intercorrelated independent variables are being used, and when these blocks differ with respect to the number of variables they contain. (See [10].) It can be shown, for example, that very small differences in total correlations with the dependent variable imply rather large differences in partial correlations and slope estimates, so that if one relies on the relative magnitudes of these partial coefficients, he can expect to find rather large differences from one sample to the next, or among replications that utilize somewhat different measuring instruments. The implication is that whenever independent variables are highly intercorrelated, it will be necessary to have *both* large samples and accurate measurement.

As a numerical example of the computation of the multiple R, let us see how much of the variation in discrimination can be explained by both per cent Negro and per cent urban. Using Eq. (19.14) we get

$$R_{1\cdot23}^2 = r_{12}^2 + r_{13\cdot2}^2(1 - r_{12}^2) = (.536)^2 + (.332)^2[1 - (.536)^2]$$

$$= .2873 + (.1102)(.7127) = .3658$$

Hence
$$R_{1\cdot23} = .605$$

Therefore, per cent urban explains very little variation over and above that explained by per cent Negro.

As a check on our computations, we note that the same result should be obtained if we allow per cent urban to operate first. We then get

$$r_{12\cdot3} = \frac{r_{12} - r_{13}(r_{23})}{\sqrt{1 - r_{13}^2}\,\sqrt{1 - r_{23}^2}} = \frac{.536 - (.139)(-.248)}{\sqrt{1 - (.139)^2}\,\sqrt{1 - (-.248)^2}} = .595$$

Thus
$$R_{1\cdot23}^2 = r_{13}^2 + r_{12\cdot3}^2(1 - r_{13}^2)$$

$$= (.139)^2 + (.595)^2[1 - (.139)^2] = .3667$$

and hence
$$R_{1\cdot23} = .605$$

The formulas for the multiple correlation coefficient can also readily be extended to any number of independent variables. In introducing a third independent variable, denoted as variable X_4, we merely add to the formula for $R^2_{1\cdot23}$ a term involving the product of the square of the partial between 1 and 4, controlling for 2 and 3, and the proportion of variation left unexplained by 2 and 3. Thus

$$R^2_{1\cdot234} = r^2_{12} + r^2_{13\cdot2}(1 - r^2_{12}) + r^2_{14\cdot23}[1 - r^2_{12} - r^2_{13\cdot2}(1 - r^2_{12})]$$

$$= R^2_{1\cdot23} + r^2_{14\cdot23}(1 - R^2_{1\cdot23}) \tag{19.19}$$

We can therefore keep adding to the proportion of variation explained as long as we control for all variables previously used and provided that we permit the new partial to work only on that portion of the variation left unexplained by those variables which have preceded it. Note, incidentally, the parallel with what we did in analysis of variance. As we shall presently see, we can make use of this fact in tests of significance for both multiple and partial correlation. If we were to add a fifth variable, we would get

$$R^2_{1\cdot2345} = R^2_{1\cdot234} + r^2_{15\cdot234}(1 - R^2_{1\cdot234})$$

We can again solve these equations for the partial coefficients. For example, from (19.19) we have

$$r^2_{14\cdot23} = \frac{R^2_{1\cdot234} - R^2_{1\cdot23}}{1 - R^2_{1\cdot23}} \tag{19.20}$$

indicating that the partial between 1 and 4, controlling for 2 and 3, can be interpreted as the ratio of the proportion of additional variation explained by 4 over and above that explained by 2 and 3 to the proportion of variation left unexplained by the latter two variables. We can also extend Eq. (19.17) to cover more variables. Thus

$$1 - R^2_{1\cdot234} = (1 - r^2_{12})(1 - r^2_{13\cdot2})(1 - r^2_{14\cdot23})$$

and, in general,

$$1 - R^2_{1\cdot234\cdots k} = (1 - r^2_{12})(1 - r^2_{13\cdot2}) \cdots (1 - r^2_{1k\cdot234\cdots(k-1)}) \tag{19.21}$$

The multiple-partial coefficient Sometimes it is desirable to compute a multiple correlation between a dependent variable and several independent variables, controlling for one or more independent variables. Suppose, for example, that one is trying to predict actual family size from a number of independent variables. Obviously, certain variables such as the duration of marriage and the age of wife at marriage need to be taken into consideration. On the other hand, they are so obvious that throwing them into the general multiple coefficient might obscure the effects of the remaining variables. Interest might therefore be focused on the variation in family size after such theoretically unimportant variables have explained all of the variation they could. Following Croxton and Cowden [6], we indicate the multiple-partial between variable 1 (dependent) and 2 and 3, controlling for 4, by $r_{1(23)\cdot 4}$. The formula in this case becomes

$$r_{1(23)\cdot 4}^2 = \frac{R_{1\cdot 234}^2 - r_{14}^2}{1 - r_{14}^2}$$

The above formula for the multiple-partial is a simple extension of the formulas we have been using for multiple and partial correlations. We first let control variable 4 do all of the explaining it can. We then note that $R_{1\cdot 234}^2$ represents the proportion of variation explained by all of the three independent variables taken together. The difference, then, must be due to variables 2 and 3. The numerator thus represents the proportion of variation explained by 2 and 3 over and above that explained by 4. But since we must work only on that variation left unexplained by the control variable, we divide by the quantity $1 - r_{14}^2$. Making use of the principle that we allow all control variables to operate first, we can write the general formula for the multiple-partial as

$$r_{i(jk\cdots n)\cdot tu\cdots w}^2 = \frac{R_{i\cdot jk\cdots w}^2 - R_{i\cdot tu\cdots w}^2}{1 - R_{i\cdot tu\cdots w}^2} \qquad (19.22)$$

For example,

$$r_{3(56)\cdot 124}^2 = \frac{R_{3\cdot 12456}^2 - R_{3\cdot 124}^2}{1 - R_{3\cdot 124}^2}$$

The multiple-partial does not seem to have been used very frequently in sociological research, perhaps because of its lack of familiarity to persons in the field. As a measure which enables one to handle both multiple and partial correlation problems simultaneously, its potential utility would seem to be great, however. In the next section we shall consider another kind of use for this measure.

19.6 Multiple Regression and Nonlinearity

All of our work thus far in the chapter has been based on the assumption of linear-type models. In the previous chapter we took up a test for nonlinearity, but we had very little to say about the form of the non-linear relationship except in the case of logarithmic transformations. In other words, we simply tested for the existence of a *departure* from linearity, but we made no test for the form of the curve. Although the general problem of nonlinearity is beyond the scope of this text, we can examine briefly how the techniques of multiple regression and least squares can be modified slightly so as to enable us to handle certain types of problems involving nonlinearity.

As pointed out in the previous chapter, there are an extremely large number of different forms a nonlinear relationship can take. Let us consider equations of the type

$$Y = a + b_1X + b_2X^2 + b_3X^3 + \cdots + b_kX^k \qquad (19.23)$$

If all of the coefficients b_2, b_3, . . . , b_k are zero, the equation is reduced to the familiar linear form. In other words, a straight line can be considered as a special case of this general type of curve which is referred to as a *polynomial*. Similarly, if all of the coefficients except a, b_1, and b_2 are zero, we get a polynomial of the second degree. The degree of the polynomial refers to the highest exponent of X which has a nonzero coefficient.

Polynomials have a very important property which enables us to tell the degree of the equation that is likely to be the most appropriate to one's data. Note that a polynomial of the first degree is a straight line with no bends in it. It turns out that a second-degree equation will have exactly one bend in it and, in fact, describes the particular curve that we call a parabola. A third-degree polynomial will have two bends, a fourth-degree three bends, and so on. If we ignore certain degenerate cases in which the "bends" do not behave properly, we can diagram these second-, third-, and fourth-degree equations as in Fig. 19.6. The direction

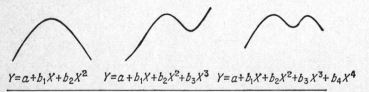

$Y = a + b_1X + b_2X^2$ $Y = a + b_1X + b_2X^2 + b_3X^3$ $Y = a + b_1X + b_2X^2 + b_3X^3 + b_4X^4$

Figure 19.6   Forms of second-, third-, and fourth-degree polynomials.

in which the parabola or higher-degree curves "open up" will depend upon the sign of the coefficients. The important point is that there will always be one less bend than is indicated by the degree of the equation.

We sometimes get empirical curves which look like one or another of these polynomials, although seldom, if ever, do we need to go higher than a third-degree equation. A simple parabola may often give a reasonably good fit to the data, especially when we realize that the curve may be quite flat and that our data need not extend far enough to complete the bend. Thus the data might be similar to those indicated in Fig. 19.7. Here, even though there may be no theoretical reason to expect scores to go down again once we have gone a certain distance along the X axis, the parabola may be a reasonable fit *within the limits of variation given in the problem*. It is thus quite possible that a least-squares parabola might fit the data much better than a straight line.

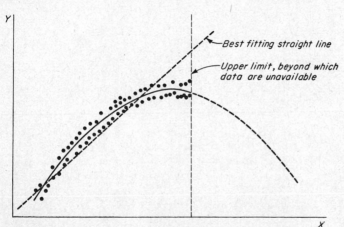

Figure 19.7 Hypothetical data best fitted by a parabola.

Suppose this turns out to be the case. How can the problem be handled? You have undoubtedly already noticed the similarity between the formula for the general polynomial and the formula for the least-squares equation involving more than one independent variable. The only difference is that in place of X_2 we have written X^2 and so on. Now suppose we were to conceptualize X^2 as a separate and distinct variable from X. As long as we are making use of abstract symbols, this is entirely possible, although admittedly this practice would make little sense in terms of a concrete variable. The mathematics of the situation turns out to be identical, however. For example, if we suspect that the relationship between discrimination and per cent Negro can be fitted by a second-degree curve, we take per cent Negro as one independent variable X_1 and (per cent

Negro)^2 as the second independent variable X_2. Therefore the second-degree equation

$$Y = a + b_1X + b_2X^2$$

which is difficult to handle by least squares is reduced to the familiar equation

$$Y = a + b_1X_1 + b_2X_2$$

To obtain a measure of goodness of fit to the parabola, we can now use the multiple correlation between Y and X_1 and X_2. The difference between the square of this multiple correlation and the square of the total r (assuming linearity) will give us a measure of the degree to which we have improved our ability to predict discrimination by using a second-degree equation rather than a straight line.

In principle, the above procedure can be extended in several ways. Third- and higher-degree equations might be used in order to obtain a somewhat better fit. Also, other variables can be added to the picture. For some of these independent variables a nonlinear model can be assumed, for others a linear one. In predicting discrimination rates from a number of independent variables, we might find that somewhat better prediction equations could be obtained assuming nonlinear models for some of the variables. In particular, perhaps the relationship between discrimination and per cent Negro may be of parabolic form, whereas all the remaining independent variables are linearly related to discrimination. The multiple least-squares equation therefore would take the form

$$Y = a + (b_1X_1 + b_2X_2) + b_3X_3 + \cdots + b_kX_k$$

where the two terms within the parentheses are needed to describe the (nonlinear) relationship between discrimination and per cent Negro. In this case also, the X_2 variable again represents (per cent Negro)^2. Conceivably, some of the other X's in the equation may also be used to indicate nonlinear relationships between discrimination and other variables.

In the above example suppose that we wished to obtain the partial correlation between discrimination and per cent Negro, controlling for the remaining variables. Since X_1 and X_2 have been used to refer to the first and second powers of per cent Negro, it would not make sense to relate Y and X_1, controlling for all of the remaining "variables," including X_2. Rather, we need to obtain the multiple correlation between Y and both X_1 and X_2, controlling for X_3, X_4, . . . , X_k. The multiple-partial coefficient can be used to accomplish this purpose.

Handling interaction as cross-products   In two-way analysis of variance, analysis of covariance (see Chap. 20), and in connection with nominal-scale dependent variables, we have conceived of statistical interaction as involving *any* departure from simple additivity. One very obvious alternative to an additive model is a multiplicative relationship of the type that would be suggested by verbal arguments to the effect that in order to have Y "present" one must have *both* X_1 and X_2 "present." When one moves beyond simple dichotomies, this idea generalizes to the notion that Y may be a multiplicative function of X_1 and X_2. One reasonably general formulation of such a relationship would be the equation

$$Y = (\alpha_1 + \beta_1 X_1)^{\gamma_1}(\alpha_2 + \beta_2 X_2)^{\gamma_2}$$

where the gamma exponents could be either positive, in which case multiplication would be implied, or negative, implying division. The function could of course be turned into an additive one by making a logarithmic transformation of all variables, and the general principle could readily be extended to more than two independent variables.

Suppose, as a reasonable approximation, we were to take both exponents to be unity so that the equation reduced to

$$Y = (\alpha_1 + \beta_1 X_1)(\alpha_2 + \beta_2 X_2) = \alpha_1\alpha_2 + \alpha_2\beta_1 X_1 + \alpha_1\beta_2 X_2 + \beta_1\beta_2 X_1 X_2$$

We immediately see that by adding a product term involving $X_1 X_2$, we can handle this kind of simple multiplicative model while retaining the additive format. We merely relabel $X_1 X_2$ as X_3, construct our measure of X_3 accordingly, and proceed. For example, we would want to measure the degree to which X_3 adds to the explained variance, and we could test for the significance of this additional term as indicated in the next section. Had we begun with three independent variables, we could have formed three product terms $X_1 X_2$, $X_1 X_3$, and $X_2 X_3$ to tap the three first-order interactions and a triple product $X_1 X_2 X_3$ to handle the higher-order interaction.

Several notes of caution are necessary. First, the use of cross-product terms is justified on the grounds that the "true" relationship is multiplicative rather than additive, whereas "nonadditivity" refers to *any* kind of departure from additivity. We thus have a somewhat more restrictive measure of interaction than was obtained in connection with analysis of variance, and it is possible that different interaction terms might have worked better (for example, $X_1 \log X_2$, $X_1 \cos X_2$, or $e^{X_1}\log X_2$). Second, if we take $X_3 = X_1 X_2$, we must realize that X_3 is an exact nonlinear function of X_1 and X_2, and therefore the product-moment correlations of X_3 with both X_1 and X_2 will ordinarily be very high. Therefore we will have a problem of multicollinearity on our hands and cannot have too much faith in our estimates of the coefficients of the X_i terms. This problem will be particularly serious when one begins with five or six independent variables and wishes to allow for all possible interactions. If the original variables are themselves highly intercorrelated, or belong in blocks, then the cross-product terms will be related to these blocks in peculiar ways. (See [1].) In such instances it may be reasonable to measure the extent to which the entire set of cross-product terms adds significantly to the explained variance by using the multiple-partial coefficient, or by comparing the multiples with and without the product terms. But the assessment of the effects of particular cross-product terms may be too risky in view of the large sampling errors involved.

There are obviously many more uses and possible extensions of multiple correlation and regression techniques than can be discussed in a general text. A sufficient number of the more elementary basic principles have been given, however, that you should be able to consult intelligently with a statistician if more complicated problems should arise.

19.7   Significance Tests and Confidence Intervals

It will of course be necessary to test both multiple and partial coefficients for significance. The null hypotheses and assumptions will be similar to those made in the case of total correlation. A random sample will as usual be assumed. The assumption of a multivariate normal distribution will assure us that each variable is normally distributed about the others, that variances are equal, and that the regression equation will have the form indicated by Eq. (19.1).[4] Having made these assumptions, we can make use of analysis-of-variance tests for the significance of various partial and multiple coefficients. We shall first take up tests for the significance of multiple correlations since these are conceptually simpler than tests for partials.

Since the square of the multiple correlation always represents the proportion of the total variation explained by the independent variables acting together, we have in effect divided this total variation into two portions, the explained and unexplained sums of squares. The analysis-of-variance table will therefore always be similar to Table 19.1.

---

[4] Once more it should be emphasized that all of the X_i need not have normal distributions, as long as the dependent variable is normally distributed about all combinations of fixed levels of the independent variables with the same variance σ^2. In other words, we assume that the disturbance term ϵ_i is normally distributed with constant variance.

Table 19.1   Analysis-of-variance test for significance of multiple correlation

	Sums of squares	Degrees of freedom	Estimate of variance	F
Total	Σx_1^2	$N - 1$		
Explained	$R^2 \Sigma x_1^2$	k	$\dfrac{R^2 \Sigma x_1^2}{k}$	$\dfrac{R^2}{1 - R^2} \dfrac{N - k - 1}{k}$
Unexplained	$(1 - R^2)\Sigma x_1^2$	$N - k - 1$	$\dfrac{(1 - R^2)\Sigma x_1^2}{N - k - 1}$	

In Table 19.1 we have indicated the dependent variable by X_1 and have let k represent the number of independent variables. If R involves, say, a dependent variable and three independent variables, there will be four parameters in the regression equation that must be estimated. We would therefore lose 4 or $(k + 1)$ degrees of freedom in using the least-squares equation to estimate the dependent variable. Thus the degrees of freedom associated with the error term will in general be

$$N - (k + 1) = N - k - 1$$

The degrees of freedom associated with the explained sum of squares can then be obtained by subtraction. Since the degrees of freedom for the explained and unexplained sums of squares will always turn out to be k and $N - k - 1$, respectively, we may write a general formula for F. Note that as was true with total correlations, the factor representing the total sum of squares cancels. We thus get as a general formula for testing for the significance of a multiple R

$$F_{k,N-k-1} = \frac{R^2}{1 - R^2} \frac{N - k - 1}{k} \tag{19.24}$$

It is not necessary, therefore, to set up the analysis-of-variance table in conventional form. Testing for the significance of the multiple correlation we obtained in explaining discrimination from per cent Negro and per cent urban (see page 457), we get

$$F_{2,147} = \frac{.3658}{1 - .3658} \frac{150 - 3}{2} = \frac{.3658}{.6342} \frac{147}{2} = 42.39$$

which is significant at the .001 level.

In testing for the significance of partial coefficients, we operate on the principle of first letting the control variables do all of the explaining they can. We then take that portion of the total sum of squares that is *unexplained* by the control variables and use this as our new total. This latter quantity is then broken into two components, the explained and the unexplained portions, and an F test made by taking the ratio of the estimates of variance based on these last two components. The process is illustrated in Table 19.2 in which we are testing for the significance of $r_{13\cdot2}$ (i.e., H_0: $\rho_{13\cdot2} = 0$).

Table 19.2   Analysis-of-variance test for significance of partial correlation $r_{13\cdot2}$

	Sums of squares	Degrees of freedom	Estimate of variance	F
Total	Σx_1^2	$N - 1$		
Explained by 2	$r_{12}^2\Sigma x_1^2$	1		
Unexplained by 2	$(1 - r_{12}^2)\Sigma x_1^2$	$N - 2$		
Explained by 3	$r_{13\cdot2}^2(1 - r_{12}^2)\Sigma x_1^2$	1	$\dfrac{r_{13\cdot2}^2(1 - r_{12}^2)\Sigma x_1^2}{(1 - r_{13\cdot2}^2)(1 - r_{12}^2)\Sigma x_1^2}$	$\dfrac{r_{13\cdot2}^2(N - 3)}{1 - r_{13\cdot2}^2}$
Unexplained by 3	$(1 - r_{13\cdot2}^2)(1 - r_{12}^2)\Sigma x_1^2$	$N - 3$	$\dfrac{}{N - 3}$	

Notice that the unexplained degrees of freedom decrease by one each time a new variable is added. Again, in the formula for F, the expression simplifies so that it becomes unnecessary to write out the whole table each time we wish to make a test. In the numerical problem we have been using (see page 438), the F value for testing for the significance of the relationship between discrimination and per cent urban, controlling for per cent Negro, becomes

$$F_{1,N-3} = \frac{r_{13\cdot2}^2}{1 - r_{13\cdot2}^2}(N - 3)$$

$$= \frac{(.332)^2}{1 - (.332)^2}(147) = 18.21 \qquad (19.25)$$

Thus the partial is significant at the .001 level.

In extending this procedure, if we wish to test for the significance of $r_{14\cdot23}$, we can take as our new total the unexplained by 2 and 3 combined. This quantity can then be broken into explained and unexplained portions and the F test made as before. Again, all quantities in both numerator and denominator of F will cancel except the factors involving the partials.

Since the degrees of freedom associated with the numerator will always be unity and since those for the denominator will be $N - k - 1$, we may write as a general formula for testing the partial $r_{ij \cdot mn \cdots t}$

$$F_{1, N-k-1} = \frac{r_{ij \cdot mn \cdots t}^2}{1 - r_{ij \cdot mn \cdots t}^2} (N - k - 1) \qquad (19.26)$$

where the total number of variables is $k + 1$.

Notice that in comparing tests for the significance of multiple and partial correlations, the final error term involving the sum of squares unexplained by all variables should be the same in both tables—provided, of course, that the same dependent and independent variables are used. We have already shown this to be the case since we know that

$$1 - R_{1 \cdot 23}^2 = (1 - r_{12}^2)(1 - r_{13 \cdot 2}^2)$$

From Tables 19.1 and 19.2 we see that these expressions are the ones that appear in the bottom rows of the respective tables.

The procedure we have described for testing partial correlations can also be used to test for the significance of the multiple-partial. By now you should be able to verify that in order to test for the significance of $r_{1(23) \cdot 45}$ we would take the sum of squares unexplained by both 4 and 5, then using the square of the multiple-partial to give the proportion of this new sum of squares which is explained by variables 2 and 3.

Confidence intervals can also be computed for both the partial and multiple coefficients by a slight modification of the z transformation procedure described in the previous chapter. We can again convert either type of coefficient into z using the table. The only change required is that the standard error of z is no longer given by

$$\sigma_z = \frac{1}{\sqrt{N - 3}}$$

Instead, we lose one more degree of freedom for each variable added so that the standard error in general becomes

$$\sigma_z = \frac{1}{\sqrt{N - k - 2}} \qquad (19.27)$$

where k represents the total number of independent variables.

We therefore obtain the 95 per cent confidence intervals for $R_{1\cdot23}$ and $r_{13\cdot2}$ as follows:

$$1.96\sigma_z = 1.96\,\frac{1}{\sqrt{146}} = .1622$$

	z	$z_l = z - 1.96\sigma_z$	$z_u = z + 1.96\sigma_z$	r_l	r_u
$R_{1\cdot23} = .605$	.7010	.5388	.8632	.492	.698
$r_{13\cdot2} = .332$	.3451	.1829	.5073	.181	.468

Thus the 95 per cent confidence interval about $R_{1\cdot23}$ goes from .492 to .698, whereas that about $r_{13\cdot2}$ ranges from .181 to .468.

Before closing the chapter, one further important point should be noted. Each time we add another variable to the least-squares equation, we lose only one more degree of freedon. We can therefore add variables with very little loss of efficiency as far as significance tests are concerned. Occasionally, the addition of more variables may lower the significance level because of the fact that they do not help explain enough additional variation to compensate for the loss in degrees of freedom. Nevertheless, we have in multiple and partial correlation a tool which, if properly used, is far more powerful than any of the methods we have previously discussed. If the number of variables used begins to approach the number of cases, however, we can expect to obtain very large multiple correlations simply because we are able to take advantage of chance fluctuations. With 15 cases and 15 variables it will be possible to pass a least-squares surface *exactly* through all of the points even if we assume a linear-type model. The multiple R will therefore automatically be unity. Therefore, like all other statistical techniques, multiple regression and correlational techniques must be used with caution. At this point it should hardly be necessary to point out that except for exploratory purposes, they should not be used unless the required assumptions are met or at least approximated.

Glossary
Beta weights
Multiple correlation
Multiple-partial correlation
Multiple regression equation
Multivariate normal distribution
Partial correlation
Polynomial equation

Exercises

1. Using the data of Exercise 1, Chap. 17:
   a. Obtain the partial correlation between moral integration and heterogeneity, controlling for mobility. Also compute the partial between moral integration and mobility, controlling for heterogeneity. (*Ans.* $-.51$; $-.63$)
   b. Obtain the multiple least-squares equation, taking moral integration as the dependent variable.
   c. What are the beta weights? How do they compare with the partials obtained in (*a*)?
   d. Compute the multiple correlation, taking moral integration as the dependent variable. How would you check your computations? (*Ans.* $R = .64$)
   e. Test for the significance of the partial and multiple correlations computed in parts (*a*) and (*d*). Place 99 per cent confidence intervals about each of these correlations.

2. Write out formulas for $r_{37 \cdot 12456}$, $R^2_{4 \cdot 1235}$, and $r^2_{5(23) \cdot 1467}$. $\left[Ans. \ (b) \ R^2_{4 \cdot 1235} = R^2_{4 \cdot 123} + r^2_{45 \cdot 123}(1 - R^2_{4 \cdot 123}) \right]$

3. Write the formulas for F which would be used in testing for the significance of each of the correlations in Exercise 2 above. $\left[Ans. \ (c) \ F = \dfrac{r^2_{5(23) \cdot 1467}}{1 - r^2_{5(23) \cdot 1467}} \ \dfrac{N-7}{2} \right]$

References

1. Althauser, R. P.: "Multicollinearity and Non-Additive Regression Models," in H. M. Blalock (ed.), *Causal Models in the Social Sciences*, Aldine Publishing Company, Chicago, 1971, chap. 26.

2. Blalock, H. M.: *Causal Inferences in Nonexperimental Research*, University of North Carolina Press, Chapel Hill, 1964, chap. 3.

3. Blalock, H. M.: "Per Cent Non-white and Discrimination in the South," *American Sociological Review*, vol. 22, pp. 677–682, 1957.

4. Christ, Carl: *Econometric Models and Methods*, John Wiley & Sons, Inc., New York, 1966, Part III.

5. Cowden, D. J.: "A Procedure for Computing Regression Coefficients," *Journal of the American Statistical Association*, vol. 53, pp. 144–150, 1958.

6. Croxton, F. E., and D. J. Cowden: *Applied General Statistics*, 3d ed., Prentice-Hall, Inc., Englewood Cliffs, N.J., 1967, chap. 21.

7. Davis, J. A.: "A Partial Coefficient for Goodman and Kruskal's Gamma," *Journal of the American Statistical Association*, vol. 62, pp. 189–193, 1967.

8. Draper, N. R., and H. Smith: *Applied Regression Analysis*, John Wiley & Sons, Inc., New York, 1966, chaps. 5–10.

9. Dwyer, P. S.: *Linear Computations*, John Wiley & Sons, Inc., New York, 1951.

10. Gordon, Robert: "Issues in Multiple Regression," *American Journal of Sociology*, vol. 73, pp. 592–616, 1968.

11. Hagood, M. J., and D. O. Price: *Statistics for Sociologists*, Henry Holt and Company, Inc., New York, 1952, chap. 25.

12. Johnston, J.: *Econometric Methods*, McGraw-Hill Book Company, New York, 1963.

13. Kendall, M. G.: *Rank Correlation Methods*, Hafner Publishing Company, Inc., New York, 1955, chap. 8.

14. Land, K. C.: "Principles of Path Analysis," in Edgar Borgatta (ed.), *Sociological Methodology 1969,* Jossey-Bass, Inc., Publishers, San Francisco, 1969, chap. 1.

15. Morris, R. N.: "Multiple Correlation and Ordinally Scaled Data," *Social Forces,* vol. 48, pp. 299–311, 1970.

16. Quade, Dana: *Nonparametric Partial Correlation,* University of North Carolina, Institute of Statistics Mimeo Series No. 526, 1967.

17. Reynolds, H. T.: *Making Causal Inferences with Ordinal Data,* University of North Carolina, Institute for Research in Social Science, Chapel Hill, 1971.

18. Simon, H. A.: "Spurious Correlation: A Causal Interpretation," *Journal of the American Statistical Association,* vol. 49, pp. 467–479, 1954.

19. Somers, R. H.: "An Approach to the Multivariate Analysis of Ordinal Data," *American Sociological Review,* vol. 33, pp. 971–977, 1968.

20. Wilson, T. P.: "A Critique of Ordinal Variables," *Social Forces,* vol. 49, pp. 432–444, 1971.

Analysis of
Covariance and
Dummy Variables

We have studied analysis of variance in which a single interval scale can be related to one or more nominal scales. In the previous chapter we saw how correlation techniques can be used to relate any number of interval scales. In analysis of covariance we combine the basic ideas of analysis of variance and correlational analysis in order to handle problems involving more than one interval scale in combination with any number of nominal scales. Thus analysis of covariance is a theoretical extension of these two procedures which ideally enables us to handle problems involving various combinations of interval and nominal scales.

Analysis of covariance computations becomes extremely tedious if carried out by hand or with a desk calculator but poses no special problems if computer programs are available. Ideally, the procedure can be extended to handle a large number of nominal and interval-scale independent variables, provided the dependent variable is an interval scale. Practically speaking, however, one is usually limited to three or four independent variables because of the fact that higher-order interactions become extremely numerous beyond that point. Analysis of covariance is formally equivalent to a procedure referred to as "dummy-variable" analysis to be discussed at the end of the chapter. The latter procedure represents a simple extension of the regression model, and a study of both procedures should provide one with a good intuitive grasp of the connection between analysis of variance and regression.

In this chapter our attention will be confined to the three-variable case in which we have one nominal and two interval scales. The basic problem which will concern us is that of relating two of these variables, controlling for the third. Although such controlling could be accomplished by taking categories of the control variable and running separate analyses within each of these classes, it is possible to achieve far greater efficiency through the use of analysis-of-covariance techniques provided interaction is not significant. In other words, the controlling can be accomplished without having to make use of an extremely large number of cases. In effect, we make use of weighted averages and adjustment procedures as we did in the case of partial correlation. In using analysis of covariance we can, however, obtain considerably more information than was possible in partial correlation since we can actually display separate correlations and slope estimates for each category of the control variable, and we can also test for interaction.

There are two types of situations with which we shall be concerned: (1) those in which we wish to relate the two interval scales, controlling for the nominal scale, and (2) those in which one of the interval scales is related to the nominal scale, the control variable being the other interval scale. Although interest will seldom be in both types of problems for any given set of data, it will be necessary to carry out most of the analysis required for the first type of problem even when interest is primarily focused on the second. For this reason, we shall proceed first with the type of problem in which the nominal-scale variable is used as the control.

20.1 Relating Two Interval Scales, Controlling for Nominal Scale

The basic methods of correlation and regression can be used to relate two or more interval scales within categories of the control variable. Having investigated each of the relationships within the separate categories, it may then be possible to pool results, obtaining average within-class correlation and least-squares coefficients, provided that it can be assumed that the relationships are the same from one category to the next. If results can be pooled, a single overall measure can be obtained which may serve as an effective summarizing measure and which will be more reliable as an estimate than any of the measures for the separate categories. The average within-class correlation coefficient so obtained can be interpreted as being directly analogous to a partial correlation coefficient since it can be used to represent the relationship between the two interval-scale variables after the control variable has been allowed to operate.

There are two significance tests we must make in this type of problem. The first is a test to see whether pooling the results for the various classes is legitimate. Here, we are essentially testing for interaction to see whether or not we can assume the same nature of relationship (as measured by the b's) in all the classes. If we cannot, then pooling will make little sense, and we must make separate analyses for each of the categories of the control variable. If pooling results does seem justified, we then go ahead and obtain an average within-class correlation, and the second test we make will be to see whether or not this coefficient is significantly different from zero.

As usual, we have to make certain assumptions about the method of sampling and the populations from which the data have been drawn, and as we might expect, these assumptions will be essentially those required by analysis of variance and correlational analysis. In broad outline, this is what we do in this first type of analysis of covariance problem. Let us now look more closely at the details of the procedures.

In order to obtain a clearer picture of what may happen when we make use of analysis of covariance to control for a nominal-scale variable, let us consider two extreme types of situations. In Fig. 20.1 we have a

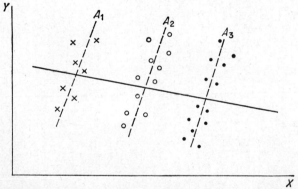

Figure 20.1 Hypothetical data indicating a weak total correlation between X and Y but stronger correlations within categories of A.

situation in which there is a slight overall or total negative correlation (indicated by the solid line) between the dependent variable Y and an independent variable X. If we look separately at each of the categories (A_1, A_2, and A_3) of the control variable A, we see that within each class there is a rather strong *positive* relationship between X and Y. In this instance, the means in X within the various categories are sufficiently different that they may obscure the basic relationship between X and Y.

If we were to superimpose the means of the three categories upon each other, we would be in effect moving the within-class equations so that they fell on top of each other, thereby obtaining a much stronger relationship between X and Y. In essence, this is what we do when we obtain an average within-class correlation coefficient. One way to visualize the process is to think in terms of our having adjusted for differences among the A categories by taking out that source of variation due to the control variable. Having adjusted for A by superimposing the means for X and Y, we can now compare the relationships between X and Y within categories by investigating the differences among within-class slopes (as indicated by the dashed lines). Superimposing the means will of course affect the a's in each of the least-squares equations but will leave these slopes and the within-class r's unchanged.

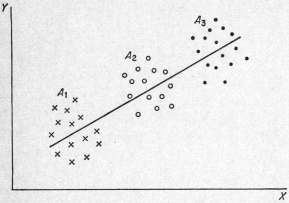

Figure 20.2 Hypothetical data indicating a strong total correlation between X and Y but weaker correlations within categories of A.

Figure 20.2 represents a contrasting situation in which there are extremely weak relationships within categories of A but where the overall relationship between X and Y is quite strong. Again, the control variable is affecting the relationship between X and Y, but this time if we were to superimpose means, we would find essentially no relationship between the two interval scales. Perhaps there is no causal relationship between X and Y at all, and the overall relationship is due to the fact that A is causing changes in both X and Y. In this instance, then, we would consider the relationship between X and Y to be spurious.

In both of these general types of situations, then, it may make very good sense to control for A. In the first case, the partial or within-class correlation will be larger in magnitude than the total; in the second it will be smaller. A carefully constructed scattergram, using different colored dots to represent the various categories of the control variable, will usually indicate whether or not it will be worth the trouble to carry out an analysis of covariance on the data at hand. If the results are similar to those indicated in either Fig. 20.1 or 20.2, it will probably prove worthwhile to proceed with the analysis. On the other hand, if the different colored dots are more or less randomly distributed on the scattergram so that the category means are not very different, it cannot be expected that analysis of covariance will produce very interesting results.

In superimposing the means of one category upon those of another we in effect control for the magnitude of these means. Actually, then, we are measuring variations and covariations about the individual category means rather than the grand means. You will remember that this is exactly what was done in analysis of variance when we divided the total sum of squares into two components. One of these components, the within-class variation, involved deviations about the category means, whereas the second component involved deviations of the class means about the total or grand mean. All that we now need to do is to extend these same procedures by breaking the total covariation, or sum of products, into explained and unexplained portions. Our reasoning will be exactly parallel to that used in connection with sums of squares. Since

$$X_{ij} - \bar{X}.. = (X_{ij} - \bar{X}._j) + (\bar{X}._j - \bar{X}..)$$

and
$$Y_{ij} - \bar{Y}.. = (Y_{ij} - \bar{Y}._j) + (\bar{Y}._j - \bar{Y}..)$$
we may write

$$(X_{ij} - \bar{X}..)(Y_{ij} - \bar{Y}..)$$
$$= [(X_{ij} - \bar{X}._j) + (\bar{X}._j - \bar{X}..)][(Y_{ij} - \bar{Y}._j) + (\bar{Y}._j - \bar{Y}..)]$$

When we sum over all cases and multiply out the factors, we obtain four terms, but the middle two will drop out. As a result we may write

$$\sum_i \sum_j (X_{ij} - \bar{X}..)(Y_{ij} - \bar{Y}..) = \sum_i \sum_j (X_{ij} - \bar{X}._j)(Y_{ij} - \bar{Y}._j)$$

Total sum of products $\quad = \quad$ within sum of products
(unexplained)

$$+ \sum_i \sum_j (\bar{X}._j - \bar{X}..)(\bar{Y}._j - \bar{Y}..)$$

$+$ between sum of products
(explained)

Again, one may use computing formulas for the total and between sums of products, obtaining the within figure by subtraction. These computing formulas turn out to be exactly analogous to those used in getting sums of squares except that a Y value replaces one of the X's so that we get cross products rather than squares. Thus we get

$$\text{Total sum of products} = \sum_i \sum_j X_{ij}Y_{ij} - \frac{\left(\sum_i \sum_j X_{ij}\right)\left(\sum_i \sum_j Y_{ij}\right)}{N}$$

(20.1)

$$\text{Between sum of products} = \sum_j \frac{\left(\sum_i X_{ij}\right)\left(\sum_i Y_{ij}\right)}{N_j}$$

$$- \frac{\left(\sum_i \sum_j X_{ij}\right)\left(\sum_i \sum_j Y_{ij}\right)}{N} \quad (20.2)$$

where N_j represents the number of cases in the jth class.

As was true for the sums of squares, the second term in both equations is the same quantity. Notice, also, that in the formula for the between sum of products the quantity in the numerator of the first term simply represents the product of the sum of the X's and the sum of the Y's for each class. The formula tells us to divide this product by the number of cases in the class and then sum over all classes.

There is one important difference between a sum of products and a sum of squares: a sum of products may be negative in value. Thus, the total covariation may be negative, but the between value may be positive. This would mean, of course, that when we subtract a positive number from a negative one, the resulting within sum of products will be a larger negative number.

PROBLEM   Before going any further, it will be helpful to introduce a numerical example and to indicate how the various computations required in analysis of covariance can be carried out in a systematic manner. Table 20.1 shows such computations for the following variables:

Y (dependent variable, interval scale): measure of educational discrimination against blacks
X (independent variable, interval scale): per cent Negro[1]
A (independent variable, nominal scale): state.

[1] In Table 20.1 per cent Negro figures have been multiplied by 10 to avoid decimals.

Data were collected for a random sample of 150 Southern counties using the 1950 census. Let us suppose, in this part of the problem, that we are interested in studying the relationship between discrimination scores and per cent Negro, controlling for the state in which the county is located.

At first glance Table 20.1 appears quite formidable, but if we examine it column by column, we see that at least the first 13 columns contain nothing really new. Columns 2, 3, 5, 7, 9, and 11 contain the basic data needed for all other computations. Columns 2 to 6 and 7 to 10 are used to obtain the total, between, and within sums of squares for the dependent and independent variables respectively. Using this computing routine, one simply works across the table, obtaining the values for each row by using the formulas as indicated at the top of each column. For example, figures in column 6 representing the sum of squares in Y are obtained by subtracting column 4 from column 5. Therefore, for Florida we have $54,989 = 3,866,409 - 3,811,420$. We thus obtain in column 6 the sum of squares within each state. When these quantities are summed, we get the within-class sum of squares so that we can enter this same quantity in the bottom row of column 6. Note that this particular computing routine differs from the one we have previously used in working analysis-of-variance problems in that we have obtained the within-class sum of squares directly, subtracting this value from the total to get the between sum of squares. Thus $1,370,555 = 2,961,762 - 1,591,207$.

In obtaining the total sum of squares we use exactly the same procedure as that used in the case of each state: we subtract column 4 from column 5. In so doing, of course, we are making use of the formula

$$\Sigma y^2 = \Sigma Y^2 - \frac{(\Sigma Y)^2}{N} = 40,399,788 - \frac{(74,938)^2}{150}$$
$$= 40,399,788 - 37,438,026 = 2,961,762$$

Here the N for the totals row is the total number of cases in the sample (150).

Notice that the totals and sums rows contain exactly the same entries in columns 3, 5, 7, 9, and 11 involving the raw scores ΣY, ΣY^2, ΣX, ΣX^2, and ΣXY. But the entries are different for columns 4, 8, and 12 involving the correction factors to be subtracted in order to obtain Σy^2, Σx^2, and Σxy. Actually, the "sums" figures are not really needed in columns 4, 8, and 12 except for the purpose of checking computations. For example, the formula $(6) = (5) - (4)$ also holds for the sums row and thus, as a

Table 20.1 Computations for analysis of covariance*

Class (1)	N_j (2)	For computing sums of squares in Y (col. 6)				For computing sums of squares in X (col. 10)			
		ΣY (3)	$(\Sigma Y)^2/N_j$ (4) = (3)²/(2)	ΣY^2 (5)	Σy^2 (6) = (5) − (4)	ΣX (7)	$(\Sigma X)^2/N_j$ (8) = (7)²/(2)	ΣX^2 (9)	Σx^2 (10) = (9) − (8)
Florida	11	6,475	3,811,420	3,866,409	54,989	2,683	654,408	744,861	90,453
Alabama	8	4,030	2,030,112	2,168,898	138,786	3,367	1,417,086	1,964,231	547,145
Arkansas	10	4,608	2,123,366	2,223,740	100,374	3,211	1,031,052	1,236,701	205,649
Georgia	33	18,911	10,837,149	11,239,451	402,302	12,707	4,892,965	5,826,629	933,664
Kentucky	9	2,724	824,464	891,102	66,638	695	53,669	63,293	9,624
Louisiana	15	7,476	3,726,038	3,926,182	200,144	5,257	1,842,403	2,025,311	182,908
North Carolina	24	9,281	3,589,040	3,862,309	273,269	7,459	2,318,195	3,266,843	948,648
Mississippi	20	12,206	7,449,322	7,586,664	137,342	10,419	5,427,778	6,043,283	615,505
South Carolina	11	5,967	3,236,826	3,371,315	134,489	4,676	1,987,725	2,367,054	379,329
Tennessee	9	3,260	1,180,844	1,263,718	82,874	1,088	131,527	229,200	97,673
Sums	150	74,938	38,808,581	40,399,788	1,591,207	51,562	19,756,808	23,767,406	4,010,598
Totals	150	74,938	37,438,026	40,399,788	2,961,762	51,562	17,724,266	23,767,406	6,043,140
Between class (explained by A)					1,370,555				2,032,542
Within class (unexplained by A)					1,591,207				4,010,598

* Adapted from [4], Table 74, pp. 486–487, with the kind permission of the publisher.

Table 20.1 Computations for analysis of covariance (Continued)

Class (1)	For computing covariations (col. 13) ΣXY (11)	(ΣX)(ΣY)/N_j (12) = (3)(7)/(2)	Σxy (13) = (11) − (12)	Slopes b = Σxy/Σx² (14) = (13)/(10)	Explained by X (Σxy)²/Σx² (15) = (13)(14)	Unexplained by X Σy² − (Σxy)²/Σx² (16) = (6) − (15)	For computing correlations r² = (Σxy)²/Σx²Σy² (17) = (15)/(6)	r (18) = ±√(17)
Florida	1,601,644	1,579,311	22,333	.24690	5,514	49,475	.10027	.317
Alabama	1,894,209	1,696,126	198,083	.36203	71,712	67,074	.51671	.719
Arkansas	1,579,758	1,479,629	100,129	.48689	48,752	51,622	.48570	.697
Georgia	7,765,621	7,281,881	483,740	.51811	250,630	151,672	.62299	.789
Kentucky	217,349	210,353	6,996	.72693	5,086	61,552	.07632	.276
Louisiana	2,700,374	2,620,089	80,285	.43894	35,240	164,904	.17607	.420
North Carolina	3,203,824	2,884,457	319,367	.33665	107,515	165,754	.39344	.627
Mississippi	6,620,545	6,358,716	261,829	.42539	111,379	25,963	.81096	.900
South Carolina	2,737,694	2,536,517	201,177	.53035	106,694	27,795	.79333	.891
Tennessee	464,348	394,098	70,250	.71924	50,527	32,347	.60968	.781
Sums	28,785,366	27,041,177	1,744,189			798,158		
Totals	28,785,366	25,759,688	3,025,678	.50068	1,514,896	1,446,866	.51148	.715
Between class (explained by A)			1,281,489			614,189		
Within class (unexplained by A)			1,744,189	b_w = .43489	758,530	832,677	.47670	.690

Table 20.1 Computations for analysis of covariance (*Continued*)

Class (1)	$\bar{X}_{.j} = \Sigma X/N_j$ $(19) = (7)/(2)$	$x = \bar{X}_{.j} - \bar{X}_{..}$ $(20) = (19) - \bar{X}_{..}$	$b_w x$ $(21) = b_w(20)$	$\bar{Y}_{.j} = \Sigma Y/N_j$ $(22) = (3)/(2)$	$\bar{Y}'_{.j} = \bar{Y}_{.j} - b_w x$ $(23) = (22) - (21)$
			For computing adjusted $\bar{Y}$'s (col. 23)		
Florida	243.909	−99.838	−43.42	588.64	632.06
Alabama	420.875	77.128	33.54	503.75	470.21
Arkansas	321.100	−22.647	−9.85	460.80	470.65
Georgia	385.060	41.313	17.97	573.06	555.09
Kentucky	77.222	−266.525	−115.91	302.67	418.58
Louisiana	350.467	6.720	2.92	498.40	495.48
North Carolina	310.792	−32.955	−14.33	386.71	401.04
Mississippi	520.950	177.203	77.06	610.30	533.24
South Carolina	425.091	81.344	35.38	542.45	507.07
Tennessee	120.889	−222.858	−96.92	362.22	459.14
Sums Totals	$\bar{X}_{..} = 343.747$			$\bar{Y}_{..} = 499.59$	

check, we note that

$$1,591,207 = 40,399,788 - 38,808,581$$

The sums figure for column 4, namely 38,808,581, was obtained by adding the results for each state, whereas the "totals" figure of 37,438,026 was obtained using the total sample size of 150. Thus

$$37,438,026 = (74,938)^2/150$$

It will be helpful at this point to work through a sufficient number of the computations in columns 2 to 6 and 7 to 10 so that you understand what is involved and become convinced that the results you obtain by this new method are exactly the same (subject to rounding errors) as those we would have obtained by the old method.

Columns 11 to 13 are used to break the covariation into component parts, in an analogous manner. As indicated above, the formulas are similar to those used for analysis of variance, except that products replace squares, and we therefore obtain column 13 by subtracting 12 from 11 as indicated by the computing formulas. Again, we compute the within sum of products directly and the between value by subtraction. Thus the total covariation is 3,025,678, and the within is 1,744,189, giving 1,281,489 as the between-class covariation. It so happens for these data that all three sums of products, as well as all state values, are positive, but this will not necessarily be the case. We now have performed the basic computations we shall need for our later work, having obtained the total, explained, and unexplained sums for y^2, x^2, and xy. Our attention can now be focused on the various tests and measures needed in carrying out the analysis. The remaining columns in Table 20.1 will be explained as we come to them.

Testing for interaction It will be remembered that in two-way analysis of variance the first test we made was for an interaction effect. The reason for making this test first was that if the two independent variables produce different results when acting in combination than would be expected on the basis of their separate effects, then it makes very little sense theoretically to study the effects of one while controlling for the other. In other words, the relationship between one independent variable and the dependent variable differs according to the value of the control variable. If this is the case, the relationship should be studied separately within each of the categories of the control variable. We face a similar problem in analysis of covariance, although instead of thinking in terms of

the additivity assumption, we shall now find ourselves explicitly comparing the slopes of the least-squares equations within each of the categories.
Let us first note the parallel between the assumption of additivity and
the assumption of equal slopes. Then we shall be in a better position to
understand the nature of the test for interaction in analysis of covariance.

In Chap. 16, which covers analysis of variance, we used the following
numerical example to illustrate additivity.

	A_1	A_2	A_3
B_1	5	10	20
B_2	10	15	25
B_3	25	30	40

It was noted that it is not necessary to assume equal differences between
the scores of B_1 and B_2, on the one hand, and B_2 and B_3 on the other.
But we did have to assume that the differences between B_1 and B_2 are
the same for each of the A categories. Now let us suppose that the variable B actually represents an interval-scale variable X that has been
categorized. We shall assume that the relationships between X and
the dependent variable Y (represented by scores in the body of the table)
are linear within each of the categories of A. A little thought will convince you that by properly locating the B categories along the X axis,
we can translate the additivity property into the statement that the three
regression lines all have the same slope. Figure 20.3 indicates this relationship. Thus we see that a test for additivity is directly analogous to a
test of the hypothesis that the within-class slopes are equal.

In testing for interaction in two-way analysis of variance, we took the
amount of variation in the dependent variable that could not be explained
by the two nominal scales when additivity was assumed. This quantity
was then broken down into two components: the amount that could be
explained by interaction, and the amount still left unexplained by the
between-column, between-row, and interaction effects. The ratio of these
last two components was used to test for interaction. In analysis of
covariance we do exactly the same thing, but as we might expect, our
procedure takes a somewhat different form. We have just seen that the
assumption of additivity is analogous to the assumption that the population slopes within each of the categories are the same. If there is a significant interaction effect, however, this will mean a different relationship
for at least some of the categories. In other words, a given change in X
will produce different changes in Y in the various classes of A. If we now
take the amount of variation in Y left unexplained by X assuming equal

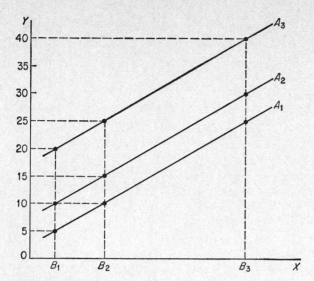

Figure 20.3 Lines with equal slopes indicating no interaction.

slopes, we can see how much additional variation we can explain through interaction. Then we can test for interaction by comparing the interaction sum of squares with the error term.

How do we determine the amount of variation we can attribute to interaction? In order to answer this question we first must ask ourselves how much variation we could possibly expect to explain using linear models within each of the categories of A. Obviously, the individual least-squares equation for each category gives us the best fit that can be expected of a straight line, and the correlation coefficient computed on the data for that particular category will give a measure of goodness of fit. We therefore can obtain figures for each category representing the amount of the variation in Y that is explained by X, using the straight line which best fits the data for that particular category. When we sum the explained variations for each of the categories, we thus obtain the amount of variation actually explained by all of the separate least-squares equations. Similarly, in summing the unexplained sums of squares we obtain the amount of variation in Y that is still left unexplained by these separate least-squares lines.

In Table 20.1 these computations have been carried out in columns 15 and 16. In the case of Florida, for example, the total variation in Y (column 6) is 54,989. Of this quantity, 5,514 is explained by the least-squares equation which best fits the Florida data, and 49,475 is left unexplained. Of the total variation in Y (2,961,762), the quantity 798,158 represents the amount left unexplained by these separate least-squares equations.

We next must ask how much variation is left unexplained when it is assumed that there is no interaction effect. If there is no interaction, then all of the slopes for the categories of A will be equal. Our best estimate of this common slope will be a pooled estimate which is a weighted average of the individual within-class slopes. These slopes have been computed in column 14. The pooled estimate or average within-class slope has also been computed in column 14 by making use of the within-class data from columns 10 and 13. Thus the value .43489 was obtained by dividing 1,744,189 by 4,010,598.

We can now compare the relative explaining abilities of the separate within-class least-squares lines, each with a different slope, and a number of straight lines drawn through the means of each category but all having the same slope, that is, the average within-class b (see Fig. 20.4). These

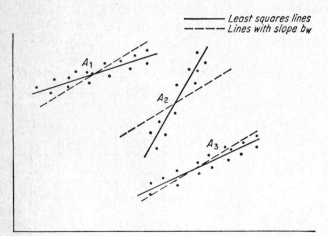

Figure 20.4 Comparison of separate least-squares lines with lines drawn through category means but all having the same slope b_w.

latter parallel lines cannot be expected to explain as much of the total variation as the actual best-fitting lines for each of the categories, but if there is in fact no interaction in the population data, the various regression equations will all have the same slope, and we can expect that the least-squares lines will not differ too markedly in slope. In other words, if there is no interaction, the series of parallel dashed lines will approximate fairly closely the actual least-squares equations for each category. Since the value of the average within-class slope will then be not too different from that of each of the separate within-class slopes, the dashed lines will have almost as high explanatory power as the solid lines.

Because of sampling fluctuations we can expect some interaction within the sample even though there may be none within the population. The solid and dashed lines will never be identical and therefore the latter lines will always leave somewhat more unexplained variation than the individual least-squares lines. The question now is whether or not the separate least-squares lines differ sufficiently among themselves, and therefore from the dashed lines, to warrant our concluding that the interaction effect is statistically significant.

In this and other analysis-of-covariance tests we must make a number of assumptions which are essentially those assumptions required by analysis of variance and regression analysis. As usual, independent random sampling must be assumed. We shall also assume bivariate normality between X and Y within each of the categories of A. Furthermore, we shall assume that the variances for both X and Y are the same within all A categories.[2] In the test for interaction our null hypothesis will of course be that each of the category relationships between X and Y involves the same slope β.

In Table 20.2 an F test is made for interaction. We take the amount of variation in Y left unexplained by X and A, assuming no interaction or

Table 20.2   Analysis-of-variance test for interaction

	Sum of squares	Degrees of freedom	Estimate of variance	F
Unexplained by X and A, assuming no interaction	832,677	$N - (k + 1) = 139$		
Explained by interaction	34,519	$k - 1 = 9$	3,835.4	<1.0
Error	798,158	$N - 2k = 130$	6,139.7	

equal population slopes. This quantity can be found from Table 20.1 by reading across the within-class row until we come to the unexplained

[2] Once more, it will not be necessary to assume the normality of the X's as long as the Y's are normally distributed (with equal variances) about fixed X scores within each category of the nominal-scale variable.

by X column (column 16). Since the figure 832,677 was obtained by using the same set of figures as was the average within-class b, we see that we have essentially assumed equal slopes in computing this unexplained sum of squares. We have also seen that the quantity 798,158 represents the amount of variation left unexplained by the separate least-squares equations. The difference between these two quantities therefore represents the amount of variation that can be attributed to interaction.

In associating degrees of freedom with each of these quantities we count up the number of coefficients that have been estimated in the respective least-squares equations. Looking first at the error term, or the amount left unexplained by the separate least-squares equations, we note that for each of these separate equations we have had to compute two coefficients (a and b). We therefore will lose $2k$ degrees of freedom, where k represents the number of categories of A. Hence the degrees of freedom associated with this term will be $N - 2k$. In making use of the dashed lines, however, we have had to compute only a single slope, the average within-class b. Since each of these lines passes through a different set of sample means, however, we have different a values for each of these lines. We therefore have lost $(k + 1)$ degrees of freedom, and the degrees of freedom associated with this term will be $N - (k + 1)$ or $N - k - 1$. We can then obtain the degrees of freedom for the interaction term by subtraction, getting

$$(N - k - 1) - (N - 2k) = k - 1$$

or one less than the number of categories. We now compute F in the usual manner and conclude that since $F_{9,130} < 1.0$, we do not have significant interaction.

Since interaction did not turn out to be significant, we are justified in throwing the small amount of sample interaction back into the error term, henceforth using the quantity 832,677 as the variation that is unexplained by both X and A. Since we are on the wrong end of the test for interaction, we must of course be somewhat cautious in doing this. With such a large N and small value of F we are undoubtedly safe in ruling out interaction in this particular problem, however.

Had interaction been significant, our next step would be to look for the state or states that are out of line with the rest. This is easily accomplished by looking at the column of b's. If several states are obviously creating the interaction effect and if good theoretical reasons can be suggested why this is the case, it may be possible to exclude these states and repeat the analysis with the remainder. If no states stand out in this manner, it may be necessary to proceed by analyzing each state separately. In such an eventuality, valuable theoretical insights might be gained by asking oneself why the relationship between discrimination and per cent Negro should differ from one state to the next.

One possible strategy that can be used whenever there are sizable differences among the slopes is to rank-order the categories (here, states) with respect to the magnitudes of the slopes and then to attempt to locate some specific *variable* that is highly correlated with this ordering. For example, perhaps when we rank the states from low to high with respect to steepness of slopes (here all positive), we might find that the states with the steepest slopes tend to be the most urbanized or most industrialized. If this were the case, we might obtain a measure Z of urbanization (or industrialization), replacing the nominal scale "state" with Z, and then utilize some specific alternative to an additive model such as the multiplicative function $Y = kX^{b_1}Z^{b_2}$. By taking logarithms of both sides, this multiplicative function could be transformed into the additive equation $\log Y = \log k + b_1 \log X + b_2 \log Z$.

The average within-class correlation   Having established that there is no significant interaction effect, we are now justified in pooling the individual within-class r's to obtain an average within-class correlation coefficient that will be analogous to the partial correlation coefficient. In other words, since we are justified in assuming a single slope for all the regression equations, we can also assume that the population correlation coefficients will all be the same and that the common value can be estimated by pooling the sample r's for the various classes. The average within-class correlation coefficient, which we can symbolize as $r_{XY \cdot A}$, is computed in the same manner as the average within-class b by using within-class data given in the bottom row of Table 20.1. (See columns 17 and 18.) The square of this coefficient can be interpreted as the proportion of variation in Y that is left unexplained by A but is explained by X. Thus

$$.47670 = (.690)^2 = \frac{758{,}530}{1{,}591{,}207}$$

If you will note the formulas used in computing each of these numbers, you will see that the interpretation follows immediately from these formulas. As a rough check on one's computations, the average within-class r should turn out to be comparable in magnitude with the various separate within-class r's. Since it is essentially a weighted average, the states with the largest numbers of counties will have the greatest effect in determining its value. If some of the b's in column (14) turn out to be negative, the comparable r's in column (18) should of course be given negative signs.

If we wished a measure analogous to a multiple R, we might take the ratio of the amount of variation explained by *both* X and A to the total sum of squares. In this problem, for example, we have explained $2,961,762 - 832,677$ or $2,129,085$. Therefore we have explained $2,129,085/2,961,762$ or 71.9 per cent of the variation. We must remember, however, that if we wish to form a multiple R by taking the square root of this value, the result will be partially a function of the average number of cases within categories of A (see Sec. 16.5).

We can test for the significance of $r_{XY \cdot A}$ in the usual manner. First we let the control variable A do all the explaining it can. We then let X go to work on the variation left unexplained by A, breaking this latter quantity into two components. The first of these components will be that portion which is explained by X, while the other will be the error term which is left unexplained by both X and A (assuming no interaction). We have already seen that the degrees of freedom for the error term will be $N - (k + 1)$. The degrees of freedom associated with the variation unexplained by A, which is found in the bottom row of column 6, will of course be $N - k$ (see Sec. 16.1). This leaves 1 degree of freedom associated with the component which is unexplained by A but explained by X. The results of this test are summarized in Table 20.3. We thus see

Table 20.3 Analysis-of-variance test for significance of average within-class correlation $(\rho_{XY \cdot A})$

	Sum of squares	Degrees of freedom	Estimate of variance	F
Unexplained by A	1,591,207	$N - k = 140$		
Unexplained by A but explained by X	758,530	1	758,530	126.6
Error (assuming no interaction)	832,677	$N - (k + 1) = 139$	5,990.5	

that the average within-class correlation is significant at the .001 level.

Before we conclude this portion of the chapter in which we have studied the relationship between two interval scales while controlling for the nominal scale, we can draw a comparison with the type of controlling accomplished by partial correlation. Controlling by means of analysis of covariance obviously involves considerably more work than the use of partial correlation. As can readily be imagined, extensions involving additional variables will begin to require so many computations that analysis of covariance will generally not be feasible without the aid of computer programs. On the other hand, analysis of covariance supplies us with more information than does partial correlation. Not only may we make a test for interaction, but we can investigate the relationships between X and Y within each of the categories of the control variables, comparing the various values of r and b. In making use of partial correlations, we obtain only the single measure which is comparable to the average within-class correlation, and we cannot make a test for interaction.

We see that analysis of covariance has a number of advantages over analyses using partial correlations, especially in studies where interaction may be expected. Thus in some instances it may be well worthwhile to convert one of the interval scales into a nominal scale and to proceed with analysis of covariance in preference to partial correlation, even though in so doing we may lose information with respect to the level of measurement.

20.2 Relating Interval and Nominal Scales, Controlling for Interval Scale

In one-way analysis of variance we related an interval scale to a single nominal scale, testing for the significance of the differences among the means of the categories of A. In order to determine the magnitude of the relationship between the two variables, we computed an intraclass correlation coefficient. We also obtained the means of the various categories which could be used for descriptive purposes to indicate the relative scores of one category compared with the others. In two-way analysis of variance we were able to control for a nominal scale and te t for interaction. We found ourselves severely limited, however, in that we were required to have equal numbers of cases in each of the subcells. In this section we shall take up situations in which we wish to relate Y and A but where the control variable is an interval scale X.

Suppose that our interest is primarily in discovering the relationship between discrimination rates and subregions of the South, as defined by the various states. Admittedly, states are not the best kinds of units to delineate subregions, but they will serve for illustrative purposes. Obviously, a variable such as per cent Negro needs to be controlled since the various Southern states differ considerably in their minority percentages. Suppose we were to categorize per cent Negro and proceed with separate analyses of variance for each of these categories. Note that we would probably not even attempt two-way analysis of variance because of the necessity of having equal subclasses. But would separate analyses of variance really solve our problem? We would immediately find that when we examined the counties with low minority percentages, we would be excluding practically all of the counties in Mississippi and Alabama and including practically all of those in Kentucky and Tennessee. On the other hand, there would be at most one or two counties from these latter states among the counties with high minority percentages. Thus in attempting to control by this method we almost do away with our problem in that only a few of the states will be represented in each of the separate analyses. The effects of subregions or states would be hopelessly confounded with per cent Negro. We cannot hold the one variable constant without at the same time decreasing the variability of the other.

Although we cannot actually hold this control variable constant, we can, by using analysis of covariance, make certain adjustments for its effects. Specifically, if we are willing to assume that the regressions of Y on X within each of the categories of A have a common slope which can be estimated by the average within-class b, we can estimate the change in Y produced by a given change in X. In other words, we can make certain predictions about what would happen to the discrimination rates in each state if the minority percentages were to change. In particular, we can ask ourselves what would happen to these rates if the various percentages of blacks were to be equalized. This sort of adjustment process yields purely hypothetical results, and this fact should be made perfectly clear. We are not getting discrimination rates for the various states with per cent Negro actually held constant; we can only predict what they would be if this were to happen and if the relationships between X and Y were as assumed. Quite conceivably, if blacks were actually to redistribute themselves more evenly among the Southern states, the particular relationships we found between X and Y would no longer hold. Nevertheless, such an adjustment procedure can often lead to useful insights.

If it can be assumed that there is no interaction effect, we have seen that we can best estimate the common slopes of the within-class regression equations by means of the average within-class b computed in Table 20.1. The adjustment procedure we shall use can now be described. We would like to adjust each of the class means $\bar{Y}_{\cdot j}$ so as to take into consideration the fact that the means in X also differ from state to state. For convenience we shall assume that all of the $\bar{X}_{\cdot j}$'s are adjusted to the grand mean of the X's. This will involve moving the mean in X for each class a distance of $(\bar{X}_{\cdot\cdot} - \bar{X}_{\cdot j})$. Figure 20.5 indicates this difference as the length of the base of the triangle. But we know that in order to obtain the amount of change in Y for a given change in X, we must multiply the change in X by the average within-class b. Therefore $\bar{Y}_{\cdot j}$ changes by the amount $b_w(\bar{X}_{\cdot\cdot} - \bar{X}_{\cdot j})$ where we have used the symbol b_w to represent the average within-class slope. The adjusted value of the means for Y can now be found by adding this increment to the original mean for Y.

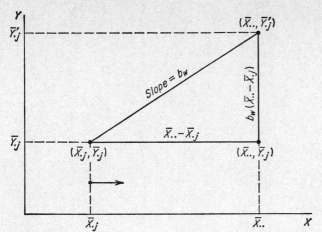

Figure 20.5 Geometrical interpretation of the computations for the adjusted Y means.

Thus, letting $\bar{Y}'_{.j}$ represent the adjusted value, we get

$$\bar{Y}'_{.j} = \bar{Y}_{.j} + b_w(\bar{X}_{..} - \bar{X}_{.j})$$

$$= \bar{Y}_{.j} - b_w(\bar{X}_{.j} - \bar{X}_{..}) \tag{20.3}$$

The second of these forms, which just involves reversing the order of $\bar{X}_{..}$ and $\bar{X}_{.j}$ and changing signs, is the form used in computing the adjusted $\bar{Y}$ in Table 20.1. Notice that the slope in this particular example is positive. Also, the change from $\bar{X}_{.j}$ to $\bar{X}_{..}$ as shown in Fig. 20.5 is positive. Exactly the same algebraic results hold in cases where the slope is negative or where the X value is decreased. By now you should be able to convince yourself that this is the case.

Figure 20.6 should help you visualize what we have done in adjusting the mean Y values. We have in effect moved each of the class means parallel to the slope of the average within-class b to a position where all of the $\bar{X}$'s are equal to the grand mean of the X's. The adjusted $\bar{Y}$'s can be found along the vertical dashed line corresponding to the grand mean of the X's. The relative magnitudes of the means in Y may be considerably altered. In Fig. 20.6 the unadjusted $\bar{Y}$ values are such that the mean of A_1 is slightly below that of A_2 which, in turn, is substantially less than that of A_3. Notice that A_1 has a very small $\bar{X}$ value, however. Since the slope is represented as positive, adjusting for X has the effect of increasing the value of Y in the case of A_1. On the other hand, the

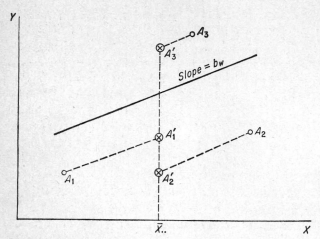

Figure 20.6 Geometrical interpretation of adjusting Y means by sliding category means parallel to line with slope b_w.

adjustment process decreases the Y values for both A_2 and A_3 since both of these categories have relatively large X values. As a result, the adjusted $\bar{Y}$ for A_1 is actually larger than that for A_2, and the value for A_3 is much closer to that of A_1.

If you will refer back to Table 20.1, columns 22 and 23, you will note the effect on discrimination rates of adjusting for per cent Negro. Florida, with its relatively low percentage of blacks, now stands out as having very high adjusted rates, whereas states such as Mississippi and Tennessee have been brought more into line with the remaining states. Note also that the overall differences among states have been considerably reduced.

It was mentioned earlier in the chapter that analysis of covariance will be useful if scattergrams show that the various class means in X are quite different in value. This can be seen from Fig. 20.6. Had the class means in X been closely bunched around the grand mean, the bases, and therefore the legs, of the triangles would have been quite short. In other words, adjusting for X would not produce a very marked effect since, in reality, there was very little adjustment actually needed. Had all the class means in X been exactly equal, we would in effect have had a control for X. It is only when the class means in X are very different that we can expect adjustment to have a noticeable effect. Put another way, there must be a fairly strong relationship between X and A, the two independent variables.

Something else is also necessary in order for adjustment to be worthwhile. Had the average within-class b been very small numerically, it would have taken a very large change in X to produce a slight change in Y. Thus, if there is little or no relationship between X and Y within classes of A, there will be no point in adjusting for X. These observations are of course consistent with common sense which tells us that there is not much to be gained in controlling for a variable which is not related to both of the variables in which we are interested. True, if X is related only to the dependent variable, it may be controlled as a disturbing influence. We can see from Fig. 20.6, however, that unless there are some variations with respect to X among categories of A, there will be little point in adjusting.

In using the average within-class b we have had to assume no interaction effect. Therefore, it will be necessary to carry out the test for interaction as well as the computations for b before going ahead with the adjustment process. If interaction turns out to be significant, the problem is more complicated and beyond the scope of this text. Under some circumstances it may be possible to adjust using the individual within-class slopes. However, the interpretation must be carefully made. Suppose, for example, that the slope for Mississippi turned out to be quite different from that of Tennessee. Could we legitimately use their individual least-squares lines to adjust Y values? This would require the assumption that Mississippi maintains essentially the same patterns of discrimination as it continues to lose blacks. The fact that other states show different relationships with per cent Negro suggests, however, that this may not be a legitimate assumption. The fact that interaction has been shown to exist should make us cautious in projecting what would happen if the X's actually were changed. On the other hand, if we find essentially the same relationship between per cent Negro and discrimination in each of the states, i.e., no interaction, we are somewhat more confident that adjustment will not lead us too far astray.

We next must raise the question as to the significance of the differences among the adjusted Y means. Differences among the unadjusted means may or may not have been significant, but this does not mean that the same result will hold for the adjusted values. Perhaps adjusting for X has had the effect of moving the Y values closer together. Or perhaps they are now farther apart. We have accomplished a descriptive task, that of actually obtaining the adjusted figures so that they may be displayed for comparative purposes. We now must test the null hypothesis that in the population the adjusted Y means are all the same. The assumptions for this test are the usual ones. We must assume independent random samples and equal variances for the adjusted $\bar{Y}$'s, and we must also make the assumptions required by regression analysis, a normal distribution of Y about fixed X's within each category of A.

Fortunately, we do not have to recompute sums of squares using the adjusted values themselves. We can carry out an analysis-of-variance test using the familiar procedure of first letting the control variable explain all of the variation it can. Since X is now our control variable, we take the amount of variation left unexplained by X as our new total sum of squares. We then break this quantity into the amount explained by A and the amount left unexplained by both variables. The degrees of freedom associated with each of these quantities have already been determined. The results of the F test are summarized in Table 20.4.

Table 20.4   Analysis-of-variance test for significance of differences among adjusted means

	Sum of squares	Degrees of freedom	Estimate of variance	F
Unexplained by X	1,446,866	$N - 2 = 148$		
Unexplained by X but explained by A	614,189	$k - 1 = 9$	68,243	
Error (assuming no interaction)	832,677	$N - (k + 1) = 139$	5,990.5	11.39

Thus we see that the adjusted differences, although smaller than the original ones, are significant at the .001 level. We conclude that, although adjusting for per cent Negro does decrease differences in discrimination rates among states, these differences are not completely wiped out in the process.

Finally, we may wish to compute a partial intraclass correlation between Y and A, controlling for X. This might be advisable in order to obtain a better indication of the degree of relationship between the two variables than can be indicated by the differences among adjusted means. Simply by looking at these adjusted differences one cannot get a very good idea of their magnitude relative to differences within the categories, and for this reason a partial intraclass correlation may be helpful. Generalizing the notion of intraclass correlation we may write

$$r_{iYA \cdot X} = \frac{V_b - V_e}{V_b + (\bar{n} - 1)V_e}$$

where V_b = between-class estimate (unexplained by X; explained by A)
V_e = error estimate (unexplained by X and A)
$\bar{n}$ = average number of cases per class, as computed in Eq. (16.12)

Here we are interested in the between-class estimate of the variance for the *adjusted* Y's. Our error estimate takes into consideration that X has already explained all it could of the variation in Y.

Numerically, we thus get

$$\bar{n} = \frac{1}{k-1}\left(\sum_{i=1}^{k} N_i - \frac{\sum_{i=1}^{k} N_i^2}{\sum_{i=1}^{k} N_i}\right)$$

$$= \frac{1}{9}\left(150 - \frac{2,858}{150}\right)$$

$$= \frac{1}{9}(150 - 19.05) = 14.55$$

and

$$r_{iYA \cdot X} = \frac{68,243 - 5,990.5}{68,243 + 13.55(5,990.5)}$$

$$= \frac{62,252.5}{149,414} = .417$$

20.3 Extensions of Covariance Analysis

The addition of a second nominal scale will complicate the analysis of covariance because of the requirement of equal-size subclasses. Practically speaking, this in effect means that this type of extension will not be feasible except in studies involving experimental designs in which such a control over the number of cases is possible. If we add one or more interval scales, however, the extension is straightforward in principle, although it will introduce a considerable number of additional computations. We shall have to add new columns to the computing table. In particular, there will be a column indicating the amount of variation left unexplained by both interval scales (X and Z) acting simultaneously. We shall thus become involved with multiple least-squares equations for each of the categories of A. In obtaining adjusted Y means, for example, we shall have to adjust for X and Z using both of the partial average within-class b's. Instead of moving category means parallel to a least-squares *line*, we shall now slide them parallel to an average within-class *plane*. In testing for the significance of the adjusted Y's we first let X and Z explain all they can of Y and let A work on the remainder.

Since we have by no means exhausted the subject of analysis of covariance, you may wish to consult some of the references listed below for other applications and extensions of this general method. In particular, whenever the number of categories in A is quite large, it is sometimes useful to investigate the regression of the category *means* in Y on the *means* in X, thereby in effect considering each category as a case. For example, in the problem we have been considering, we might wish to study the relationship between X and Y using states rather than counties as units, treating the $\bar{X}$'s and $\bar{Y}$'s for each state as single scores. In most problems of interest to sociologists, the number of categories in A will be too small to justify such an analysis, however, and for this reason the topic has not been discussed in the present chapter.

20.4 Dummy Variable Analysis

In both analysis of variance and analysis of covariance our explicit focus of attention has been on the process of splitting up sums of squares and sums of products into various components, whereas in regression analysis our focus was more on estimating the coefficients in an equation. It should now be useful to put the two ideas together by showing how a combination of interval and nominal scales can also be handled in terms of the regression format. Recall that in the case of two-way analysis of variance it was pointed out that we may utilize an additive model of the form

$$Y_{ijk} = \mu + \alpha_i + \beta_j + \gamma_{ij} + \epsilon_{ijk}$$

whereas in multiple regression we used equations of the form

$$Y_i = \alpha + \beta_1 X_1 + \beta_2 X_2 + \cdots + \beta_k X_k + \epsilon_i$$

Apart from differences in notation, which need not bother us, we see two apparent differences in these equations: (1) we did not allow for interaction in the simplified regression model, and (2) the additive model in the case of analysis of variance does not contain any X's representing interval scales. We noted, however, that the regression model does not place any restrictions on the X's in terms of their frequency distributions, although we *may* assume them to be normally distributed. In particular, some or all of the X's could be assigned scores of 1 or 0, and we shall see that this will enable us to handle nominal scales as special cases. Also, we have noted that certain of the X's may be products of other X's (e.g., we may set $X_3 = X_1 X_2$), and through this device we may handle interaction terms in the regression context. Therefore, the analysis of variance procedure may be considered as a special case of regression analysis, and vice versa, and both may be subsumed under a single more general mathematical model.

In order to keep the argument from becoming too abstract, let us assume that we are dealing with an interval-scale dependent variable Y, two interval-scale independent variables X_1 and X_2, and a single nominal scale consisting of four categories. Suppose Y is one's income at age 35; X_1 his years of schooling; X_2 his occupational status score; and Z_i a variable (to be described below) representing the region in which the respondent resides. If there are four regions (Northeast, South, Midwest, and West), we may utilize three Z_i as follows:

$$Z_1 = 1 \text{ if respondent resides in the Northeast}$$

$$= 0 \text{ otherwise}$$

$$Z_2 = 1 \text{ if respondent resides in the South}$$

$$= 0 \text{ otherwise}$$

and $\qquad Z_3 = 1$ if respondent resides in the Midwest

$$= 0 \text{ otherwise}$$

The "variable" Z_i is referred to as a *dummy variable* since simple scores of 1 and 0 have been arbitrarily assigned. Actually, we might have used a different set of scores, but the use of 1's and 0's will keep the analysis as simple as possible. Notice that we do not need to utilize a Z_4 that takes on the value 1 in the case of the West and 0 otherwise, because if we know the values of Z_1, Z_2, and Z_3, we know the value of Z_4 with certainty. In particular, respondents from the West will be given scores of zero for all three Z's. As long as we are dealing with a single nominal scale, and as long as we do not insert a constant α into the equation, it will be possible to include Z_4 in the equation. But if we also retain α, or if we have more than one nominal scale and attempt to retain Z's for all categories, we shall find that least-squares procedures will break down owing to the fact that, in this example, Z_4 is a perfect linear function of Z_1, Z_2, and Z_3. In fact we can see that $Z_4 = 1 - (Z_1 + Z_2 + Z_3)$. In practice, then, if we form the habit of always "suppressing" one category of each nominal scale, we shall be in a position to apply ordinary least squares under the usual assumptions. It will turn out that the suppressed category (here the West) will form the basis of comparison with the remaining categories.

We may now form an estimating equation as follows:

$$Y = a + b_1X_1 + b_2X_2 + c_1Z_1 + c_2Z_2 + c_3Z_3$$

Let us interpret this equation before introducing a more complex model that allows for interaction. Suppose we are dealing with a respondent from the West, for whom $Z_1 = Z_2 = Z_3 = 0$. In his case the equation reduces to

$$Y = a + b_1X_1 + b_2X_2$$

If we now compare this individual with one from the Northeast, for whom $Z_1 = 1$ and $Z_2 = Z_3 = 0$, we see that for the latter individual the equation will contain an additional term $c_1Z_1 = c_1(1) = c_1$ which we may consider to be added onto the intercept a. Thus for the respondent from the Northeast we have

$$Y = (a + c_1) + b_1X_1 + b_2X_2$$

and we see that c_1 can be interpreted as the *difference* in the intercepts between the two regions. Similarly, c_2 can be interpreted as the difference between the intercepts for individuals from the South as compared with those from the West. In this sense, the suppressed category represents a comparison group for all three remaining categories. In causal terms, the c_i might be interpreted as the increments or decrements in income that would be incurred if the individuals were to migrate from the West to the other regions.

Let us next consider the case where we wish to allow for interactions between region and either X_1 or X_2. For simplicity we shall confine our attention to X_1, dropping X_2 from the equation. In the case of analysis of covariance we saw that interaction showed up as a difference among the within-category slopes. This can be handled in terms of the dummy-variable formulation by introducing terms of the form $d_{ij}X_iZ_j$. In the case of a single independent variable X_1 and three Z_i our equation will be

$$Y = a + b_1X_1 + c_1Z_1 + c_2Z_2 + c_3Z_3 + d_{11}X_1Z_1 + d_{12}X_1Z_2 + d_{13}X_1Z_3$$

In the case of a respondent from the West, for whom $Z_1 = Z_2 = Z_3 = 0$, the equation reduces to $Y = a + b_1X_1$. But for the individual from the Northeast the equation will be

$$Y = a + b_1X_1 + c_1Z_1 + d_{11}X_1Z_1 = (a + c_1) + (b_1 + d_{11})X_1$$

since $Z_1 = 1$ for all persons from this region. If we compare the equation for Northeasterners with our "standard" equation for the Westerners, we now not only have an intercept difference c_1 but a difference in slopes as well. Thus we may interpret d_{11} as the increment (or decrement) that we add to the slope of the relationship between X_1 and Y when individuals move from the West to the Northeast. Similar interpretations can be given to d_{12} and d_{13}, and if these coefficients depart significantly from zero, we infer that there is interaction present in the population. Furthermore, an examination of the magnitudes of the d_{ij} may be useful in explaining such interactions.

We have just considered the case where there is one interval and one nominal independent variable, and the results of such an analysis will give us results that are identical to those obtained using covariance analysis. The computational procedures are very simple, provided one has available computer programs capable of handling multiple regression problems. One merely utilizes the scores of interval-scale variables as they stand, converting his nominal scales into dummy variables Z_i. These are then analyzed exactly as would be done in connection with multiple regression procedures. Each of the coefficients b_i, c_j, and d_{ij} may be tested for significance. One may obtain multiple and partial correlations, and so forth. For example, if one wanted to measure the explanatory power of region, with a control for all of the X_i, he would first see if interactions could be ruled out. If this were possible, he could then compute a multiple-partial relating Y to all of the Z_j taken together, with a control for all of the X_i.

If one wishes to utilize two or more nominal scales, two rather simple alternative strategies are available. One possibility is to combine the two scales into a single nominal scale and proceed as above. For example, if one wanted to study race and sex interactions with education X_1 as these affect income Y, he might utilize the four combinations black-male (Z_1), black-female (Z_2), white-female (Z_3), and white-male (suppressed), thereby comparing the three remaining race-sex combinations with white males as a standard group. The second alternative is to utilize two different dummy variables, one for sex and the other for race. If we let $Z_1 = 1$ for all blacks and $W_1 = 1$ for all females, then we could explicitly introduce first-order interactions between X_1 and race by adding a term involving the product X_1Z_1, and we could similarly utilize the term X_1W_1 to tap the income-sex interaction. But we could also handle a race-sex interaction term by utilizing the product W_1Z_1 which would be unity only for black females. Furthermore, we could handle the higher-order interaction of race-sex-income through a product term $X_1W_1Z_1$.

If there are two nominal scales with r and c categories, respectively, then there will be $(r - 1)$ and $(c - 1)$ nonsuppressed categories, and we will need $(r - 1)(c - 1)$ product terms to handle all of the two-factor interactions. We may therefore express Y as a function of the main effects of the row variable, the main effects of a column variable, and a series of interaction terms. We thus may treat two-way analysis of variance as a special case of dummy-variable analysis, and furthermore we do not need to assume equal numbers of cases in all of the subcells since we are allowing for intercorrelations among the independent variables. As was true in the case of regression analysis, however, we must pay the price of a theoretical ambiguity resulting from an overlap in the variation that will be "explained" by the two correlated independent variables. As an exercise, it may be useful to the reader to reconceptualize problems discussed in the analysis of variance chapter in terms of the dummy-variable framework.

20.5 Concluding Remarks

We have covered a number of statistical approaches to multivariate analysis, though a number of more specialized topics have not been treated. Perhaps the fundamental problem of multivariate analysis, in the absence of well-specified theories that dictate precisely the steps to be followed, is that of finding relatively systematic ways of dealing with various kinds of complications. The basic task is that of eliminating as many such complications as possible, but only after having permitted ourselves to discover their existence and to assess their importance. The general strategy is to set up a roughly ordered set of priorities and to attempt to eliminate first those potential complications in which one is least interested, gradually working toward a more intensive analysis involving those that are most central to one's theoretical interests and that are empirically the most important.

There are several kinds of complexities that have been mentioned only briefly in passing. These include the realistic possibility of both random and nonrandom measurement errors of various kinds. As noted, the former have received some degree of attention in the statistical literature, whereas the latter have been virtually ignored until only very recently. A second kind of complexity is encountered in realistic nonexperimental research where one needs to allow for reciprocal causation. We have assumed that the choice of dependent variable is not problematic and that there is no feedback effect from dependent to independent variables. Although we have allowed for the possibility of intercorrelated independent variables, we have not considered models that attempt to account for these intercorrelations by taking some of the "independent" variables as functions of the others. These topics will be treated in a later volume and have been studied in considerable detail by econometricians in connection with simultaneous-equation models. (See especially, Christ [2] and Johnston [6].)

One type of complication that has been studied involves the addition of explanatory variables to an equation, variables which, as we have just noted, may be intercorrelated. It has been noted that whenever these intercorrelations are high relative to correlations with the dependent variable(s), it will be especially difficult to separate out their component effects. Therefore, one form that simplifications must always take is that of the reduction of the number of explanatory variables to a reasonable number. This may be achieved through a number of devices. One of these is to separate the variables into "blocks" and to treat only the blocks as distinct. One may either construct a single score for the entire block (e.g., socioeconomic status), or he may use measures such as the multiple-partial correlation coefficient to assess the effects of the entire block. Along with these purely statistical operations must be included a careful theoretical conceptualization concerning the nature of the particular blocks of variables that have been formed. Techniques such as multiple factor analysis, cluster analysis, latent structure analysis, multiple classification analysis, and canonical correlation may be used for this purpose.

As a general rule, it is usually the case that an investigator will be able to delineate his independent variables into several sets according to his theoretical interests. First, there will be those variables in which he is primarily interested. Second, there will be a set of independent variables that he intends to use as control variables. These are variables that he fully expects to have major impacts on the variables of primary interest, but in terms of his own research design they are being considered as "nuisance variables." They cannot be ignored, but they are of little theoretical interest. Finally, there will be a set of variables—perhaps a very large one—that are thought to have relatively minor importance, or that have been suggested as variables that one may fall back upon in case the remainder are found to have little explanatory value. In exploratory studies it makes very good sense to include these variables, since theoretical guidelines are usually very vague. The basic strategy of analysis, in the case of this third type of variable, is to begin by seeing how many of this set can be eliminated right away. Those that cannot may then be transferred to the second set. The general point is that, in trying to narrow down the scope of the analysis, one should work from the outside inwards, so to speak. Try to eliminate the complications first. In this instance, this implies getting rid of variables with marginal explanatory power. As a general rule, unless there are ample research funds available, many such variables may automatically be eliminated if the zero-order correlations with the dependent variables are negligible *or* if the variables are extremely highly associated with other independent variables of more fundamental interest.

Possible nonlinearities are another form of complexity that should always be investigated in the case of interval scales, and that can sometimes be crudely assessed in the case of ordinal data. Certainly, all bivariate relationships (including those among the independent variables) should be routinely tested for nonlinearity by comparing E^2 with r^2. If this difference is statistically significant but yet numerically small (owing to a very large sample), then one will have to decide whether the increment explained by allowing for nonlinearity is worth the price of the added complexity. This will depend on one's research priorities and the centrality of the particular relationship for the subsequent analysis. For example, if a nonlinear relationship is embedded in a complex set of relationships among as many as ten or fifteen variables, then it is probably not worth the price of the added complexity. But if there are only three or four major variables, and if the dependent variable is one's primary focus of attention, then such a refinement may be deemed worthwhile. If so, then one should attempt to specify a reasonably simple mathematical function (e.g., logarithmic, parabolic, or exponential function) that will explain almost as much of the variance as the completely unrestricted nonlinear function (i.e., with no restrictions on the category means), the explanatory power of which is measured by E^2. In other words, it is not sufficient to indicate that an important relationship is *nonlinear*. A specific form should be indicated and a test made to see whether this form (e.g., a parabola) gives a significantly better fit than a straight line. The further possibility that the particular form of the relationship may also vary with the level of other variables (implying interaction) should also be investigated. For example, a relationship may be logarithmic for males but linear for females. With as many as eight or ten variables in a system, the number of possible complexities of this type increases geometrically as each variable is added. Usually, however, most of the potential complexities will not materialize.

Finally, one should always investigate the possibility of interactions or nonadditive relationships among the independent variables. With multiple independent variables, there will be numerous higher-order interactions that are practically always ignored in analysis. One reasonable strategy is to look for all possible two-variable interactions. Most of these will undoubtedly be found to be negligible. Tests may be made for the significance of entire sets of interactions through the use of multiple-partial coefficients. Suppose, for example, that one has four independent variables X_1, X_2, X_3, and X_4. One could add to the regression equation all the possible $X_i X_j$ cross-products, testing to see whether this entire set of variables added significantly to the explained variance. If not, all the interactions could be omitted. If they did, then perhaps at least some could be eliminated.

Whenever one finds a reasonably large number of two-variable interactions that are significant, then there would be motivation to proceed to look for higher-order interactions. The assumption, here, is that higher-order interactions will not appear in the absence of lower-order interactions. The theoretical grounds for making such an assumption may not be entirely clear, but in purely empirical terms the assumption appears to be entirely reasonable. Certainly, if one were to find large third- and fourth-order interactions in the absence of first-order interactions, the theoretical explanation would be indeed difficult to derive. Perhaps a somewhat weaker case would have to be made in favor of ignoring two-factor interactions in the absence of main effects, but at least in the case of low-priority variables, near-zero main effects will ordinarily justify one in neglecting the study of interactions involving these variables. Occasionally, of course, one may be led astray, but shortcuts must inevitably be taken in multivariate analyses.

The main point that should be emphasized is that the search for interactions (and nonlinearities) should be both systematic and routine. It should not be a "hit-or-miss" matter of looking for only a select subset of the possible interactions while neglecting the rest. Rarely are sociological (or other social-science) theories explicit and precise enough to specify and predict such interactions (especially higher-order interactions) in advance of data collection. The main pitfall of this kind of "dragnet" approach to data analysis is, of course, that a certain number of statistically significant interactions will appear merely by chance. Therefore, when they are found, it should be ascertained whether or not they are *patterned* in systematic ways. Do they tend to involve only two or three of the variables, for example?

The above remarks imply that there is always a danger of overanalyzing one's data, particularly in instances where the number of parameters to be estimated begins to approach the total sample size, or whenever there are a very large number of complexities of rather minor importance. There is inevitably a strain between the need for simplicity, on the one hand, and for increased explanatory power, on the other. There can be no hard and fast rules of thumb for deciding between the two, particularly in view of the fact that there are many different types of complexity.

Studies vary considerably in terms of the degree to which they are primarily exploratory and to which they are guided by a well-defined theory. They also vary with respect to the quality of measurements, as previously implied. Where measurement has been crude and theory is weak, but where there are large numbers of potential explanatory variables, there are routinized computer procedures for carrying out exploratory analyses. (Sonquist and Morgan [8].) Where a more explicit theory is available, simultaneous-equation techniques are recommended. Whenever the sample size is sufficient, a recommended strategy is to split the sample (randomly) in half, or even in thirds. One may then conduct a purely exploratory study with the first subsample, using these data to develop his theoretical explanations, which may then be tested using the remaining data. In such a way, multivariate statistical techniques may be adapted very flexibly to the occasion and may be used for the purposes of developing one's theories as well as testing them.

Exercises

1. Check as many of the computations in Table 20.1 as is necessary in order for you to understand how these figures were obtained.

2. Take the data for Exercise 1, Chap. 17, and collapse the heterogeneity index into the following categories: 10.0–14.9, 15.0–19.9, 20.0–24.9, 25.0–29.9, and 30.0–49.9. Referring to moral integration as Y, mobility as X, and heterogeneity as A:
    a. Test for interaction. (*Ans.* $F = 2.17$)
    b. Obtain $r_{XY \cdot A}$ and test for significance. (*Ans.* $F = 13.6$)
    c. Adjust the category means in Y for differences with respect to X.
    d. Test for the significance of differences among the adjusted $\bar{Y}$'s. (*Ans.* $F = 2.71$)
    e. Obtain the partial intraclass correlation $r_{iYA \cdot X}$.

3. Conduct a dummy-variable analysis on the data of Exercise 2, allowing for interaction, and compare your results with those of the covariance analysis.

References

    1. Boyle, R. P.: "Path Analysis and Ordinal Data," *American Journal of Sociology*, vol. 75, pp. 461–480, 1970.
    2. Christ, Carl: *Econometric Models and Methods*, John Wiley & Sons, Inc., New York, 1966, Part III.
    3. Dixon, W. J., and F. J. Massey: *Introduction to Statistical Analysis*, 3d ed., McGraw-Hill Book Company, New York, 1969, chap. 12.
    4. Hagood, M. J., and D. O. Price: *Statistics for Sociologists*, Henry Holt and Company, Inc., New York, 1952, chap. 24.
    5. Johnson, P. O.: *Statistical Methods in Research*, Prentice-Hall, Inc., Englewood Cliffs, N.J., 1949, chaps. 10 and 11.
    6. Johnston, J.: *Econometric Methods*, McGraw-Hill Book Company, New York, 1963.
    7. Schuessler, Karl: "Covariance Analysis in Sociological Research," in Edgar Borgatta (ed.), *Sociological Methodology 1969*, Jossey-Bass, Inc., Publishers, San Francisco, 1969, chap. 7.
    8. Sonquist, J. A., and J. N. Morgan: *The Detection of Interaction Effects*, Institute for Social Research, University of Michigan, Ann Arbor, 1964.
    9. Suits, Daniel: "The Use of Dummy Variables in Regression Equations," *Journal of the American Statistical Association*, vol. 52, pp. 548–551, 1957.

Sampling

Part 5

Sampling 21

All the tests we have considered, as well as the procedures used for obtaining confidence intervals, have required the assumption of random sampling. In fact, you may have formed the impression that random sampling is the only respectable kind of sampling used by the statistician. This is far from the case. There are four basic types of probability sampling which will be discussed in this chapter: random sampling, systematic sampling, stratified sampling, and cluster sampling. As we shall see, it is possible to make use of statistical inference with each of these four types of probability samples, although it is unfortunately true that at the present time we are quite restricted as to the number of different types of tests that can be used with nonrandom probability samples. Especially in the case of cluster samples, our formulas also become much more complicated. In a general text such as this, it will therefore be impossible to do much more than to indicate some general considerations of strategy in choosing the type of sampling that will be most appropriate in a given situation.

We have indicated that there are four basic types of probability sampling, one of which is random sampling. What, then, is a probability sample? The distinguishing characteristic of a probability sample is that every individual must have a *known* probability of being included in the sample. In a random sample we have seen that all combinations of individuals have an equal chance of appearing. In making statistical inferences it is not absolutely necessary that all probabilities be equal, since, if the probability of selection is known, it will be possible to adjust for unequal probabilities by a weighting procedure of some kind. It is essential, however, that probabilities be known in order to arrive at the proper weights. If probabilities are unknown, it will be impossible to make legitimate use of statistical inference. With nonprobability sampling we may actually obtain a very representative sample, but we shall not be in a position to evaluate the risks of error involved. After describing and comparing each of the four types of probability sampling, we shall discuss briefly certain instances where nonprobability samples are likely to be obtained.

21.1 Simple Random Sampling

It has been emphasized that in random sampling not only must each individual have an equal chance of being selected but that all combinations must be equally probable. We have also indicated that it is usually more convenient to sample without replacement. Sampling specialists usually refer to such a sample as a "simple random sample." Notice that on each successive draw the probability of an individual's being selected is slightly increased because of the fact that there will be fewer and fewer individuals left unselected from the population. If, on any given draw, the probabilities of all remaining individuals being selected are equal regardless of the individuals previously selected, then we have a simple random sample. In effect, we have independence from one draw to the next except for the fact that no individual can be selected twice.

By what mechanical procedures are random samples selected? It is sometimes erroneously thought that almost any "hit-or-miss" method of sampling will yield a random sample. This is far from the case. Such methods almost invariably lead to a biased sample because of the human element involved. In order to assure ourselves that all individuals, including those who are atypical or difficult to locate, do in fact have an equal chance of appearing, we must ordinarily go to great lengths to draw our sample. First, we must be sure that each individual in the population is listed once and only once. We can then associate a number with each position on the list and make use of some mechanical procedure, such as that used in a bingo game, to assure equal probabilities of selection. Let us first examine certain problems that may be encountered in obtaining such lists, or what sampling specialists refer to as a "sampling frame."

It might be thought that obtaining a list is usually a simple matter. In most practical research problems this is not the case. Often, there are no lists at all. For example, there is no list of residents of the United States or the state of Michigan. There will almost certainly be no list of blacks or Japanese-Americans living in a given community. If no list exists, it may be extremely expensive to compile one. If this is the case, other methods of probability sampling may be preferable to simple random sampling. Lists may exist, but they may be out of date. Some individuals may not have been included while others on the list may no longer be members of the population concerned. City directories, which at first thought appear to be the ideal source for one who wishes to draw a random sample of residents, may be so out of date by the time they are published as to be highly misleading. Those individuals who have recently arrived will be excluded from the list and therefore will have no chance of being selected in the sample. To the degree that such persons differ from the rest of the population with respect to whatever characteristics are being studied, the researcher will obtain a biased sample and misleading results. Other lists, such as telephone directories or motor-vehicle registration lists, may be biased in that lower-income groups are especially likely to be underrepresented. It is safe to say, therefore, that no matter how accurate a list appears to be, you should always carefully investigate its adequacy. A poor list can be worse than none at all if it leads to an unusually biased sample.

What can we do if the list is inadequate? If the list is complete but involves duplications, the problem is relatively simple provided, of course, that the duplications are readily spotted. For example, if the list consists of all children at a particular school and we wish to select a random sample of *parents*, we would undoubtedly discover that some parents had more than one child attending school. Therefore if we gave each *child's* card an equal chance of being selected, certain parents would have a better chance than others of being included. In order to remedy this situation we could simply remove the cards of all but one sibling, or we could select a parent only if his oldest child's card were chosen, rejecting him if any of his other children's cards have been included.

It should be noted that if Jones's second or third child were selected and we therefore did not include Jones in the sample, it would not be legitimate to substitute for Jones the parent whose child appeared next to Jones's on the list. If this were done, persons with cards next to parents with more than one child would have a higher probability of being selected. The correct procedure would be to omit Jones and go on to the next card selected by probability methods. Another alternative which is theoretically possible but which may create additional problems for analysis would be to include Jones if any of his children's cards are selected but to give him relatively less weight in the analysis. Thus, if he has three children and therefore three times the ordinary probability of selection, his scores would be given one-third the weight of the parent with only one child.

In most problems it is more likely that the list will be incomplete or that it will include names of individuals who are no longer members of the population. Here it will again be possible to purify the list until it is correct. If this is not feasible, it may be desirable to redefine the population slightly to conform to the list. Suppose a list of employees is known to be complete and accurate as of the first of the year. Rather than obtain the names of all persons hired since that date, it may be possible to confine one's attention to employees who were with the company prior to this time and who are presently employed there. Any persons included in the sample but found to have left the company can then be ignored. Notice, however, that the population studied would not be *all* present employees, and any reader should be made well aware of this fact.

Having obtained an accurate list, it is a relatively simple matter to draw a random sample. Theoretically, a number of mechanical devices could be used to assure equal probabilities of selection. One could use a well-shuffled deck of cards or draw numbers from a hat. Perhaps a round cage containing balls with numbers on them would give more reliable results because of the tendency of cards or slips of paper to stick together when shuffled or mixed. Actually, the researcher need not go through such an involved process since tables of random numbers have already been constructed for this purpose. Such tables can be constructed by using mechanical devices such as indicated above or by using electronic devices. For example, one could place an equal number of balls with the digits 0, 1, 2, . . . , 9 in a basket and proceed to select balls, each time replacing and shuffling thoroughly. The resulting digits could then be used to form a table of random numbers such as that given in Table B of Appendix 2.

In using a table of random numbers it makes no difference whether we go down columns or across rows, nor does it matter which column or row we start with—as long as our decision is made prior to examining the data. To illustrate the use of a table of random numbers, let us suppose that one wishes to draw a sample of size 100 from a population consisting of 736 individuals. Since the number 736 consists of three digits, we shall find it convenient to select three adjacent columns (any three), selecting another three columns when we come to the bottom of the page. Suppose, for example, that we decide to use the first three columns of the first page of Table B. As the first case in the sample, we select the first number between 001 and 736 which appears. This number is 100. In other words, the one hundredth individual will be in the sample. We proceed down columns 1 to 3, obtaining the numbers 375 and 084. We then come to the number 990. This would correspond to the 990th individual in the population, but since there is no such individual we move on to the next number which is 128.

After a while, we begin to run into numbers which have already been selected. Since we are sampling without replacement, we must omit these repetitions until we have finally selected 100 cases. This is all there is to it. The reason that the process is so simple and that arbitrary decisions can be made as to the use of columns or rows is, of course, that the numbers appearing in the table are completely random. As a matter of fact, it is almost impossible to use such a table incorrectly unless columns (or rows) are repeated or unless one cheats by deciding he wants the 219th case in the sample and deliberately looks for a column containing this number.

Correction for sampling without replacement   It was mentioned in Chap. 9, which was on probability, that when we sample without replacement, we violate the assumption of independence and that, strictly speaking, we therefore must modify our formulas to take this fact into consideration. Usually the problem is not a serious one since the sample selected is a small fraction of the population, and therefore the chances of any given individual's being selected two or more times is rather slight. If the sample size is as much as one-fifth the size of the population, however, it may be desirable to introduce correction factors whenever these factors are known. Unfortunately, exact correction factors are known only in the simplest kinds of problems. This fact is seldom a disturbing one, however, since, if we were going to select a sample which is one-third or one-half the size of the population, we would probably be in a position to select the entire population anyway. The use of a correction factor for formulas involving the standard error of the mean will be discussed below. In more complicated cases you should refer to a standard text on sampling, although you will probably not find any discussion of correction factors for the various nonparametric tests in such texts. Such tests, however, are most applicable for small samples where the replacement problem is of minor importance.

The formula we should actually use for the standard error of the mean if we have sampled without replacement is

$$\sigma_{\bar{X}} = \sqrt{1 - f}\,\frac{\sigma}{\sqrt{N}} \tag{21.1}$$

where f represents the *sampling fraction* or the ratio of the number of cases in the sample to the size of the population. If we refer to the sample size as N and the population size as M, we may rewrite the correction factor as

$$\sqrt{1 - \frac{N}{M}}$$

It can immediately be seen that if the sample size is relatively small as compared with M, the value of the correction factor becomes approxi-

mately unity, and there is little or no point in using it. Thus, if a sample
of 500 is selected from a population of 10,000, the sampling fraction is
$\frac{1}{20}$ and the value of the correction factor becomes .975. Notice that
since the correction factor must be less than 1 for finite populations, the
corrected value of the standard error will always be less than the uncorrected figure. Thus, if we desire a small standard error, as is usually the
case, we shall be on the conservative side if we do not make use of the
correction. Unless the sampling fraction is of the order of one-fifth or
more, we seldom bother to use it.

This same correction factor can be used in other formulas involving
standard errors of means or proportions. Thus, if an estimate is to be
used, we would use the formula

$$\hat{\sigma}_{\bar{x}} = \sqrt{1 - f}\left(\frac{\hat{\sigma}}{\sqrt{N}}\right) = \sqrt{1 - f}\left(\frac{s}{\sqrt{N - 1}}\right) \qquad (21.2)$$

In a difference-of-means test there would be two sampling fractions, and
the basic formula for the estimate of the standard error of the difference
of means would be

$$\hat{\sigma}_{\bar{X}_1 - \bar{X}_2} = \sqrt{(1 - f_1)\frac{\hat{\sigma}_1{}^2}{N_1} + (1 - f_2)\frac{\hat{\sigma}_2{}^2}{N_2}} \qquad (21.3)$$

21.2   Systematic Sampling

Another type of sampling used quite frequently is likely to be confused
with simple random sampling and, in fact, is often used interchangeably
with simple random sampling. In systematic sampling, instead of using a
table of random numbers, we simply go down a list taking every kth
individual, starting with a randomly selected case among the first k individuals. Thus, if we wanted to select a sample of 90 persons from a list
of 1,800, we would take every twentieth in the list. Our first choice,
however, must be determined by some random process such as the use of
a table of random numbers. Suppose the eleventh person were selected.
The sample would then consist of individuals numbered 11, 31, 51, 71,
91, ...

Systematic sampling is obviously much simpler than random sampling
whenever a list is extremely long or whenever a large sample is to be
drawn. If a telephone directory or city directory could legitimately be
used, we can imagine the difficulty in locating the 512th, 1,078th, and
15,324th individuals. If the ordering used in compiling the list can be
considered to be essentially random with respect to the variables being

measured, a systematic sample will be equivalent to a simple random sample. For example, most lists are given in alphabetical order. Surnames, of course, are not random. A husband and wife listed separately would have practically no chance of appearing together in the sample unless their name were an extremely common one. Certain ethnic groups may have an undue proportion of names beginning with the same letter (O'Brien, O'Neil, etc.). Actually, in the case of alphabetical lists we have something approximating a stratified sample (see below) in which ethnic groups may have a tendency to be grouped together. Taking every kth individual is thus likely to give a proper representation of each group. In practice since alphabetical ordering is essentially irrelevant to most variables studied, we are usually safe in considering a systematic sample as equivalent to a simple random sample. Special formulas have been developed, however, which make use of somewhat different assumptions. In most cases they will hardly be worth the extra trouble.

There are two types of situations in which systematic sampling may cause serious biases. Fortunately, neither occurs very frequently in sociological problems. *First,* the individuals may have been ordered so that a trend occurs. If persons have been listed according to office, prestige, or seniority, the position of the random start may affect the results. Suppose, for example, that the sampling fraction is $\frac{1}{30}$. Two persons may draw systematic samples with very different random starts. A random start of two would yield a considerably higher average score (if individuals were ranked from high to low) than a start of 27 since each individual in the first sample will be 25 ranks ahead of the comparable person in the second sample. If a trend of this sort is noticed, the list may have to be shuffled somewhat or a "middle start" used (e.g., start with the fifteenth or sixteenth individual).

The *second* type of situation to be avoided is that in which the list has some periodic or cyclical characteristic which corresponds to the sampling fraction. For example, in a housing development or apartment house every eighth dwelling unit may be a corner unit. If it is somewhat larger than the others its occupants can be expected to differ as well. If the sampling fraction also happens to be $\frac{1}{8}$, one could obtain a sample with either all corner units or no corner units depending on the random start. To avoid this pitfall, one could change the sampling fraction slightly to $\frac{1}{7}$ or $\frac{1}{9}$, or he could make use of several different random starts. After selecting ten households, he could pick another random number and go to ten more residences, draw a third number, and so on.

Systematic sampling, in combination with other designs, is often used in social surveys because of its simplicity. An untrained interviewer can much more easily be told to go to every third house in a block than to use a table of random numbers. As with simple random sampling, how-ever, the listing must be complete and accurate. If the interviewer were to miss the smaller apartments or certain residences in back alleys, serious errors could result. It is important to realize that in all types of proba-bility sampling there must be both some element of randomization and some sort of a complete listing. As we shall see presently, however, the nature of the required lists may differ from one design to the next, some being much simpler to obtain than others. It will always be necessary for the researcher to examine his list carefully and to know how it has been constructed and the nature of its defects.

21.3 Stratified Sampling

While the differences between simple random and systematic sampling are usually relatively minor in terms of cost saving or problems of analysis, the remaining two basic types of sampling differ in certain fundamental respects from the preceding two already discussed. As we shall see, both stratified and cluster sampling can be used, under certain circumstances, to improve the efficiency of the sampling design. In other words, they may be designed to yield greater accuracy for the same cost or, if you prefer, to involve less cost for the same accuracy. It will also be found that both these designs require different formulas from any we have previously used.

In a stratified sample we first divide all individuals into groups or categories and then select independent samples within each group or stratum. It is important that the strata be defined in such a way that each individual appears in one and only one stratum. In the simplest and most frequently used types of stratified sampling, we take either a simple random sample or a systematic sample within each stratum. The sampling fractions for each stratum may be equal, in which case we speak of *proportional* stratified sampling, or we may have *disproportional* strati-fied sampling.

One reason that we often stratify a sample is that different sampling methods or lists may have been used for each stratum. For example, the strata may consist of separate factories, schools, or dormitories, each of which was studied at a different time and by different persons. It might have been completely unfeasible to combine the lists for all strata and then to select a single random sample. Another important reason for stratifying rather than taking a random sample is to reduce the number of cases required in order to achieve a given degree of accuracy. To the degree that the strata are homogeneous with respect to the vari-ables being studied, we can improve the efficiency of the design. By discussing proportional and disproportional stratified sampling, we shall see more specifically some of the advantages of stratified sampling over simple random sampling.

Proportional stratified sampling Proportional stratified sampling is often used to assure a more representative sample than might be expected under simple random or systematic sampling. Suppose, for example, that it is known that there are 600 Protestants, 300 Catholics, and 100 Jews in a given population. If a random sample of size 100 were drawn, we would certainly not expect to get exactly 60 Protestants, 30 Catholics, and 10 Jews. The proportion of Jews, especially, might be relatively either too large or too small. Now suppose we were interested in studying some variable, such as church attendance, which is closely related to denomination. Suppose, also, that we wished to estimate the mean number of times persons in the population attended church. It is easy to see intuitively that a proportional stratified sample in which the sampling fractions for all three strata were $\frac{1}{10}$ (i.e., consisting of 60 Protestants, 30 Catholics, and 10 Jews) would ordinarily yield more reliable results than a simple random sample.

We have here, in effect, a problem analogous to analysis of variance. In a random sample there are two sources of variation. There may be sampling errors *within* each stratum, and there may be errors *between* strata with respect to the relative numbers selected. Not only might we select very atypical Jews or Catholics but also might select too many or too few of each type. In stratified sampling we have eliminated this type of between-stratum variation and are left only with the within variation. If the strata were completely homogeneous, proportional stratified sampling would always yield exactly correct results whereas simple random sampling would not. On the other hand, if the strata were as heterogeneous as would be expected by chance, we would gain nothing by stratifying. In other words, if the differences between groups are small as compared with the within differences, stratification is of no help. Thus the gain from stratifying is roughly proportional to the intraclass correlation between the two variables. If the criterion for stratifying is highly related to the variable studied, the gain may therefore be considerable. In gaining control over the number of cases in each stratum, something that was not possible in random sampling, we can assure ourselves of more accuracy for a given size sample.

You should not come to expect too much of proportional stratified sampling. If the size of the sample is relatively large, we expect, of course, that chance factors alone will assure us of approximately the correct proportions from each of the strata. Since problems of analysis are not made too complex by proportional stratified sampling, there is little to lose by stratifying, however. It is usually neither essential nor feasible to hunt around to obtain a single "best" criterion for stratifying. To obtain a proportional stratified sample the sizes of the population strata must be known, and it will, of course, only be possible to stratify according to variables for which information is given from the listing at the time the sample is drawn. This often means that one is confined to such simple variables as sex, age, occupation, or area of residence. Several of these variables may even be used in combination, if so desired, although it will seldom be advantageous to stratify by more than two or three variables simultaneously. Since stratification is such a simple procedure, however, its possibilities should always be explored.

Disproportional stratified sampling   In disproportional stratified sampling we make use of different sampling fractions to manipulate the number of cases selected in order to improve still further the efficiency of the design. There are several types of situations in which this type of sampling is desirable. Often our interest may center primarily on the separate subpopulations represented by the strata rather than on the entire population. Suppose, for example, that we wished to *compare* the three major religious groups with respect to their attendance. Obviously, both simple random sampling and proportional stratified sampling would give us too few Jews in the sample to make meaningful comparisons. We might therefore decide to select equal numbers from each group, thereby giving each Jewish person a probability of selection equal to three times that of each Catholic and six times that of any given Protestant. If we were to select 50 from each group, the respective sampling fractions would thus be $\frac{1}{12}$, $\frac{1}{6}$, and $\frac{1}{2}$. If we then wished to generalize to the entire population in order to estimate the mean attendance figure, we would have to weight the means of the three strata to compensate for the fact that Jews have been oversampled. This weighting procedure will be described below.

Even if our goal is to generalize to the entire population rather than to compare different subpopulations, it may still be desirable to make use of disproportional stratified sampling if either (1) the standard deviations within the separate strata differ considerably among themselves or (2) the cost of gathering data varies substantially from stratum to stratum. There will always be some *optimum allocation* for which the sampling design will have maximum efficiency. In other words, there will be a certain set of sampling fractions that will yield the smallest sampling error for a given cost. We can obtain such an optimum allocation if we *make the sampling fraction for each stratum directly proportional to the standard deviation within the stratum and inversely proportional to the square root of cost of each case within the stratum.* Let us see intuitively why this is the case, looking first at the question of the standard deviations.

If one particular stratum is unusually homogeneous with respect to the variable being studied, it will be unnecessary to select a very large sample from this stratum in order to obtain a given degree of accuracy. On the other hand, it will be advisable to take a much larger sample from a very heterogeneous stratum. Since our overall accuracy will be determined primarily by the degree of accuracy in the weakest link in the chain, so to speak, it is important that we do not have one or two strata with large sampling errors. This is especially true if these strata happen to be large ones. It would be pointless to have perfect accuracy in several of the smaller strata but a very large sampling error in another stratum. Therefore, if we take relatively more cases from the heterogeneous strata and fewer from the homogeneous ones, we can get by with fewer cases. As it turns out mathematically, the desired sampling fractions are actually proportional to the relative standard deviations rather than the variances.

A word of caution is necessary at this point. A particular stratum may be very homogeneous with respect to one variable being studied and yet heterogeneous with respect to another. Since a research project usually involves a study of more than one variable, it therefore may be extremely difficult to find allocations which are optimal, or nearly optimal, for more than one variable. In fact, a design that is very efficient for one variable may be extremely inefficient for another. Therefore, it is best to consult a sampling specialist and to be well aware of the important variables before using disproportional allocation. When in doubt, proportional allocation would be much safer.

Thus far, cost considerations have never been discussed because we have been implicitly assuming equal costs for gathering data on all individuals. Suppose, however, that this is not the case and that certain strata involve higher costs than others. Various administrators, for example, may permit different data-collecting techniques, or perhaps the physical layouts in the different strata are such that more time is consumed interviewing in one stratum than in the rest. Other factors being equal, it would obviously be less expensive to select a relatively larger number of cases from the cheapest strata. It can be shown mathematically that optimum allocation will be attained if sampling fractions are taken inversely proportional to the square root of the cost factors.

Notice that in the special case where all costs are equal and where all within-stratum standard deviations are equal, the sampling fractions will also be equal and we have the situation in which proportional stratification gives us optimum allocation. In general, it is usually wise to follow the rule of using proportional stratification unless cost differentials are very great or unless stratum standard deviations are substantially different. As will be seen below, the use of disproportional sampling tends to complicate problems of analysis and should therefore not be used unless it is clearly to one's advantage to do so.

We have not as yet faced up to an extremely important question. How can we make use of cost calculations and the relative standard deviations when these are unknown at the time the sample is drawn? The obvious answer is that they have to be estimated, just as we had to make enlightened guesses as to the values of certain parameters before we could estimate the size of the sample we would need. We must realize, however, that the kinds of estimates we need are not the kinds of estimates we make from sample statistics. Of course, it would be possible to do a pilot study in order to obtain such estimates, but unless the study is to be an extremely large and expensive one, such an outlay of money will probably not be feasible. Our estimates must therefore be based on the experience of experts or on the results of previous studies. The situation is not quite as bad as it sounds, however, since it is usually possible to obtain very good approximations to optimum allocation with very crude guesses as to costs and standard deviations. In other words, if there is reason to suspect large differences among strata with respect to either of these factors, an enlightened guess is likely to yield a design which is almost as efficient as would be obtained with exact values.

Computations with stratified samples When computing estimates of means and estimating standard errors from stratified samples, we must compute values separately for each of the strata and then weight them according to the relative size of the stratum in the population. If we let W_i indicate the weight of the ith stratum in the population and if we set $\Sigma W_i = 1$, thereby reducing the weights to proportions, we can write the formula for estimating the population mean as

$$\bar{X} = \sum_{i=1}^{k} W_i \bar{X}_i$$

where the $\bar{X}_i$ are the sample means for each of the k strata. This formula is as we would expect. It simply says that if one stratum is three times as large as a second, its mean should receive three times as much weight.

If *proportional* stratified sampling has been used and if we let N_i and M_i, respectively, indicate the sizes of the sample and population for the ith stratum, then by definition all N_i/M_i will be equal to N/M. But since for the ith stratum

$$\bar{X}_i = \frac{\sum\limits_{j=1}^{N_i} X_{ij}}{N_i}$$

and also

$$W_i = \frac{M_i}{M} = \frac{N_i}{N}$$

we have

$$\bar{X} = \sum_{i=1}^{k} \frac{N_i}{N} \frac{\sum\limits_{j=1}^{N_i} X_{ij}}{N_i} = \frac{1}{N} \sum_{i=1}^{k} \sum_{j=1}^{N_i} X_{ij}$$

This double summation simply means that we have summed all of the X's. Since we then divide this sum by the total number of cases to get $\bar{X}$, we thus see that in the case of proportional stratified sampling we could have obtained the estimate of μ exactly as we did in the case of simple random samples. For this reason we refer to proportional stratification as being *self-weighting*. In other words, each stratum has received its proper weight. If stratification has been disproportional, we must multiply each $\bar{X}_i$ by the weight of that stratum *in the population*.

In estimating the standard error of the mean, our computations cannot be worked out so easily. We must first estimate the standard error for each stratum and then pool the results as was done in the difference-of-means test and in analysis of variance. It will be remembered that instead of summing standard deviations, we worked with the variances and sums of squares. We also have to square the weights W_i. The formula for the estimated *variance* of the mean using stratified sampling can thus be written

$$\hat{\sigma}_{\bar{X}}^2 = \Sigma W_i^2 \hat{\sigma}_{\bar{X}_i}^2$$

where $\hat{\sigma}_{\bar{X}_i}^2$ indicates an estimate of the variance of the mean within the ith stratum. We can obtain the estimated standard error of the mean by taking the square root of the above expression and can then compute the t statistic as before.

Suppose, for example, that there are three counties and that data for the counties can be summarized as in Table 21.1. Notice that we have obtained a disproportionate sample since unequal sampling fractions have been used. Let us assume that simple random sampling was used within

Table 21.1 Data for computing parameter estimates from stratified samples

	County			Total
	1	2	3	
Size of county (M_i)	10,000	15,000	25,000	50,000 (= M)
Weight (W_i)	.20	.30	.50	1.00
Size of sample (N_i)	50	50	50	150 (= N)
Sample mean ($\bar{X}_i$)	3,100	4,300	3,800	
Sample standard deviation (s_i)	500	400	300	

each stratum and that samples were independently drawn. The estimated standard errors, ignoring the factor $1 - f$, are

County 1:
$$\frac{s_1}{\sqrt{N_1 - 1}} = \frac{500}{\sqrt{49}} = 71.4$$

County 2:
$$\frac{s_2}{\sqrt{N_2 - 1}} = \frac{400}{\sqrt{49}} = 57.1$$

County 3:
$$\frac{s_3}{\sqrt{N_3 - 1}} = \frac{300}{\sqrt{49}} = 42.9$$

The estimated mean and variance will therefore be

$$\bar{X} = .20(3,100) + .30(4,300) + .50(3,800) = 3,810$$

and $\quad \hat{\sigma}_{\bar{X}}^2 = (.20)^2(71.4)^2 + (.30)^2(57.1)^2 + (.50)^2(42.9)^2$

$$= 957.5$$

Although computations for means and proportions are straightforward in the case of stratified samples, it should be recognized that one cannot legitimately use the various nonparametric tests, tests for the significance of correlation, analysis of covariance, etc., without substantial modification. Unfortunately, you will generally not find discussions of these problems in textbooks on sampling. We know how to handle complicated statistical problems if we can assume the simplest kind of sampling, random sampling. With the more complicated sampling designs we can handle the simplest of statistical problems such as estimating means or proportions, computing confidence intervals for means and proportions, and making tests of differences of means. There is a gap, however, when it comes to more sophisticated statistical techniques with complicated sample designs.

21.4 Cluster Sampling

In stratified sampling we divided our population into groups that we called strata, and we sampled from *every* stratum. Sometimes it is advantageous to divide the population into a large number of groups, called clusters, and to sample *among* the clusters. For example, we might divide a city into several hundred census tracts and then select 40 tracts for our sample. Such a sampling design is referred to as cluster sampling and is frequently used in social surveys in order to cut down on the cost of gathering data. As we shall see presently, the aim in cluster sampling is to select clusters that are as *heterogeneous* as possible but that are small enough to cut down on such expenses as listing costs and travel costs involved in interviewing.

In cluster sampling we do not sample our elements directly. Instead, we sample clusters or groups of elements. In the simplest of cluster designs we might use random selection among clusters and then select every individual within those clusters included in the sample of clusters. Such a design is often referred to as a single-stage cluster design since sampling occurs only once in the process. In multistage sampling, on the other hand, the design may be much more complicated. We might first take a simple random sample of census tracts within the city. Then within each tract we might take a simple random sample of blocks (smaller clusters). Finally, the interviewer might be instructed to select every third dwelling unit within those blocks included and to interview every second adult within each of these households. Thus, sampling procedures may enter the selection process at a number of points. It is essential in probability sampling, of course, that there be some element of randomness in the procedure. Sampling fractions may be computed which produce unbiased samples so that every individual in the population has an equal chance of appearing in the sample. It will not be possible to ensure independence of selection by this method, however. Persons within the same cluster will generally have a better chance of appearing together in the same sample than members of different clusters. In fact, the whole purpose of cluster sampling is to ensure that this will occur.

It will be instructive to compare cluster sampling with both simple random sampling and stratified sampling. In order to simplify the argument, let us suppose that we are using a single-stage cluster design in which clusters are selected randomly and then every individual within the sampled clusters is used in the total sample. How does cluster sampling differ from stratified sampling? Notice that although both involve dividing the population into groups, they in a sense involve opposite sampling operations. In stratified sampling we *sample individuals* within every stratum. We are therefore sure that every stratum is represented by a certain number of cases. Our sampling errors involve variability *within* the strata. We therefore want the strata to be as homogeneous as possible and as different as possible from each other.

In (single-stage) cluster sampling, on the other hand, we have no source of sampling error within a cluster because every case is being used. Since we are only taking a sample of clusters our error now involves variability *between* the clusters. If the cluster means differ considerably as compared to the variability within clusters, we run the risk of obtaining some very unusual clusters in our sample of clusters. If this should occur, and if the clusters are homogeneous, our sampling error could be considerable. But if the clusters are heterogeneous as compared with differences among clusters, we can get by with relatively few large clusters. Suppose, in the extreme, that every cluster were heterogeneous and that, by comparison, differences among cluster means were insignificant. We could then simply select one very large cluster and obtain an excellent sample. However, if the clusters were completely homogeneous, we would need only one case in each cluster. We thus attempt to obtain homogeneous strata but heterogeneous clusters, the reason for the variance in strategy being the difference in the point at which the sample is drawn.

Let us now compare cluster sampling with simple random sampling. In practically all examples you will encounter, cluster samples will be *less efficient* (i.e., will yield greater sampling errors) than simple random samples *of the same size.* As we shall see shortly, however, it may *cost* considerably less to obtain cluster samples. Our problem will be essentially that of balancing cost and efficiency. How, then, do we compare the relative efficiencies of the two designs? Efficiency is most conveniently measured in terms of the size of the standard error of the estimate, a small standard error indicating high efficiency. As we have seen, it is desirable to obtain clusters that are as heterogeneous as possible. This intuitive notion can be translated into a formula involving the intraclass correlation coefficient. It can be shown that the ratio of the variances of the estimates of μ for cluster and for simple random samples of the same size is approximately

$$\frac{\sigma \bar{X}_C{}^2}{\sigma \bar{X}_R{}^2} = 1 + \rho_i(\bar{N} - 1)$$

where $\sigma \bar{X}_C{}^2$ and $\sigma \bar{X}_R{}^2$ represent the variances of the means for cluster and simple random samples, respectively, ρ_i represents the population intraclass correlation, and $\bar{N}$ is the mean number of cases selected from each of the clusters.

Notice that the ratio of variances will ordinarily be greater than unity, indicating larger variances (and hence standard errors) for cluster sampling. The expression will be greater than unity unless either $\bar{N} = 1$ or $\rho_i \leq 0$. Obviously, if $\bar{N} = 1$ the cluster sample reduces to the special case of a random sample since each cluster consists of a single case. Intraclass correlation is, of course, a measure of homogeneity. If the cluster is more homogeneous than would be expected by chance, ρ_i will be greater than zero, and the more homogeneous the cluster, the larger the value of ρ_i. Conceivably, ρ_i can be negative. But this would require that the cluster be more heterogeneous than would be expected by chance. As it turns out, the kinds of clusters which we ordinarily select for practical purposes will nearly always be at least as homogeneous as expected by chance.

We see that if $\rho_i > 0$, the larger the number of cases $\bar{N}$ selected from the cluster, the greater the ratio of variances and therefore the lower the relative efficiency of the cluster design. This can be seen intuitively. If a cluster is quite homogeneous, we do not need very many cases to obtain an accurate estimate of its mean. We might very well have taken a small sample from within the cluster and used the money saved to study additional clusters. There are thus two factors which determine the relative efficiency of the cluster design: the degree of homogeneity within the clusters and the size of the cluster itself. We want to select only a small number of cases from homogeneous clusters; if clusters are heterogeneous, we can afford to take more cases within each cluster without seriously impairing the efficiency.

As we have already indicated, cluster sampling is usually more economical than simple random sampling. Suppose, for example, that one wished to obtain a nationwide sample to study voting or fertility behavior. In the first place, no list of adults would be available. The cost of compiling such a list would be prohibitive. Lists of counties are available, however. It would certainly be much less expensive to draw a random (or systematic or stratified) sample of counties and then to work only within those counties actually selected. Even within each county a simple random sample would probably not be feasible. There is another obvious cost-saving factor. It will be much less expensive to send interviewers into, say, 50 counties than to scatter them all over the countryside. In a simple random sample, there might be only ten persons selected in the state of Montana. With cluster sampling, local interviewers can be efficiently trained, and each can obtain a relatively large number of interviews without incurring huge travel expenses. Cluster samplings taken on state, county, or city levels would all have the same advantages, although to a lesser degree to be sure.

There are a number of costs involved in any sample survey. It is these costs, not the number of cases, that set the limits to the study. There are certain fixed costs which will be independent of the sample design and the number of cases selected. For our purposes, these can be ignored since they can simply be subtracted from the total funds available. Then there will be costs involved in actually listing the units to be sampled. As we have seen, cluster sampling often reduces listing costs considerably. Certain costs will be directly proportional to the number of cases ultimately selected. The salary paid to the interviewer while he is actually talking to the respondent, costs of coding the data, and certain computational costs fall into this category.

Other costs may be proportional to the number of *clusters* selected, however. Most travel costs, including costs of call-backs, are of this variety. It will be more economical to send an individual into a given county for several days and then into a second county than to have him travel all over the state only to find that certain respondents are not home the first time he calls. Generally speaking, whenever travel costs and other costs which depend on the number of *clusters* selected are quite large as compared with costs which vary directly with the number of cases, cluster sampling will be more economical than simple random sampling. For example, in a survey of a large area involving very short interviews, cluster sampling may be appropriate. If interviews last for several hours or more, simple random sampling may be more sensible provided listing costs are not prohibitive.

Thus, when deciding which design to use, one must balance cost considerations against the efficiency of the design. Whichever method will yield a smaller standard error for a given cost should be used. Since it is not necessary to take every individual within sampled clusters, a mult-stage sample may provide a satisfactory compromise. We then have the complicated problem of selecting an optimum design in which we must determine the number of stages in which sampling is to be used, the number of clusters to be used, and the number of cases to be selected from within each cluster. The problem is further complicated by the fact that most studies will undoubtedly involve not one but a number of variables and that the clusters will not be of the same size. Lest there be any doubt in your mind at this point, it is therefore always wise to consult a sampling specialist before making a decision as to design. Careful planning not only can be economical but also can result in fewer problems when it comes to analysis of the data.

Before closing this section on cluster sampling, it is again necessary to inject a word of caution. Formulas found in this text cannot be used with cluster sampling. As has been pointed out, errors introduced by using simple-random-sampling formulas for data collected from cluster samples can be extremely serious. These errors are not of the order of magnitude of errors introduced by using the normal table instead of the *t* table. They may be much greater. Instead of having significance at the .05 level, the true level (as obtained by correct cluster-sample formulas) may be as high as .50 (see [3]). If we wish to reject a null hypothesis, we shall seldom if ever be on the conservative side using random-sampling formulas for clustered data. It will be remembered that cluster samples are less efficient than simple random samples of the same size. Therefore, simple-random-sample formulas will *underestimate* the true standard errors. Put differently, a cluster sample of a given size may be equivalent in terms of efficiency to a much smaller random sample. A cluster sample of size 800 may be equivalent in efficiency to a simple random sample of 500. If simple-random-sample formulas are used with an N of 800, therefore, we are more likely to obtain significance than if the correct procedures were used.

One should thus be extremely cautious in analyzing data from clustered samples. He should not make use of statistics such as chi square unless the sampling specialist can help him introduce appropriate correction factors. The problem is not quite so serious with stratified samples because, if anything, stratified samples are more efficient than simple random samples. A stratified sample of a given size may be equivalent in efficiency to a somewhat larger random sample, and the researcher will generally be on the conservative side in rejecting a null hypothesis using random-sample formulas. This is not always the case, however, and proper caution should therefore be exercised.

21.5 Nonprobability Sampling

We turn, briefly, to certain situations in which nonprobability sampling has been used. The major disadvantage of nonprobability sampling is that we can obtain no valid estimate of our risks of error. Therefore, statistical inference is not legitimate and should not be used. This does not mean that nonprobability sampling is never appropriate. In exploratory studies, the main goal of which is to obtain valuable insights which ultimately may lead to testable hypotheses, probability sampling either may be too expensive or lead to fewer such insights. One may wish to interview persons who are in especially good positions to supply information, for example. Or he may wish to interview extreme cases, those who will provide him with most striking differences. If this is done, of course, he has no legitimate right to test for the significance of differences among extremes unless he is attempting to generalize to a population made up entirely of such persons. The fact that you can undoubtedly think of studies in which statistical tests have been made on such extreme cases does not mean that such a procedure is legitimate. This is not to deny, however, that useful insights may be obtained by such a comparison.

Nonprobability methods are sometimes used when the purpose is to make generalizations about a population sampled. Such methods invariably either make use of the interviewer's judgment as to the individuals to be included or permit an individual sampled to be selected out of the study on some nonrandom basis.

Quota samples sometimes used in public opinion surveys seem, on the surface, to be similar to stratified samples. An interviewer is given certain "quotas" he must fill. He must have so many females over 40, so many persons with an income of less than $3,000, or a certain percentage of Catholics. But it is left up to his discretion *which* females over 40 or which Catholics to interview. Being only human, he is likely to select those persons who are most conveniently located. If he goes to their home, he may select only those persons who are at home at the time. Even if he is consciously aware of such a selective tendency, it will be difficult for him to correct for it exactly. An extremely conscientious interviewer might even oversample persons who are seldom at home or lower-class individuals who might be missed by other interviewers. Perhaps a well-trained person may become quite expert in the use of his judgment. But it will be difficult if not impossible to tell. If any group which is either under- or oversampled happens to differ markedly from others with respect to the variable being studied, the sample may be seriously biased. What is more, we have no way of estimating just how biased it may be.

Whenever lists are incomplete or whenever a large percentage of persons must be considered as nonrespondents, we have in effect another example of nonprobability sampling. If one receives a 50 per cent return on a mailed questionnaire, serious biases can be introduced owing to the fact that nonrespondents may differ significantly from those who returned the questionnaire. Thus, even though pains may have been taken initially to obtain a probability sample, certain individuals actually have no probability of being included in the ultimate sample because they have selected themselves out by refusing to answer. It is for this reason that it is of utmost importance to follow up a mailed questionnaire with one or more postcards in order to obtain a higher percentage return. Likewise, the interviewer must learn to be persistent and must expect to make several call-backs in order to get a sufficient return. Obviously, a substantial bias cannot be compensated for by a large sample.

21.6 Nonsampling Errors and Sample Size

Even if one has been extremely careful to design a study that meets all of the requirements of good sampling, there will always be certain nonsampling errors involved. Probability theory enables us to evaluate the risks of sampling errors, i.e., those errors introduced by virtue of the fact that samples vary from one to the next. Nonsampling errors, on the other hand, are errors of measurement. In a study involving an interview or questionnaire there will always be response errors. In some cases, such as persons' ages, for example, there may be a consistency of errors leading to a definite bias. In other examples, response errors may be more or less random. An interviewer's own biases may color his results.

In this text we cannot go into a detailed discussion of the kinds of nonsampling errors possible. One extremely important point is worth mentioning, however. There is nothing to be gained in reducing sampling errors below a certain point as compared with nonsampling errors. If these two types of errors can be assumed to be independent of each other, we can diagram the situation as in Fig. 21.1. The total error is thus a

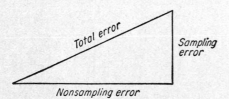

Figure 21.1 Relationship between total error and sampling and nonsampling errors.

function of two independent sources of error and cannot be substantially reduced unless both types are simultaneously controlled. If nonsampling mistakes such as response or interviewing errors are large, there is no point in taking a huge sample in order to reduce the standard error of the estimate since the total error will be primarily determined by the length of the base of the triangle. Likewise, if one is willing to go to great pains to reduce nonsampling errors to a minimum, it will be foolish for him to make use of a small sample, thereby having a large sampling error. A proper balance between sampling and nonsampling errors should therefore be maintained. Research accuracy limits effective sample size and vice versa. Unfortunately, nonsampling errors are usually difficult to estimate. If errors can be estimated, however, the most effective total design would be one for which both legs of the triangle are equal. It is well to keep this fact in mind.

Glossary
Cluster sample
Sampling fraction
Simple random sample
Stratified sample
Systematic sample

References
1. Cochran, W. G.: *Sampling Techniques*, John Wiley & Sons, Inc., New York, 1953.
2. Hansen, M. H., W. N. Hurwitz, and W. G. Madow: *Sample Survey Methods and Theory*, vol. 1, John Wiley & Sons, Inc., New York, 1953.
3. Kish, L.: "Confidence Intervals for Clustered Samples," *American Sociological Review*, vol. 22, pp. 154–165, April, 1957.
4. Kish, L.: "Selection of the Sample," in L. Festinger and D. Katz (eds.), *Research Methods in the Behavioral Sciences*, The Dryden Press, Inc., New York, 1953, chap. 5.
5. Kish, L.: *Survey Sampling*, John Wiley & Sons, Inc., New York, 1965.
6. Lazerwitz, Bernard: "Sampling Theory and Procedures," in H. M. Blalock and Ann B. Blalock (eds.), *Methodology in Social Research*, McGraw-Hill Book Company, New York, 1968, chap. 8.

Appendix

Review of Algebraic Operations

Since most students will have forgotten much of what they learned in elementary algebra, a brief review of some of the basic algebraic operations may be helpful. Some of these rules will be stated below in very concise form. If a more extensive review is needed, you should consult an algebra text.

One of the basic things to remember about arithmetic and algebraic manipulations is that the *order* in which the operation is carried out is of extreme importance. Generally speaking, whenever there is a fairly complicated expression, one works from the inside out. The following rules should be more or less committed to memory.

1. *Expanding a Squared Sum or Difference*

$$(a + b)^2 = a^2 + 2ab + b^2 \neq a^2 + b^2$$

$$(a - b)^2 = a^2 - 2ab + b^2 \neq a^2 - b^2$$

The converse holds when dealing with square roots

$$\sqrt{a^2 + 2ab + b^2} = \sqrt{(a + b)^2} = a + b$$

It is definitely *not true* that

$$\sqrt{a^2 + b^2} = a + b$$

2. *Dividing by a Sum or Difference*   Although it is true that

$$\frac{a + b}{c} = \frac{a}{c} + \frac{b}{c}$$

we cannot so readily simplify the expressions

$$\frac{a}{b+c} \quad \text{or} \quad \frac{a}{b-c}$$

For example, $\qquad\qquad \dfrac{a}{b+c} \neq \dfrac{a}{b} + \dfrac{a}{c}$

3. *Dividing by a Fraction* If the denominator is itself a fraction, we can place the denominator of the denominator in the numerator as follows:

$$\frac{a}{b/c} = a\frac{c}{b} = \frac{ac}{b}$$

Likewise, $\qquad\qquad \dfrac{a/b}{c/d} = \dfrac{a}{b}\dfrac{d}{c} = \dfrac{ad}{bc}$

and $\qquad\qquad \dfrac{a}{b/(c+d)} = a\dfrac{c+d}{b} = \dfrac{a(c+d)}{b}$

4. *Multiplication of Powers* If we have the product of a number raised to the power a and the *same number* raised to the power b, we may *add* exponents. Thus

$$X^a X^b = X^{a+b} \quad \text{and} \quad X^3 X^2 = X^5$$

But $\quad X^a + X^b \neq X^{a+b} \quad$ and $\quad X^3 + X^2 = X^2(X+1) \neq X^5$

Similarly, we subtract exponents when dividing

$$\frac{X^a}{X^b} = X^{a-b} \quad \text{and} \quad \frac{X^3}{X^2} = X^1 = X$$

In particular, $\qquad \dfrac{X^a}{X^a} = X^{a-a} = X^0 = 1$

Thus, any real number (other than 0) when raised to the 0 power is 1.

5. *Negative Exponents* A number raised to a negative power may be written as the reciprocal of that number raised to the same positive power. Thus

$$X^{-a} = \frac{1}{X^a} \quad \text{and} \quad X^{-2} = \frac{1}{X^2}$$

6. *Removing or Adding Parentheses* Here we follow the rule of working from the inside out. A negative sign before a parenthesis means that each term within the parentheses must have its sign changed if the parentheses are removed. Thus

$$a(b - c) = ab - ac$$

and $\quad -[a - (b - c)] = -[a - b + c] = -a + b - c$

and $\quad a - [b - (c - d)^2] = a - [b - (c^2 - 2cd + d^2)]$

$$= a - [b - c^2 + 2cd - d^2]$$

$$= a - b + c^2 - 2cd + d^2$$

Similarly, we must change all signs of quantities placed inside any parentheses we may introduce if the parentheses are preceded by a negative sign. Thus

$$a - b - c = a - (b + c)$$

and $\quad a - b + c - d = (a - b) + (c - d) = -(b - a) - (d - c)$

Use of summation signs In statistics it is frequently necessary to make use of formulas involving sums of numerous quantities. As a shorthand substitute for writing out each of these sums at length, we make use of the Greek letter Σ (capital sigma) to indicate a summation. As a general rule, whenever this symbol appears, it means that all quantities appearing to the right of it should be summed. Rather than using completely different letters for each quantity being summed (for example, a, b, c, d, e, f, . . .), we ordinarily make use of a single letter (usually X, Y, or Z), together with a subscript i, j, or k which can take on any numerical values we please. Usually, although not always, the first score will be symbolized by X_1, the second by X_2, and so forth. We then make use of Σ as follows:

$$\sum_{i=1}^{N} X_i = X_1 + X_2 + X_3 + \cdots + X_N$$

The notations above and below Σ are used to indicate that i takes on the successive values 1, 2, 3, up to N. Similarly, we might write

$$\sum_{i=3}^{8} X_i = X_3 + X_4 + X_5 + X_6 + X_7 + X_8$$

In this latter case, we have been instructed to add the scores of the third through the eighth observations.

If we follow the usual rules of algebra, we may derive certain rules that must apply to summations. Most of these rules will be stated with little or no explanation, since many obviously follow from the definition of Σ and very simple rules of algebra.

1. $\displaystyle\sum_{i=}^{N} X_i^2 = X_1^2 + X_2^2 + X_3^2 + \cdots + X_N^2$

2. $\displaystyle\sum_{i=1}^{N} X_i Y_i = X_1 Y_1 + X_2 Y_2 + X_3 Y_3 + \cdots + X_N Y_N$

3. $\displaystyle\sum_{i=1}^{N} (X_i + Y_i) = (X_1 + Y_1) + (X_2 + Y_2) + \cdots + (X_N + Y_N)$

$$= (X_1 + X_2 + \cdots + X_N)$$
$$+ (Y_1 + Y_2 + \cdots + Y_N)$$
$$= \sum_{i=1}^{N} X_i + \sum_{i=1}^{N} Y_i$$

4. $\displaystyle\sum_{i=1}^{N} (X_i - Y_i) = \sum_{i=1}^{N} X_i - \sum_{i=1}^{N} Y_i \qquad \text{(see 3)}$

5. $\displaystyle\sum_{i=1}^{N} (X_i + Y_i)^2 = \sum_{i=1}^{N} (X_i^2 + 2X_i Y_i + Y_i^2)$

$$= \sum_{i=1}^{N} X_i^2 + \sum_{i=1}^{N} 2X_i Y_i + \sum_{i=1}^{N} Y_i^2$$
$$\neq \sum_{i=1}^{N} X_i^2 + \sum_{i=1}^{N} Y_i^2$$

Note: The factor 2 can be taken out of the second term, giving $2 \sum\limits_{i=1}^{N} X_i Y_i$ (see 6)

6. If k is a constant

$$\sum_{i=1}^{N} kX_i = kX_1 + kX_2 + \cdots + kX_N$$

$$= k(X_1 + X_2 + \cdots + X_N) = k \sum_{i=1}^{N} X_i$$

7. $\left(\sum\limits_{i=1}^{N} X_i \right)^2 = (X_1 + X_2 + \cdots + X_N)^2$

$$= X_1{}^2 + X_2{}^2 + \cdots + X_N{}^2 + 2X_1 X_2$$
$$+ 2X_1 X_3 + \cdots + 2X_{N-1} X_N$$
$$\neq X_1{}^2 + X_2{}^2 + \cdots + X_N{}^2$$

In other words, we must distinguish between

$$\sum_{i=1}^{N} X_i{}^2 \quad \text{and} \quad \left(\sum_{i=1}^{N} X_i \right)^2$$

We may also sometimes find it convenient to express a sum in terms of a double summation over two indices i and j. Each quantity to be summed can be written with a double subscript (ij). The quantity $\sum\limits_{i=1}^{N} \sum\limits_{j=1}^{M} X_{ij}$ means that we first sum the second subscript j from 1 to M, and then, working outwards, we sum i from 1 to N. Thus

$$\sum_{i=1}^{N} \sum_{j=1}^{M} X_{ij} = \sum_{i=1}^{N} (X_{i1} + X_{i2} + X_{i3} + \cdots + X_{iM})$$

$$= (X_{11} + X_{12} + \cdots + X_{1M}) + (X_{21} + X_{22} + \cdots + X_{2M})$$
$$+ \cdots + (X_{N1} + X_{N2} + \cdots + X_{NM})$$

Similarly,

$$\sum_{i=1}^{N} \left(\sum_{j=1}^{M} X_{ij} \right)^2 = \sum_{i=1}^{N} (X_{i1} + X_{i2} + \cdots + X_{iM})^2$$

Working with very large or very small numbers In working with very large or small numbers, especially when taking squares or square roots, it is often convenient to make use of powers of 10. Since $10^1 = 10$, $10^2 = 100$, $10^3 = 1,000$ and so forth, by counting the number of places to either the right or left of the decimal we can write any figure as a number between 0 and 10 times a certain power of 10. Thus

$$13 = 1.3(10) = 1.3 \times 10^1$$

$$138 = 1.38(100) = 1.38 \times 10^2$$

$$1,382 = 1.382(1,000) = 1.382 \times 10^3$$

$$1,382,461 = 1.382461 \times 10^6$$

$$.13 = \frac{1.3}{10} = 1.3 \times 10^{-1}$$

$$.013 = \frac{1.3}{100} = \frac{1.3}{10^2} = 1.3 \times 10^{-2}$$

$$.00013 = \frac{1.3}{10,000} = 1.3 \times 10^{-4}$$

If we wish to square the quantity 1,382, we get

$$1,382^2 = (1.382 \times 10^3)^2 = 1.382^2 \times 10^6$$

$$= 1.9099 \times 1,000,000 = 1,909,900$$

It is thus much easier to keep track of the decimal point.

In taking square roots it is simplest to make use of even powers of ten.

Since $\sqrt{100} = \sqrt{10^2} = 10$ and $\sqrt{10,000} = \sqrt{10^4} = 10^2 = 100$

and, in general, $\sqrt{10^{2k}} = 10^k$

whereas

$$\sqrt{1,000} = \sqrt{10^3} = 10\sqrt{10} \quad \text{and} \quad \sqrt{100,000} = \sqrt{10^5} = 100\sqrt{10}$$

we see that it is always possible to take even powers of 10 out from under the radical, although this is not possible with odd powers. When taking a square root, therefore, we can count the number of *pairs* of digits to the right or left of the decimal and can express the original quantity as a number between 1 and 100 times an even power of 10.

$$13 = 1.3(10) = 1.3 \times 10^1$$
$$138 = 1.38(100) = 1.38 \times 10^2$$
$$1{,}382 = 1.382(1{,}000) = 1.382 \times 10^3$$
$$1{,}382{,}461 = 1.382461 \times 10^6$$
$$.13 = \frac{1.3}{10} = 1.3 \times 10^{-1}$$
$$.013 = \frac{1.3}{100} = \frac{1.3}{10^2} = 1.3 \times 10^{-2}$$
$$.00013 = \frac{1.3}{10{,}000} = 1.3 \times 10^{-4}$$

Tables 2

Table A Table of squares and square roots

Number	Square	Square root	Number	Square	Square root
1	1	1.0000	31	9 61	5.5678
2	4	1.4142	32	10 24	5.6569
3	9	1.7321	33	10 89	5.7446
4	16	2.0000	34	11 56	5.8310
5	25	2.2361	35	12 25	5.9161
6	36	2.4495	36	12 96	6.0000
7	49	2.6458	37	13 69	6.0828
8	64	2.8284	38	14 44	6.1644
9	81	3.0000	39	15 21	6.2450
10	1 00	3.1623	40	16 00	6.3246
11	1 21	3.3166	41	16 81	6.4031
12	1 44	3.4641	42	17 64	6.4807
13	1 69	3.6056	43	18 49	6.5574
14	1 96	3.7417	44	19 36	6.6332
15	2 25	3.8730	45	20 25	6.7082
16	2 56	4.0000	46	21 16	6.7823
17	2 89	4.1231	47	22 09	6.8557
18	3 24	4.2426	48	23 04	6.9282
19	3 61	4.3589	49	24 01	7.0000
20	4 00	4.4721	50	25 00	7.0711
21	4 41	4.5826	51	26 01	7.1414
22	4 84	4.6904	52	27 04	7.2111
23	5 29	4.7958	53	28 09	7.2801
24	5 76	4.8990	54	29 16	7.3485
25	6 25	5.0000	55	30 25	7.4162
26	6 76	5.0990	56	31 36	7.4833
27	7 29	5.1962	57	32 49	7.5498
28	7 84	5.2915	58	33 64	7.6158
29	8 41	5.3852	59	34 81	7.6811
30	9 00	5.4772	60	36 00	7.7460

SOURCE: H. Sorenson, *Statistics for Students of Psychology and Education*, McGraw-Hill Book Company, New York, 1936, table 72, pp. 347–359, with the kind permission of the author.

Table A   Table of squares and square roots (*Continued*)

Number	Square	Square root	Number	Square	Square root
61	37 21	7.8102	101	1 02 01	10.0499
62	38 44	7.8740	102	1 04 04	10.0995
63	39 69	7.9373	103	1 06 09	10.1489
64	40 96	8.0000	104	1 08 16	10.1980
65	42 25	8.0623	105	1 10 25	10.2470
66	43 56	8.1240	106	1 12 36	10.2956
67	44 89	8.1854	107	1 14 49	10.3441
68	46 24	8.2462	108	1 16 64	10.3923
69	47 61	8.3066	109	1 18 81	10.4403
70	49 00	8.3666	110	1 21 00	10.4881
71	50 41	8.4261	111	1 23 21	10.5357
72	51 84	8.4853	112	1 25 44	10.5830
73	53 29	8.5440	113	1 27 69	10.6301
74	54 76	8.6023	114	1 29 96	10.6771
75	56 25	8.6603	115	1 32 25	10.7238
76	57 76	8.7178	116	1 34 56	10.7703
77	59 29	8.7750	117	1 36 89	10.8167
78	60 84	8.8318	118	1 39 24	10.8628
79	62 41	8.8882	119	1 41 61	10.9087
80	64 00	8.9443	120	1 44 00	10.9545
81	65 61	9.0000	121	1 46 41	11.0000
82	67 24	9.0554	122	1 48 84	11.0454
83	68 89	9.1104	123	1 51 29	11.0905
84	70 56	9.1652	124	1 53 76	11.1355
85	72 25	9.2195	125	1 56 25	11.1803
86	73 96	9.2736	126	1 58 76	11.2250
87	75 69	9.3274	127	1 61 29	11.2694
88	77 44	9.3808	128	1 63 84	11.3137
89	79 21	9.4340	129	1 66 41	11.3578
90	81 00	9.4868	130	1 69 00	11.4018
91	82 81	9.5394	131	1 71 61	11.4455
92	84 64	9.5917	132	1 74 24	11.4891
93	86 49	9.6437	133	1 76 89	11.5326
94	88 36	9.6954	134	1 79 56	11.5758
95	90 25	9.7468	135	1 82 25	11.6190
96	92 16	9.7980	136	1 84 96	11.6619
97	94 09	9.8489	137	1 87 69	11.7047
98	96 04	9.8995	138	1 90 44	11.7473
99	98 01	9.9499	139	1 93 21	11.7898
100	1 00 00	10.0000	140	1 96 00	11.8322

Table A   Table of squares and square roots (*Continued*)

Number	Square	Square root	Number	Square	Square root
141	1 98 81	11.8743	181	3 27 61	13.4536
142	2 01 64	11.9164	182	3 31 24	13.4907
143	2 04 49	11.9583	183	3 34 89	13.5277
144	2 07 36	12.0000	184	3 38 56	13.5647
145	2 10 25	12.0416	185	3 42 25	13.6015
146	2 13 16	12.0830	186	3 45 96	13.6382
147	2 16 09	12.1244	187	3 49 69	13.6748
148	2 19 04	12.1655	188	3 53 44	13.7113
149	2 22 01	12.2066	189	3 57 21	13.7477
150	2 25 00	12.2474	190	3 61 00	13.7840
151	2 28 01	12.2882	191	3 64 81	13.8203
152	2 31 04	12.3288	192	3 68 64	13.8564
153	2 34 09	12.3693	193	3 72 49	13.8924
154	2 37 16	12.4097	194	3 76 36	13.9284
155	2 40 25	12.4499	195	3 80 25	13.9642
156	2 43 36	12.4900	196	3 84 16	14.0000
157	2 46 49	12.5300	197	3 88 09	14.0357
158	2 49 64	12.5698	198	3 92 04	14.0712
159	2 52 81	12.6095	199	3 96 01	14.1067
160	2 56 00	12.6491	200	4 00 00	14.1421
161	2 59 21	12.6886	201	4 04 01	14.1774
162	2 62 44	12.7279	202	4 08 04	14.2127
163	2 65 69	12.7671	203	4 12 09	14.2478
164	2 68 96	12.8062	204	4 16 16	14.2829
165	2 72 25	12.8452	205	4 20 25	14.3178
166	2 75 56	12.8841	206	4 24 36	14.3527
167	2 78 89	12.9228	207	4 28 49	14.3875
168	2 82 24	12.9615	208	4 32 64	14.4222
169	2 85 61	13.0000	209	4 36 81	14.4568
170	2 89 00	13.0384	210	4 41 00	14.4914
171	2 92 41	13.0767	211	4 45 21	14.5258
172	2 95 84	13.1149	212	4 49 44	14.5602
173	2 99 29	13.1529	213	4 53 69	14.5945
174	3 02 76	13.1909	214	4 57 96	14.6287
175	3 06 25	13.2288	215	4 62 25	14.6629
176	3 09 76	13.2665	216	4 66 56	14.6969
177	3 13 29	13.3041	217	4 70 89	14.7309
178	3 16 84	13.3417	218	4 75 24	14.7648
179	3 20 41	13.3791	219	4 79 61	14.7986
180	3 24 00	13.4164	220	4 84 00	14.8324

Table A   Table of squares and square roots (*Continued*)

Number	Square	Square root	Number	Square	Square root
221	4 88 41	14.8661	261	6 81 21	16.1555
222	4 92 84	14.8997	262	6 86 44	16.1864
223	4 97 29	14.9332	263	6 91 69	16.2173
224	5 01 76	14.9666	264	6 96 96	16.2481
225	5 06 25	15.0000	265	7 02 25	16.2788
226	5 10 76	15.0333	266	7 07 56	16.3095
227	5 15 29	15.0665	267	7 12 89	16.3401
228	5 19 84	15.0997	268	7 18 24	16.3707
229	5 24 41	15.1327	269	7 23 61	16.4012
230	5 29 00	15.1658	270	7 29 00	16.4317
231	5 33 61	15.1987	271	7 34 41	16.4621
232	5 38 24	15.2315	272	7 39 84	16.4924
233	5 42 89	15.2643	273	7 45 29	16.5227
234	5 47 56	15.2971	274	7 50 76	16.5529
235	5 52 25	15.3297	275	7 56 25	16.5831
236	5 56 96	15.3623	276	7 61 76	16.6132
237	5 61 69	15.3948	277	7 67 29	16.6433
238	5 66 44	15.4272	278	7 72 84	16.6733
239	5 71 21	15.4596	279	7 78 41	16.7033
240	5 76 00	15.4919	280	7 84 00	16.7332
241	5 80 81	15.5242	281	7 89 61	16.7631
242	5 85 64	15.5563	282	7 95 24	16.7929
243	5 90 49	15.5885	283	8 00 89	16.8226
244	5 95 36	15.6205	284	8 06 56	16.8523
245	6 00 25	15.6525	285	8 12 25	16.8819
246	6 05 16	15.6844	286	8 17 96	16.9115
247	6 10 09	15.7162	287	8 23 69	16.9411
248	6 15 04	15.7480	288	8 29 44	16.9706
259	6 20 01	15.7797	289	8 35 21	17.0000
250	6 25 00	15.8114	290	8 41 00	17.0294
251	6 30 01	15.8430	291	8 46 81	17.0587
252	6 35 04	15.8745	292	8 52 64	17.0880
253	6 40 09	15.9060	293	8 58 49	17.1172
254	6 45 16	15.9374	294	8 64 36	17.1464
255	6 50 25	15.9687	295	8 70 25	17.1756
256	6 55 36	16.0000	296	8 76 16	17.2047
257	6 60 49	16.0312	297	8 82 09	17.2337
258	6 65 64	16.0624	298	8 88 04	17.2627
259	6 70 81	16.0935	299	8 94 01	17.2916
260	6 76 00	16.1245	300	9 00 00	17.3205

Table A   Table of squares and square roots (*Continued*)

Number	Square	Square root	Number	Square	Square root
301	9 06 01	17.3494	341	11 62 81	18.4662
302	9 12 04	17.3781	342	11 69 64	18.4932
303	9 18 09	17.4069	343	11 76 49	18.5203
304	9 24 16	17.4356	344	11 83 36	18.5472
305	9 30 25	17.4642	345	11 90 25	18.5742
306	9 36 36	17.4929	346	11 97 16	18.6011
307	9 42 49	17.5214	347	12 04 09	18.6279
308	9 48 64	17.5499	348	12 11 04	18.6548
309	9 54 81	17.5784	349	12 18 01	18.6815
310	9 61 00	17.6068	350	12 25 00	18.7083
311	9 67 21	17.6352	351	12 32 01	18.7350
312	9 73 44	17.6635	352	12 39 04	18.7617
313	9 79 69	17.6918	353	12 46 09	18.7883
314	9 85 96	17.7200	354	12 53 16	18.8149
315	9 92 25	17.7482	355	12 60 25	18.8414
316	9 98 56	17.7764	356	12 67 36	18.8680
317	10 04 89	17.8045	357	12 74 49	18.8944
318	10 11 24	17.8326	358	12 81 64	18.9209
319	10 17 61	17.8606	359	12 88 81	18.9473
320	10 24 00	17.8885	360	12 96 00	18.9737
321	10 30 41	17.9165	361	13 03 21	19.0000
322	10 36 84	17.9444	362	13 10 44	19.0263
323	10 43 29	17.9722	363	13 17 69	19.0526
324	10 49 76	18.0000	364	13 24 96	19.0788
325	10 56 25	18.0278	365	13 32 25	19.1050
326	10 62 76	18.0555	366	13 39 56	19.1311
327	10 69 29	18.0831	367	13 46 89	19.1572
328	10 75 84	18.1108	368	13 54 24	19.1833
329	10 82 41	18.1384	369	13 61 61	19.2094
330	10 89 00	18.1659	370	13 69 00	19.2354
331	10 95 61	18.1934	371	13 76 41	19.2614
332	11 02 24	18.2209	372	13 83 84	19.2873
333	11 08 89	18.2483	373	13 91 29	19.3132
334	11 15 56	18.2757	374	13 98 76	19.3391
335	11 22 25	18.3030	375	14 06 25	19.3649
336	11 28 96	18.3303	376	14 13 76	19.3907
337	11 35 69	18.3576	377	14 21 29	19.4165
338	11 42 44	18.3848	378	14 28 84	19.4422
339	11 49 21	18.4120	379	14 36 41	19.4679
340	11 56 00	18.4391	380	14 44 00	19.4936

Table A   Table of squares and square roots (*Continued*)

Number	Square	Square root	Number	Square	Square root
381	14 51 61	19.5192	421	17 72 41	20.5183
382	14 59 24	19.5448	422	17 80 84	20.5426
383	14 66 89	19.5704	423	17 89 29	20.5670
384	14 74 56	19.5959	424	17 97 76	20.5913
385	14 82 25	19.6214	425	18 06 25	20.6155
386	14 89 96	19.6469	426	18 14 76	20.6398
387	14 97 69	19.6723	427	18 23 29	20.6640
388	15 05 44	19.6977	428	18 31 84	20.6882
389	15 13 21	19.7231	429	18 40 41	20.7123
390	15 21 00	19.7484	430	18 49 00	20.7364
391	15 28 81	19.7737	431	18 57 61	20.7605
392	15 36 64	19.7990	432	18 66 24	20.7846
393	15 44 49	19.8242	433	18 74 89	20.8087
394	15 52 36	19.8494	434	18 83 56	20.8327
395	15 60 25	19.8746	435	18 92 25	20.8567
396	15 68 16	19.8997	436	19 00 96	20.8806
397	15 76 09	19.9249	437	19 09 69	20.9045
398	15 84 04	19.9499	438	19 18 44	20.9284
399	15 92 01	19.9750	439	19 27 21	20.9523
400	16 00 00	20.0000	440	19 36 00	20.9762
401	16 08 01	20.0250	441	19 44 81	21.0000
402	16 16 04	20.0499	442	19 53 64	21.0238
403	16 24 09	20.0749	443	19 62 49	21.0476
404	16 32 16	20.0998	444	19 71 36	21.0713
405	16 40 25	20.1246	445	19 80 25	21.0950
406	16 48 36	20.1494	446	19 89 16	21.1187
407	16 56 49	20.1742	447	19 98 09	21.1424
408	16 64 64	20 1990	448	20 07 04	21.1660
409	16 72 81	20.2237	449	20 16 01	21.1896
410	16 81 00	20.2485	450	20 25 00	21.2132
411	16 89 21	20.2731	451	20 34 01	21.2368
412	16 97 44	20.2978	452	20 43 04	21.2603
413	17 05 69	20.3224	453	20 52 09	21.2838
414	17 13 96	20.3470	454	20 61 16	21.3073
415	17 22 25	20.3715	455	20 70 25	21.3307
416	17 30 56	20.3961	456	20 79 36	21.3542
417	17 38 89	20.4206	457	20 88 49	21.3776
418	17 47 24	20.4450	458	20 97 64	21.4009
419	17 55 61	20.4695	459	21 06 81	21.4243
420	17 64 00	20.4939	460	21 16 00	21.4476

Table A   Table of squares and square roots (*Continued*)

Number	Square	Square root	Number	Square	Square root
461	21 25 21	21.4709	501	25 10 01	22.3830
462	21 34 44	21.4942	502	25 20 04	22.4054
463	21 43 69	21.5174	503	25 30 09	22.4277
464	21 52 96	21.5407	504	25 40 16	22.4499
465	21 62 25	21.5639	505	25 50 25	22.4722
466	21 71 56	21.5870	506	25 60 36	22.4944
467	21 80 89	21.6102	507	25 70 49	22.5167
468	21 90 24	21.6333	508	25 80 64	22.5389
469	21 99 61	21.6564	509	25 90 81	22.5610
470	22 09 00	21.6795	510	26 01 00	22.5832
471	22 18 41	21.7025	511	26 11 21	22.6053
472	22 27 84	21.7256	512	25 21 44	22.6274
473	22 37 29	21.7486	513	26 31 69	22.6495
474	22 46 76	21.7715	514	26 41 96	22.6716
475	22 56 25	21.7945	515	26 52 25	22.6936
476	22 65 76	21.8174	516	22 62 56	22.7156
477	22 75 29	21.8403	517	26 72 89	22.7376
478	22 84 84	21.8632	518	26 83 24	22.7596
479	22 94 41	21.8861	519	26 93 61	22.7816
480	23 04 00	21.9089	520	27 04 00	22.8035
481	23 13 61	21.9317	521	27 14 41	22.8254
482	23 23 24	21.9545	522	27 24 84	22.8473
483	23 32 89	21.9773	523	27 35 29	22.8692
484	23 42 56	22.0000	524	27 45 76	22.8910
485	23 52 25	22.0227	525	27 56 25	22.9129
486	23 61 96	22.0454	526	27 66 76	22.9347
487	23 71 69	22.0681	527	27 77 29	22.9565
488	23 81 44	22.0907	528	27 87 84	22.9783
489	23 91 21	22.1133	529	27 98 41	23.0000
490	24 01 00	22.1359	530	28 09 00	23.0217
491	24 10 81	22.1585	531	28 19 61	23.0434
492	24 20 64	22.1811	532	28 30 24	23.0651
493	24 30 49	22.2036	533	28 40 89	23.0868
494	24 40 36	22.2261	534	28 51 56	23.1084
495	24 50 25	22.2486	535	28 62 25	23.1301
496	24 60 16	22.2711	536	28 72 96	23.1517
497	24 70 09	22.2935	537	28 83 69	23.1733
498	24 80 04	22.3159	538	28 94 44	23.1948
499	24 90 01	22.3383	539	29 05 21	23.2164
500	25 00 00	22.3607	540	29 16 00	23.2379

Table A **Table of squares and square roots** (*Continued*)

Number	Square	Square root	Number	Square	Square root
541	29 26 81	23.2594	581	33 75 61	24.1039
542	29 37 64	23.2809	582	33 87 24	24.1247
543	29 48 49	23.3024	583	33 98 89	24.1454
544	29 59 36	23.3238	584	34 10 56	24.1661
545	29 70 25	23.3452	585	34 22 25	24.1868
546	29 81 16	23.3666	586	34 33 96	24.2074
547	29 92 09	23.3880	587	34 45 69	24.2281
548	30 03 04	23.4094	588	34 57 44	24.2487
549	30 14 01	23.4307	589	34 69 21	24.2693
550	30 25 00	23.4521	590	34 81 00	24.2899
551	30 36 01	23.4734	591	34 92 81	24.3105
552	30 47 04	23.4947	592	35 04 64	24.3311
553	30 58 09	23.5160	593	35 16 49	24.3516
554	30 69 16	23.5372	594	35 28 36	24.3721
555	30 80 25	23.5584	595	35 40 25	24.3926
556	30 91 36	23.5797	596	35 52 16	24.4131
557	31 02 49	23.6008	597	35 64 09	24.4336
558	31 13 64	23.6220	598	35 76 04	24.4540
559	31 24 81	23.6432	599	35 88 01	24.4745
560	31 36 00	23.6643	600	36 00 00	24.4949
561	31 47 21	23.6854	601	36 12 01	24.5153
562	31 58 44	23.7065	602	36 24 04	24.5357
563	31 69 69	23.7276	603	36 36 09	24.5561
564	31 80 96	23.7487	604	36 48 16	24.5764
565	31 92 25	23.7697	605	36 60 25	24.5967
566	32 03 56	23.7908	606	36 72 36	24.6171
567	32 14 89	23.8118	607	36 84 49	24.6374
568	32 26 24	23.8328	608	36 96 64	24.6577
569	32 37 61	23.8537	609	37 08 81	24.6779
570	32 49 00	23.8747	610	37 21 00	24.6982
571	32 60 41	23.8956	611	37 33 21	24.7184
572	32 71 84	23.9165	612	37 45 44	24.7385
573	32 83 29	23.9374	613	37 57 69	24.7588
574	32 94 76	23.9583	614	37 69 96	24.7790
575	33 06 25	23.9792	615	37 82 25	24.7992
576	33 17 76	24.0000	616	37 94 56	24.8193
577	33 29 29	24.0208	617	38 06 89	24.8395
578	33 40 84	24.0416	618	38 19 24	24.8596
579	33 52 41	24.0624	619	38 31 61	24.8797
580	33 64 00	24.0832	620	38 44 00	24.8998

Table A Table of squares and square roots (*Continued*)

Number	Square	Square root	Number	Square	Square root
621	38 56 41	24.9199	661	43 69 21	25.7099
622	38 68 84	24.9399	662	43 82 44	25.7294
623	38 81 29	24.9600	663	43 95 69	25.7488
624	38 93 76	24.9800	664	44 08 96	25.7682
625	39 06 25	25.0000	665	44 22 25	25.7876
626	39 18 76	25.0200	666	44 35 56	25.8070
627	39 31 29	25.0400	667	44 48 89	25.8263
628	39 43 84	25.0599	668	44 62 24	25.8457
629	39 56 41	25.0799	669	44 75 61	25.8650
630	39 69 00	25.0998	670	44 89 00	25.8844
631	39 81 61	25.1197	671	45 02 41	25.9037
632	39 94 24	25.1396	672	45 15 84	25.9230
633	40 06 89	25.1595	673	45 29 29	25.9422
634	40 19 56	25.1794	674	45 42 76	25.9615
635	40 32 25	25.1992	675	45 56 25	25.9808
636	40 44 96	25.2190	676	45 69 76	26.0000
637	40 57 69	25.2389	677	45 83 29	26.0192
638	40 70 44	25.2587	678	45 96 84	26.0384
639	40 83 21	25.2784	679	46 10 41	26.0576
640	40 96 00	25.2982	680	46 24 00	26.0768
641	41 08 81	25.3180	681	46 37 61	26.0960
642	41 21 64	25.3377	682	46 51 24	26.1151
643	41 34 49	25.3574	683	46 64 89	26.1343
644	41 47 36	25.3772	684	46 78 56	26.1534
645	41 60 25	25.3969	685	46 92 25	26.1725
646	41 73 16	25.4165	686	47 05 96	26.1916
647	41 86 09	25.4362	687	47 19 69	26.2107
648	41 99 04	25.4558	688	47 33 44	26.2298
649	42 12 01	25.4755	689	47 47 21	26.2488
650	42 25 00	25.4951	690	47 61 00	26.2679
651	42 38 01	25.5147	691	47 74 81	26.2869
652	42 51 04	25.5343	692	47 88 64	26.3059
653	42 64 09	25.5539	693	48 02 49	26.3249
654	42 77 16	25.5734	694	48 16 36	26.3439
655	42 90 25	25.5930	695	48 30 25	26.3629
656	43 03 36	25.6125	696	48 44 16	26.3818
657	43 16 49	25.6320	697	48 58 09	26.4008
658	43 29 64	25.6515	698	48 72 04	26.4197
659	43 42 81	25.6710	699	48 86 01	26.4386
660	43 56 00	25.6905	700	49 00 00	26.4575

Table A   Table of squares and square roots (*Continued*)

Number	Square	Square root	Number	Square	Square root
701	49 14 01	26.4764	741	54 90 81	27.2213
702	49 28 04	26.4953	742	55 05 64	27.2397
703	49 42 09	26.5141	743	55 20 49	27.2580
704	49 56 16	26.5330	744	55 35 36	27.2764
705	49 70 25	26.5518	745	55 50 25	27.2947
706	49 84 36	26.5707	746	55 65 16	27.3130
707	49 98 49	26.5895	747	55 80 09	27.3313
708	50 12 64	26.6083	748	55 95 04	27.3496
709	50 26 81	26.6271	749	56 10 01	27.3679
710	50 41 00	26.6458	750	56 25 00	27.3861
711	50 55 21	26.6646	751	56 40 01	27.4044
712	50 69 44	26.6833	752	56 55 04	27.4226
713	50 83 69	26.7021	753	56 70 09	27.4408
714	50 97 96	26.7208	754	56 85 16	27.4591
715	51 12 25	26.7395	755	57 00 25	27.4773
716	51 26 56	26.7582	756	57 15 36	27.4955
717	51 40 89	26.7769	757	57 30 49	27.5136
718	51 55 24	26.7955	758	57 45 64	27.5318
719	51 69 61	26.8142	759	57 60 81	27.5500
720	51 84 00	26.8328	760	57 76 00	27.5681
721	51 98 41	26.8514	761	57 91 21	27.5862
722	52 12 84	26.8701	762	58 06 44	27.6043
723	52 27 29	26.8887	763	58 21 69	27.6225
724	52 41 76	26.9072	764	58 36 96	27.6405
725	52 56 25	26.9258	765	58 52 25	27.6586
726	52 70 76	26.9444	766	58 67 56	27.6767
727	52 85 29	26.9629	767	58 82 89	27.6948
728	52 99 84	26.9815	768	58 98 24	27.7128
729	53 14 41	27.0000	769	59 13 61	27.7308
730	53 29 00	27.0185	770	59 29 00	27.7489
731	53 43 61	27.0370	771	59 44 41	27.7669
732	53 58 24	27.0555	772	59 59 84	27.7849
733	53 72 89	27.0740	773	59 75 29	27.8029
734	53 87 56	27.0924	774	59 90 76	27.8209
735	54 02 25	27.1109	775	60 06 25	27.8388
736	54 16 96	27.1293	776	60 21 76	27.8568
737	54 31 69	27.1477	777	60 37 29	27.8747
738	54 46 44	27.1662	778	60 52 84	27.8927
739	54 61 27	27.1846	779	60 68 41	27.9106
740	54 76 00	27.2029	780	60 84 00	27.9285

Table A Table of squares and square roots (*Continued*)

Number	Square	Square root	Number	Square	Square root
781	60 99 61	27.9464	821	67 40 41	28.6531
782	61 15 24	27.9643	822	67 56 84	28.6705
783	61 30 89	27.9821	823	67 73 29	28.6880
784	61 46 56	28.0000	824	67 89 76	28.7054
785	61 62 25	28.0179	825	68 06 25	28.7228
786	61 77 96	28.0357	826	68 22 76	28.7402
787	61 93 69	28.0535	827	68 39 29	28.7576
788	62 09 44	28.0713	828	68 55 84	28.7750
789	62 25 21	28.0891	829	68 72 41	28.7924
790	62 41 00	28.1069	830	68 89 00	28.8097
791	62 56 81	28.1247	831	69 05 61	28.8271
792	62 72 64	28.1425	832	69 22 24	28.8444
793	62 88 49	28.1603	833	69 38 89	28.8617
794	63 04 36	28.1780	834	69 55 56	28.8791
795	63 20 25	28.1957	835	69 72 25	28.8964
796	63 36 16	28.2135	836	69 88 96	28.9137
797	63 52 09	28.2312	837	70 05 69	28.9310
798	63 68 04	28.2489	838	70 22 44	28.9482
799	63 84 01	28.2666	839	70 39 21	28.9655
800	64 00 00	28.2843	840	70 56 00	28.9828
801	64 16 01	28.3019	841	70 72 81	29.0000
802	64 32 04	28.3196	842	70 89 64	29.0172
803	64 48 09	28.3373	843	71 06 49	29.0345
804	64 64 16	28.3549	844	71 23 36	29.0517
805	64 80 25	28.3725	845	71 40 25	29.0689
806	64 96 36	28.3901	846	71 57 16	29.0861
807	65 12 49	28.4077	847	71 74 09	29.1033
808	65 28 64	28.4253	848	71 91 04	29.1204
809	65 44 81	28.4429	849	72 08 01	29.1376
810	65 61 00	28.4605	850	72 25 00	29.1548
811	65 77 21	28.4781	851	72 42 01	29.1719
812	65 93 44	28.4956	852	72 59 04	29.1890
813	66 09 69	28.5132	853	72 76 09	29.2062
814	66 25 96	28.5307	854	72 93 16	29.2233
815	66 42 25	28.5482	855	73 10 25	29.2404
816	66 58 56	28.5657	856	73 27 36	29.2575
817	66 74 89	28.5832	857	73 44 49	29.2746
818	66 91 24	28.6007	858	73 61 64	29.2916
819	67 07 61	28.6082	859	73 78 81	29.3087
820	67 24 00	28.6356	860	73 96 00	29.3258

Table A   Table of squares and square roots (*Continued*)

Number	Square	Square root	Number	Square	Square root
861	74 13 21	29.3428	901	81 18 01	30.0167
862	74 30 44	29.3598	902	81 36 04	30.0333
863	74 47 69	29.3769	903	81 54 09	30.0500
864	74 64 96	29.3939	904	81 72 16	30.0666
865	74 82 25	29.4109	905	81 90 25	30.0832
866	74 99 56	29.4279	906	82 08 36	30.0998
867	75 16 89	29.4449	907	82 26 49	30.1164
868	75 34 24	29.4618	908	82 44 64	30.1330
869	75 51 61	29.4788	909	82 62 81	30.1496
870	75 69 00	29.4958	910	82 81 00	30.1662
871	75 86 41	29.5127	911	82 99 21	30.1828
872	76 03 84	29.5296	912	83 17 44	30.1993
873	76 21 29	29.5466	913	83 35 69	30.2159
874	76 38 76	29.5635	914	83 53 96	30.2324
875	76 56 25	29.5804	915	83 72 25	30.2490
876	76 73 76	29.5973	916	83 90 56	30.2655
877	76 91 29	29.6142	917	84 08 89	30.2820
878	77 08 84	29.6311	918	84 27 24	30.2985
879	77 26 41	29.6479	919	84 45 61	30.3150
880	77 44 00	29.6648	920	84 64 00	30.3315
881	77 61 61	29.6816	921	84 82 41	30.3480
882	77 79 24	29.6985	922	85 00 84	30.3645
883	77 96 89	29.7153	923	85 19 29	30.3809
884	78 14 56	29.7321	924	85 37 76	30.3974
885	78 32 25	29.7489	925	85 56 25	30.4138
886	78 49 96	29.7658	926	85 74 76	30.4302
887	78 67 69	29.7825	927	85 93 29	30.4467
888	78 85 44	29.7993	928	86 11 84	30.4631
889	79 03 21	29.8161	929	86 30 41	30.4795
890	79 21 00	29.8329	930	86 49 00	30.4959
891	79 38 81	29.8496	931	86 67 61	30.5123
892	79 56 64	29.8664	932	86 86 24	30.5287
893	79 74 49	29.8831	933	87 04 89	30.5450
894	79 92 36	29.8998	934	87 23 56	30.5614
895	80 10 25	29.9166	935	87 42 25	30.5778
896	80 28 16	29.9333	936	87 60 96	30.5941
897	80 46 09	29.9500	937	87 79 69	30.6105
898	80 64 04	29.9666	938	87 98 44	30.6268
899	80 82 01	29.9833	939	88 17 21	30.6431
900	81 00 00	30.0000	940	88 36 00	30.6594

Table A   Table of squares and square roots (*Continued*)

Number	Square	Square root	Number	Square	Square root
941	88 54 81	30.6757	971	94 28 41	31.1609
942	88 73 64	30.6920	972	94 47 84	31.1769
943	88 92 49	30.7083	973	94 67 29	31.1929
944	89 11 36	30.7246	974	94 86 76	31.2090
945	89 30 25	30.7409	975	95 06 25	31.2250
946	89 49 16	30.7571	976	95 25 76	31.2410
947	89 68 09	30.7734	977	95 45 29	31.2570
948	89 87 04	30.7896	978	95 64 84	31.2730
949	90 06 01	30.8058	979	95 84 41	31.2890
950	90 25 00	30.8221	980	96 04 00	31.3050
951	90 44 01	30.8383	981	96 23 61	31.3209
952	90 63 04	30.8545	982	96 43 24	31.3369
953	90 82 09	30.8707	983	96 62 89	31.3528
954	91 01 16	30.8869	984	96 82 56	31.3688
955	91 20 25	30.9031	985	97 02 25	31.3847
956	91 39 36	30.9192	986	97 21 96	31.4006
957	91 58 49	30.9354	987	97 41 69	31.4166
958	91 77 64	30.9516	988	97 61 44	31.4325
959	91 96 81	30.9677	989	97 81 21	31.4484
960	92 16 00	30.9839	990	98 01 00	31.4643
961	92 35 21	31.0000	991	98 20 81	31.4802
962	92 54 44	31.0161	992	98 40 64	31.4960
963	92 73 69	31.0322	993	98 60 49	31.5119
964	92 92 96	31.0483	994	98 80 36	31.5278
965	93 12 25	31.0644	995	99 00 25	31.5436
966	93 31 56	31.0805	996	99 20 16	31.5595
967	93 50 89	31.0966	997	99 40 09	31.5753
968	93 70 24	31.1127	998	99 60 04	31.5911
969	93 89 61	31.1288	999	99 80 01	31.6070
970	94 09 00	31.1448	1000	100 00 00	31.6228

Table B   Random numbers

10 09 73 25 33	76 52 01 35 86	34 67 35 48 76	80 95 90 91 17	39 29 27 49 45
37 54 20 48 05	64 89 47 42 96	24 80 52 40 37	20 63 61 04 02	00 82 29 16 65
08 42 26 89 53	19 64 50 93 03	23 20 90 25 60	15 95 33 47 64	35 08 03 36 06
99 01 90 25 29	09 37 67 07 15	38 31 13 11 65	88 67 67 43 97	04 43 62 76 59
12 80 79 99 70	80 15 73 61 47	64 03 23 66 53	98 95 11 68 77	12 17 17 68 33
66 06 57 47 17	34 07 27 68 50	36 69 73 61 70	65 81 33 98 85	11 19 92 91 70
31 06 01 08 05	45 57 18 24 06	35 30 34 26 14	86 79 90 74 39	23 40 30 97 32
85 26 97 76 02	02 05 16 56 92	68 66 57 48 18	73 05 38 52 47	18 62 38 85 79
63 57 33 21 35	05 32 54 70 48	90 55 35 75 48	28 46 82 87 09	83 49 12 56 24
73 79 64 57 53	03 52 96 47 78	35 80 83 42 82	60 93 52 03 44	35 27 38 84 35
98 52 01 77 67	14 90 56 86 07	22 10 94 05 58	60 97 09 34 33	50 50 07 39 98
11 80 50 54 31	39 80 82 77 32	50 72 56 82 48	29 40 52 42 01	52 77 56 78 51
83 45 29 96 34	06 28 89 80 83	13 74 67 00 78	18 47 54 06 10	68 71 17 78 17
88 68 54 02 00	86 50 75 84 01	36 76 66 79 51	90 36 47 64 93	29 60 91 10 62
99 59 46 73 48	87 51 76 49 69	91 82 60 89 28	93 78 56 13 68	23 47 83 41 13
65 48 11 76 74	17 46 85 09 50	58 04 77 69 74	73 03 95 71 86	40 21 81 65 44
80 12 43 56 35	17 72 70 80 15	45 31 82 23 74	21 11 57 82 53	14 38 55 37 63
74 35 09 98 17	77 40 27 72 14	43 23 60 02 10	45 52 16 42 37	96 28 60 26 55
69 91 62 68 03	66 25 22 91 48	36 93 68 72 03	76 62 11 39 90	94 40 05 64 18
09 89 32 05 05	14 22 56 85 14	46 42 75 67 88	96 29 77 88 22	54 38 21 45 98
91 49 91 45 23	68 47 92 76 86	46 16 28 35 54	94 75 08 99 23	37 08 92 00 48
80 33 69 45 98	26 94 03 68 58	70 29 73 41 35	53 14 03 33 40	42 05 08 23 41
44 10 48 19 49	85 15 74 79 54	32 97 92 65 75	57 60 04 08 81	22 22 20 64 13
12 55 07 37 42	11 10 00 20 40	12 86 07 46 97	96 64 48 94 39	28 70 72 58 15
63 60 64 93 29	16 50 53 44 84	40 21 95 25 63	43 65 17 70 82	07 20 73 17 90
61 19 69 04 46	26 45 74 77 74	51 92 43 37 29	65 39 45 95 93	42 58 26 05 27
15 47 44 52 66	95 27 07 99 53	59 36 78 38 48	82 39 61 01 18	33 21 15 94 66
94 55 72 85 73	67 89 75 43 87	54 62 24 44 31	91 19 04 25 92	92 92 74 59 73
42 48 11 62 13	97 34 40 87 21	16 86 84 87 67	03 07 11 20 59	25 70 14 66 70
23 52 37 83 17	73 20 88 98 37	68 93 59 14 16	26 25 22 96 63	05 52 28 25 62
04 49 35 24 94	75 24 63 38 24	45 86 25 10 25	61 96 27 93 35	65 33 71 24 72
00 54 99 76 54	64 05 18 81 59	96 11 96 38 96	54 69 28 23 91	23 28 72 95 29
35 96 31 53 07	26 89 80 93 54	33 35 13 54 62	77 97 45 00 24	90 10 33 93 33
59 80 80 83 91	45 42 72 68 42	83 60 94 97 00	13 02 12 48 92	78 56 52 01 06
46 05 88 52 36	01 39 09 22 86	77 28 14 40 77	93 91 08 36 47	70 61 74 29 41
32 17 90 05 97	87 37 92 52 41	05 56 70 70 07	86 74 31 71 57	85 39 41 18 38
69 23 46 14 06	20 11 74 52 04	15 95 66 00 00	18 74 39 24 23	97 11 89 63 38
19 56 54 14 30	01 75 87 53 79	40 41 92 15 85	66 67 43 68 06	84 96 28 52 07
45 15 51 49 38	19 47 60 72 46	43 66 79 45 43	59 04 79 00 33	20 82 66 95 41
94 86 43 19 94	36 16 81 08 51	34 88 88 15 53	01 54 03 54 56	05 01 45 11 76

SOURCE: The RAND Corporation, *A Million Random Digits*, Free Press, Glencoe, Ill., 1955, pp. 1–3, with the kind permission of the publisher.

Table B Random numbers (*Continued*)

```
98 08 62 48 26   45 24 02 84 04   44 99 90 88 96   39 09 47 34 07   35 44 13 18 80
33 18 51 62 32   41 94 15 09 49   89 43 54 85 81   88 69 54 19 94   37 54 87 30 43
80 95 10 04 06   96 38 27 07 74   20 15 12 33 87   25 01 62 52 98   94 62 46 11 71
79 75 24 91 40   71 96 12 82 96   69 86 10 25 91   74 85 22 05 39   00 38 75 95 79
18 63 33 25 37   98 14 50 65 71   31 01 02 46 74   05 45 56 14 27   77 93 89 19 36

74 02 94 39 02   77 55 73 22 70   97 79 01 71 19   52 52 75 80 21   80 81 45 17 48
54 17 84 56 11   80 99 33 71 43   05 33 51 29 69   56 12 71 92 55   36 04 09 03 24
11 66 44 98 83   52 07 98 48 27   59 38 17 15 39   09 97 33 34 40   88 46 12 33 56
48 32 47 79 28   31 24 96 47 10   02 29 53 68 70   32 30 75 75 46   15 02 00 99 94
69 07 49 41 38   87 63 79 19 76   35 58 40 44 01   10 51 82 16 15   01 84 87 69 38

09 18 82 00 97   32 82 53 95 27   04 22 08 63 04   83 38 98 73 74   64 27 85 80 44
90 04 58 54 97   51 98 15 06 54   94 93 88 19 97   91 87 07 61 50   68 47 66 46 59
73 18 95 02 07   47 67 72 52 69   62 29 06 44 64   27 12 46 70 18   41 36 18 27 60
75 76 87 64 90   20 97 18 17 49   90 42 91 22 72   95 37 50 58 71   93 82 34 31 78
54 01 64 40 56   66 28 13 10 03   00 68 22 73 98   20 71 45 32 95   07 70 61 78 13

08 35 86 99 10   78 54 24 27 85   13 66 15 88 73   04 61 89 75 53   31 22 30 84 20
28 30 60 32 64   81 33 31 05 91   40 51 00 78 93   32 60 46 04 75   94 11 90 18 40
53 84 08 62 33   81 59 41 36 28   51 21 59 02 90   28 46 66 87 95   77 76 22 07 91
91 75 75 37 41   61 61 36 22 69   50 26 39 02 12   55 78 17 65 14   83 48 34 70 55
89 41 59 26 94   00 39 75 83 91   12 60 71 76 46   48 94 97 23 06   94 54 13 74 08

77 51 30 38 20   86 83 42 99 01   68 41 48 27 74   51 90 81 39 80   72 89 35 55 07
19 50 23 71 74   69 97 92 02 88   55 21 02 97 73   74 28 77 52 51   65 34 46 74 15
21 81 85 93 13   93 27 88 17 57   05 68 67 31 56   07 08 28 50 46   31 85 33 84 52
51 47 46 64 99   68 10 72 36 21   94 04 99 13 45   42 83 60 91 91   08 00 74 54 49
99 55 96 83 31   62 53 52 41 70   69 77 71 28 30   74 81 97 81 42   43 86 07 28 34

33 71 34 80 07   93 58 47 28 69   51 92 66 47 21   58 30 32 98 22   93 17 49 39 72
85 27 48 68 93   11 30 32 92 70   28 83 43 41 37   73 51 59 04 00   71 14 84 36 43
84 13 38 96 40   44 03 55 21 66   73 85 27 00 91   61 22 26 05 61   62 32 71 84 23
56 73 21 62 34   17 39 59 61 31   10 12 39 16 22   85 49 65 75 60   81 60 41 88 80
65 13 85 68 06   87 64 88 52 61   34 31 36 58 61   45 87 52 10 69   85 64 44 72 77

38 00 10 21 76   81 71 91 17 11   71 60 29 29 37   74 21 96 40 49   65 58 44 96 98
37 40 29 63 97   01 30 47 75 86   56 27 11 00 86   47 32 46 26 05   40 03 03 74 38
97 12 54 03 48   87 08 33 14 17   21 81 53 92 50   75 23 76 20 47   15 50 12 95 78
21 82 64 11 34   47 14 33 40 72   64 63 88 59 02   49 13 90 64 41   03 85 65 45 52
73 13 54 27 42   95 71 90 90 35   85 79 47 42 96   08 78 98 81 56   64 69 11 92 02

07 63 87 79 29   03 06 11 80 72   96 20 74 41 56   23 82 19 95 38   04 71 36 69 94
60 52 88 34 41   07 95 41 98 14   59 17 52 06 95   05 53 35 21 39   61 21 20 64 55
83 59 63 56 55   06 95 89 29 83   05 12 80 97 19   77 43 35 37 83   92 30 15 04 98
10 85 06 27 46   99 59 91 05 07   13 49 90 63 19   53 07 57 18 39   06 41 01 93 62
39 82 09 89 52   43 62 26 31 47   64 42 18 08 14   43 80 00 93 51   31 02 47 31 67
```

Table B   Random numbers (*Continued*)

59 58 00 64 78	75 56 97 88 00	88 83 55 44 86	23 76 80 61 56	04 11 10 84 08
38 50 80 73 41	23 79 34 87 63	90 82 29 70 22	17 71 90 42 07	95 95 44 99 53
30 69 27 06 68	94 68 81 61 27	56 19 68 00 91	82 06 76 34 00	05 46 26 92 00
65 44 39 56 59	18 28 82 74 37	49 63 22 40 41	08 33 76 56 76	96 29 99 08 36
27 26 75 02 64	13 19 27 22 94	07 47 74 46 06	17 98 54 89 11	97 34 13 03 58
91 30 70 69 91	19 07 22 42 10	36 69 95 37 28	28 82 53 57 93	28 97 66 62 52
68 43 49 46 88	84 47 31 36 22	62 12 69 84 08	12 84 38 25 90	09 81 59 31 46
48 90 81 58 77	54 74 52 45 91	35 70 00 47 54	83 82 45 26 92	54 13 05 51 60
06 91 34 51 97	42 67 27 86 01	11 88 30 95 28	63 01 19 89 01	14 97 44 03 44
10 45 51 60 19	14 21 03 37 12	91 34 23 78 21	88 32 58 08 51	43 66 77 08 83
12 88 39 73 43	65 02 76 11 84	04 28 50 13 92	17 97 41 50 77	90 71 22 67 69
21 77 83 09 76	38 80 73 69 61	31 64 94 20 96	63 28 10 20 23	08 81 64 74 49
19 52 35 95 15	65 12 25 96 59	86 28 36 82 58	69 57 21 37 98	16 43 59 15 29
67 24 55 26 70	35 58 31 65 63	79 24 68 66 86	76 46 33 42 22	26 65 59 08 02
60 58 44 73 77	07 50 03 79 92	45 13 42 65 29	26 76 08 36 37	41 32 64 43 44
53 85 34 13 77	36 06 69 48 50	58 83 87 38 59	49 36 47 33 31	96 24 04 36 42
24 63 73 87 36	74 38 48 93 42	52 62 30 79 92	12 36 91 86 01	03 74 28 38 73
83 08 01 24 51	38 99 22 28 15	07 75 95 17 77	97 37 72 75 85	51 97 23 78 67
16 44 42 43 34	36 15 19 90 73	27 49 37 09 39	85 13 03 25 52	54 84 65 47 59
60 79 01 81 57	57 17 86 57 62	11 16 17 85 76	45 81 95 29 79	65 13 00 48 60
03 99 11 04 61	93 71 61 68 94	66 08 32 46 53	84 60 95 82 32	88 61 81 91 61
38 55 59 55 54	32 88 65 97 80	08 35 56 08 60	29 73 54 77 62	71 29 92 38 53
17 54 67 37 04	92 05 24 62 15	55 12 12 92 81	59 07 60 79 36	27 95 45 89 09
32 64 35 28 61	95 81 90 68 31	00 91 19 89 36	76 35 59 37 79	80 86 30 05 14
69 57 26 87 77	39 51 03 59 05	14 06 04 06 19	29 54 96 96 16	33 56 46 07 80
24 12 26 65 91	27 69 90 64 94	14 84 54 66 72	61 95 87 71 00	90 89 97 57 54
61 19 63 02 31	92 96 26 17 73	41 83 95 53 82	17 26 77 09 43	78 03 87 02 67
30 53 22 17 04	10 27 41 22 02	39 68 52 33 09	10 06 16 88 29	55 98 66 64 85
03 78 89 75 99	75 86 72 07 17	74 41 65 31 66	35 20 83 33 74	87 53 90 88 23
48 22 86 33 79	85 78 34 76 19	53 15 26 74 33	35 66 35 29 72	16 81 86 03 11
60 36 59 46 53	35 07 53 39 49	42 61 42 92 97	01 91 82 83 16	98 95 37 32 31
83 79 94 24 02	56 62 33 44 42	34 99 44 13 74	70 07 11 47 36	09 95 81 80 65
32 96 00 74 05	36 40 98 32 32	99 38 54 16 00	11 13 30 75 86	15 91 70 62 53
19 32 25 38 45	57 62 05 26 06	66 49 76 86 46	78 13 86 65 59	19 64 09 94 13
11 22 09 47 47	07 39 93 74 08	48 50 92 39 29	27 48 24 54 76	85 24 43 51 59
31 75 15 72 60	68 98 00 53 39	15 47 04 83 55	88 65 12 25 96	03 15 21 92 21
88 49 29 93 82	14 45 40 45 04	20 09 49 89 77	74 84 39 34 13	22 10 97 85 08
30 93 44 77 44	07 48 18 38 28	73 78 80 65 33	28 59 72 04 05	94 20 52 03 80
22 88 84 88 93	27 49 99 87 48	60 53 04 51 28	74 02 28 46 17	82 03 71 02 68
78 21 21 69 93	35 90 29 13 86	44 37 21 54 86	65 74 11 40 14	87 48 13 72 20

Table B   Random numbers (*Continued*)

```
41 84 98 45 47    46 85 05 23 26    34 67 75 83 00    74 91 06 43 45    19 32 58 15 49
46 35 23 30 49    69 24 89 34 60    45 30 50 75 21    61 31 83 18 55    14 41 37 09 51
11 08 79 62 94    14 01 33 17 92    59 74 76 72 77    76 50 33 45 13    39 66 37 75 44
52 70 10 83 37    56 30 38 73 15    16 52 06 96 76    11 65 49 98 93    02 18 16 81 61
57 27 53 68 98    81 30 44 85 85    68 65 22 73 76    92 85 25 58 66    88 44 80 35 84

20 85 77 31 56    70 28 42 43 26    79 37 59 52 20    01 15 96 32 67    10 62 24 83 91
15 63 38 49 24    90 41 59 36 14    33 52 12 66 65    55 82 34 76 41    86 22 53 17 04
92 69 44 82 97    39 90 40 21 15    59 58 94 90 67    66 82 14 15 75    49 76 70 40 37
77 61 31 90 19    88 15 20 00 80    20 55 49 14 09    96 27 74 82 57    50 81 69 76 16
38 68 83 24 86    45 13 46 35 45    59 40 47 20 59    43 94 75 16 80    43 85 25 96 93

25 16 30 18 89    70 01 41 50 21    41 29 06 73 12    71 85 71 59 57    68 97 11 14 03
65 25 10 76 29    37 23 93 32 95    05 87 00 11 19    92 78 42 63 40    18 47 76 56 22
36 81 54 36 25    18 63 73 75 09    82 44 49 90 05    04 92 17 37 01    14 70 79 39 97
64 39 71 16 92    05 32 78 21 62    20 24 78 17 59    45 19 72 53 32    83 74 52 25 67
04 51 52 56 24    95 09 66 79 46    48 46 08 55 58    15 19 11 87 82    16 93 03 33 61

83 76 16 08 73    43 25 38 41 45    60 83 32 59 83    01 29 14 13 49    20 36 80 71 26
14 38 70 63 45    80 85 40 92 79    43 52 90 63 18    38 38 47 47 61    41 19 63 74 80
51 32 19 22 46    80 08 87 70 74    88 72 25 67 36    66 16 44 94 31    66 91 93 16 78
72 47 20 00 08    80 89 01 80 02    94 81 33 19 00    54 15 58 34 36    35 35 25 41 31
05 46 65 53 06    93 12 81 84 64    74 45 79 05 61    72 84 81 18 34    79 98 26 84 16

39 52 87 24 84    82 47 42 55 93    48 54 53 52 47    18 61 91 36 74    18 61 11 92 41
81 61 61 87 11    53 34 24 42 76    75 12 21 17 24    74 62 77 37 07    58 31 91 59 97
07 58 61 61 20    82 64 12 28 20    92 90 41 31 41    32 39 21 97 63    61 19 96 79 40
90 76 70 42 35    13 57 41 72 00    69 90 26 37 42    78 46 42 25 01    18 62 79 08 72
40 18 82 81 93    29 59 38 86 27    94 97 21 15 98    62 09 53 67 87    00 44 15 89 97

34 41 48 21 57    86 88 75 50 87    19 15 20 00 23    12 30 28 07 83    32 62 46 86 91
63 43 97 53 63    44 98 91 68 22    36 02 40 09 67    76 37 84 16 05    65 96 17 34 88
67 04 90 90 70    93 39 94 55 47    94 45 87 42 84    05 04 14 98 07    20 28 83 40 60
79 49 50 41 46    52 16 29 02 86    54 15 83 42 43    46 97 83 54 82    59 36 29 59 38
91 70 43 05 52    04 73 72 10 31    75 05 19 30 29    47 66 56 43 82    99 78 29 34 78
```

Table C   Areas under the normal curve

Fractional parts of the total area (10,000) under the normal curve, corresponding to distances between the mean and ordinates which are Z standard-deviation units from the mean.

Z	.00	.01	.02	03	04	.05	06	.07	.08	.09
0.0	0000	0040	0080	0120	0159	0199	0239	0279	0319	0359
0.1	0398	0438	0478	0517	0557	0596	0636	0675	0714	0753
0.2	0793	0832	0871	0910	0948	0987	1026	1064	1103	1141
0.3	1179	1217	1255	1293	1331	1368	1406	1443	1480	1517
0.4	1554	1591	1628	1664	1700	1736	1772	1808	1844	1879
0.5	1915	1950	1985	2019	2054	2088	2123	2157	2190	2224
0.6	2257	2291	2324	2357	2389	2422	2454	2486	2518	2549
0.7	2580	2612	2642	2673	2704	2734	2764	2794	2823	2852
0.8	2881	2910	2939	2967	2995	3023	3051	3078	3106	3133
0.9	3159	3186	3212	3238	3264	3289	3315	3340	3365	3389
1.0	3413	3438	3461	3485	3508	3531	3554	3577	3599	3621
1.1	3643	3665	3686	3718	3729	3749	3770	3790	3810	3830
1.2	3849	3869	3888	3907	3925	3944	3962	3980	3997	4015
1.3	4032	4049	4066	4083	4099	4115	4131	4147	4162	4177
1.4	4192	4207	4222	4236	4251	4265	4279	4292	4306	4319
1.5	4332	4345	4357	4370	4382	4394	4406	4418	4430	4441
1.6	4452	4463	4474	4485	4495	4505	4515	4525	4535	4545
1.7	4554	4564	4573	4582	4591	4599	4608	4616	4625	4633
1.8	4641	4649	4656	4664	4671	4678	4686	4693	4699	4706
1.9	4713	4719	4726	4732	4738	4744	4750	4758	4762	4767
2.0	4773	4778	4783	4788	4793	4798	4803	4808	4812	4817
2.1	4821	4826	4830	4834	4838	4842	4846	4850	4854	4857
2.2	4861	4865	4868	4871	4875	4878	4881	4884	4887	4890
2.3	4893	4896	4898	4901	4904	4906	4909	4911	4913	4916
2.4	4918	4920	4922	4925	4927	4929	4931	4932	4934	4936
2.5	4938	4940	4941	4943	4945	4946	4948	4949	4951	4952
2.6	4953	4955	4956	4957	4959	4960	4961	4962	4963	4964
2.7	4965	4966	4967	4968	4969	4970	4971	4972	4973	4974
2.8	4974	4975	4976	4977	4977	4978	4979	4980	4980	4981
2.9	4981	4982	4983	4984	4984	4984	4985	4985	4986	4986
3.0	4986.5	4987	4987	4988	4988	4988	4989	4989	4989	4990
3.1	4990.0	4991	4991	4991	4992	4992	4992	4992	4993	4993
3.2	4993.129									
3.3	4995.166									
3.4	4996.631									
3.5	4997.674									
3.6	4998.409									
3.7	4998.922									
3.8	4999.277									
3.9	4999.519									
4.0	4999.683									
4.5	4999.966									
5.0	4999.997133									

SOURCE: Harold O. Rugg, *Statistical Methods Applied to Education*, Houghton Mifflin Company, Boston, 1917, appendix table III, pp. 389–390, with the kind permission of the publisher.

Table D   Distribution of *t*

df	Level of significance for one-tailed test					
	.10	.05	.025	.01	.005	.0005
	Level of significance for two-tailed test					
	.20	.10	.05	.02	.01	.001
1	3.078	6.314	12.706	31.821	63.657	636.619
2	1.886	2.920	4.303	6.965	9.925	31.598
3	1.638	2.353	3.182	4.541	5.841	12.941
4	1.533	2.132	2.776	3.747	4.604	8.610
5	1.476	2.015	2.571	3.365	4.032	6.859
6	1.440	1.943	2.447	3.143	3.707	5.959
7	1.415	1.895	2.365	2.998	3.499	5.405
8	1.397	1.860	2.306	2.896	3.355	5.041
9	1.383	1.833	2.262	2.821	3.250	4.781
10	1.372	1.812	2.228	2.764	3.169	4.587
11	1.363	1.796	2.201	2.718	3.106	4.437
12	1.356	1.782	2.179	2.681	3.055	4.318
13	1.350	1.771	2.160	2.650	3.012	4.221
14	1.345	1.761	2.145	2.624	2.977	4.140
15	1.341	1.753	2.131	2.602	2.947	4.073
16	1.337	1.746	2.120	2.583	2.921	4.015
17	1.333	1.740	2.110	2.567	2.898	3.965
18	1.330	1.734	2.101	2.552	2.878	3.922
19	1.328	1.729	2.093	2.539	2.861	3.883
20	1.325	1.725	2.086	2.528	2.845	3.850
21	1.323	1.721	2.080	2.518	2.831	3.819
22	1.321	1.717	2.074	2.508	2.819	3.792
23	1.319	1.714	2.069	2.500	2.807	3.767
24	1.318	1.711	2.064	2.492	2.797	3.745
25	1.316	1.708	2.060	2.485	2.787	3.725
26	1.315	1.706	2.056	2.479	2.779	3.707
27	1.314	1.703	2.052	2.473	2.771	3.690
28	1.313	1.701	2.048	2.467	2.763	3.674
29	1.311	1.699	2.045	2.462	2.756	3.659
30	1.310	1.697	2.042	2.457	2.750	3.646
40	1.303	1.684	2.021	2.423	2.704	3.551
60	1.296	1.671	2.000	2.390	2.660	3.460
120	1.289	1.658	1.980	2.358	2.617	3.373
∞	1.282	1.645	1.960	2.326	2.576	3.291

SOURCE: Table D is abridged from Table III of R. A. Fisher and F. Yates, *Statistical Tables for Biological, Agricultural and Medical Research* (1948 ed.), published by Oliver & Boyd, Ltd., Edinburgh and London, by permission of the authors and publishers.

Table E   Critical values of r in the runs test, $P = .05$

For the two-sample runs test any value of r which is equal to or less than that shown in the body of the table is significant at the .05 level with direction not predicted, or at the .025 level with direction predicted.

N_1 \ N_2	2	3	4	5	6	7	8	9	10	11	12	13	14	15	16	17	18	19	20
4		2																	
5		2	2	3															
6		2	3	3	3														
7		2	3	3	4	4													
8	2	2	3	3	4	4	5												
9	2	2	3	4	4	5	5	6											
10	2	3	3	4	5	5	6	6	6										
11	2	3	3	4	5	5	6	6	7	7									
12	2	3	4	4	5	6	6	7	7	8	8								
13	2	3	4	4	5	6	6	7	8	8	9	9							
14	2	3	4	5	5	6	7	7	8	8	9	9	10						
15	2	3	4	5	6	6	7	8	8	9	9	10	10	11					
16	2	3	4	5	6	6	7	8	8	9	10	10	11	11	11				
17	2	3	4	5	6	7	7	8	9	9	10	10	11	11	12	12			
18	2	3	4	5	6	7	8	8	9	10	10	11	11	12	12	13	13		
19	2	3	4	5	6	7	8	8	9	10	10	11	12	12	13	13	14	14	
20	2	3	4	5	6	7	8	9	9	10	11	11	12	12	13	13	14	14	15

SOURCE: F. S. Swed and C. Eisenhart, "Tables for Testing Randomness of Grouping in a Sequence of Alternatives," *Annals of Mathematical Statistics*, vol. 14, pp. 83–86, 1943, with the kind permission of the authors and publisher.

Table F Table of probabilities associated with values as small as observed values of U in the Mann-Whitney test (with direction predicted)*

U \ N_1	1	2	3		U \ N_1	1	2	3	4
	$N_2 = 3$					$N_2 = 4$			
0	.250	.100	.050		0	.200	.067	.028	0.14
1	.500	.200	.100		1	.400	.133	.057	.029
2	.750	.400	.200		2	.600	.267	.114	.057
3		.600	.350		3		.400	.200	.100
4			.500		4		.600	.314	.171
5			.650		5			.429	.243
					6			.571	.343
					7				.443
					8				.557

U \ N_1	1	2	3	4	5		U \ N_1	1	2	3	4	5	6
	$N_2 = 5$							$N_2 = 6$					
0	.167	.047	.018	.008	.004		0	.143	.036	.012	.005	.002	.001
1	.333	.095	.036	.016	.008		1	.286	.071	.024	.010	.004	.002
2	.500	.190	.071	.032	.016		2	.428	.143	.048	.019	.009	.004
3	.667	.286	.125	.056	.028		3	.571	.214	.083	.033	.015	.008
4		.429	.196	.095	.048		4		.321	.131	.057	.026	.013
5		.571	.286	.143	.075		5		.429	.190	.086	.041	.021
6			.393	.206	.111		6		.571	.274	.129	.063	.032
7			.500	.278	.155		7			.357	.176	.089	.047
8			.607	.365	.210		8			.452	.238	.123	.066
9				.452	.274		9			.548	.305	.165	.090
10				.548	.345		10				.381	.214	.120
11					.421		11				.457	.268	.155
12					.500		12				.545	.331	.197
13					.579		13					.396	.242
							14					.465	.294
							15					.535	.350
							16						.409
							17						.469
							18						.531

SOURCE: H. B. Mann and D. R. Whitney, "On a Test of Whether One of Two Random Variables Is Stochastically Larger than the Other," *Annals of Mathematical Statistics*, vol. 18, pp. 52–54, 1947, with the kind permission of the authors and publisher.

* If direction has not been predicted in advance, probabilities may be doubled.

Table F   Table of probabilities associated with values as small as observed values of U **in the Mann Whitney test** (*Continued*)

$$N_2 = 7$$

U \ N₁	1	2	3	4	5	6	7
0	.125	.028	.008	.003	.001	.001	.000
1	.250	.056	.017	.006	.003	.001	.001
2	.375	.111	.033	.012	.005	.002	.001
3	.500	.167	.058	.021	.009	.004	.002
4	.625	.250	.092	.036	.015	.007	.003
5		.333	.133	.055	.024	.011	.006
6		.444	.192	.082	.037	.017	.009
7		.556	.258	.115	.053	.026	.013
8			.333	.158	.074	.037	.019
9			.417	.206	.101	.051	.027
10			.500	.264	.134	.069	.036
11			.583	.324	.172	.090	.049
12				.394	.216	.117	.064
13				.464	.265	.147	.082
14				.538	.319	.183	.104
15					.378	.223	.130
16					.438	.267	.159
17					.500	.314	.191
18					.562	.365	.228
19						.418	.267
20						.473	.310
21						.527	.355
22							.402
23							.451
24							.500
25							.549

Table F Table of probabilities associated with values as small as observed values of U in the Mann-Whitney test (*Continued*)

$$N_2 = 8$$

U \ N_1	1	2	3	4	5	6	7	8
0	.111	.022	.006	.002	.001	.000	.000	.000
1	.222	.044	.012	.004	.002	.001	.000	.000
2	.333	.089	.024	.008	.003	.001	.001	.000
3	.444	.133	.042	.014	.005	.002	.001	.001
4	.556	.200	.067	.024	.009	.004	.002	.001
5		.267	.097	.036	.015	.006	.003	.001
6		.356	.139	.055	.023	.010	.005	.002
7		.444	.188	.077	.033	.015	.007	.003
8		.556	.248	.107	.047	.021	.010	.005
9			.315	.141	.064	.030	.014	.007
10			.387	.184	.085	.041	.020	.010
11			.461	.230	.111	.054	.027	.014
12			.539	.285	.142	.071	.036	.019
13				.341	.177	.091	.047	.025
14				.404	.217	.114	.060	.032
15				.467	.262	.141	.076	.041
16				.533	.311	.172	.095	.052
17					.362	.207	.116	.065
18					.416	.245	.140	.080
19					.472	.286	.168	.097
20					.528	.331	.198	.117
21						.377	.232	.139
22						.426	.268	.164
23						.475	.306	.191
24						.525	.347	.221
25							.389	.253
26							.433	.287
27							.478	.323
28							.522	.360
29								.399
30								.439
31								.480
32								.520

Table G   Table of critical values of U in the Mann-Whitney test

Critical values of U at $\alpha = .001$ with direction predicted or at $\alpha = .002$ with direction not predicted

N_1 \ N_2	9	10	11	12	13	14	15	16	17	18	19	20
1												
2												
3									0	0	0	0
4		0	0	0	1	1	1	2	2	3	3	3
5	1	1	2	2	3	3	4	5	5	6	7	7
6	2	3	4	4	5	6	7	8	9	10	11	12
7	3	5	6	7	8	9	10	11	13	14	15	16
8	5	6	8	9	11	12	14	15	17	18	20	21
9	7	8	10	12	14	15	17	19	21	23	25	26
10	8	10	12	14	17	19	21	23	25	27	29	32
11	10	12	15	17	20	22	24	27	29	32	34	37
12	12	14	17	20	23	25	28	31	34	37	40	42
13	14	17	20	23	26	29	32	35	38	42	45	48
14	15	19	22	25	29	32	36	39	43	46	50	54
15	17	21	24	28	32	36	40	43	47	51	55	59
16	19	23	27	31	35	39	43	48	52	56	60	65
17	21	25	29	34	38	43	47	52	57	61	66	70
18	23	27	32	37	42	46	51	56	61	66	71	76
19	25	29	34	40	45	50	55	60	66	71	77	82
20	26	32	37	42	48	54	59	65	70	76	82	88

SOURCE: D. Auble, "Extended Tables for the Mann-Whitney Statistic," *Bulletin of the Institute of Educational Research at Indiana University*, vol, 1, no. 2, Tables 1, 3, 5 and 7, 1953, with the kind permission of the publisher; as adapted in S. Siegel *Nonparametric Statistics*, McGraw-Hill Book Company, New York, 1956, table K.

Table G Table of critical values of U in the Mann-Whitney test (*Continued*)

Critical values of U at $\alpha = .01$ with direction predicted or at $\alpha = .02$ with direction not predicted

N_1 \ N_2	9	10	11	12	13	14	15	16	17	18	19	20
1												
2					0	0	0	0	0	0	1	1
3	1	1	1	2	2	2	3	3	4	4	4	5
4	3	3	4	5	5	6	7	7	8	9	9	10
5	5	6	7	8	9	10	11	12	13	14	15	16
6	7	8	9	11	12	13	15	16	18	19	20	22
7	9	11	12	14	16	17	19	21	23	24	26	28
8	11	13	15	17	20	22	24	26	28	30	32	34
9	14	16	18	21	23	26	28	31	33	36	38	40
10	16	19	22	24	27	30	33	36	38	41	44	47
11	18	22	25	28	31	34	37	41	44	47	50	53
12	21	24	28	31	35	38	42	46	49	53	56	60
13	23	27	31	35	39	43	47	51	55	59	63	67
14	26	30	34	38	43	47	51	56	60	65	69	73
15	28	33	37	42	47	51	56	61	66	70	75	80
16	31	36	41	46	51	56	61	66	71	76	82	87
17	33	38	44	49	55	60	66	71	77	82	88	93
18	36	41	47	53	59	65	70	76	82	88	94	100
19	38	44	50	56	63	69	75	82	88	94	101	107
20	40	47	53	60	67	73	80	87	93	100	107	114

Table G   Table of critical values of U in the Mann-Whitney test (*Continued*)

Critical values of U at $\alpha = .025$ with direction predicted or
at $\alpha = .05$ with direction not predicted

N_1 \ N_2	9	10	11	12	13	14	15	16	17	18	19	20
1												
2	0	0	0	1	1	1	1	1	2	2	2	2
3	2	3	3	4	4	5	5	6	6	7	7	8
4	4	5	6	7	8	9	10	11	11	12	13	13
5	7	8	9	11	12	13	14	15	17	18	19	20
6	10	11	13	14	16	17	19	21	22	24	25	27
7	12	14	16	18	20	22	24	26	28	30	32	34
8	15	17	19	22	24	26	29	31	34	36	38	41
9	17	20	23	26	28	31	34	37	39	42	45	48
10	20	23	26	29	33	36	39	42	45	48	52	55
11	23	26	30	33	37	40	44	47	51	55	58	62
12	26	29	33	37	41	45	49	53	57	61	65	69
13	28	33	37	41	45	50	54	59	63	67	72	76
14	31	36	40	45	50	55	59	64	67	74	78	83
15	34	39	44	49	54	59	64	70	75	80	85	90
16	37	42	47	53	59	64	70	75	81	86	92	98
17	39	45	51	57	63	67	75	81	87	93	99	105
18	42	48	55	61	67	74	80	86	93	99	106	112
19	45	52	58	65	72	78	85	92	99	106	113	119
20	48	55	62	69	76	83	90	98	105	112	119	127

Table G Table of critical values of U in the Mann-Whitney test (*Continued*)

Critical values of U at $\alpha = .05$ with direction predicted or at $\alpha = .10$ with direction not predicted

N_1 \ N_2	9	10	11	12	13	14	15	16	17	18	19	20
1											0	0
2	1	1	1	2	2	2	3	3	3	4	4	4
3	3	4	5	5	6	7	7	8	9	9	10	11
4	6	7	8	9	10	11	12	14	15	16	17	18
5	9	11	12	13	15	16	18	19	20	22	23	25
6	12	14	16	17	19	21	23	25	26	28	30	32
7	15	17	19	21	24	26	28	30	33	35	37	39
8	18	20	23	26	28	31	33	36	39	41	44	47
9	21	24	27	30	33	36	39	42	45	48	51	54
10	24	27	31	34	37	41	44	48	51	55	58	62
11	27	31	34	38	42	46	50	54	57	61	65	69
12	30	34	38	42	47	51	55	60	64	68	72	77
13	33	37	42	47	51	56	61	65	70	75	80	84
14	36	41	46	51	56	61	66	71	77	82	87	92
15	39	44	50	55	61	66	72	77	83	88	94	100
16	42	48	54	60	65	71	77	83	89	95	101	107
17	45	51	57	64	70	77	83	89	96	102	109	115
18	48	55	61	68	75	82	88	95	102	109	116	123
19	51	58	65	72	80	87	94	101	109	116	123	130
20	54	62	69	77	84	92	100	107	115	123	130	138

Table H   Table of critical values of T in the Wilcoxon matched-pairs signed-ranks test

N	Level of significance, direction predicted		
	.025	.01	.005
	Level of significance, direction not predicted		
	.05	.02	.01
6	0	—	—
7	2	0	—
8	4	2	0
9	6	3	2
10	8	5	3
11	11	7	5
12	14	10	7
13	17	13	10
14	21	16	13
15	25	20	16
16	30	24	20
17	35	28	23
18	40	33	28
19	46	38	32
20	52	43	38
21	59	49	43
22	66	56	49
23	73	62	55
24	81	69	61
25	89	77	68

SOURCE: F. Wilcoxon, *Some Rapid Approximate Statistical Procedures*, American Cyanamid Company, New York, 1949, table I, p. 13, with the kind permission of the author and publisher; as adapted in S. Siegel, *Nonparametric Statistics*, McGraw-Hill Book Company, New York, 1956, table G.

Table I Distribution of χ^2

Probability

df	.99	.98	.95	.90	.80	.70	.50	.30	.20	.10	.05	.02	.01	.001
1	.0³157	.0³628	.00393	.0158	.0642	.148	.455	1.074	1.642	2.706	3.841	5.412	6.635	10.827
2	.0201	.0404	.103	.211	.446	.713	1.386	2.408	3.219	4.605	5.991	7.824	9.210	13.815
3	.115	.185	.352	.584	1.005	1.424	2.366	3.665	4.642	6.251	7.815	9.837	11.341	16.268
4	.297	.429	.711	1.064	1.649	2.195	3.357	4.878	5.989	7.779	9.488	11.668	13.277	18.465
5	.554	.752	1.145	1.610	2.343	3.000	4.351	6.064	7.289	9.236	11.070	13.388	15.086	20.517
6	.872	1.134	1.635	2.204	3.070	3.828	5.348	7.231	8.558	10.645	12.592	15.033	16.812	22.457
7	1.239	1.564	2.167	2.833	3.822	4.671	6.346	8.383	9.803	12.017	14.067	16.622	18.475	24.322
8	1.646	2.032	2.733	3.490	4.594	5.527	7.344	9.524	11.030	13.362	15.507	18.168	20.090	26.125
9	2.088	2.532	3.325	4.168	5.380	6.393	8.343	10.656	12.242	14.684	16.919	19.679	21.666	27.877
10	2.558	3.059	3.940	4.865	6.179	7.267	9.342	11.781	13.442	15.987	18.307	21.161	23.209	29.588
11	3.053	3.609	4.575	5.578	6.989	8.148	10.341	12.899	14.631	17.275	19.675	22.618	24.725	31.264
12	3.571	4.178	5.226	6.304	7.807	9.034	11.340	14.011	15.812	18.549	21.026	24.054	26.217	32.909
13	4.107	4.765	5.892	7.042	8.634	9.926	12.340	15.119	16.985	19.812	22.362	25.472	27.688	34.528
14	4.660	5.368	6.571	7.790	9.467	10.821	13.339	16.222	18.151	21.064	23.685	26.873	29.141	36.123
15	5.229	5.985	7.261	8.547	10.307	11.721	14.339	17.322	19.311	22.307	24.996	28.259	30.578	37.697
16	5.812	6.614	7.962	9.312	11.152	12.624	15.338	18.418	20.465	23.542	26.296	29.633	32.000	39.252
17	6.408	7.255	8.672	10.085	12.002	13.531	16.338	19.511	21.615	24.769	27.587	30.995	33.409	40.790
18	7.015	7.906	9.390	10.865	12.857	14.440	17.338	20.601	22.760	25.989	28.869	32.346	34.805	42.312
19	7.633	8.567	10.117	11.651	13.716	15.352	18.338	21.689	23.900	27.204	30.144	33.687	36.191	43.820
20	8.260	9.237	10.851	12.443	14.578	16.266	19.337	22.775	25.038	28.412	31.410	35.020	37.566	45.315
21	8.897	9.915	11.591	13.240	15.445	17.182	20.337	23.858	26.171	29.615	32.671	36.343	38.932	46.797
22	9.542	10.600	12.338	14.041	16.314	18.101	21.337	24.939	27.301	30.813	33.924	37.659	40.289	48.268
23	10.196	11.293	13.091	14.848	17.187	19.021	22.337	26.018	28.429	32.007	35.172	38.968	41.638	49.728
24	10.856	11.992	13.848	15.659	18.062	19.943	23.337	27.096	29.553	33.196	36.415	40.270	42.980	51.179
25	11.524	12.697	14.611	16.473	18.940	20.867	24.337	28.172	30.675	34.382	37.652	41.566	44.314	52.620
26	12.198	13.409	15.379	17.292	19.820	21.792	25.336	29.246	31.795	35.563	38.885	42.856	45.642	54.052
27	12.879	14.125	16.151	18.114	20.703	22.719	26.336	30.319	32.912	36.741	40.113	44.140	46.963	55.476
28	13.565	14.847	16.928	18.939	21.588	23.647	27.336	31.391	34.027	37.916	41.337	45.419	48.278	56.893
29	14.256	15.574	17.708	19.768	22.475	24.577	28.336	32.461	35.139	39.087	42.557	46.693	49.588	58.302
30	14.953	16.306	18.493	20.599	23.364	25.508	29.336	33.530	36.250	40.256	43.773	47.962	50.892	59.703

For larger values of df, the expression $\sqrt{2\chi^2} - \sqrt{2df - 1}$ may be used as a normal deviate with unit variance, remembering that the probability for χ^2 corresponds with that of a single tail of the normal curve.

SOURCE: Table I is reprinted from Table IV of R. A. Fisher and F. Yates, *Statistical Tables for Biological, Agricultural and Medical Research* (1948 ed.), published by Oliver & Boyd Ltd,., Edinburgh and London, by permission of the authors and publishers.

Table J   Distribution of F

$$p = .05$$

n_1 n_2	1	2	3	4	5	6	8	12	24	∞
1	161.4	199.5	215.7	224.6	230.2	234.0	238.9	243.9	249.0	254.3
2	18.51	19.00	19.16	19.25	19.30	19.33	19.37	19.41	19.45	19.50
3	10.13	9.55	9.28	9.12	9.01	8.94	8.84	8.74	8.64	8.53
4	7.71	6.94	6.59	6.39	6.26	6.16	6.04	5.91	5.77	5.63
5	6.61	5.79	5.41	5.19	5.05	4.95	4.82	4.68	4.53	4.36
6	5.99	5.14	4.76	4.53	4.39	4.28	4.15	4.00	3.84	3.67
7	5.59	4.74	4.35	4.12	3.97	3.87	3.73	3.57	3.41	3.23
8	5.32	4.46	4.07	3.84	3.69	3.58	3.44	3.28	3.12	2.93
9	5.12	4.26	3.86	3.63	3.48	3.37	3.23	3.07	2.90	2.71
10	4.96	4.10	3.71	3.48	3.33	3.22	3.07	2.91	2.74	2.54
11	4.84	3.98	3.59	3.36	3.20	3.09	2.95	2.79	2.61	2.40
12	4.75	3.88	3.49	3.26	3.11	3.00	2.85	2.69	2.50	2.30
13	4.67	3.80	3.41	3.18	3.02	2.92	2.77	2.60	2.42	2.21
14	4.60	3.74	3.34	3.11	2.96	2.85	2.70	2.53	2.35	2.13
15	4.54	3.68	3.29	3.06	2.90	2.79	2.64	2.48	2.29	2.07
16	4.49	3.63	3.24	3.01	2.85	2.74	2.59	2.42	2.24	2.01
17	4.45	3.59	3.20	2.96	2.81	2.70	2.55	2.38	2.19	1.96
18	4.41	3.55	3.16	2.93	2.77	2.66	2.51	2.34	2.15	1.92
19	4.38	3.52	3.13	2.90	2.74	2.63	2.48	2.31	2.11	1.88
20	4.35	3.49	3.10	2.87	2.71	2.60	2.45	2.28	2.08	1.84
21	4.32	3.47	3.07	2.84	2.68	2.57	2.42	2.25	2.05	1.81
22	4.30	3.44	3.05	2.82	2.66	2.55	2.40	2.23	2.03	1.78
23	4.28	3.42	3.03	2.80	2.64	2.53	2.38	2.20	2.00	1.76
24	4.26	3.40	3.01	2.78	2.62	2.51	2.36	2.18	1.98	1.73
25	4.24	3.38	2.99	2.76	2.60	2.49	2.34	2.16	1.96	1.71
26	4.22	3.37	2.98	2.74	2.59	2.47	2.32	2.15	1.95	1.69
27	4.21	3.35	2.96	2.73	2.57	2.46	2.30	2.13	1.93	1.67
28	4.20	3.34	2.95	2.71	2.56	2.44	2.29	2.12	1.91	1.65
29	4.18	3.33	2.93	2.70	2.54	2.43	2.28	2.10	1.90	1.64
30	4.17	3.32	2.92	2.69	2.53	2.42	2.27	2.09	1.89	1.62
40	4.08	3.23	2.84	2.61	2.45	2.34	2.18	2.00	1.79	1.51
60	4.00	3.15	2.76	2.52	2.37	2.25	2.10	1.92	1.70	1.39
120	3.92	3.07	2.68	2.45	2.29	2.17	2.02	1.83	1.61	1.25
∞	3.84	2.99	2.60	2.37	2.21	2.09	1.94	1.75	1.52	1.00

Values of n_1 and n_2 represent the degrees of freedom associated with the larger and smaller estimates of variance respectively.

SOURCE: Table J is abridged from Table V of R. A. Fisher and F. Yates, *Statistical Tables for Biological, Agricultural and Medical Research* (1948 ed.), published by Oliver & Boyd, Ltd., Edinburgh and London, by permission of the authors and publishers.

Table J   Distribution of F **(***Continued***)**

$$p = .01$$

n_1 n_2	1	2	3	4	5	6	8	12	24	∞
1	4052	4999	5403	5625	5764	5859	5981	6106	6234	6366
2	98.49	99.01	99.17	99.25	99.30	99.33	99.36	99.42	99.46	99.50
3	34.12	30.81	29.46	28.71	28.24	27.91	27.49	27.05	26.60	26.12
4	21.20	18.00	16.69	15.98	15.52	15.21	14.80	14.37	13.93	13.46
5	16.26	13.27	12.06	11.39	10.97	10.67	10.27	9.89	9.47	9.02
6	13.74	10.92	9.78	9.15	8.75	8.47	8.10	7.72	7.31	6.88
7	12.25	9.55	8.45	7.85	7.46	7.19	6.84	6.47	6.07	5.65
8	11.26	8.65	7.59	7.01	6.63	6.37	6.03	5.67	5.28	4.86
9	10.56	8.02	6.99	6.42	6.06	5.80	5.47	5.11	4.73	4.31
10	10.04	7.56	6.55	5.99	5.64	5.39	5.06	4.71	4.33	3.91
11	9.65	7.20	6.22	5.67	5.32	5.07	4.74	4.40	4.02	3.60
12	9.33	6.93	5.95	5.41	5.06	4.82	4.50	4.16	3.78	3.36
13	9.07	6.70	5.74	5.20	4.86	4.62	4.30	3.96	3.59	3.16
14	8.86	6.51	5.56	5.03	4.69	4.46	4.14	3.80	3.43	3.00
15	8.68	6.36	5.42	4.89	4.56	4.32	4.00	3.67	3.29	2.87
16	8.53	6.23	5.29	4.77	4.44	4.20	3.89	3.55	3.18	2.75
17	8.40	6.11	5.18	4.67	4.34	4.10	3.79	3.45	3.08	2.65
18	8.28	6.01	5.09	4.58	4.25	4.01	3.71	3.37	3.00	2.57
19	8.18	5.93	5.01	4.50	4.17	3.94	3.63	3.30	2.92	2.49
20	8.10	5.85	4.94	4.43	4.10	3.87	3.56	3.23	2.86	2.42
21	8.02	5.78	4.87	4.37	4.04	3.81	3.51	3.17	2.80	2.36
22	7.94	5.72	4.82	4.31	3.99	3.76	3.45	3.12	2.75	2.31
23	7.88	5.66	4.76	4.26	3.94	3.71	3.41	3.07	2.70	2.26
24	7.82	5.61	4.72	4.22	3.90	3.67	3.36	3.03	2.66	2.21
25	7.77	5.57	4.68	4.18	3.86	3.63	3.32	2.99	2.62	2.17
26	7.72	5.53	4.64	4.14	3.82	3.59	3.29	2.96	2.58	2.13
27	7.68	5.49	4.60	4.11	3.78	3.56	3.26	2.93	2.55	2.10
28	7.64	5.45	4.57	4.07	3.75	3.53	3.23	2.90	2.52	2.06
29	7.60	5.42	4.54	4.04	3.73	3.50	3.20	2.87	2.49	2.03
30	7.56	5.39	4.51	4.02	3.70	3.47	3.17	2.84	2.47	2.01
40	7.31	5.18	4.31	3.83	3.51	3.29	2.99	2.66	2.29	1.80
60	7.08	4.98	4.13	3.65	3.34	3.12	2.82	2.50	2.12	1.60
120	6.85	4.79	3.95	3.48	3.17	2.96	2.66	2.34	1.95	1.38
∞	6.64	4.60	3.78	3.32	3.02	2.80	2.51	2.18	1.79	1.00

Values of n_1 and n_2 represent the degrees of freedom associated with the larger and smaller estimates of variance respectively.

Table J   Distribution of F (Continued)

$$p = .001$$

n_1 / n_2	1	2	3	4	5	6	8	12	24	∞
1	405284	500000	540379	562500	576405	585937	598144	610667	623497	636619
2	998.5	999.0	999.2	999.2	999 3	999.3	999.4	999.4	999.5	999.5
3	167.5	148.5	141.1	137.1	134.6	132.8	130.6	128.3	125.9	123.5
4	74.14	61.25	56.18	53.44	51.71	50.53	49.00	47.41	45.77	44.05
5	47.04	36.61	33.20	31.09	29.75	28.84	27.64	26.42	25.14	23.78
6	35.51	27.00	23.70	21.90	20.81	20.03	19.03	17.99	16.89	15.75
7	29.22	21.69	18.77	17.19	16.21	15.52	14.63	13.71	12.73	11.69
8	25.42	18.49	15.83	14.39	13.49	12.86	12.04	11.19	10.30	9.34
9	22.86	16.39	13.90	12.56	11.71	11.13	10.37	9.57	8.72	7.81
10	21.04	14.91	12.55	11.28	10.48	9.92	9.20	8.45	7.64	6.76
11	19.69	13.81	11.56	10.35	9.58	9.05	8.35	7.63	6.85	6.00
12	18.64	12.97	10.80	9.63	8.89	8.38	7 71	7.00	6.25	5.42
13	17.81	12.31	10.21	9.07	8.35	7.86	7.21	6.52	5.78	4.97
14	17.14	11.78	9.73	8.62	7.92	7.43	6.80	6.13	5.41	4.60
15	16.59	11.34	9.34	8.25	7.57	7.09	6.47	5.81	5.10	4.31
16	16.12	10.97	9.00	7.94	7.27	6.81	6.19	5.55	4.85	4.06
17	15.72	10.66	8.73	7.68	7.02	6.56	5.96	5.32	4.63	3.85
18	15.38	10.39	8.49	7.46	6.81	6.35	5.76	5.13	4.45	3.67
19	15.08	10.16	8.28	7.26	6.61	6.18	5.59	4.97	4.29	3.52
20	14.82	9.95	8.10	7.10	6.46	6.02	5.44	4.82	4.15	3.38
21	14.59	9.77	7.94	6.95	6.32	5.88	5.31	4.70	4.03	3.26
22	14.38	9.61	7.80	6.81	6.19	5.76	5.19	4.58	3.92	3.15
23	14.19	9.47	7.67	6.69	6.08	5.65	5.09	4.48	3.82	3.05
24	14.03	9.34	7.55	6.59	5.98	5.55	4.99	4.39	3.74	2 97
25	13.88	9.22	7.45	6.49	5.88	5.46	4.91	4.31	3.66	2.89
26	13.74	9.12	7.36	6.41	5.80	5.38	4.83	4.24	3.59	2.82
27	13.61	9.02	7.27	6.33	5.73	5.31	4.76	4.17	3.52	2.75
28	13.50	8.93	7.19	6.25	5.66	5.24	4.69	4.11	3.46	2.70
29	13.39	8.85	7.12	6.19	5.59	5.18	4.64	4.05	3.41	2.64
30	13.29	8.77	7.05	6.12	5.53	5.12	4.58	4.00	3.36	2.59
40	12.61	8.25	6.60	5.70	5.13	4.73	4.21	3.64	3.01	2.23
60	11.97	7.76	6.17	5.31	4.76	4.37	3.87	3.31	2.69	1.90
120	11.38	7.31	5.79	4.95	4.42	4.04	3.55	3.02	2.40	1.56
∞	10.83	6.91	5.42	4.62	4.10	3.74	3.27	2.74	2.13	1.00

Values of n_1 and n_2 represent the degrees of freedom associated with the larger and smaller estimates of variance respectively.

Table K Values of z for given values of r

r	.000	.001	.002	.003	.004	.005	.006	.007	.008	.009
.000	.0000	.0010	.0020	.0030	.0040	.0050	.0060	.0070	.0080	.0090
.010	.0100	.0110	.0120	.0130	.0140	.0150	.0160	.0170	.0180	.0190
.020	.0200	.0210	.0220	.0230	.0240	.0250	.0260	.0270	.0280	.0290
.030	.0300	.0310	.0320	.0330	.0340	.0350	.0360	.0370	.0380	.0390
.040	.0400	.0410	.0420	.0430	.0440	.0450	.0460	.0470	.0480	.0490
.050	.0501	.0511	.0521	.0531	.0541	.0551	.0561	.0571	.0581	.0591
.060	.0601	.0611	.0621	.0631	.0641	.0651	.0661	.0671	.0681	.0691
.070	.0701	.0711	.0721	.0731	.0741	.0751	.0761	.0771	.0782	.0792
.080	.0802	.0812	.0822	.0832	.0842	.0852	.0862	.0872	.0882	.0892
.090	.0902	.0912	.0922	.0933	.0943	.0953	.0963	.0973	.0983	.0993
.100	.1003	.1013	.1024	.1034	.1044	.1054	.1064	.1074	.1084	.1094
.110	.1105	.1115	.1125	.1135	.1145	.1155	.1165	.1175	.1185	.1195
.120	.1206	.1216	.1226	.1236	.1246	.1257	.1267	.1277	.1287	.1297
.130	.1308	.1318	.1328	.1338	.1348	.1358	.1368	.1379	.1389	.1399
.140	.1409	.1419	.1430	.1440	.1450	.1460	.1470	.1481	.1491	.1501
.150	.1511	.1522	.1532	.1542	.1552	.1563	.1573	.1583	.1593	.1604
.160	.1614	.1624	.1634	.1644	.1655	.1665	.1676	.1686	.1696	.1706
.170	.1717	.1727	.1737	.1748	.1758	.1768	.1779	.1789	.1799	.1810
.180	.1820	.1830	.1841	.1851	.1861	.1872	.1882	.1892	.1903	.1913
.190	.1923	.1934	.1944	.1954	.1965	.1975	.1986	.1996	.2007	.2017
.200	.2027	.2038	.2048	.2059	.2069	.2079	.2090	.2100	.2111	.2121
.210	.2132	.2142	.2153	.2163	.2174	.2184	.2194	.2205	.2215	.2226
.220	.2237	.2247	.2258	.2268	.2279	.2289	.2300	.2310	.2321	.2331
.230	.2342	.2353	.2363	.2374	.2384	.2395	.2405	.2416	.2427	.2437
.240	.2448	.2458	.2469	.2480	.2490	.2501	.2511	.2522	.2533	.2543
.250	.2554	.2565	.2575	.2586	.2597	.2608	.2618	.2629	.2640	.2650
.260	.2661	.2672	.2682	.2693	.2704	.2715	.2726	.2736	.2747	.2758
.270	.2769	.2779	.2790	.2801	.2812	.2823	.2833	.2844	.2855	.2866
.280	.2877	.2888	.2898	.2909	.2920	.2931	.2942	.2953	.2964	.2975
.290	.2986	.2997	.3008	.3019	.3029	.3040	.3051	.3062	.3073	.3084
.300	.3095	.3106	.3117	.3128	.3139	.3150	.3161	.3172	.3183	.3195
.310	.3206	.3217	.3228	.3239	.3250	.3261	.3272	.3283	.3294	.3305
.320	.3317	.3328	.3339	.3350	.3361	.3372	.3384	.3395	.3406	.3417
.330	.3428	.3439	.3451	.3462	.3473	.3484	.3496	.3507	.3518	.3530
.340	.3541	.3552	.3564	.3575	.3586	.3597	.3609	.3620	.3632	.3643
.350	.3654	.3666	.3677	.3689	.3700	.3712	.3723	.3734	.3746	.3757
.360	.3769	.3780	.3792	.3803	.3815	.3826	.3838	.3850	.3861	.3873
.370	.3884	.3896	.3907	.3919	.3931	.3942	.3954	.3966	.3977	.3989
.380	.4001	.4012	.4024	.4036	.4047	.4059	.4071	.4083	.4094	.4106
.390	.4118	.4130	.4142	.4153	.4165	.4177	.4189	.4201	.4213	.4225
.400	.4236	.4248	.4260	.4272	.4284	.4296	4308	.4320	.4332	.4344
.410	.4356	.4368	.4380	.4392	.4404	.4416	.4429	.4441	.4453	.4465
.420	.4477	.4489	.4501	.4513	.4526	.4538	.4550	.4562	.4574	.4587
.430	.4599	.4611	.4623	.4636	.4648	.4660	.4673	.4685	.4697	.4710
.440	.4722	.4735	.4747	.4760	.4772	.4784	.4797	.4809	.4822	.4835
.450	.4847	.4860	.4872	.4885	.4897	.4910	.4923	.4935	.4948	.4961
.460	.4973	.4986	.4999	.5011	.5024	.5037	.5049	.5062	.5075	.5088
.470	.5101	.5114	.5126	.5139	.5152	.5165	.5178	.5191	.5204	.5217
.480	.5230	.5243	.5256	.5279	.5282	.5295	.5308	.5321	.5334	.5347
.490	.5361	.5374	.5387	.5400	.5413	.5427	.5440	.5453	.5466	.5480

SOURCE: Albert E. Waugh, *Statistical Tables and Problems*, McGraw-Hill Book Company, New York, 1952, table A11, pp. 40–41, with the kind permission of the author and publisher.

Table K   Values of z for given values of r (Continued)

r	.000	.001	.002	.003	.004	.005	.006	.007	.008	.009
.500	.5493	.5506	.5520	.5533	.5547	.5560	.5573	.5587	.5600	.5614
.510	.5627	.5641	.5654	.5668	.5681	.5695	.5709	.5722	.5736	.5750
.520	.5763	.5777	.5791	.5805	.5818	.5832	.5846	.5860	.5874	.5888
.530	.5901	.5915	.5929	.5943	.5957	.5971	.5985	.5999	.6013	.6027
.540	.6042	.6056	.6070	.6084	.6098	.6112	.6127	.6141	.6155	.6170
.550	.6184	.6198	.6213	.6227	.6241	.6256	.6270	.6285	.6299	.6314
.560	.6328	.6343	.6358	.6372	.6387	.6401	.6416	.6431	.6446	.6460
.570	.6475	.6490	.6505	.6520	.6535	.6550	.6565	.6579	.6594	.6610
.580	.6625	.6640	.6655	.6670	.6685	.6700	.6715	.6731	.6746	.6761
.590	.6777	.6792	.6807	.6823	.6838	.6854	.6869	.6885	.6900	.6916
.600	.6931	.6947	.6963	.6978	.6994	.7010	.7026	.7042	.7057	.7073
.610	.7089	.7105	.7121	.7137	.7153	.7169	.7185	.7201	.7218	.7234
.620	.7250	.7266	.7283	.7299	.7315	.7332	.7348	.7364	.7381	.7398
.630	.7414	.7431	.7447	.7464	.7481	.7497	.7514	.7531	.7548	.7565
.640	.7582	.7599	.7616	.7633	.7650	.7667	.7684	.7701	.7718	.7736
.650	.7753	.7770	.7788	.7805	.7823	.7840	.7858	.7875	.7893	.7910
.660	.7928	.7946	.7964	.7981	.7999	.8017	.8035	.8053	.8071	.8089
.670	.8107	.8126	.8144	.8162	.8180	.8199	.8217	.8236	.8254	.8273
.680	.8291	.8310	.8328	.8347	.8366	.8385	.8404	.8423	.8442	.8461
.690	.8480	.8499	.8518	.8537	.8556	.8576	.8595	.8614	.8634	.8653
.700	.8673	.8693	.8712	.8732	.8752	.8772	.8792	.8812	.8832	.8852
.710	.8872	.8892	.8912	.8933	.8953	.8973	.8994	.9014	.9035	.9056
.720	.9076	.9097	.9118	.9139	.9160	.9181	.9202	.9223	.9245	.9266
.730	.9287	.9309	.9330	.9352	.9373	.9395	.9417	.9439	.9461	.9483
.740	.9505	.9527	.9549	.9571	.9594	.9616	.9639	.9661	.9684	.9707
.750	.9730	.9752	.9775	.9799	.9822	.9845	.9868	.9892	.9915	.9939
.760	.9962	.9986	1.0010	1.0034	1.0058	1.0082	1.0106	1.0130	1.0154	1.0179
.770	1.0203	1.0228	1.0253	1.0277	1.0302	1.0327	1.0352	1.0378	1.0403	1.0428
.780	1.0454	1.0479	1.0505	1.0531	1.0557	1.0583	1.0609	1.0635	1.0661	1.0688
.790	1.0714	1.0741	1.0768	1.0795	1.0822	1.0849	1.0876	1.0903	1.0931	1.0958
.800	1.0986	1.1014	1.1041	1.1070	1.1098	1.1127	1.1155	1.1184	1.1212	1.1241
.810	1.1270	1.1299	1.1329	1.1358	1.1388	1.1417	1.1447	1.1477	1.1507	1.1538
.820	1.1568	1.1599	1.1630	1.1660	1.1692	1.1723	1.1754	1.1786	1.1817	1.1849
.830	1.1870	1.1913	1.1946	1.1979	1.2011	1.2044	1.2077	1.2111	1.2144	1.2178
.840	1.2212	1.2246	1.2280	1.2315	1.2349	1.2384	1.2419	1.2454	1.2490	1.2526
.850	1.2561	1.2598	1.2634	1.2670	1.2708	1.2744	1.2782	1.2819	1.2857	1.2895
.860	1.2934	1.2972	1.3011	1.3050	1.3089	1.3129	1.3168	1.3209	1.3249	1.3290
.870	1.3331	1.3372	1.3414	1.3456	1.3498	1.3540	1.3583	1.3626	1.3670	1.3714
.880	1.3758	1.3802	1.3847	1.3892	1.3938	1.3984	1.4030	1.4077	1.4124	1.4171
.890	1.4219	1.4268	1.4316	1.4366	1.4415	1.4465	1.4516	1.4566	1.4618	1.4670
.900	1.4722	1.4775	1.4828	1.4883	1.4937	1.4992	1.5047	1.5103	1.5160	1.5217
.910	1.5275	1.5334	1.5393	1.5453	1.5513	1.5574	1.5636	1.5698	1.5762	1.5825
.920	1.5890	1.5956	1.6022	1.6089	1.6157	1.6226	1.6296	1.6366	1.6438	1.6510
.930	1.6584	1.6659	1.6734	1.6811	1.6888	1.6967	1.7047	1.7129	1.7211	1.7295
.940	1.7380	1.7467	1.7555	1.7645	1.7736	1.7828	1.7923	1.8019	1.8117	1.8216
.950	1.8318	1.8421	1.8527	1.8635	1.8745	1.8857	1.8972	1.9090	1.9210	1.9333
.960	1.9459	1.9588	1.9721	1.9857	1.9996	2.0140	2.0287	2.0439	2.0595	2.0756
.970	2.0923	2.1095	2.1273	2.1457	2.1649	2.1847	2.2054	2.2269	2.2494	2.2729
.980	2.2976	2.3223	2.3507	2.3796	2.4101	2.4426	2.4774	2.5147	2.5550	2.5988
.990	2.6467	2.6996	2.7587	2.8257	2.9031	2.9945	3.1063	3.2504	3.4534	3.8002

r	z
.9999	4.95172
.99999	6.10303

Name Index

Subject Index